Contents

v

List of Figures and Tables

FIGURES

TABLES

Foreword

It is now almost 170 years since the tsetse flies of Africa were first described and assigned to the genus *Glossina* named by Wiedemann, yet they continue to demand our attention to this day. Few groups of insects have attracted the interest of successive generations of amateur and professional scientists for so long. The reasons for this are numerous: tsetse are relatively large flies and easy to recognize; they are widespread in Africa, occurring in a variety of habitats, many with attractive scenic and wilderness characteristics; they feed on the blood of numerous species of exotic African vertebrates, which themselves are a subject of great interest; and they exhibit a remarkable range of host preferences and other behavioural characteristics. Above all, however, tsetse flies transmit the African pathogenic trypanosomes which cause sleeping sickness in man and trypanosomal disease in domestic livestock. Historically, many thousands of people and millions of head of livestock have died in major trypanosomosis epidemics, and the presence of tsetse flies has frequently led to the abandonment of settlements and efforts to use animals for transport, traction and livestock farming. The flies have thus been traditionally classified by parasitologists as insects of major medical and veterinary importance. More up-to-date terminology now emphasizes their adverse influence on social and economic development and their role in environmental protection and agricultural sustainability, but these aspects have almost always also been considered in past tsetse control planning.

Many methods have been devised with variable degrees of success to control tsetse populations and trypanosomosis in view of their deleterious effects on human and animal health. An immense body of work on these topics has been recorded in an extensive archive literature, and in many scholarly books, as indicated in the text of this new volume. Regrettably, however, tsetse and trypanosomosis problems still plague elements of the populations of some 40 African nations. Efforts to improve existing control measures and to identify new control techniques have therefore been intensified, notably during the past three or four decades. For example, techniques for the aerial application of insecticides have now become a fine art and a revival of interest in tsetse attractants and traps has let to improvements and innovations in design which have been rewarded by greatly increased catching efficiency. Outstanding original research on the behaviour of tsetse populations has underpinned success in these areas and resulted in the introduction of insecticide-impregnated targets which, when well-managed, have proved to be as effective as traps. Great care is being taken to avoid contamination or destruction of the environment and to ensure sound land use planning. Research on trypanotolerant livestock and the application of special formulations of insecticide directly to the skin of cattle also holds promise of further improved control technology for the future.

Much of the experience gained in the use of the new techniques and their integration with traditional control approaches has not yet been fully evaluated or collated with older knowledge and a thorough stocktaking is considerably overdue. This applies equally to the many advances in knowledge accruing from research on the physiology and genetics of tsetse populations. In 1982, Stephen Leak joined the scientific staff of the International Laboratory for Research on Animal Diseases (ILRAD) in Kenya to provide entomological support to scientific groups at ILRAD and the International Livestock Centre for Africa (ILCA) studying tsetse challenge in relation to the epidemiology, diagnosis and control of trypanosomosis, in resistant and susceptible ruminant livestock. He had previously worked for 6 years on aspects of tsetse control in Zambia. Since 1982, his research has included studies on tsetse species of all of the major ecological groups, supervision of trials of new tsetse traps and insecticide-impregnated targets, and the utility of 'pour-on' insecticide formulations. His work has been done at sites in at least ten countries in Africa. From an early stage of his appointment by ILRAD he was also encouraged to note and evaluate developments in other areas of tsetse biology and ecology, frequently reported only in the informal literature of departmental, local meeting and consultancy reports. He has made good use of these opportunities and has now produced an outstanding account of recent developments in these fields. He has cited approximately 2000 publications from an even greater body of relevant reports and it has given me considerable pleasure to see how commendably well he has managed the task of integrating established and new findings.

Some 30 years ago there were suggestions that increasing human population pressure would so alter rural environments that tsetse flies would be greatly reduced or even eliminated. While this sequence of events has undoubtedly occurred in certain densely populated areas, animal trypanosomosis is still an adverse economic force requiring attention in huge areas in Africa and there are reports that human trypanosomosis is currently a greater problem than it was in the 1930s. Civil wars, massive population movements and financial stringencies continue to result in failures of administrative infrastructure and provide the circumstances favouring uncontrolled advances in distribution of tsetse flies and the spread of disease. However, where some civil order prevails it is still possible to mount successful tsetse and trypanosomosis control schemes. Aerial applications of insecticide and deployment of insecticide-treated targets have been used recently in a major control scheme involving four countries in southeast Africa. The sterile insect technique, and tsetse traps and targets have been used successfully on significant scales in several countries in eastern and western Africa. The future will almost certainly see further tsetse and trypanosomosis control schemes, since recent projections forecast continuation of tsetse problems well into the next century. Major agencies of the United Nations organization continue to conduct tsetse and trypanosomosis control programmes, and substantial new projects are under discussion involving groups of countries in various regions in Africa. These projected programmes of pest control and management will need the involvement of further generations of scientific and administrative personnel for success. The newcomers will surely find Stephen Leak's account of new developments in tsetse biology and ecology, and experience of their application in Africa during the past 30 years, invaluable, as they approach their tasks.

A.R. Gray
Duns, Scotland
May 1998

Acknowledgements

It is my pleasure to thank Dr A.R. Gray, the former Director General of the International Laboratory for Research on Animal Diseases (ILRAD), for encouraging me with the writing of this book at an early stage and for agreeing to write the foreword. Dr Bill Thorpe also provided valuable advice and encouragement in the early stages. Peter de Leeuw, an ecologist formerly working with the International Livestock Centre for Africa (ILCA), deserves particular thanks for going through large sections of the manuscript and providing valuable editorial advice as well as for useful discussion about tsetse, trypanosomosis and livestock production in Africa generally. I thank Dr A.M. Jordan for reading through the whole manuscript in an editorial and technical capacity; any errors remain my responsibility. There are a number of people who have supplied information, commented on parts of the text, or with whom I have had useful discussions. These include Drs Pierre Cattand, Mark Eisler, Guy Hendrickx, Russ Kruska, Andrew Peregrine, Robin Reid, John Rowlands and Professor Ian Maudlin. I wish to express my thanks to the Trustees of the Natural History Museum, London, for permission to reproduce Figs 2.1 and 2.2. I thank Dr Guy D'Ieteren and my employers at the International Livestock Research Institute, and formerly ILRAD, for providing me with the facilities which made it possible to write this book; the staff of the Graphic Arts Unit of those two institutes, Dave Elsworth, Joel Mwaura and Francis Shikhubari for assistance with the figures; and the library staff for supplying me with the many references.

I dedicate this book to my daughter, Melissa.

Part I

Tsetse Biology and Ecology

Chapter 1

Introduction

Domestic livestock are of importance in Africa not only as a source of milk and meat, adding protein to human diets which are often deficient, but also, and perhaps more importantly, as a source of animal traction, enabling farmers to cultivate larger areas with crops providing staple foods. Trypanosomosis, a disease caused by protozoan parasites of the genus *Trypanosoma*, transmitted cyclically by the tsetse fly (*Glossina* spp.) (Fig. 1.1), is arguably still the main constraint to livestock production on the continent, preventing full use of the land to feed the rapidly increasing human population (Murray *et al.*, 1991). Sleeping sickness, the disease caused in humans by species of *Trypanosoma* (Fig. 1.2), is an important yet neglected disease which poses a threat to millions of people in tsetse-infested areas. Often wrongly thought of as a disease of the past, the prevalence of human sleeping sickness is currently increasing in many areas.

Although only eradication of the tsetse fly vector can remove the threat of the disease, eradication is regarded as an impractical goal for much of the 10 million km^2 of tsetse-infested Africa. Control of the vector and of the disease, rather than eradication, is seen as a more realistic approach. Various methods to control trypanosomosis are being investigated, including immunological approaches, use of chemotherapy and exploitation of the trypanotolerance trait.

Because of the economic importance of the tsetse fly as a vector of trypanosomosis, a large amount of literature has been produced. This review is intended to summarize that literature and highlight some of the areas concerning epidemiology, which are relevant to current control techniques. Aspects requiring future research are also discussed. Basic biology and anatomy have been described in some of the standard texts on tsetse and trypanosomosis and therefore no attempt is made to cover these aspects in detail. Instead, more recent findings are covered and some of the basic anatomy and biology is described where there is relevance to ecological or epidemiological research topics. Rather than attempt to review thoroughly subjects that have already been reviewed by experts in their

3

Fig. 1.1. Adult *Glossina brevipalpis* – mating (Dave Elsworth, ILRI).

fields, some of the pertinent points are summarized here and the reader is referred to recent reviews for further details. A number of authoritative reviews have been published on aspects of tsetse ecology or biology in recent years. Among the recently published books (and not so recent) are: *Trypanosomiasis Control and African Rural Development* (Jordan, 1986); *Trypanosomiasis – a Veterinary Perspective* (Stephen, 1986); and *The Role of the Trypanosomiases in African Ecology* (Ford, 1971). These publications reviewed the literature on tsetse and trypanosomosis control, animal trypanosomosis and the ecology of human and animal trypanosomosis in Africa, respectively. *The African Trypanosomiases,* edited by Mulligan (1970), is also a useful source book despite being published over 25 years ago. The last two named titles are now out of print, however. The five-volume series of manuals for tsetse control produced by the Food and Agriculture Organization (FAO) of the United Nations are also valuable source books. In this review I mainly refer to literature from 1970 onwards, and summarize information from the large number of publications not readily accessible to many tsetse/trypanosomosis workers. I also draw upon some earlier literature in order to provide the necessary background.

Of the 31 species and subspecies of tsetse fly, only a small number have been studied in much detail, those being the ones considered most

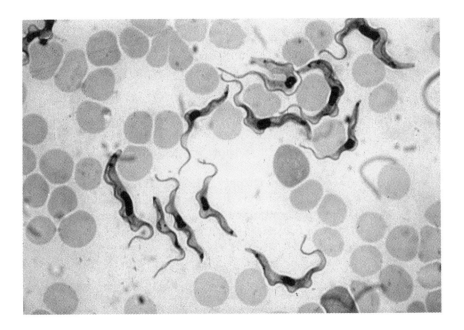

Fig. 1.2. *Trypanosoma b. brucei* trypanosomes (Dave Elsworth, ILRI).

important from an epidemiological and economical viewpoint. The importance of those species may result from their being most widespread, in terms of abundance and distribution, or from the closeness of their contact with humans and/or domestic livestock. Because of their epidemiological significance, and as more is known about them, those species will be discussed more fully than some of the lesser-known species, particularly of the *fusca* group. This will also avoid some repetition. Species regarded as of particular importance are: *Glossina morsitans, G. pallidipes, G. swynnertoni* and *G. austeni* from the *morsitans* group; and *G. tachinoides, G. palpalis* and *G. fuscipes* and their subspecies from the *palpalis* group. *G. brevipalpis* and *G. longipennis* of the *fusca* group are discussed in more detail because of their anomalous position in that group. Their different ecology compared with other *fusca* group flies results in closer contact with domestic livestock.

There is currently a much greater awareness of the need to evaluate the effects of successful control of tsetse and trypanosomosis on the environment and to integrate control schemes into natural resource management. This is necessary in order to achieve the objectives of increased food production, to keep pace with the growing human population in Africa, whilst avoiding detrimental environmental effects. The direct effects (for example, on non-target organisms) have been shown to be relatively small and transient and it is the longer-term changes in land use, which could result in environmental degradation, that require more detailed investigation. A major difficulty of those sorts of analyses is to determine to what extent effects can be attributed to tsetse or trypanosomosis control rather than to other factors, such as policy changes affecting the sensitive issues of land tenure and use, associated with human population growth and land availability. Clearly, it is bad land-use practices rather than successful tsetse control that can result in harmful effects such as overgrazing. Will human population growth naturally result in eradication of tsetse in the foreseeable future and make discussion of possible harmful effects of their control irrelevant?

One of the most significant changes in biological research since the last books on tsetse and trypanosomosis were written has resulted from advances in molecular biology and molecular genetics, which have opened up new fields of study. These include the study of genetic variation in tsetse populations – microsatellite markers; the production of novel, improved diagnostics of trypanosome infections in tsetse, such as DNA probes and PCR. There are opportunities for the development of new techniques for analysing the origin of tsetse bloodmeals. Similar possibilities have arisen in trypanosomosis research, for studying genetic variation, in particular the importance of clonal or interbreeding populations and their epidemiological significance. These studies also contribute to our understanding of the evolution and phylogeny of trypanosome species. Again, the same types of novel diagnostics have been or are being developed.

Finally, molecular genetics is allowing the rapid mapping of the bovine genome and identification of genes responsible for the trypanotolerance trait. This will enable marker-assisted selection of trypanotolerance or offer the possibility of introgressing trypanotolerance genes into productive livestock.

Chapter 2

Classification and Anatomy

2.1 CLASSIFICATION

Tsetse flies are classified into one genus, *Glossina*, of the family Glossinidae, Order *Diptera* – the two-winged flies. There are 31 species and subspecies identified at present (Table 2.1). One of the first descriptions of tsetse flies was published by Westwood (1851), regarding specimens sent to him from South Africa by Frank Vardon in 1850. The genus *Glossina* had already been established by Wiedemann (1830) but the flies sent to Westwood were a new species, *G. morsitans* Westw. The most recently identified species of tsetse fly was *G. frezili*, which inhabits mangrove swamps in the Congo (Gouteux, 1987a).

The economic and biological importance of the genus *Glossina* lies in the fact that its species are cyclical vectors of protozoan parasites belonging to the genus *Trypanosoma*. Species of *Trypanosoma* transmitted by tsetse cause diseases of economic importance in humans and domestic animals in sub-Saharan Africa. Biting flies can also transmit trypanosomes mechanically, both in Africa and in other parts of the world. A trypanosome found in South America, *Trypanosoma cruzi*, causes the disease of humans known as Chagas disease, which is transmitted cyclically by reduviid bugs, *Triatoma* and *Rhodnius* spp.

Evidence of an earlier, wider distribution of tsetse, and of their evolutionary age, arose from the discovery of fossil flies in Florissant shales in Colorado, North America. Four fossil tsetse species found in those shales were named: *G. osborni*, *G. oligocaena*, *G. veterna* and *G. armatipes*. These fossils are from the Oligocene period (Cockerell, 1907, 1909, 1919). Thus, it appears that tsetse flies originated at least 40 million years ago and inhabited parts of the United States of America. During that period, tsetse may have fed on giant terrestrial reptiles, and Cockerell, who described some of these fossils, suggested that tsetse may have had a role in the disappearance of some tertiary mammals in America (Cockerell, 1907, 1909).

Table 2.1. Tsetse species and subspecies (by groups or subgenera).

Morsitans (*Glossina*)	Palpalis (*Nemorhina*)	Fusca (*Austenina*)
G. morsitans submorsitans Newstead 1910	G. palpalis palpalis Rob.-Desvoidy 1830	G. fusca fusca Walker 1849
G. morsitans centralis Machado 1970	G. palpalis gambiensis Vanderplank 1949	G. nigrofusca nigrofusca Newstead 1910
G. morsitans morsitans Westwood 1850	G. fuscipes fuscipes Newstead 1910	G. nigrofusca hopkinsi van Emden 1944
G. austeni Newstead 1912	G. fuscipes quanzensis Pires 1948	G. medicorum Austen 1911
G. pallidipes Austen 1903	G. fuscipes martinii Zumpt 1935	G. tabaniformis Westwood 1850
G. swynnertoni Austen 1923	G. tachinoides Westwood 1850	G. brevipalpis Newstead 1910
G. longipalpis Wiedemann 1830	G. pallicera pallicera Bigot 1891	G. longipennis Corti 1895
	G. pallicera Newstead Austen 1929	G. frezili Gouteux 1987
	G. caliginea Austen 1911	G. severini Newstead 1913
		G. haningtoni Newstead & Evans 1922
		G. fuscipleuris Austen 1911
		G. vanhoofi Henrard 1952

A stercorarian trypanosome, *Trypanosoma grayi*, which is a parasite of *Crocodilus niloticus*, usually transmitted by *G. fuscipes*, may be a survivor from the Mesozoic period as its life cycle is considered more primitive than the salivarian trypanosomes. The example of this parasite is used to support theories of the evolutionary process resulting in transmission of trypanosomes via tsetse flies to vertebrate hosts.

The classification of tsetse is based largely on morphological differences in the structure of the genitalia. Three subgenera are identified based on these morphological differences and can also be broadly differentiated from ecological characteristics. Differences in structure of the superior claspers of the male genitalia (Fig. 2.1) can easily be used to distinguish between the groups (Newstead, 1911a,b). Within the subgenera, males of the *palpalis* group are identified mainly from the morphology of the inferior claspers (Fig. 2.2), *morsitans* group flies from the superior claspers, and *fusca* group females most easily from the signum, which is a chitinized plate at the anterior of the oviduct. Descriptions of anatomical features of the genitalia and keys for identification can be found in Mulligan (1970), updated by Jordan (1993); the latter publication includes a key to the identification of tsetse puparia. More recently, investigations into sex pheromones of tsetse flies (shown to be hydrocarbons present in the cuticle) resulted in a method of classifying tsetse based on cuticular alkanes (Carlson *et al.*, 1993). This allows the genetic distance between species to be judged, which may provide some insight into ecological differences between species of the same group.

In addition to standard conventional keys for identification of tsetse, a computer program for identification of *Glossina* spp. was recently developed, combining morphological features with information on distribution. This program also provides brief background information for each species (Brunhes *et al.*, 1994).

Pollock (1971, 1974) reviewed the classification of tsetse flies and their evolution, mainly in relation to the male genitalia (Fig. 2.2). He discussed the question of *Glossina* being put in the family Muscidae or, as Brues *et al.* (1954) suggested, in a separate family, Glossinidae. Pollock concluded that *Glossina* were not related to *Stomoxys* and should not be regarded as muscids. He further concluded that the genus *Glossina* originated between the Palaeocene and the Oligocene, as indicated from the fossil flies referred to earlier.

Tsetse flies may be closely related to species of another section of the higher Diptera, the Pupipara. The Pupipara consists of the rather specialized families of parasitic flies: Hippoboscidae, Streblidae and Nycteribiidae; however, the family Glossinidae does not belong to that section. Comparisons of the genitalia and abdominal segmentation of *Glossina, Gasterophilus* and sheep keds (*Melophagus* spp.), of the family Hippoboscidae, supported Pollock's theory that tsetse and Hippoboscidae evolved from flies resembling Gasterophilidae (Pollock, 1973). Pollock postulated that ancestral adult

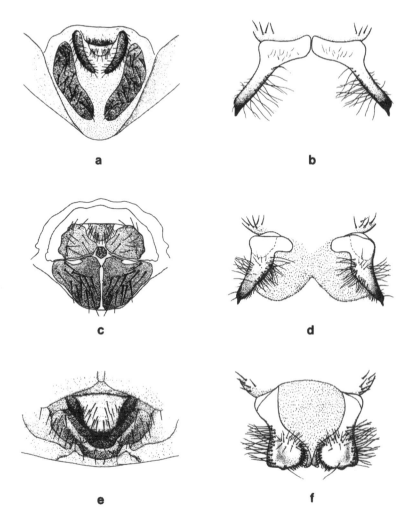

Fig. 2.1. Distinguishing features of *Glossina* genitalia for subgeneric classification. (a) Female external armature – *fusca* group; (b) male superior claspers – *fusca* group; (c) female external armature – *palpalis* group; (d) male superior claspers – *palpalis* group; (e) female external armature – *morsitans* group; (f) male superior claspers – *morsitans* group. (From Jordan, 1993.)

Gasterophilidae must have had functional mouthparts. The reproductive process in Hippoboscidae also involves the hatching of an egg within the uterus, where it is nourished through the larval stages before being deposited. In common with tsetse flies, they deposit third-stage larvae ready to pupate, and have milk glands for nourishing the larva *in utero* (Lenoble and Denlinger, 1982). Pollock questioned established views regarding the

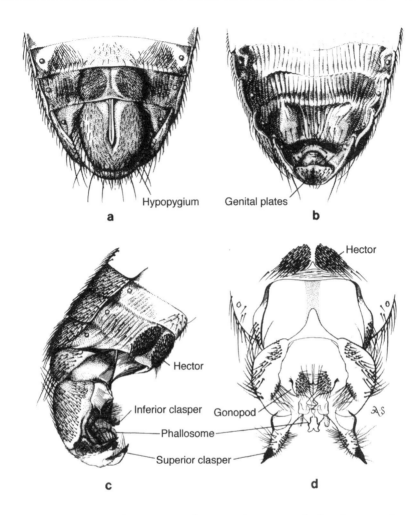

Fig. 2.2. Genitalia of *Glossina*. (a) Male external armature; (b) female external armature; (c) and (d) male genitalia, side and ventral views. (From Jordan, 1993.)

classification of tsetse flies, and other higher Diptera, suggesting the formation of a superfamily Gasterophiloidea to contain the three related families, Gasterophilidae, Hippoboscidae and Glossinidae. Pollock's hypotheses have not found widespread acceptance amongst systematists (e.g. Griffiths, 1976). Lane and Crosskey (1993), in an editors' footnote to Jordan (1993), stated that they 'followed the classification of Wood (1987) and McAlpine (1989) under which the "gasterophilids" are treated as a sub-family of Oestridae ... and the Glossinidae treated as a family of the Hippoboscoidea (= erstwhile "Pupipara" ... We consider the relationships of Glossinidae essentially settled.'

2.1.1 Relationships between subgeneric classification and habitat

In the remote past a dispersal of the genus *Glossina* probably took place, leading to its separation into a number of phylogenetic groups. It is generally accepted that the forest habitat was the original one from which tsetse evolved and that the *fusca* group (predominantly forest flies) is the most primitive (Newstead, 1911b; Bursell, 1958a; Potts, 1970). The *morsitans* group tsetse might have evolved from the more primitive *fusca* group in order to adapt to the appearance of savanna vegetation (Bursell, 1958a). This would have involved changes in the structure of the puparia, rendering them increasingly impermeable to water and thus allowing them to survive desiccation. The vegetation changes took place mainly in the tertiary period, continuing into the quaternary. They resulted principally from the upwarping of the land surface between Lake Victoria and the western rift valley. One result of this was that rivers which drained westwards before the Pleistocene were reversed and, for part of their course, then drained eastwards to the lake or the Nile. In contrast with the accepted theory, Pollock (1974) suggested that evolution of *Glossina* involved a progressive invasion from an original semi-arid grassy woodland habitat of, first, riverine and, later, forest habitats. Pollock believed that *fusca* group tsetse were the most recently evolved because the genitalia of males appeared to be more specialized than those of the *morsitans* group.

The anomalous distributions of *G. brevipalpis* and *G. fuscipleuris* could be evidence of a former eastward extension of forests. Bursell (1958a) noted that the forest flies generally cannot survive in the savanna, whilst the savanna flies cannot, or do not, penetrate the forest. He did not discuss the evolution of the *palpalis* group flies. Three main habitat divisions were recognized by Bursell:

1. Hygrophytic habitats – closely resembling ancestral habitat and characterized by evergreen forest.
2. Semi-arid habitats – in which the breeding places are protected from the action of grass fires.
3. Xerophytic habitats – in which there is a prolonged dry season, when breeding sites are exposed to grass fires.

Machado (1959) also disputed the theory that the *fusca* group represents most closely the ancestral *Glossina* as he believed that there was insufficient evidence that the Oligocene fossil *Glossina* from Colorado was especially closely related to the modern *fusca* group. However, studies of the genetic distance between different tsetse species (Gooding *et al.*, 1991) and results of polymerase chain reaction (PCR) analysis of tsetse endosymbionts (Aksoy *et al.*, 1995) provide further evidence that *fusca* group tsetse are evolutionarily the most ancient. The endosymbionts of *G. brevipalpis* are most divergent, followed by *G. morsitans* and then the *palpalis* group.

Results of molecular genetic analysis of tsetse enzymes, based on poly-acrylamide gel electrophoresis (PAGE), indicate that *G. austeni* is closer to the *palpalis* group flies than to the *morsitans* group and that the classification should be amended (Gooding, 1982). Interestingly, in the initial description of *G. austeni*, Newstead (1912) had stated that it belonged to the *palpalis* group. The superior claspers of the male genitalia do, however, more closely resemble those of *morsitans* group species. Autosomal and X-banding patterns of *G. austeni* chromosomes are identical to those of the *palpalis* group despite its genetic similarity to *G. m. morsitans*. Travassos Santos Dias (1987) suggested that *G. austeni* be placed in a new, monotypic subgenus *Machadomyia*, but it generally continues to be classed in the *morsitans* group.

Speciation of the *palpalis* group tsetse may have arisen from climatic changes at the end of the tertiary period (Machado, 1954). These changes consisted of alternating wet and dry periods, which resulted in alternate regression and extension of forest areas of West Africa. This caused geographical separation of tsetse, following which subspeciation took place (Challier, 1973). Pollock (1974) discussed tsetse evolution in relation to morphological differences in the male genitalia (Fig. 2.2).

2.2 ANATOMY

A brief outline of the anatomy of tsetse is given here; for a complete and detailed description, refer to Mulligan (1970). Figure 2.3 shows the basic anatomy of female tsetse. The first published description of the internal anatomy of the tsetse fly was by Minchin (1905). Despite being published almost 100 years ago, this is still a good description, but it is not readily available. Not long afterwards, Evans (1919) described the external genitalia specifically with regard to improving the classification of female tsetse flies. Some of the specialized features of tsetse anatomy compared with other higher Diptera are mentioned in this chapter.

The exoskeleton and outer integument (skin) of all insects is formed from a cuticle, the main component of which is chitin – a nitrogenous poly-saccharide, which is fairly inert to water, dilute acids and alkalis, and organic solvents. The main non-chitinous constituent of the cuticle is protein; sclerotin, a brownish tanned protein, forms the hard parts of the cuticle. In a newly emerged fly the cuticle is still soft, but hardens after a few hours (Wigglesworth, 1972). Cuticle lines the tracheae, which are branched tubules supplying air throughout the body of the fly, and some other internal organs.

The head
The structure of the head of *Glossina* is typical of the family Muscidae, but the proboscis is long and slender. The mouthparts start from a bulb-like

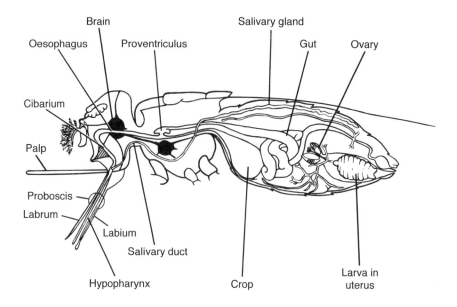

Fig. 2.3. Outline anatomy of *Glossina* (female).

swelling at the base of the head, and are normally ensheathed between the long maxillary palps. The haustellum of the proboscis arises from a relatively small, membranous rostrum which is usually swung back, allowing the bulbous base of the haustellum to lie firmly against the lower part of the head. The haustellum is composed of the labium, the labrum and the hypopharynx. It is the base of the labium that forms the bulb. The labial gutter engages with the labrum and encloses the hypopharynx to form the food canal up which blood is sucked. The cibarial pump forms the sucking apparatus and lies within the rostrum of the proboscis. When the haustellum is retracted the pump is pushed up into the head. Super-contracting muscles of resilin, an elastic structural protein, form part of the functional structure of the cibarial pump for pushing blood out of the pump into the oesophagus (Rice, 1970a, b). Contractions of the cibarial pump are coordinated via stretch receptors on each side of the pump's anterior wall (Rice, 1970c).

The head of the tsetse larva has a reduced skeletal structure compared with other higher Diptera, and has become specialized for its mode of nutrition; thus the mandibular hooks found in other Cyclorrhaphan Diptera are absent (Roberts, 1971).

Male genitalia

The male genitalia (Fig. 2.2) and the process of spermatogenesis of tsetse flies have been described in detail by Itard (1970), Roberts (1972) and Pollock (1974).

Characteristic structures of the male genitalia are the inferior and superior claspers. These are used to engage with the female genitalia during copulation and are useful taxonomical features for species identification. Morphological differences of the head of the inferior claspers are particularly important for identifying subspecies of *G. palpalis*, and variations of these organs have been described from a number of sites in West Africa including Côte d'Ivoire and the Congo (Gouteux and Dagnogo, 1985; Challier and Dejardin, 1987). The length of the head of the inferior claspers was homogeneous in different populations from Côte d'Ivoire but different from those in the Congo. Challier and Dejardin discussed these observations in relation to the taxonomy of subspecies of *G. palpalis*.

Female genitalia

Genitalia of *fusca* group females have an external armature consisting of five plates – one pair dorsal, one pair lateral and a single median sternal plate (Fig. 2.1). There is no mediodorsal plate and in most species of the group there is a well-developed signum. The signum is a chitinized structure on the inner surface of the wall of the uterus, towards the anterior end, the function of which is unclear. It is a useful feature for identification of *fusca* group females.

Palpalis group flies have six external plates; in addition to those present in the *fusca* group flies they also have a small mediodorsal plate but there is no signum. *Morsitans* group flies have only a pair of fused anal plates and a median sternal plate; dorsal plates are generally absent and there is no signum.

Internal female reproductive system

In the abdomen of female tsetse flies are pouches of ectodermal origin, lined with a cuticle. These are called spermathecae, and their function is to store spermatozoa received from the male. The spermathecae are spherical, yellowish-brown in colour, and are connected by spermathecal ducts to the uterus. The sperm stored in the spermathecae will usually be sufficient to fertilize all the ova produced by the female during an average life span. Sperm passes down these ducts to fertilize each ovum as ovulation takes place and the ovum enters the uterus.

A cellular layer consisting of secretory cells surrounds the cuticular intima of the spermathecae (Jordan, 1972). Extracellular ducts lead from the porous head of each secretory cell into the lumen of the spermathecae. Jordan believed that the secretions passing down these ducts, probably containing nutrients, would be the medium in which the spermatozoa are maintained while in the spermathecae. Kokwaro *et al.* (1981) largely confirmed this as they found that the substance in the lumen of the spermathecae was a mixture of carbohydrate, probably neutral muco-polysaccharides, and proteins. These materials accumulate in cavities within the epithelial secretory cells and are then transferred to the main lumen of the receptacle in which sperm are also stored. Kokwaro *et al.* believed these products to be utilized by the sperm stored within the spermathecae.

External marks termed 'cicatrices' on the abdomen of female tsetse have been reported for several species and are believed to be caused by accidents and unsuccessful predation. They are more frequent in older flies and were discussed by Ryan *et al.* (1982a); their epidemiological significance, if any, is not clear.

Chapter 3

Biology

In this chapter, the biology and life cycle of tsetse are described. Some differences in biology of tsetse, compared with other Diptera, and features relevant to epidemiology and control are referred to.

3.1 LIFE CYCLE

The life cycle of tsetse is outlined in Fig. 3.1. The adult female produces a single egg, which hatches to a first-stage (instar) larva in the uterus. After a period of development and moulting, a third-stage larva is deposited on the ground. Females produce one full-grown larva every 9–10 days, which then pupates in light or sandy soil. The adult fly will emerge after a puparial period that varies according to temperature but may be around 30 days at 24°C. Consequently, tsetse flies have a very low rate of reproduction, closer to that of small mammals than to most other insects. Insects such as this are termed *k* strategists and they differ from most insect species, which produce large numbers of eggs and are termed *r* strategists. The reproductive method of tsetse, in which the single egg hatches and develops to a third-stage larva in the uterus of the female fly, where it is supplied with nutrients, is known as adenotrophic viviparity. The maternal 'care' given to each larva enables a higher degree of survival of each offspring than would be the case for *r* strategists, which rely on the numerous offspring produced to offset the greater mortality to which they will be subjected. Generations of tsetse flies overlap, although there is some temperature-determined seasonality in the emergence of new flies. In colonies, female tsetse can each produce at least ten offspring in their reproductive lives, although it is likely that fewer would be produced in the wild. The optimum temperature for the reproductive cycle of female tsetse in laboratory colonies is about 25°C; a constant temperature of 30°C can cause sterility in *G. f. fuscipes* (Mellanby, 1936).

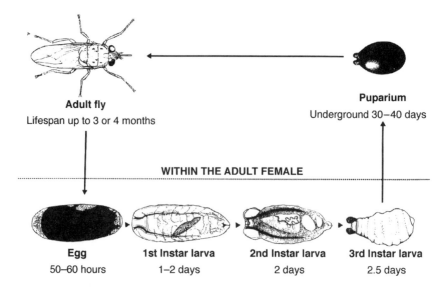

Fig. 3.1. Life cycle of the tsetse fly (*Glossina* spp.).

3.2 SPERMATOGENESIS, COPULATION AND FERTILIZATION

Spermatogenesis only occurs during the puparial stage of males and ceases once the male fly has emerged (Curtis, 1968a; Itard, 1970). During copulation, sperm are transferred from the male to the uterus of the female. The sperm are contained in a spermatophore in most, and probably all, tsetse species. Spermatophores, first described by Pollock (1970), are prominent colourless objects found in the uterus of recently mated females, and enclose a single, closely packed sperm mass. Pollock's work indicates that the spermatophore is formed in the female uterus during lengthy copulation and that the sperm are ejaculated into it during the jerking phase at the end of copulation. The spermatophore wall comprises two layers, similar in components to secretions of the male accessory glands (Kokwaro and Odhiambo, 1980). It is thought that these glands produce spermatophore precursor secretions that protect the sperm mass whilst migrating to the spermathecae (Pollock, 1974). The secretory proteins of the male accessory reproductive gland and the spermatophore share common immunological characteristics, different from those of the testes (Kokwaro *et al.*, 1986).

Under laboratory conditions a single male is capable of inseminating up to 15 females, and male *Glossina morsitans* may copulate with females even before they are mature enough to inseminate them. The period of maturation is approximately 3 days and does not depend solely upon a bloodmeal being taken. *G. austeni* males were less sexually responsive than *G. morsitans* shortly after emergence but nearly always inseminated when they did copulate. There was less increase in copulation performance in flies starved for 4 days but copulatory behaviour did not decrease. At emergence, accessory glands are larger in male *G. austeni* than in *G. morsitans* and in both species they cease to grow after day 3. The size of the accessory glands and the duration of copulation were related to insemination success, but the latter was quite variable. Sperm transfer apparently occurs at the very end of copulation.

As the egg of a female fly passes from the ovary into the uterus it is fertilized by sperm, which pass down the duct from the spermathecae (where they are stored after copulation) to the uterus. Fertilization only takes place when the egg has completely descended into the uterus, and its micropyle comes to rest opposite the opening of the spermathecal ducts. The supply of sperm is then controlled by spermathecal valves (Roberts, 1973a). The stored sperm will normally be sufficient for the reproductive life of the female. Therefore, as Glasgow (1967) put it, 'A mean male age at death of nine days would enable males to discharge the reproductive function.' It is likely, however, that in most areas of tsetse infestation the life span of male flies would be significantly longer than this.

3.3   LARVAL DEVELOPMENT

A description of the eclosion of *G. swynnertoni* larvae (Jackson, 1948a) is likely to be typical for all tsetse species. The mouth of the first-instar larva is anterodorsal, blocked by a chitinized median tooth about 0.05 mm long. This larval 'egg-tooth', which is lost with the skin of the first instar, is used to puncture the chorion, or eggshell, which then splits along a line of weakness on the dorsal side, allowing the first-instar larva to hatch. Formerly, muscles in a structure called the choriothete were thought to aid removal of the chorion after its puncture by the larval egg-tooth. The choriothete, which supports the developing embryo, was also thought to be involved in removal of the cast skins of the moulting larva but no evidence of this was subsequently found (Tobe and Davey, 1971; Roberts, 1973b).

The choriothete of *G. austeni* is a highly modified invagination of the antero-ventral wall of the uterus, forming a tongue-like organ attached to the uterine floor (Tobe and Davey, 1971). Secretory cells in the choriothete epithelium produce a mucoprotein that serves as a cement to fasten the chorion to its surface (Roberts, 1973b).

The third-instar larvae have respiratory lobes called polypneustic lobes, the surfaces of which are covered in narrow slits called stigmata ('supernumerary stigmata', Newstead *et al.*, 1924). The polypneustic lobes are hard, dark, sclerotized structures, which differ morphologically between subgenera (Zdárek *et al.*, 1996). Air enters through the stigmata, which open into outer air chambers, three in each lobe, and communicate with inner air chambers that are continuous with the lateral tracheal trunks of the larva.

The chambers are believed to act as filters (Zdárek *et al.*, 1996). Contraction of respiratory muscles causes intake of air but this mechanism is only functional during the last third of intra-uterine life (Bursell, 1955). Bursell believed that there were valves in the air tubes, which opened and closed the stigmata, but Zdárek *et al.* (1996) found no evidence for this, or of sensory innervation of the lobes, which had previously been suspected.

3.4   LARVIPOSITION AND PUPARIATION

The behaviour of the larvae of tsetse, and other flies, during metamorphosis and pupariation was the subject of reviews by Zdárek and Denlinger (1993) and by Denlinger and Zdárek (1994). These authors pointed out that unlike most dipteran larvae, which make a commitment to metamorphose early in the third instar, the third-instar larva of tsetse is only competent to pupariate after attaining full size (33.6 mg for *G. morsitans*). This is disputed by Jordan (A.M. Jordan, Nairobi, 1998, personal communication), who claims that much smaller larvae can also pupariate.

Shortly after deposition by the female, the larva attempts to burrow. Uric acid from the anus then spreads over the larva and is believed to protect it from predation. It is unclear whether the uric acid is secreted or is simply excretion of accumulated waste from intra-uterine life. Whether the larva succeeds in burrowing, and the depth to which it can burrow, depend on particle size, moisture content of the soil and possibly soil temperature. Larvae have difficulty in burrowing in soil of too fine a texture. For *G. m. submorsitans* and *G. tachinoides,* the optimum particle size is between 0.2 and 1 mm, while larvae of *G. palpalis* burrowed best in forest soil of particle size 1.8–2.5 mm (Parker, 1956a). Larvae of *G. austeni* and *G. brevipalpis* burrow deeper in damp sand (8.5 cm) than in dry sand (2–3 cm) (Stuhlmann, 1907) although pupariation is delayed for several hours by contact with water (Zdárek and Denlinger, 1991). Larvae exhibit a strong photonegative response, but this is not essential for stimulating burrowing into the soil.

In laboratory experiments involving temperature changes, regular cycles of sudden increases in temperature of about 6°C for 2 h caused a reduction in larviposition during each period of increased temperature and a significant increase before the heat impulse. When the temperature was allowed to fall rapidly, larviposition rate doubled in the following 15 min (Robinson *et al.*, 1985).

3.4.1   Control of pupariation

The neuroendocrine system of tsetse larvae produces ecdysone, a hormone which mediates the tanning of the larval cuticle and formation of the puparium. If the larvae are ligatured, so that the brain is separated from the posterior end, this results in pupariation of the posterior part (Langley, 1967a). (This was not the case in similar experiments, with other Muscidae, in which ligaturing prevented pupariation of the posterior part.) Langley's explanation for this was that a sufficient quantity of ecdysone must be present in tsetse larvae at the time of larviposition to mediate puparium formation and tanning. However, the central nervous system could inhibit the action of ecdysone in response to sensory stimulation; this would delay pupariation until the larva reaches suitable environmental conditions. Ligaturing the larva cuts off the influence of the nervous system and prevents inhibition of ecdysone in the posterior part. A less likely alternative explanation was that the ecdysone was produced in the posterior part. Neural regulation of pupariation was investigated further by Zdárek and Denlinger (1992b) who concluded that it was only partially controlled by hormonal factors.

3.4.2 Puparial metabolism and duration of the interlarval period

In the laboratory, the interlarval period of tsetse is related to temperature, although non-linearly (Glasgow, 1963). In field studies, Dransfield *et al.* (1989) detected a 9-day interlarval period independent of temperature. Mark–release–recapture experiments using flies of known age led Hargrove (1994) to conclude that there was a relationship between temperature and interlarval period that resulted from varying larval growth rate. In addition to temperature, the interlarval period of *G. morsitans* could significantly increase as a result of food deprivation during the adult fly's pregnancy (Saunders, 1972). Timing of larval deposition could also be affected by food availability (Langley and Stafford, 1990).

Fat is the main source of energy during puparial development, and as its utilization is temperature dependent, predictions of puparial duration can be made. Puparial duration can vary widely between tsetse species, localities and seasons, from 20 to 90 days (Phelps and Burrows, 1969a). Experimentally, the optimum incubation temperature for *G. m. morsitans* puparia is 23°C; in a study by Potts (1930a), puparia did not survive beyond 13 days when kept at 35°C. In their predictive model, Phelps and Burrows assumed that rates of development were constant throughout puparial life, under constant temperature. Nonetheless, they recognized from other studies (Rajagopal and Bursell, 1965) that there might be variation in metabolic rates. Although higher temperatures generally increase the rate of metabolism and thus shorten the puparial duration, very high but sublethal temperatures can cause a delay between completion of development and eclosion. The experimental data suggested that this source of error was not great. Unfortunately, although accurate predictions could be made in the laboratory, large variations between temperature regimes in different breeding sites made precise prediction in the field difficult. Consequently, only average puparial durations, across a range of sites in an area, were meaningful. Predictions could otherwise be made for just one site and not extrapolated to others. The formula for determining puparial duration of *G. morsitans* kept at alternating temperatures of 25°C and 20°C for 24-h periods, derived by Phelps and Burrows (1969a), is:

$$d^* = [(n + 1)/n]\, x_{max}$$

The puparial duration for puparia collected from the field could be determined from this formula by adjusting the observed time taken by the last adult to emerge (x_{max}) by a function of the sample size, where $d^* =$ estimated duration and $n =$ number of adults emerging.

Formulae were also derived for determining the puparial period and interlarval period at a constant temperature (Glasgow, 1963). Of course temperatures are not constant under natural conditions for these lengths of time; but for certain seasons, and as temperatures are more stable in the soil, it can be assumed that variation would not be great enough to invalidate calculations based on these formulae. The formulae, from Glasgow (1970), are:

$$\text{Puparial period in days} = 1/[0.0323 + 0.0028\,(t - 24)] \qquad (3.1)$$

$$\text{First larva born on day} = 1/[0.0661 + 0.0035\,(t - 24)] \qquad (3.2)$$

$$\text{Interlarval period} = 1/[0.0859 + 0.0069\,(t - 24)] \qquad (3.3)$$

Hargrove (1994) pointed out that Glasgow published no data supporting his derivation of a formula relating interlarval period to temperature. Furthermore, there are differences in puparial duration between the sexes and possibly between species (Saunders, 1962). In laboratory colonies of tsetse, female flies emerge 1–2 days before males (Birkenmeyer and Dame, 1975). Progress is being made in utilizing this differential in emergence times for the automation of mass-rearing facilities for the sterile male tsetse control technique (Chapter 18, section 18.4).

In experiments at two sites in Zimbabwe, in which wild and marked flies were released 12 h after emergence, the rate of growth of oocytes increased with temperature for *G. m. morsitans* but not for *G. pallidipes*, and was always higher than in laboratory studies. For both species, at one site, where the mean temperature was 22°C, the first larva was produced at about day 18 and subsequent larvae at 11-day intervals. For flies released at the second site, the intervals decreased with temperature by about 0.5 days per 1°C and were 2 days shorter than at the first site for any given temperature (measured in a Stevenson screen).

Metabolic activity continues in puparia maintained at a constant 16°C but adults do not emerge. Glasgow (1970) stated that below 17°C and above 32°C emergence would not occur as fat reserves were exhausted before development was completed, but qualified this by referring to Rajagopal and Bursell (1965) who indicated wider temperature limits for completion of development. The median lethal temperature for 26-day-old puparia for

6-h exposures is 40°C (Phelps and Burrows, 1969b). Small variations from these experimental observations may occur in flies from different areas in Africa that may have adapted to local conditions. Evidence from Zimbabwe suggests that *G. m. morsitans* lives at temperatures 2–6°C lower than the average Stevenson screen temperature (Hargrove and Coates, 1990).

In the first 5 days of puparial life, glycogen and protein are formed from free amino acids. Subsequently, the protein content falls until the final period of puparial development when the developing exoskeleton is apparent. At that time glycogen and free sugars are depleted, probably because they contribute to chitin synthesis (Stafford, 1973). Survival of successfully emerging adults depends upon sufficient fat reserves to support them until their first bloodmeal.

There is some evidence that relative humidity, as well as temperature, affects the breeding of tsetse and larviposition site selection. In Botswana, the rate of reproduction of *G. morsitans* was lowest during the hot dry months, when holes in the ground were important breeding sites (Atkinson, 1971a). During the hot dry season such sites were significantly more humid than tree holes, fallen logs, or under leaf litter (Atkinson, 1971b). The greater proportion of puparia deposited in holes in the dry season was believed to reflect a behavioural change, which avoided increased mortality in that season. Survival of puparia is dependent upon relative humidity, and in laboratory experiments 11 out of 12 puparia hatched at 98% relative humidity, compared with only two out of 15 at 11% (Buxton, 1936). This effect was not seen during the hot dry season in the field in Nigeria, possibly because the soil in breeding sites had a high moisture-holding capacity, thus providing a suitable microclimate (Buxton, 1936). A mortality of up to 50% was observed in puparia of *G. morsitans* collected in natural breeding sites (Potts, 1930a).

3.4.3 Eclosion of adults from puparia

The adult fly emerges from the puparium, and from the ground, with the help of a structure at the front of the head called the ptilinum. This structure, found in certain groups of Diptera, takes the form of a sac folded in a cavity of the head. It is evaginated at the time of eclosion and rhythmic contractions of the thoracic and abdominal musculature allow the adult fly to break out of the puparium and reach the soil surface.

The peak emergence of tsetse is in the mid-afternoon, predominantly between 15 h and 18 h for *G. morsitans*, depending on temperature. This is in contrast with most Diptera, which emerge at dawn (Dean *et al.*, 1968). Again in contrast with most Diptera, the periodicity of eclosion is temperature dependent, rather than daylight dependent (Denlinger and Zdárek, 1994). Adult *G. austeni* emerging from their puparia have significantly larger fat reserves than *G. m. morsitans*, even though the ratio of fat to dry weight is similar (Langley, 1971). As there were no differences in the rate of oxygen consumption, Langley (1971) concluded that some substrate other than fat might be used by *G. austeni.*

Immediately after eclosion the wings, abdomen and thorax are highly compressed, although the legs are already full sized. The newly emerged adult tsetse therefore has to expand by about 90%, compared with an increase in body length of about 30% for blowflies (*Sarcophaga* spp.). This expansion takes place as a result of air taken into the gut, by means of the cibarial pump. The adult must then feed in order to complete growth of flight muscles and the endocuticle (Hargrove, 1975a).

Sex ratios of tsetse flies emerging from puparia are close to unity but, as female flies live longer than males, more females are generally found in a representative adult population sample. In unbiased samples, females would comprise between 70 and 80% of an average population.

Chapter 4

Physiology

How long tsetse flies live and how far they can fly are simple questions on which research has been conducted, seeking to answer the related questions: How far will a fly's energy reserves allow it to fly? What are these energy reserves? How are these energy reserves managed efficiently for optimum survival of the species? How often must a fly feed to maintain necessary energy reserves for flight, survival and reproduction? Finally, complex questions may be asked relating to the way in which a tsetse fly's behaviour is affected by its energy requirements. Many of these questions are still controversial and require a good understanding of the physiological processes associated with the uptake and metabolism of a bloodmeal.

Professor Einar Bursell pioneered research into tsetse physiology and made significant contributions to the pool of knowledge (e.g. Bursell, 1963b, 1966a, 1975a). He was followed more recently by Langley (e.g. 1965, 1966, 1970), whilst Rogers and Randolph have tried to link an understanding of tsetse physiology with ecology (e.g. Randolph and Rogers, 1978, 1981; Rogers and Randolph, 1978b, 1986a). The physiology of tsetse was thoroughly reviewed by Langley (1977). These and more recent advances in this area are discussed here.

Tsetse physiology is closely linked to behaviour and, thus, is a significant component influencing the ecology of tsetse and trypanosomosis epidemiology. Tsetse behaviour also has important implications for control techniques. For example, Colvin and Gibson (1992), in a review of tsetse host-seeking behaviour, showed how this behaviour made them vulnerable to control using baits. This vulnerability was associated with their being obligate blood feeders and having a low reproductive rate. Some aspects of behavioural physiology are discussed in Chapters 8 and 15 to 18, on ecology and tsetse control, respectively.

4.1 PROCESSING OF THE BLOODMEAL

Processes of metabolism of the bloodmeal in tsetse were reviewed by Bursell *et al.* (1974). Some of that information, together with more recent research findings, is summarized here.

4.1.1 The role of saliva

Tsetse flies discharge saliva whilst probing the host for a bloodmeal. As saliva may contain trypanosomes, parasites can be exuded each time the fly probes before successfully obtaining a meal. The epidemiological significance of this led to investigations into the frequency of probing by tsetse flies before successfully obtaining a bloodmeal. The effect of trypanosome infections in tsetse on feeding behaviour and success are discussed in Chapter 11 (section 11.7). Tsetse saliva, like that of many other bloodsucking insects, contains a powerful anticoagulant enzyme. The anticoagulant function of tsetse saliva was first suggested by Stuhlmann (1907), later demonstrated by Yorke and Macfie (1924) and identified as an antithrombin (Lester and Lloyd, 1928). Vector saliva might play an important role in arthropod disease transmission by creating a suitable environment for the disease agent in both the host and vector (Titus and Ribeiro, 1990). Saliva has vasodilatory and anticoagulant properties that make it easier for the fly to find and obtain its bloodmeal. Two platelet aggregation inhibitors have been identified from saliva of *G. morsitans* (Mant and Parker, 1981). In a saline extract from salivary glands, these inhibitors completely abolish thrombin-induced platelet aggregation. Saliva may also have immunosuppressive, anti-inflammatory properties that would prevent an adverse reaction in the host.

Tsetse saliva does not break down bloodmeal erythrocytes, but saliva of *G. morsitans* exhibits cholinesterase activity, similar to enzyme activity of saliva in most insects in its reaction to inhibitors and ability to hydrolyse a wide range of substrates (Golder and Patel, 1982). The quantity of saliva secreted by tsetse appears to increase as they become hungrier, but there is no evidence that infected tsetse salivate more copiously than non-infected ones (Youdeowei, 1976).

4.1.2 The role of digestive enzymes

A number of studies have been carried out to investigate the enzymes involved in digestion of the bloodmeal. In addition to improving the under-standing of digestive processes in tsetse, the mechanisms by which trypanosomes survive and develop, or alternatively are controlled in the hostile environment of the tsetse alimentary tract, have been elucidated.

Stimulated by the feeding process, trypsin begins to be secreted from the anterior and posterior portions of the midgut. Some questions arose regarding the stimuli for trypsin or protease production following the conflicting observations of Akov (1972), Langley (1966) and Langley and Abasa (1970). Laboratory experiments showed that trypsin secretion could be stimulated by feeding flies on goats and on various blood components, except for washed bovine erythrocytes (Gooding, 1974). The quantity of secretion was correlated with the amounts of protein and carboxypeptidase B in the posterior midgut (Gooding, 1974, 1977a), which led Gooding to conclude that the amount of protein, rather than the size of the meal ingested, determined the level of trypsin in the posterior midgut. This hypothesis agreed with Akov's observations rather than that of Langley. Other proteolytic enzymes (carboxypeptidases, a chymotrypsin-like enzyme and aminopeptidases) have been isolated from *G. morsitans* and subspecies of *G. palpalis* (Cheeseman and Gooding, 1985). The level of activity of these enzymes varies with fly species and age. Figure 4.1 shows the sequence of digestive processes diagrammatically.

Two fibrinolysin proteases from the midgut of adult female *G. m. centralis* have been purified and characterized (Endege *et al.*, 1989). These were serine proteases similar to trypsin-like enzymes detected in *G. m. morsitans*. An assay developed by Stiles *et al.* (1991) showed that although proteases were present in the posterior midgut of *G. p. palpalis*, the anterior midgut possessed no detectable proteases but did contain subspecies-specific protease (trypsin) inhibitors. A protease inhibitor in the anterior midgut of *G. m. morsitans* is produced cyclically in relation to the host feeding cycle (Houseman, 1980). This inhibits some activity of protease VI (a trypsin-like enzyme) and trypsin, but does not affect protease VI hydrolysis of haemoglobin.

An increase in enzyme activity in teneral adults suggests that enzyme production involves neuroendocrine control, which is mediated by the stomatogastric nervous system and responds to inflation of the crop with air during expansion of the adult (Langley, 1967a). No nervous connection exists between the crop and the midgut, and protease production was there-fore thought to be controlled humorally through the distension of the crop acting on the neuroendocrine system (Langley, 1966). A haemolytic agent, haemolysin, which lyses erythrocytes, has been detected in the digestive section of *G. morsitans* midguts (not in the anterior non-digestive section) (Gooding, 1977c).

Female tsetse, in common with other female bloodsucking insects, ingest larger bloodmeals and have higher levels of digestive proteases to deal with it than do males (Gooding, 1977a). Protein in the tsetse bloodmeal stimulates enzyme activity in the midgut. Specifically, trypsin inhibitors found in bloodmeal serum (termed 'secretogogues' by Gooding, 1977a) are believed to stimulate production of trypsin and carboxypeptidase B, in the fly midgut.

Digestion of blood proteins appears to take place only in the posterior section of the midgut, not the anterior, and involves the six enzymes identified by Gooding (1977a,b), which convert proteins to peptides and free

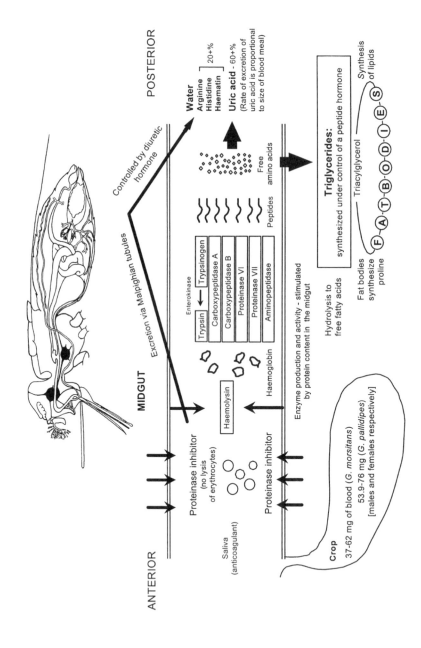

Fig. 4.1. Digestive processes in tsetse flies (*Glossina* spp.).

amino acids. The rate of digestion in wild flies is higher than in laboratory-reared flies, probably due to an intrinsically lower rate of secretory activity of the neuroendocrine system, which controls midgut protease production (Bursell *et al.*, 1974). The neuroendocrine system of tsetse secretes continuously, in contrast with other insects in which there are cycles of activity (Langley, 1967b).

Interactions between digestive enzymes and trypanosome infections
An enterokinase digestive enzyme in adult *G. m. morsitans* activates bovine trypsinogen to trypsin (Vundla and Whitehead, 1985). An analogous trypsin enzyme in *Aedes aegypti* mosquitoes is responsible for destroying ookinetes of *Plasmodium gallinaceum in vitro* (Gass and Yeates, 1979). This led Vundla and Whitehead (1985) to speculate that digestive enzymes, such as enterokinase, may retard the establishment of *T. b. brucei* in tsetse. *Trypanosoma b. brucei* causes a concentration-dependent decrease in trypsin, or trypsin-like enzymes, in crude midgut homogenates of *G. m. morsitans*, suggesting that trypanosomes might overcome the hostile tsetse-fly midgut barrier by inhibition of this enzyme activity (Imbuga *et al.*, 1992). Experimentally, D-glucosamine inhibits the activity of midgut trypsin in *G. morsitans* and could play a role in susceptibility of tsetse to trypanosome infection (Osir *et al.*, 1993). A midgut lectin–trypsin complex found in *G. longipennis* also showed specific activity for glucosamine (Osir *et al.*, 1995). Effects of trypanosome infection on longevity, reproductive performance and feeding behaviour, and influence of enzymes and enzyme inhibitors on the establishment of trypanosome infections of tsetse flies, are discussed further in Chapter 11.

4.1.3   Water balance and excretion

The large increase in weight of tsetse flies after feeding causes difficulty in flying, and increases their vulnerability to predators. It is therefore important for them to lose some of the weight of the bloodmeal rapidly. This is done by rapid excretion of excess water from the bloodmeal via the Malpighian tubules whilst maintaining the water balance and without risking desiccation. Water loss is a function of size and can be estimated from its relationship to residual dry weight. Size difference between species of tsetse can be striking. One of the biggest species, *G. brevipalpis*, is about 5 times as heavy as the smallest, such as *G. austeni* or *G. tachinoides*.

One means of reducing water loss in resting flies is the closure of spiracles. The rate of water loss by *G. m. morsitans* decreases in the course of desiccation due to a progressive increase in the degree of spiracular control (Bursell, 1957a). The rate of water loss of flies whose spiracles had been blocked, or of flies whose spiracles were kept open by exposure to carbon dioxide, was an inverse linear function of humidity. Normal flies showed an asymptotic curve with relatively low rates of loss in dry air. This is a result of spiracular control: the drier the air, the greater the degree of closure of the spiracles. The thoracic spiracles are normally closed at 25°C, and open in response to carbon dioxide, or high temperature, resulting in a slight decrease in body temperature, possibly due to evaporation of body moisture in dry air (Edney and Barrass, 1962). In the puparial and pharate adult stages the supernumerary stigmata are blocked by remnants of the third-instar larva tracheal apparatus and no water is lost through them. The size of a fly could theoretically be expected to show some relationship to its habitat and the need to reduce water loss; however, Bursell (1959a) found no correlation between resistance to desiccation and habitat. He tentatively concluded that the water balance of adult tsetse was no obstacle to the invasion of semi-arid habitats.

The water content of a tsetse bloodmeal is reduced from 79% to about 55% within 3 h after feeding (Bursell, 1960a). This requires an active transport mechanism in which water is absorbed from the bloodmeal, across the wall of the anterior midgut, into the haemocoel. The transport mechanism involves movement of sodium ions and is dependent upon sodium concentration. This is helped by a favourable osmotic gradient between the gut contents and the haemolymph in the presence of a minimal (0.05 M) NaCl concentration (Langley and Pimley, 1973). Water extracted from the bloodmeal rehydrates the tsetse tissues and the excess is excreted. Tsetse can reabsorb sodium from the primary urine produced by the Malpighian tubules during diuresis. This reabsorption of ions may lead to an increased circulation of water through the excretory system (Gee, 1977). A variety of anions are able to support diuresis, at rates which are related to the size of the anion (Peacock, 1986). Bicarbonate is an exception and can support a high rate of diuresis. The water content of faecal material varies according to the relative humidity of the tsetse's environment during the hunger cycle.

The large amount of nitrogen from amino acids that the tsetse fly's food contains must also be excreted in order to avoid water loss and desiccation. Uric acid is the main excretory product, representing the final product of nitrogen metabolism in most insects, and makes up more than 60% of the dry weight of tsetse faeces. A further 20% or more is excreted as arginine, histidine and haematin, and their elimination is a characteristic feature of tsetse excretory physiology (Moloo, 1978). Elimination of excess water containing those amino acids, as well as salts (Gee, 1975a), occurs about 30 min after feeding, via the Malpighian tubules (Bursell, 1960b; Moloo and Kutuza, 1970), and is controlled by a diuretic hormone (Gee, 1975a). The hormone, produced in response to distension of the abdomen, can apparently be destroyed in the Malpighian tubules, indicating that they, too, could play a role in control of diuresis (Gee, 1975b). The rate of uric acid excretion is proportional to the bloodmeal size. During puparial development the level of uric acid rises to reach a plateau after the first 5 days and does not drop until emergence of the adult fly (Brown *et al.*, 1973), at which time the uric acid, together with other excretory products, are excreted in the meconium. It would appear that the rise in uric acid levels during the first 5 days results from metabolism of free amino acids, and no longer rises thereafter because no free amino acids remain.

4.1.4 Metabolism, nutrition and digestion – adaptation of the metabolism of the bloodmeal to the reproductive cycle of female tsetse flies

Studies on bloodmeal digestion in tsetse, and subsequent metabolism of digestive products, have been carried out to determine the optimum frequency of feeding and the theoretical daily capacity for flight, and to link these physiological capabilities to tsetse behaviour. The possible effects of these processes on establishment of trypanosome infections in tsetse have also been investigated as all of these aspects may play a role in the epidemiology and control of trypanosomosis.

Table 4.1 summarizes estimates of the mean weight of bloodmeals for three tsetse species obtained for wild and laboratory-reared flies. Laboratory-reared tsetse, which are generally offered food at frequent intervals, are likely to take smaller bloodmeals than wild flies, which feed less frequently.

The bloodmeal initially passes into the midgut, but back-pressure then causes it to enter the crop, where it remains until it is actively emptied back into the midgut. This is regulated by peristaltic movements of the upper digestive system and by sphincter muscles. Large differences were observed between the rate of digestion of captive flies and wild flies as measured by the precipitin test (Weitz and Buxton, 1953). Only 28% of the bloodmeals of wild *G. swynnertoni* could be identified after 3 days, compared with 90–100% of bloodmeals from captive flies 3 days after feeding. There is some evidence that increasing the general level of activity of tsetse can increase the rate of digestion. Stages in bloodmeal metabolism are shown diagrammatically in Fig. 4.2.

Female tsetse have a significant capacity to store nutrients destined for larval development. A bloodmeal taken in early pregnancy provides these

Table 4.1. Estimates of the mean size of tsetse bloodmeals.

Tsetse species	Mean wet mass (mg)		Reference
	Male	Female	
G. morsitans	37.3	62.3	Taylor, 1976 (wild and laboratory-reared flies)
G. pallidipes	53.9	76.3	Taylor, 1976 (wild and laboratory-reared flies)
G. brevipalpis	113 (range 80–147)	–	Moloo and Kutuza, 1970 (wild flies)
G. brevipalpis	60 (range 34–111)	–	Moloo and Kutuza, 1970 (laboratory-reared flies)

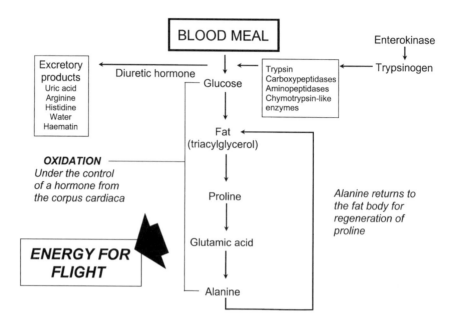

Fig. 4.2. Metabolism of the bloodmeal of tsetse flies.

nutrients for the developing larva, which is particularly important during the late stage of pregnancy when the larva is developing rapidly. During the first half of the 9-day reproductive cycle, lipids – mainly in the form of triglycerides resulting from digestion of a bloodmeal – are synthesized and stored as triacylglycerol in the female's fat-body (Pimley, 1985). Normal growth of the larva is a function of optimum feeding throughout the pregnancy cycle (Moloo, 1976a,b).

Most lipid synthesis takes place 18 h after feeding. During the second half of the reproductive cycle, lipids are transported from the fat-body to the larva in the uterus. The host's blood affects neither the rate of synthesis of the lipids nor their composition, although the rate of synthesis does change during the reproductive and feeding cycles. Synthesis of lipids in tsetse seems to be under the control of a peptide hormone that is produced in the medial neurosecretory cells of the midbrain (Pimley and Langley, 1981, 1982). This hormone is stored in the corpus cardiaca prior to release into the haemolymph during the fly's reproductive cycle. The role of the hormone may be to reduce metabolic demands on the fat-body during the latter half of the pregnancy cycle (Pimley and Langley, 1982). In this way, nutrients obtained at that time could be passed on to the larva without competition from maternal fat cells. Pimley and Langley's studies suggested that the hormone stimulates synthesis of the amino acid, proline, from alanine, and hydrolysis of triglycerides to free fatty acids. Nutrients can also be synthesized from glucose in the third-instar larva, shortly after formation of the puparium. Tsetse larvae synthesize large amounts of glutamic acid, whereas in adults, proline is the main product synthesized (Moloo, 1976a,b).

Relating digestive metabolism to energy needs, Bursell (1965) estimated that for every 1 mg dry weight of blood ingested by a tsetse fly almost 0.5 mg must be excreted in order to dispose of the nitrogen. Considering that synthesis of a complex molecule like uric acid would involve expenditure of some energy, Bursell (1965) estimated that considerably less than 50% of the bloodmeal would become available as a source of metabolic energy.

The respiratory rate of tsetse increases after feeding, possibly reflecting the transformation of the bloodmeal into lipid and proline food reserves as well as into uric acid (Rajagopal and Bursell, 1966). The amount of energy required for this transformation was estimated from the oxygen consumption of resting flies, which increases with increasing temperature. Depletion of fat reserves, before adult emergence, could be a major cause of mortality from starvation, particularly at the limits of tsetse distribution, such as in parts of Zimbabwe (Rajagopal and Bursell, 1966).

4.1.5 The role of symbiotic organisms

There is a giant-cell zone, termed the mycetome, in the anterior midgut epithelium (Stuhlmann, 1907). This zone contains intracellular rod-like structures, sometimes called bacteroids, recognized by Stuhlmann as symbiotic microorganisms that were thought to be important in the process of digestion and possibly in providing the tsetse with additional nutrients. Wigglesworth (1929) reviewed the process of digestion in tsetse, and of the enzymes involved, but at that time the function of the so-called bacteroids remained unclear. Bacteroids are common in many insects and may confer some advantage. In insects with restricted diets, such as tsetse feeding on blood or aphids feeding on plant sap, specialized mycetomes containing these endosymbionts are well developed. Although their function remains unclear, they are probably responsible for synthesis of vitamins (folic acid and vitamin B complex metabolites), contributing to tsetse nutrition (Wink, 1979; Nogge, 1981). Folic acid may also be utilized by trypanosomes infecting tsetse, which could be of significance to their transmission. Endosymbionts may also be necessary to support intra-uterine larval development in female tsetse. If tsetse are fed on blood containing antibiotics (e.g. oxytetracycline and sulphaquinoxaline) or lysozyme, their reproduction is adversely affected because the symbionts are killed or damaged, leading to secondary sterility (Nogge, 1976). Furthermore, offspring from females treated in this way were sterile and usually died within 10 days (Aksoy *et al.*, 1995). Similarly, tsetse fed on rabbits immunized with symbionts became aposymbiotic and their fecundity was drastically reduced, but without affecting their longevity (Nogge, 1978). The number of symbionts in tsetse may be directly related to metabolic processes in the fly (Nogge and Ritz, 1982).

The ultrastructure of tsetse bacteroids is similar to Gram-negative bacteria, rather than rickettsia, which are much smaller (Reinhardt *et al.*, 1972).

DNA probes and polymerase chain reaction (PCR) assays have been developed to identify these tsetse symbionts and three types have now been demonstrated. In addition to the larger bacteroids in the mycetome, there are smaller ones found in the midgut epithelium, ovaries and early embryos. Bacteria in the midgut belong to the gamma subdivision of *Proteobacteria*, whereas ovarian symbionts are of the alpha subdivision and are probably *Rickettsia* (O'Neill *et al.*, 1993). The presence of two distinct types of organism, previously inferred from their differing morphology, was confirmed. Not all tsetse species carry both types of bacteria; thus, O'Neill *et al.* (1993) found the rickettsial *Wolbachia* genus in *G. austeni*, *G. brevipalpis* and *G. m. morsitans*, but not in *G. fuscipes*, *G. palpalis* or *G. tachinoides*. This has implications for the vector competence of those species, which are discussed in Chapter 11.

Wolbachia bacteria are trans-ovarially transmitted through the milk glands (Pell and Southern, 1975). This was suspected by Ma and Denlinger (1974) after they found them in the larval midgut of *G. palpalis* and extra-cellularly, in the lumen of the milk glands. Intracellular bacteroids were also found in the ovaries of *G. austeni* (Huebner and Davey, 1974).

The rickettsial *Wolbachia* bacteria in the ovaries confer some repro-ductive advantage to infected tsetse, suggesting possibilities for control based on hybrid sterility. A mating between an uninfected female tsetse and an infected male will not result in fertile offspring due to cytoplasmic incompatibility, whilst other combinations (for example, an infected female and an uninfected male) can successfully reproduce (Beard *et al.*, 1993a). The symbionts found in the mycetome area of the midgut are classed as primary or *p*-bacteria (endosymbionts); those found further down the midgut, associated with midgut epithelial cells, are secondary or *s*-bacteria (endosymbionts) (Aksoy, 1995). Phylogenic studies have shown that these bacteria are related to the primary symbionts of aphids. Furthermore, there appears to be a relationship between the phylogeny of tsetse and symbiont subgenera, indicating that the symbionts were acquired before speciation of tsetse, as there is no horizontal transmission of symbionts between species. PCR assays have been used to describe mycetome endosymbionts and their phylogenetic lineages (Aksoy *et al.*, 1995).

4.2   METABOLISM – ADAPTATION TO THE ENVIRONMENT AND MODE OF LIFE

The haematophagous and viviparous habits of tsetse flies require a large musculature relative to that of other insects. This, in turn, implies a high energy consumption requiring a specialized metabolism adapted to their way of life. Aspects of this metabolism are discussed here.

Products of bloodmeal digestion are stored as lipids, which constitute the main food reserve of tsetse (Jack, 1939). However, fats cannot be quickly mobilized for the needs of flight metabolism. Initially, carbohydrate was thought to sustain the early phases of flight (as in most Diptera), but no glycogen reserves were found in tsetse, while amino acid content decreased as flies became hungry (Bursell, 1960b). Only minute amounts of carbohydrates can be found in tsetse, and enzyme systems for their oxidation are insufficiently developed to allow significant flight energy to be produced (Norden and Paterson, 1969). Activity of enzymes to break down carbohydrates is far lower in tsetse than in blowflies (*Sarcophaga* spp.), in which carbohydrates are present in considerable amounts. These and subsequent studies (Birtwisle, 1974; Hargrove, 1976a) showed that metabolites other than carbohydrates provided most of the flight energy for tsetse flies. Glycogen, which occurs in the proventriculus of tsetse, may be used for supplying energy for secretory processes, but not for flight (D'Costa *et al.*, 1973).

Bursell (1963a) hypothesized that proline was the source of energy for initial flight activity, subsequently sustained by utilization of lipid reserves. Using radioactive ^{14}C amino acids, Bursell demonstrated that proline was a substrate of oxidative metabolism providing energy for muscles both during rest and during flight (Bursell, 1963a, 1966a, 1975a). As proline reserves are depleted during flight and cannot rapidly be replenished, flight performance declines quickly during individual flights. Thus, tsetse are limited to short periods of high-speed flight, estimated at between 6.5 and 7.5 m s^{-1} (about 0.4 km min^{-1}) in open country (Gibson and Brady, 1985, 1988).

4.2.1 Proline metabolism

As the major source of energy, the metabolic processes for proline utilization in tsetse are of interest. While amino acid content varies considerably in different developmental stages of the fly, proline occurs in the greatest concentration in the haemolymph of *G. m. morsitans* (Cunningham and Slater, 1974). Proline is synthesized mainly from lipids in the fat-bodies, and is oxidized to alanine in the muscles. In resting flies proline content is high and alanine content low, but during flight the proline drops sharply and alanine content increases proportionately (Bursell, 1963a). Proline, providing 0.52 mol ATP g^{-1} compared with 0.18 mol ATP g^{-1} for carbohydrates, is particularly important during the first seconds of flight.

Bursell and Slack (1968) found that the content of proline, alanine and glutamate in the thorax of tsetse flies (*G. morsitans*) caught by two different sampling methods differed. Flies caught on a stationary bait animal had significantly less proline and significantly more alanine and glutamate than flies caught on a fly-round. They suggested that these differences occurred because flies from the bait sample had been more active than those caught on the fly-round sample immediately preceding capture.

During flight, proline is converted to glutamic acid and subsequently, by partial oxidation, to alanine. Re-synthesis of proline from alanine is enhanced during flight and alanine is then transported from thoracic flight muscles to fat-bodies for proline regeneration (Bursell, 1963a, 1966b, 1967; McCabe and Bursell, 1975a,b; Hoek *et al.,* 1976). However, proline synthesis in the fat-body is probably negligible compared with the capacity of flight muscle for its oxidation (Bursell, 1977a). The extra-mitochondrial enzymes, oxaloacetic carboxylase and alanine-oxoglutarate amino-transferase, identified in *G. morsitans* and *G. austeni,* were thought to be involved in proline catabolism, but Bursell and Slack (1976) suspected that an unidentified mitochondrial enzyme must also be involved. About 20% of the glutamate produced by proline oxidation is oxidized by glutamic dehydrogenase in tsetse flight muscle mitochondria (Bursell, 1975b; Bursell and Slack, 1976). The remaining 80% undergoes transamination by reacting with pyruvate to form alanine, and is eventually regenerated to proline. Phospholipid composition of flight muscle sarcosomes is generally similar to that of other dipterans although mitochondrial metabolism of tsetse may differ (D'Costa and Rutesasira, 1973a). The enzyme initiating the proline metabolism pathway in flight muscle mitochondria is proline dehydro-genase, which is activated by ADP and ATP (Norden and Venturas, 1972). Calcium and phosphate play a lesser role. In the presence of phosphate, ADP controls proline oxidation by flight muscle mitochondria (Bursell and Slack, 1976) acting at the level of the respiratory chain, rather than at the level of dehydrogenases. A malic enzyme occurs in insect flight muscle (Hoek *et al.,* 1976) but appears particularly important in the proline oxida-tion pathway of tsetse, in which it exhibits higher activity. Confirmation of this arose from experiments showing that the enzyme, produced by flight muscle mitochondria, appears to play a role in replenishment of pyruvate (Norden and Matanganyidze, 1977). This facilitates alanine production during the transamination process. Regenerated proline is then transported from the fat-body, through the haemolymph, to flight muscle. The metabolic pathway and enzymes involved in re-synthesis of proline in fat-bodies have been well described (McCabe and Bursell, 1975a,b; Bursell, 1977a; Konji *et al.,* 1984). According to Konji *et al.* (1988), isocitrate stimulates the synthesis of proline.

A hormone, stored and released by the corpora cardiaca, controls proline oxidation and appears to be dependent upon the presence of calcium ions (Pimley, 1985).

A newly emerged, unfed tsetse fly, with partially developed thoracic flight muscles, is termed a teneral fly. A teneral fly can be recognized from the soft feel of its thorax (due to the incompletely developed thoracic muscles), and from the ability to extrude the ptilinum by gently squeezing the head. The short flight duration of teneral tsetse might be due to the low proline levels and absence of a residual bloodmeal, although flight can continue for 4–7 min after thoracic proline is used up (Hargrove, 1976a).

The incomplete development of flight muscles also restricts flight duration in teneral flies.

Residual bloodmeal amino acids appear to be involved in early stages of flight in non-teneral flies by significantly stimulating proline uptake (Njagi *et al.*, 1992). This takes place in flight muscle mitochondria, which contain a proline transport mechanism, possibly located in a hydrophobic part of the mitochondrial membrane. The mitochondrial transport mechanism seems to be energy dependent, although ATP does not appear to be the immediate source of energy (Njagi *et al.*, 1992). Furthermore, the dependence of proline transport on temperature is consistent with a membrane transport process rather than simple diffusion.

4.2.2 Maturation of the adult – development of thoracic musculature

Wingbeat frequency of adult tsetse increases during maturation and is associated with increased amounts of thoracic muscle, cuticle and proline. In males, wingbeat frequency also rises more or less linearly with increasing temperature up to 36°C. The increase in frequency is specifically associated with increases in the mass of contractile protein and in the amount of mitochondrial substance per unit of contractile material (shown by Anderson and Finlayson, 1973; Bursell, 1973a). In other words, as the fly matures, it develops more muscle and a greater density of mitochondria in that muscle, as a result of which it produces enough energy to increase wingbeat frequency. During flight, wingbeat frequency declines as the proline is utilized and not significantly replenished (Hargrove, 1980). The rate of decline of wingbeat frequency is much greater than in other insects that have been studied and is greater in teneral than non-teneral flies, but during maturation of the thoracic musculature there is a marked increase in lift (Hargrove, 1975b). Under experimental conditions tsetse are capable of producing very high lift during initial stages of flight and Hargrove (1975b) attributed this ability to the combination of large wings and a large thoracic musculature. He suggested that the ability of tsetse to produce a large amount of lift was a necessity imposed by their haematophagous and viviparous mode of life.

In nature, thoracic muscle development in male *G. m. morsitans* occurs primarily during the first two hunger cycles, and the muscles increase in size by 2–3 mg after the second or third bloodmeal. The rate of development, measured as thoracic residual dry weight, is roughly halved in laboratory-reared flies, in which flight activity is also lower (Dame *et al.*, 1968). These authors concluded that irreversible behavioural and physiological inhibition in flight and in flight musculature occurred within the first few hours of adult life as a result of confinement. Langley (1970) disagreed, arguing that the rate, rather than the extent, of muscle development was affected in colony flies, and that the decreased activity of released flies was behavioural rather than physiological. Langley's view was supported by evidence that if laboratory-reared flies are induced to fly soon after they have taken their first bloodmeal, the rate of development can be increased to normal levels (Bursell *et al.*, 1972). Wild flies digest bloodmeals more rapidly than those maintained in a laboratory (Bursell *et al.*, 1974).

4.2.3   Energy reserves for flight and survival

Both thoracic cuticular residual dry weight and thoracic proline reserves increase during maturation. The increase in weight of the cuticle is triggered by the first bloodmeal (Hargrove, 1975a). As Bursell (1973a) did not account for this increase in cuticular weight in his estimation of thoracic musculature, he overestimated the increases in mitochondrial and contractile protein during maturation. The increases of these proteins are, none the less, still substantial and are correlated with improvements in flight performance. Delayed maturation of the adult may be an adaptation to the viviparous mode of reproduction (Bursell, 1973a).

The amount of energy available to female *G. m. morsitans* limits the duration of individual flights to a few minutes and restricts the total daily flying time (Bursell and Taylor, 1980). For males, daily flight duration is estimated at about 15 min in the hot season and more than twice as long in the cold season.

Mature male flies use most of each bloodmeal for lipid production, with virtually no increase in residual dry weight (RDW), whilst mature females use it both for conversion to fat and for the regular production of full-grown larvae, resulting in an increase in both fat and RDW (Randolph and Rogers, 1981). In Zimbabwe, the mean size, and the proportion of fat in newly emerged *G. m. morsitans*, was greatest when development took place at 24°C. At temperatures lower or higher than this, size and fat content diminished comparatively rapidly.

When fat reserves fall to about 6% of a fly's total dry body-mass, it risks death from starvation (Rogers *et al.*, 1994). The lethal lower limit of fat content is estimated at about 0.5 mg for *G. swynnertoni* and this species is closer to starvation in the dry season than in the wet season because availability of water restricts the distribution of vertebrate hosts. Nutritional stress is also associated with production of small puparia (Glasgow and Bursell, 1961).

4.3   REPRODUCTIVE PHYSIOLOGY

4.3.1   The reproductive cycle

Tobe and Langley (1978) reviewed the reproductive physiology of *Glossina*. Adenotrophic viviparity involves cyclical production of eggs, which hatch in the uterus, and in tsetse there are numerous specific adaptations to this mode of reproduction. Each ovary consists of two meroistic–polytrophic ovarioles. The ovaries are enclosed in a common epithelial sheath and are served by a system of trachea and tracheoles. Each ovariole comprises a germarium and vitellarium with a single developing follicle. The anterior part of the germarium contains oogonia and pre-follicular cells. One of these oogonia undergoes four consecutive and apparently synchronous divisions to produce a cyst of 16 daughter nuclei contained in a syncytial cytoplasm, which soon becomes surrounded by pre-follicular tissue (Huebner *et al.*, 1975). The primary follicle develops and is eventually pushed into the vitellarium by the division and growth of the germarial cells (Moloo, 1971). Protein for development of the oocytes may be transferred via the nurse cells. Huebner *et al.* obtained results contrary to Moloo's regarding transfer of protein and RNA to tsetse oocytes; they thought that follicle cells might be involved in protein synthesis and that haemolymph contributed little protein for oogenesis.

4.3.2   The regulation of ovulation – hormonal and nervous control

The regulation of ovulation is a complex process involving secretion of hormones and regulatory compounds by the nervous centres of the endocrine system in the head, such as neurosecretory cells of the corpus cardiacum and corpus allatum (Fig. 4.3). Additional stimuli occur in the reproductive organs of the female fly, and there are also physical and chemical stimuli from the male during copulation. Various studies over the years have substantially elucidated the processes controlling ovulation in tsetse flies (e.g. Saunders and Dodd, 1972). Normal cyclical ovulation only occurs following mating, which stimulates the endocrine system, and virgin unmated females do not ovulate (Dodd, 1971; Odhiambo, 1971; Ejezie and Davey, 1977; Chaudhury *et al.*, 1980; Wall, 1989a) – in contrast with earlier observations that they simply ovulated late (Mellanby, 1937; Vanderplank, 1947a).

Ejezie (1983) reviewed the role of hormones in controlling reproduction of tsetse. Median neurosecretory cells in the head of *G. austeni* are involved in ovulation (Foster, 1972), and during the pregnancy cycle they undergo cyclical changes of net synthesis and release of hormones, correlated with ovulation and larviposition. The corpus allatum also undergoes cyclical changes, correlated with changes in volume of the milk gland (Ejezie and Davey, 1974). Ovulation appears to be regulated by neurosecretory products released at neuromuscular junctions in the ovaries and oviducts (Robert *et al.*, 1984). Control of parturition is more complicated; a first neurohormone appears to be produced in the nerve centres and is then released in neurohaemal areas in the vicinity of the uterus, enhancing its contraction. A second neurohormone may be released in the corpora cardiaca and the median perisympathetic organ.

There is evidence that one neural factor (originating from the uterus during mating) acts as an afferent stimulus, whilst another (originating from

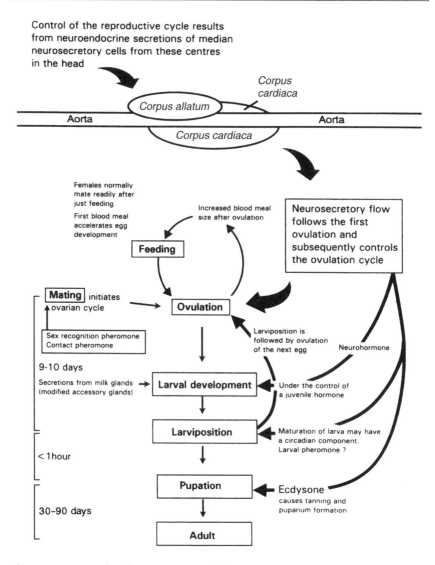

Fig. 4.3. Hormonal and nervous control of the reproductive cycle in female tsetse flies.

the ovary during oocyte maturation) regulates a cerebral hormone controlling ovulation (Chaudhury *et al.*, 1980). Earlier observations of Chaudhury and Dhadialla (1976) on mechanisms controlling ovulation had led them to the following conclusions:

● Ovulation is dependent on the physical act of copulation.
● Ovulation is independent of insemination and the transfer of male accessory gland secretion or seminal fluid.

- The duration of mating is critical in inducing ovulation.
- The corpus cardiacum is important in the regulation of ovulation.
- A blood-borne factor in the mated female is ultimately responsible for induction of ovulation.
- Ovarian development may have a role in the control mechanism.

It is likely that the median neurosecretory cells are activated first by nervous impulses during mating, and subsequently by ovarian distension. These cells then release an ovulation-stimulating hormone into the haemolymph, probably through the corpus cardiacum. The stimulus of mating appears to lead to an increase in bloodmeal size in female tsetse (Ejezie and Davey, 1977). Unmated females do not ovulate and their mature eggs eventually disintegrate.

In *G. m. morsitans* kept at 25°C, and probably for all tsetse species, ovulation occurs about 1 h after the previous larviposition (Dodd, 1971; Foster, 1974). The egg hatches on day 3.8, ecdysis to the second instar occurs on day 4.9, the third-instar cuticle is formed on day 6.8 and parturition occurs on day 9.0. Parturition depends on the larva having reached a suitable stage of development, but also shows a daily rhythm with peak parturition in late afternoon, at least for some species. Laboratory experiments indicate that parturition follows a circadian pattern, peaking 9 h after lighting is put on, and is associated with pulses of haemolymph pressure (Denlinger and Ma, 1974; Denlinger and Zdárek, 1992; Zdárek *et al.*, 1992).

An important vector of trypanosomes to livestock is *G. pallidipes,* and its reproduction has been closely studied. In the laboratory, ovulation in unmated adult *G. pallidipes* is delayed until an age of at least 15 days (Wall, 1989a). However, in the field, inseminated females were first detected at 5–6 days old and most became inseminated at 8–10 days of age. It was therefore concluded that mating takes place at or just before ovulation. Wall (1989a) rejected the concept of gross differences in *G. pallidipes* mating behaviour between geographical areas as a result of his field studies in Kenya and Zimbabwe.

4.3.3   Larval nutrition

Secretions from highly modified accessory glands, known as milk glands, nourish the developing larva (Ma and Denlinger, 1974). Rapid protein synthesis is required in pregnant females to provide nourishment to the larva, and their protein-secreting glands are highly efficient and well adapted to this purpose (Tobe and Davey, 1972). The milk glands undergo a cycle of secretion and release of the uterine milk during each pregnancy, with a peak of secretion in early pregnancy. A study of the secretory cells of the uterine gland during the second pregnancy cycle of *G. m. morsitans* showed that quantitative morphological changes of the cells were functionally correlated with development of the larva in the uterus and physiological events during pregnancy (Moloo, 1983). Stimulation of milk synthesis by the uterine gland appears to be controlled by a neuroendocrine hormone (Pimley, 1983). Some of the components for 'milk' production probably originate from secretions of the fat-bodies; during most of the pregnancy cycle, protein synthesis in the fat-body is low compared with that of the milk gland (Riddiford and Dhadialla, 1990).

Uterine milk contains mostly protein and lipid in roughly equal proportions, estimated at 48% of dry weight as lipid (Cmelik *et al.*, 1969). Langley and Pimley (1979a) estimated protein content at about 55% and, in contrast with observations by Ma *et al.* (1975) and Moloo (1976b), milk composition remained relatively constant throughout the reproductive cycle. At eclosion of the first-instar larva the milk is rich in acidic lipids, which are later replaced by proteins (Ma *et al.*, 1975). The free lipids in the milk are phospholipids, cholesterol, triglycerides and a number of other unidentified ones. The milk also contains large quantities of tyrosine, which may be necessary for the tanning of larval and adult cuticles.

During early pregnancy, the rate of milk synthesis exceeds the rate of uptake by the larva (Denlinger and Ma, 1974). To accommodate the excess, secretory reservoirs expand, a process that reverses after day 6. Uterine gland secretion and larval feeding appear to continue throughout the intra-uterine life of the larva (Langley and Pimley, 1974). In laboratory experiments, free, U-^{14}C labelled amino acids are excreted by pregnant adult females during diuresis, suggesting that they have little capacity for storing nutrient substrates for their developing larvae (Langley and Pimley, 1974).

Assuming a 48 h digestion time, a large bloodmeal taken on about the 5th day of a 9-day interlarval period would provide the bulk of the nutrients to synthesize lipids for larval growth (Langley and Pimley, 1974). This assumption fits with field observations, confirming the importance of this large bloodmeal (Langley and Pimley, 1979b). Milk secretion continues from the time the first-instar larva hatches until parturition (Roberts, 1971; Tobe *et al.*, 1973), disproving earlier suggestions that milk gland secretions were confined to a short period after ovulation (Hoffmann, 1954; Bursell and Jackson, 1957).

During early pregnancy the fat-body synthesizes triglyceride, but this ability diminishes before the larva hatches and is followed by a massive transfer of triglyceride from the fat-body to the uterine gland where 'larval milk' is synthesized (Langley *et al.*, 1981a). This transfer happens at the time of most rapid larval growth (Langley and Bursell, 1980) and occurs via haemolymph diglycerides, which form a lipid transport mechanism. Lipid utilization during reproduction is outlined in Fig. 4.4. Diglycerides are an important means of lipid transport in insect haemolymph, and may provide a substantial proportion of energy consumed during flight in insects other than tsetse. Fat reserves accumulating in the tsetse larva provide most of the energy required for puparial development (D'Costa and Rutesasira, 1973b).

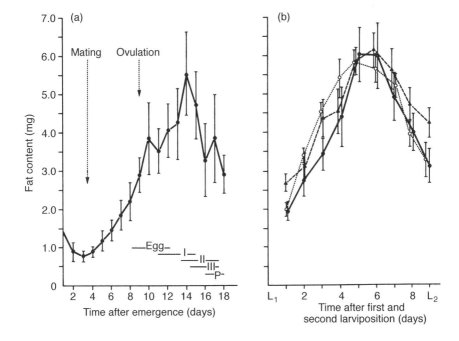

Fig. 4.4. Lipid utilization by *Glossina*. Vertical lines represent fiducial limits of means. (a) Net daily change in lipid reserves (mg) of fertilized female *G. morsitans* from emergence to the 18th day of adult life and fed on pig blood. Horizontal bars represent the time range for different developmental events *in vitro*; I, II, III, larval instars; P = parturition. (b) Net daily change in lipid reserves (mg) of fertilized female *G. morsitans* between the first and second larvipositions: ●—●, fed on pig blood; ▲—▲, fed on cow blood; ○—○, fed on live goat. (From Langley and Pimley, 1979b.)

Unused fat reserves remaining after puparial development form the food reserves for the subsequently emerging adult fly prior to its first bloodmeal (Bursell, 1958a) and are crucial to its survival.

Temperature is the main factor affecting fat consumption during the puparial period and the optimum is 25°C (Bursell, 1960c); estimates for fat consumption at low temperatures were inaccurate, as there is a non-linear relationship between the logarithm of oxygen consumption by tsetse puparia and temperature (Rajagopal and Bursell, 1965; Phelps, 1973). As 1 mg of fat needs 2056 mm^3 of oxygen for consumption, Phelps was able to calculate survival time of resting tsetse (based on temperature and fat reserves) that ranged from 81.2 h for a fly with 0.3 mg of fat at 16°C, to 28.2 h for a fly with 0.5 mg of fat at 32°C. In Zimbabwe, field-caught flies survived for longer and there was a marked seasonal variation in size, with largest puparia being found at the time of lower temperatures.

Abortion

Starvation of pregnant flies or death of an embryo or developing larva can result in abortion (Langley, 1977). This is also an important consequence of sublethal contact with insecticides, particularly the synthetic pyrethroids. Table 4.2 summarizes estimates of abortion rates in wild populations of tsetse.

4.3.4 Temperature and larviposition

Temperature influences the larviposition rhythm of tsetse. In the laboratory, at constant temperature and humidity, peak larviposition of *G. m. morsitans* occurs 8–9 h after lights are turned on (Robinson *et al.*, 1985). Sudden increases in temperature of 6°C for 2 h during reproductive cycles significantly reduce larviposition during the periods of increased temperature. If temperature is allowed to fall rapidly, larviposition rate doubles in the following 15 min (Robinson *et al.*, 1985). Tsetse may, therefore, respond to seasonal cycles of changing temperature in a way that allows them to deposit larvae in the most favourable temperature conditions for their survival. Female tsetse tend to choose darker resting sites (a photonegative response) as pregnancy progresses and when temperatures increase. Parturition is to some extent dependent on neurosecretions from the mother; therefore, nervous control may also play a major part in enabling tsetse to find suitable places where larvae can pupariate. Cyclic nucleotides are important regulatory compounds in most biological systems. In tsetse, cyclic AMP and cyclic GMP levels change during pregnancy and at parturition, and may stimulate ovulation and parturition (Denlinger *et al.*, 1984; Denlinger and Zdárek, 1992; Zdárek and Denlinger, 1993).

There is a transfer of substances from the male to the female haemolymph during copulation (shown by radiolabelling experiments) (Gillott and Langley, 1981). These substances are probably synthesized in the male accessory glands and stimulate receptivity of female tsetse. Unfed females are refractory to copulation and subsequently receptivity declines with age or after mating. In the laboratory, synthesis of accessory gland secretions begins before adult emergence and continues uninterrupted, independent of the corpus allatum (Samaranayaka-Ramasamy, 1981).

Table 4.2. Estimates of abortion rates in wild tsetse populations.

Tsetse species	Location	Abortion rate (%)	Reference
G. m. morsitans	Zambia	0.04	Harley, 1966a
G. m. morsitans	Zambia	8.72	Madubunyi, 1975
G. m. morsitans	Zambia	7.0	Madubunyi, 1978
G. pallidipes	Kenya	1–5	Dransfield and Brightwell, 1989
G. brevipalpis	Uganda	0–12	Okiwelu, 1977a

4.4 PHEROMONES AND HORMONES

Pheromones are substances released by individuals of a species, usually to initiate a behavioural response in other individuals of the same species. They have been identified for many insect species and in some cases have been used as important tools for the monitoring and control of insect pests (e.g. 'Gyplure', for control of the gypsy moth, *Lymantria dispar*, a pest of deciduous trees and shrubs in America) (Van Den Bosch and Messenger, 1973). They are often produced to attract members of the opposite sex for mating and can be effective at long distances. Others are short-range or contact pheromones, which usually initiate specific behaviour. Hormones, on the other hand, are directly active biological compounds affecting the individual in which the hormone is produced. The manipulation of both hormones and pheromones have been investigated as possible means to control tsetse flies. Substances produced by one species which produce a behavioural response in another have been termed kairomones. Host odours and substances on the skin, such as sebum, are in this category and are discussed in Chapter 10.

4.4.1 Pheromones

Sex pheromone

Sex-recognition pheromones of tsetse were first detected following gas chromatographic analysis of cuticular hydrocarbons and are saturated methyl-branched hydrocarbons (Langley *et al.*, 1975). Sex pheromones isolated from the cuticle of female *G. m. morsitans* trigger mating behaviour in the male at ultra-short range, or upon contact with baited decoys (Carlson *et al.*, 1978). The male mating response only occurred if decoys were approximately the same size and shape as a female tsetse. Each tsetse species appears to possess a unique sex-recognition pheromone (Langley *et al.*, 1975) and some of these are shown in Table 4.3.

The sex pheromone of *G. pallidipes* elicits increasing responses with increasing doses (Carlson *et al.*, 1984). At about the same time as their first identification, a sex-recognition pheromone in *G. p. palpalis*, which elicited

Table 4.3. Sex pheromones identified in tsetse.

Tsetse species	Pheromone	Reference
G. m. morsitans	15,19,23-trimethylheptatriacontane	Carlson *et al.*, 1978; Huyton *et al.*, 1980a
G. austeni	15,19-dimethyltritriacontane	Langley, 1979
G. pallidipes	13,23-dimethylpentatriacontane	McDowell *et al.*, 1981; Carlson *et al.*, 1984*

*McDowell *et al.* (1981) and Carlson *et al.* (1978, 1984) identified different isomers of the same compound.

a sexual response in the male, was reported (Offor *et al.*, 1981). Similar long-chain branched hydrocarbons were examined for three other species of tsetse (Nelson and Carlson, 1986). These varied considerably, but were simpler in males than in females and it appeared likely that they were species-specific or even subspecies-specific sex pheromones.

Natural biosynthesis of the sex pheromone takes place in cells closely associated with the female abdominal cuticle (Langley and Carlson, 1983). It is then believed to be released from the cuticle (at least in *G. m. morsitans*) via unicellular glands located on the fly's legs (Schlein *et al.*, 1980). From these cells, it spreads over the external body surface by diffusion and grooming behaviour. Natural pheromone seems to be a component of surface cuticular waxes and appears on the pharate adult female about 2 days before emergence from the puparium and remains throughout its life. In the laboratory, the pheromone does not stimulate males unless it is placed on a three-dimensional object; thus, both mechano- and chemoreceptors probably have to be stimulated for mating behaviour to be initiated. The chemoreceptors are thought to be basiconic sensillae on the tibiae and tarsi (Coates and Langley, 1982a). In contrast, Den Otter and Saini (1985) believed that *G. m. morsitans* males perceived the pheromone only through olfactory receptors on the antennae. These conflicting views were resolved after Langley *et al.* (1987) demonstrated that chemoreceptors on the tarsi and tibiae of the legs were stimulated by pheromone on contact, or at ultra-short range. Removal of the antennae did not change responses; thus, antennal olfactory receptors were not involved. They suggested that Den Otter and Saini's observations were due to some contaminant of their pheromone preparation. Chemoreceptors on the legs of *G. tachinoides* are more common on males than on females, which is compatible with their likely role in pheromone perception (D'Amico *et al.*, 1992).

A combination of transmission electron microscopy and electrophysiology enabled Deportes *et al.* (1994) to detect hair shaped chemo- and mechanoreceptor sensillae on the front side of the wings of *G. f. fuscipes*, which were stimulated experimentally by sex pheromones.

Male responsiveness to decoys increased to a maximum 3–4 days after emergence in regularly fed *G. m. morsitans* (Huyton *et al.*, 1980b), while in the field male tsetse responded with a standard behavioural pattern after landing (Hall, 1988). Firstly, the pheromone was perceived; secondly, tsetse oriented themselves on the decoy and engaged their genitalia; and thirdly, a long period of quiescence corresponded to copulation. The probability of a fly leaving before copulating decreased during the sequence, but was initially dependent on the dose of pheromone used: the lower the dose, the higher the rate of leaving. These results suggested that the role of the pheromone is to initiate copulation (Hall, 1988).

Larval pheromone

There is some evidence for the existence of a larval pheromone released from the anal orifice of tsetse larvae (Nash *et al.*, 1976). In *G. m. morsitans* the larval orifice appears to open either at parturition, or soon after, and the secretion contains uric acid, indicating that it is a product of the Malpighian tubules. The pheromone seems to affect choice of larviposition site, resulting in aggregation of larvae in breeding sites, although Rowcliffe and Finlayson (1981) disputed this. Flies larviposited significantly more frequently in an experimental incubator chamber if it had a water- or ether-soluble extract of larval excretion in addition to any pupariating larvae. This supports the existence of a larval pheromone for attracting females to breeding sites and could explain aggregations of puparia that have some-times been reported during the dry season (Leonard and Saini, 1993). An analogous aggregation pheromone has been identified for species of *Simulium damnosum* black flies, which lay their eggs in batches, often with many flies ovipositing simultaneously at a single site. In this case, the pheromone is released from eggs immediately after oviposition and attracts other gravid females to the same site (McCall *et al.*, 1997).

4.4.2   Hormones

Juvenile hormone
Topical application of a juvenile-hormone analogue to young puparia of *G. m. morsitans* causes developmental abnormalities during metamorphosis (Langley *et al.*, 1988a). The use of these juvenile-hormone mimics in attempts to control tsetse (Langley *et al.*, 1990a) is described in Chapter 18. Experimentally, a plant-derived ecdysterone injected into the thorax of adult female tsetse and a juvenile-hormone analogue applied to the abdomen have been used to disrupt the pregnancy cycle (Denlinger, 1975). A sufficient dose of either the ecdysterone or the juvenile-hormone analogue caused abortion in current and sometimes subsequent pregnancies.

Combinations of sex pheromones with chemosterilants or hormone mimics for tsetse control have been investigated for field use (Langley and Hall, 1986; Langley *et al.*, 1981a, 1982a,b, 1988a, 1990a). This subject is discussed in more detail in Chapter 18.

4.5   THE NERVOUS SYSTEM

The brain, neuroendocrine and stomatogastric nervous systems of tsetse were described in some detail by Langley (1965) and the reader is referred to this publication for descriptive details.

4.5.1   Neurosecretions

There are two neurosecretory elements in the corpus cardiacum of tsetse, and of other insects; these are the neurosecretory axon terminals from the brain and the intrinsic neurosecretory cells themselves (Awiti *et al.*, 1985).

Neurosecretory granules in the nervus corporis cardiacum (ncc) axons of *G. m. morsitans* are small compared with the large granules synthesized in the corpus cardiacum. The secretory granules cluster into small vesicles, and secretions from both the ncc axons and the corpus cardiacum appear to be released into the haemolymph (Awiti *et al.*, 1985). Awiti and colleagues found evidence of a neurotransmitter function of neurohormones secreted from the ncc axons. Electron microscope studies led them to conclude that the aortic wall has taken over the neurohaemal function of the corpus cardiacum, the function of which otherwise remains that of synthesizing an intrinsic hormone. Activity of the corpus cardiacum to synthesize and/or release its secretion is most likely controlled in the aortic wall via a neuro-transmitter substance.

4.5.2   Sense organs

Using scanning electron microscopy, D'Amico *et al.* (1991) showed the morphology of proprioceptive hairs of the prothoracic organ and the existence of hairs, probably mechanoreceptors, on the ptilinum. Chemoreceptors were later identified on the dorsal surface of the costal vein of the wings of adult flies and described in more detail (Baldet *et al.*, 1992).

The involvement of the nervous system in the feeding processes and the sensory and motor innervation of the tsetse gut were described by Rice (1972). The rapid transfer of blood from the crop to the midgut for elimination of excess fluid is probably controlled by multi-terminal neurones associated with the proventriculus. Midgut neurones are thought to monitor the state of distension of the anterior midgut (Finlayson and Rice, 1972).

A number of sensillae are found on the cibarium and on the labrum and labium of tsetse mouthparts, and are likely to provide information regarding odours, temperature and tactile stimulation associated with the feeding process. Mechanoreceptors on the labium may monitor changes resulting from eversion of the labella during probing, in addition to which there are gustatory chemoreceptors (Rice *et al.*, 1973a). Together, these sensillae coordinate the piercing mechanisms of the proboscis. Additional sensillae on the labrum and cibarium consist of mechanoreceptors, which are presumed to monitor movement of saliva and blood flow. The limited number of gustatory (taste) chemoreceptors are believed to detect erythro-cytes, whilst others act as flow meters and generally coordinate cibarial pumping of the bloodmeal and saliva (Rice *et al.*, 1973b). Functions of these sensillae are discussed further, in relation to feeding and trypanosome infections, in Chapter 11.

There is little information on sensory organs of tsetse larvae. They have few sensillae on their body surface, probably because their free-living period is brief. They apparently possess photoreceptors, and there are chemoreceptors and mechanoreceptors at the anterior end of the antenno-maxillary processes (Finlayson, 1972). These are probably a combination of tactile sensillae and a hygroreceptor sensillum whose functions are to discriminate between dry and humid conditions within the microclimate, enabling the larvae to find suitable sites for pupariation and to initiate this pupariation in the short free-living period after larviposition.

4.6   PTERIDINES

Biopterin and its derivatives affect the growth of crithidial flagellate parasites of mosquitoes. Consequently, Harmsen (1970) investigated their presence in tsetse in order to determine their role, if any, in the development of *Trypanosoma* species. As in other Diptera, pteridines are found in the heads of tsetse and he suggested that they could play a role in trypanosome growth and morphogenesis. The accumulation of pteridines in the head has formed the basis of a method for ageing tsetse flies described in Chapter 26. No role for pteridines in trypanosome development has yet been demonstrated.

Chapter 5

Genetics of Tsetse Flies

Gooding (1984a) published a comprehensive review of tsetse genetics. Most data are available for *G. m. morsitans* and species of the *palpalis* group whilst little information is available for *fusca* group species. Initial difficulties in the study of tsetse chromosomes and genetics arose, firstly, from the preparation of material for cytological examination and secondly, interpretation of the data obtained. Itard (1973) described methods for their study in a review of tsetse cytogenetics.

5.1   CHROMOSOME NUMBER AND POLYMORPHISM

Chromosomes of tsetse were first demonstrated in *G. morsitans centralis* by Vanderplank (1948), who observed that it had three pairs. Later, Itard (1966a) found six somatic chromosomes in *G. tachinoides* and ten in *G. m. morsitans*. Riordan (1968) reported that *G. palpalis* has three pairs, and suggested that six chromosomes may be the diploid number for most, or all, species of *Glossina*. Subsequently, however, species of the *fusca* group that have been examined were shown to have the highest number of chromosomes, up to a diploid number of 16 (Maudlin, 1970), whilst *palpalis* group species have the simplest chromosome formula. Maudlin (1970) found eight pairs of chromosomes in the *fusca* group tsetse, *G. brevipalpis*. *Glossina pallidipes* has four pairs of chromosomes: two long, one medium and one short (Hulley, 1968). The three larger pairs were metacentric, in contrast with Itard's observations for the species that he studied. Itard (1974) found six chromosomes in *G. p. gambiensis*, of which four are autosomal and two are sex chromosomes. Table 5.1 summarizes the chromosome number of tsetse species that have been studied.

   The small chromosomes found in *morsitans* group tsetse are termed supernumerary, or B chromosomes (Itard, 1974). Although Maudlin (1970) disagreed with this opinion at the time, it is now accepted that tsetse of the *morsitans* group, but not the *palpalis* group (Davies and Southern, 1976), do

Table 5.1. Chromosome number of tsetse.

Group	Tsetse species	Diploid chromosome number	Supernumerary chromosomes
Morsitans	G. pallidipes	8	Variable
	G. longipalpis	8	
	G. m. centralis	6	Variable
	G. m. morsitans	10	Variable
	G. m. submorsitans	8	Variable
	G. swynnertoni	8	Probably variable
	G. austeni	6	Variable
Palpalis	G. palpalis	6	Absent
	G. p. gambiensis	6	Absent
	G. tachinoides	6	Absent
	G. f. fuscipes	6	Absent
Fusca	G. brevipalpis	16 (probably the same for most fusca group species)	
	G. f. congolensis	22	

have supernumerary B chromosomes (Jones and Rees, 1982; Bushrod, 1984). Analyses of *fusca* group tsetse have been insufficient to confirm that supernumerary chromosomes are absent from that group. Southern *et al.* (1972) found that the chromosome number in *G. austeni* varied between individuals, the variation being restricted to the supernumerary Bgroup chromosomes. The number of these chromosomes varies both between populations (stocks) of the same species of tsetse and seasonally in wild populations, although within-species differences do not necessarily interfere with successful interbreeding between two populations. Supernumerary chromosomes apparently affect fitness, or may be a correlate of fitness, in some tsetse populations (Warnes and Maudlin, 1992). Although their precise role is not known, in other species B chromosomes are often deleterious to individual survival. In a comparison between a population of *G. pallidipes* from Zimbabwe, which possessed supernumerary (B) chromosomes, and a population from Uganda that did not, Langley *et al.* (1984) found that the Uganda population had a higher reproductive rate than the Zimbabwe population, but the offspring were slightly smaller.

Elsen *et al.* (1989) claimed to have shown polytene chromosomes in the nuclei of aortic cells of adult tsetse for the first time, although they were previously reported from tissues of larvae, puparia and pharate adults within puparia (Burchard and Baldry, 1970; Southern *et al.*, 1973; Southern and Pell, 1974, 1981).

Meiosis in male *G. m. morsitans* was described by Southern *et al.* (1972), who concluded that this species represents an example of an achiasmate system. Chiasmata are found during female meiosis but only rarely in males of *G. m. morsitans*, *G. austeni* and *G. f. fuscipes*, suggesting that genetic recombination is more frequent in females. Meiosis in males is completed

9–10 days after larviposition, whilst in females it occurs throughout adult life. The diploid chromosome number in male *G. m. morsitans* is $2n = 8 \pm 1$ autosomes and the XY sex pair. No recognizable diplotene or diakinesis stage was observed. Southern (1980) observed that meiosis in *morsitans* group hybrid males proceeds normally but in hybrid males of the *palpalis* group the chromosomes tend to fragment. Genetic material may, therefore, be passed from one taxon to another and genetic recombination could produce completely new chromosomes in descendants of hybrids. Vanderplank (1948) observed that hybrid male *palpalis* flies were fertile and could successfully transfer sperm, which Gooding (1984a) found inconsistent with Southern's observations of chromosome fragmentation. Hybrid females of both *palpalis* and *morsitans* groups are, to a degree, fertile, but almost all hybrid male crosses between *G. p. gambiensis* and *G. p. palpalis* were sterile (Gooding, 1988a). A marker gene indicated that the X chromosome was involved in this sterility. The possibility of tsetse control through hybrid sterility was suggested by Potts (1944), following hybridization of *G. swynnertoni* and *G. m. centralis*.

5.2   SEX DETERMINATION

Male tsetse of all species that have been studied are heterogametic. As with *Drosophila*, the sex of tsetse is determined by X chromosome dosage; thus, males can be XY or XYY and females may be XX, XXY or XXXY (Maudlin, 1979). The number of Y chromosomes can vary but has no effect on the sex phenotype of the fly, although it is required for production of motile sperm (Southern, 1980) and a correlation has been detected between the number of Y chromosomes and the duration of the puparial period for *G. p. palpalis* (Gooding, 1984a). Sex ratio distortion genes are likely to exist in the tsetse genome (Maudlin, 1979), and a significant sex ratio distortion has been observed in two colonies of *G. m. submorsitans* (Gooding, 1984a; Rawlings and Maudlin, 1984). The sex ratio distortion is thought to be caused by 'distorter males' either passing on only X-bearing sperm to their mates or passing on non-functional Y-bearing sperm (Rawlings and Maudlin, 1984). The sex ratio distortion observed in the laboratory could be affected by environmental factors such as temperature or feeding frequency; a decreased feeding frequency altered the sex ratio towards a more normal situation (Rawlings and Maudlin, 1984).

5.3   GENETIC VARIATION IN TSETSE POPULATIONS

Potentially, tsetse populations could show genetic intraspecific variation related to vectorial capacity or behaviour, for example trap avoidance, or avoidance of insecticide-treated cattle, which could be important for epidemiological investigations or for control schemes. It was suggested that tsetse at Nguruman, Kenya, have been selected for trap resistance (ICIPE,

1992) although this was strongly refuted by Brightwell *et al.* (1997). The most commonly used method to examine genetic variation has been iso-enzyme analysis, which shows differences in enzyme allele frequencies within or between populations.

There is some evidence of within-species geographical variation in behaviour and host preferences, but there has been little work to establish the genetic basis of such variation. Demonstrated behavioural differences within a tsetse species, which appear to have a genetic basis, include different feeding frequencies of female *G. pallidipes* from different locations in Kenya (Van Etten, 1982a), and activity patterns and responses to temperature, also in *G. pallidipes* (Van Etten, 1982b). Genetic variation in tsetse populations from Zambia (Gooding, 1989), Burkina Faso (Gooding, 1981), Tanzania (Gooding *et al.*, 1993) and Kenya (Kence *et al.*, 1995) have been investigated. In Zambia, gene–enzyme systems of two populations of *G. m. centralis* were examined using polyacrylamide gel electrophoresis (PAGE) to separate enzymatic proteins. Although no significant genetic differences were found between flies caught by different sampling methods, there were differences in allele frequencies at loci for three enzymes in the two populations. Gooding (1989) concluded from this study that there was little gene flow between the two populations.

The same PAGE technique was used to examine genetic polymorphism of three tsetse species in Burkina Faso (Gooding, 1981). There was little or no variation between populations and the average frequency of poly-morphic populations per locus was small. The study indicated that *G. p. gambiensis* and *G. tachinoides* are more closely related to each other than either species is to *G. m. submorsitans*. There was little genetic varia-tion in enzyme loci for those three species collected within 150 km of Bobo Dioulasso. In Tanzania, a very low level of genetic variation was found in a population of *G. swynnertoni*, compared with other tsetse populations, suggesting that the sample may have come from a small, inbred population (Gooding *et al.*, 1993). Elsewhere, significant differences in heterozygosities of geographically separated populations of *G. pallidipes* have been demon-strated. Examination of allele frequencies of subsets of a tsetse population with differing trypanosome infection rates indicated that the populations in those subsets were not genetically substructured (Gooding, 1992).

In a genetic analysis of samples of *G. pallidipes* from Kenya, Zimbabwe and Mozambique, Krafsur *et al.* (1997) found greater gene diversity in the southern African populations than in the Kenyan population. This was interpreted as being a result of the smaller, discontinuous populations in Kenya providing a smaller gene pool than in that of the larger, continuously distributed population of southern Africa. There was a higher degree of genetic drift in the Kenyan population, which may have resulted from post-mating reproductive isolation, or because *G. pallidipes* was less mobile than it is generally considered to be (Krafsur *et al.*, 1997).

Isoenzyme analyses have revealed little polymorphism in tsetse populations and an alternative approach has been to identify microsatellite markers. So far these have been identified for *G. p. gambiensis* and will be used to estimate gene flows within that species (Solano *et al.*, 1997).

Two laboratory stocks of *G. m. morsitans* were found to have differing reproductive performance and chromosome band differences (Jordan *et al.*, 1977). To determine whether this was related to genetic differences between the two strains, Gooding and Rolseth (1981) analysed their enzyme allele frequencies. This confirmed that there was genetic variation between the strains, consistent with the differences in reproductive performance, but did not provide any genetic basis for the differences.

A low value of heritability (h^2) of adult weight has been estimated for colony-bred *G. m. morsitans* (Gooding and Hollebone, 1976) and this reflects selection for size in tsetse populations, discussed in relation to mortality in Chapter 9.

Comparing allele frequency for enzymes of laboratory-reared and field-caught tsetse, a phenogram was constructed (Fig. 5.1) by Gooding *et al.* (1991), which indicates tentatively that the subgenus *Austenina* (*fusca* group) is the oldest of the three subgenera within the genus *Glossina*.

5.3.1   Visible traits

Only a few visible mutants have been observed, either in wild or in laboratory-bred tsetse. Mutations controlling body (*ocra*) or eye colour (*salmon*) have been reported and *salmon* eye colour mutations could theoretically be used for genetic control. The allele for *salmon* is pleiotropic, with its locus on the X chromosome (Gooding, 1979). It is lethal at temperatures of 23–25°C and has been investigated because of its potential as a genetic means of controlling tsetse populations. Computer models have shown that effective control could be achieved if *salmon* flies behave the same as wild-type flies in the field. On the other hand, they are more susceptible to trypanosome infections so it would be hard to justify releases of those flies in the field. The possibility of using this mutation as a tsetse control agent seems remote. *Salmon* type flies were over 100 times as sensitive to light as wild-type flies over much of their spectral sensitivity range (Davis and Gooding, 1983). The *ocra* mutant was found in *G. m. morsitans* and is easily recognizable as the fly's general appearance is yellowish rather than dark brown. The mutation is believed to reside on the differential part of the X chromosome and is thought to be a sex-linked recessive as it is only found in F_2 males (Bolland *et al.*, 1974).

Other visible mutants that have been observed are for long scutellar bristles in females (Gooding, 1984b).

5.4   GENETIC INFLUENCE ON VECTORIAL CAPACITY

There is ample evidence of genetic variation in vector competence between species and, unsurprisingly, there is some experimental evidence for a role

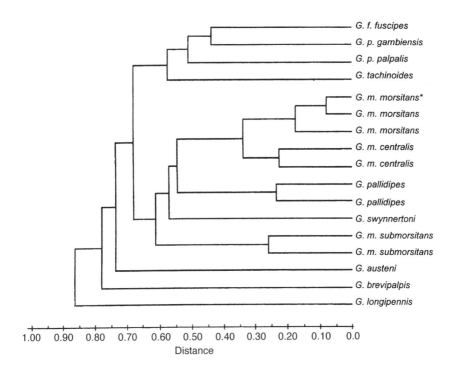

Fig. 5.1. Phenogram of the genetic distance for 12 species and subspecies of tsetse flies, using arc distance (Cavelli-Sforza and Edwards, 1967) based upon allelic frequencies at structural loci for ten enzymes. (From Gooding *et al.*, 1991.) The genetic distance between any two species is a measure of the genetic differences that have accumulated in the two species since they diverged.
* Tsetse subspecies shown more than once have different origins.

of the tsetse genome in the cyclical development of trypanosomes and determination of vectorial capacity within species (Jordan, 1974). Maudlin (1982) subsequently produced colonies at Bristol with different susceptibility to infection. *Salmon* mutants of *G. m. morsitans* appear to be better vectors of *Trypanasoma congolense* and *T. brucei brucei* (Distelmans *et al.*, 1985; Makumyaviri *et al.*, 1984a,b). Higher mature and procyclic infection rates can be achieved under experimental conditions in mutant compared with normal *G. m. morsitans* (Distelmans *et al.*, 1985). There may be a biochemical link between these observed phenomena and tryptophan, which is an essential amino acid metabolized by trypanosomes. *Salmon* mutants have a fault in their tryptophan metabolism resulting in its accumulation, which seems to predispose tsetse to trypanosome infection. A mutation in *G. p. palpalis*, also affecting tryptophan metabolism, does not result in its accumulation and these mutants do not differ significantly from wild flies in their vectorial capacity (D'Haeseleer *et al.*, 1987). In haematophagous insects, susceptibility to vertebrate pathogens is controlled by autosomal genes, sex-linked recessive genes or maternally inherited factors. According to Milligan *et al.* (1995), Gooding (1992) made errors in analyses which led him to conclude that there was an association between trypanosome infection maturation and X-linked eye colour mutants, but in fact there is no such association.

Experimentally, male tsetse usually acquire higher salivary gland infection rates for *Trypanozoon* trypanosomes than females. Until recently, no genetic mechanism had been demonstrated for the control of establishment of trypanosomes in the tsetse fly midgut, or maturation of infective stages. Refractoriness or susceptibility to *Trypanozoon* midgut infection is maternally inherited (Maudlin *et al.*, 1986), and a single sex-linked gene model can be fitted to data on salivary gland infection rates (Maudlin *et al.*, 1991). Initially, Maudlin and colleagues thought that maturation of *T. b. gambiense* infections established in the midgut was a phenomenon associated with trypanosome genotype (Maudlin *et al.*, 1986). Subsequently, evidence was found for the involvement of maternally inherited rickettsia-like organisms (RLOs) in the establishment of trypanosomes in tsetse and interactions with lectins affecting maturation (see Chapter 11). The RLOs appear to influence trypanosome establishment by producing an endochitinase that generates glucosamine in the midgut. This inhibits the trypanosomicidal activity of midgut lectins. Midgut infection rates of *T. b. rhodesiense* in *G. m. morsitans* show a dose-related increase when flies are fed *N*-acetyl-D-glucosamine in the infective meal and for 4 subsequent days (Welburn *et al.*, 1993). The lectins, however, have a dual role and also stimulate maturity in those trypanosomes that succeed in establishing themselves in the midgut. The number of RLOs in the midgut is influenced by the temperature at which puparia are maintained and increases rapidly after flies take bloodmeals. It is still not known whether there are genetic mechanisms in tsetse which regulate the number of midgut RLOs or whether this is randomly dependent on the numbers transmitted from the female to her offspring. The influence of environmental factors is also unclear.

Maximum-likelihood estimates of the gene frequencies in four tsetse species suggest that maturation of *T. b. brucei* infections may be under the control of a sex-linked recessive gene (Maudlin *et al.*, 1991). Welburn and Maudlin (1992) showed that maternally inherited susceptibility to midgut infection with *T. congolense* is a phenomenon restricted to the teneral state of the fly.

Other issues related to studies of tsetse genetics are the potential for the development of insecticide resistance and even of trap avoidance, although the latter is less likely. The probability of resistance developing would be greater if the resistance gene was dominant, or co-dominant (additive) to its wild-type allele.

Chapter 6

Sampling Tsetse Populations

The sampling of tsetse populations may be carried out in order to study their population dynamics, providing an understanding of their ecology (which is essential for carrying out effective control), or to estimate the level of trypanosomosis risk (or tsetse challenge) for a given area, or to evaluate control measures. For these purposes it is necessary to sample a representative portion of a tsetse population, or at least to know which part of the population is being sampled, particularly for ecological studies.

In this chapter, following a description of techniques for population sampling, factors that affect sample size and composition, causes of variability and factors affecting efficiency of sampling methods are reviewed. Finally, methods for estimating actual population sizes from sample data are described.

Because all available sampling methods are biased in respect of the portion of the population they sample, much work has been undertaken to determine what these biases are and how the sample relates to the whole population. Considerable day-to-day variability occurs in sample sizes of tsetse populations from the same area, for reasons which are still unclear. Samples of tsetse populations caught in traps, or by other techniques, will depend upon activity as well as abundance of the flies. Therefore, unless something is known about the flies' activity, relative abundance cannot be accurately estimated; similarly, if the abundance is not known, the level of activity of the population cannot be properly assessed. These problems were discussed by Vale (1993a).

Despite these difficulties, trap catches are commonly used to estimate apparent abundance in the absence of alternative and easily used means to do so. For this purpose, it has to be assumed that seasonal or annual variations in abundance are not due to seasonal or annual differences in activity. This may be a reasonable assumption; at Nguruman, Kenya, trap catches of *G. pallidipes* were well correlated with estimates of absolute population size with no evidence of seasonal change in trap efficiency (Brightwell *et al.*, 1997). This relationship may not hold for all tsetse species or all locations.

6.1 SAMPLING METHODS

6.1.1 Fly-rounds

In the early days of tsetse research, man fly-rounds were commonly used for assessing and studying the ecology of tsetse populations, particularly in anglophone countries of Africa (see Chapter 9, section 9.1). This method consisted of marking a route through the area to be surveyed, along which two or more people would walk carrying hand-nets (Fig. 6.1). At intervals along the route, sometimes marked on trees with permanent signs, the team would stop and catch as many as possible of the tsetse flies observed. The flies would frequently alight on the backs of the moving people and take off when they stopped.

A number of modifications of this sampling method were tested, ranging from spiral routes through the bush to the use of electric back-packs to electrocute alighting flies. The electrocuted flies would then fall into a collecting device. Another modification, found useful for low-density tsetse populations and for species particularly unwilling to come to humans, was the use of bait oxen led through the bush along a marked route. The ox was stopped at regular intervals whilst tsetse flies landing on it or flying around it were caught. More flies (*G. morsitans* and *G. pallidipes*) could be caught from a dark ox than from a light one (Dean *et al.*, 1968).

One of the major problems with fly-round samples was the repellency of humans to many species of tsetse (see section 6.3); catches of some important species of the *morsitans* group consisted mainly of males and very hungry flies, and consequently gave misleading information. This observation led to further development of other sampling techniques such as traps.

6.1.2 Traps

A large number of trap designs have been tested for sampling or control of tsetse populations, but most were either rather inefficient or too cumbersome, and fly-rounds remained the commonest sampling technique until the development of the biconical trap in the 1970s (Challier and Laveissière, 1973). Figures 6.2 and 6.3 show early trap designs and traps based on the biconical trap, respectively. Traps assumed a new importance since the possibility arose of using them in conjunction with odours for control, particularly of *morsitans* group tsetse (with the exception of *G. austeni*) and also for some *palpalis* group flies, although there has been less success in identifying potent odour attractants for *palpalis* group tsetse. Attraction to such traps depends almost entirely on the distant perception of odours so that the most important feature of the trap is its ability to capture those flies attracted to its immediate vicinity.

Fig. 6.1. Two fly-catchers traversing a fly-round in Zambia.

Prominent examples of traps used for sampling, rather than just for control, are the F3 or 'Box trap' developed by Flint in Zimbabwe, and the Epsilon trap, also developed in Zimbabwe (Flint, 1985). The NGU trap, developed by Brightwell *et al.* (1987, 1991) at Nguruman in Kenya, was primarily designed for tsetse control, but has also been used for population monitoring.

Cuisance (1989) reviewed the use of traps and screens for both the control and sampling of tsetse populations and gave details of their designs and functions. Whilst some of the more significant aspects of the development of traps are summarized here, the reader is referred to this publication for more detailed information. A list of the main tsetse traps either currently or previously used, with references for their design, is given in Table 6.1.

Some of the early traps, in addition to being somewhat inefficient, were bulky and difficult to use (Fig. 6.2), and their large-scale deployment would have been costly. Subsequent developments have therefore been directed at producing cheaper, lighter, more efficient and more easily managed traps.

(a)

Harris trap

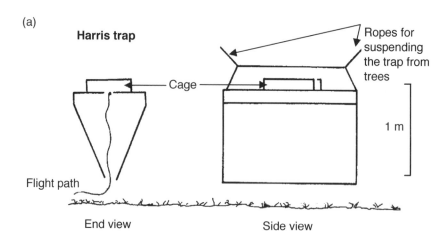

Cage

Ropes for
suspending
the trap from
trees

1 m

Flight path

End view　　　　　　　　　　Side view

Swynnerton trap

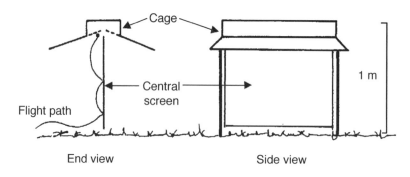

Cage

Central
screen

Flight path

1 m

End view　　　　　　　　　　Side view

Morris trap (animal trap)

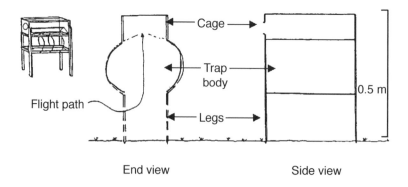

Cage

Trap
body

Flight path

Legs

0.5 m

End view　　　　　　　　　　Side view

(b) **Langridge trap**

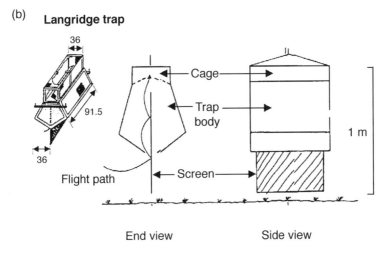

End view   Side view

Moloo trap

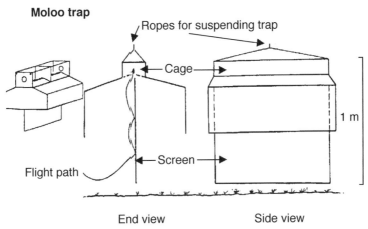

End view   Side view

Fig. 6.2. Trap designs – early models.

The biconical trap

The biconical trap (Challier and Laveissière, 1973) became the most widely used tsetse trap in Africa in the 1970s and 1980s, and several more recent traps have been based on its design (Fig. 6.3). Some examples of these modifications are discussed below. Apart from modifications using different coloured cloth (e.g. Takken, 1984), an automatic biconical trap was designed to reduce personnel requirements for ecological surveys and for carrying out periodicity surveys (Ryan *et al.*, 1981). This trap, which was a combination of the biconical trap with an electrocuting grid in the cone and a suction trap segregator, automatically separated flies caught at different time intervals.

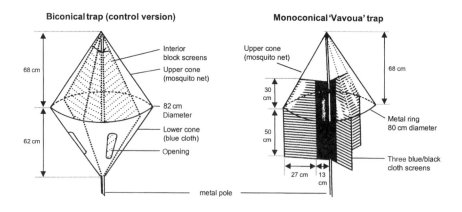

Biconical trap (control version)

68 cm

62 cm

- Interior block screens
- Upper cone (mosquito net)
- 82 cm Diameter
- Lower cone (blue cloth)
- Opening

Monoconical 'Vavoua' trap

Upper cone (mosquito net)

30 cm

50 cm

27 cm   13 cm

68 cm

- Metal ring 80 cm diameter
- Three blue/black cloth screens

metal pole

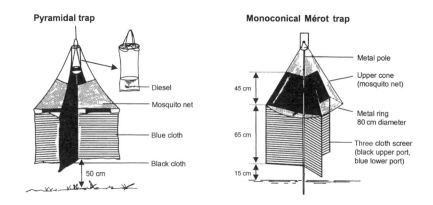

Pyramidal trap

- Diesel
- Mosquito net
- Blue cloth
- Black cloth

50 cm

Monoconical Mérot trap

45 cm

65 cm

15 cm

- Metal pole
- Upper cone (mosquito net)
- Metal ring 80 cm diameter
- Three cloth screen (black upper port, blue lower port)

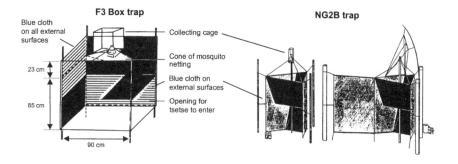

F3 Box trap

Blue cloth on all external surfaces

23 cm

85 cm

90 cm

- Collecting cage
- Cone of mosquito netting
- Blue cloth on external surfaces
- Opening for tsetse to enter

NG2B trap

Fig. 6.3. Trap designs – the biconical trap and its modifications.

Table 6.1. Trap designs for tsetse sampling and control.

Trap type	Purpose	Reference
Currently used		
Biconical	Sampling, control	Challier and Laveissière, 1973
Ngu – variations	Control	Brightwell *et al.*, 1987; Dransfield *et al.*, 1990
Monoconical (Lancien)	Control	Lancien, 1981
Vavoua	Control	Laveissière and Grebaut, 1990
Bipyramidal	Control used without insecticide impregnation	Gouteux *et al.*, 1991
Pyramidal	Control	Gouteux and Lancien, 1986
Epsilon trap	Sampling	Flint, 1985; Vale, 1988
F3	Sampling	Flint, 1985; Green and Flint, 1986
Sticky trap	Sampling	Ryan and Molyneux, 1981; Vreysen *et al.*, 1996
No longer used		
Harris trap	Control and sampling	Harris, 1938
Morris trap	Sampling, visual no bait	Morris and Morris, 1949
Langridge	Sampling/control	Langridge, 1975
Moloo – screen awning	Sampling	Moloo, 1973

Sticky traps

Sticky traps using non-setting adhesives have been used for catching tsetse (Ryan and Molyneux, 1981). One of the best adhesives tested was Oecotak®, which is a polybutene compound. Other similar products are Tanglefoot® and Stickem®. Sticky traps are not commonly used for monitoring tsetse populations although they were found useful for monitoring populations of *G. austeni* on Zanzibar, where it occurred at low densities and was difficult to sample with other types of trap. An early sticky panel was a three-dimensional design, but others used on Zanzibar simply consist of a monopanel or 'legpanel', to the lower part of which a sticky material is applied. The panels measure 60 cm by 70 cm and are coloured 'baby blue'; they are coated with sticky materials, of which polyisobutylene LMW was the best (Vreysen *et al.*, 1996).

Other traps and sampling devices

Gouteux and Lancien (1986) described a pyramidal trap, used for the capture and control of tsetse in the Congo. This was 2–5 times more efficient than the biconical trap. In a more recent study, the biconical trap, Mérot's monoconical trap, F3 trap, Vavoua trap and screen trap were compared for the capture of *G. tachinoides* in Burkina Faso. The biconical trap was best whilst the Mérot and F3 traps were poor (Amsler *et al.*, 1994b).

Artificial refuges can be useful for sampling fed and resting flies; a variety of artificial refuges used experimentally were described by Vale (1971). These are simply designed to imitate the natural shady resting sites or larviposition sites of tsetse, such as warthog (*Phacochoerus africanus*) burrows. The devices used by Vale (1971) were box and horizontal pipe refuges and ventilated pits. Hale Carpenter (1923) also described artificial refuges for collecting tsetse puparia; these consisted of low sloping thatched roofs built over suitable loose soil near water on the shores of Lake Victoria.

Puparial searches

For some species of tsetse, searching for puparia can be a useful and complementary way of determining their presence and breeding sites, although it is time-consuming, and may not be very productive for certain species. This technique may be of particular use for species that do not readily come to traps, such as *G. austeni*. Abedi and Miller (1963) described a method for obtaining large numbers of puparia for starting a colony of *G. palpalis*. They poured soil from potential breeding sites into a 4-gallon (32 l) drum, filled with water. Puparia could then be separated out on the water surface. If the soil was poured slowly, almost all the puparia were set free to float; otherwise the soil needed to be broken up gently by hand. Puparia were removed, dried on filter paper and stored in a glass tube with a little sand. They collected over 5000 puparia by this method. Various other pupae likely to be encountered whilst searching for tsetse puparia have been described (Smith and Baldry, 1968).

6.2   TRAP EFFICIENCY – ELECTRIC NETS

The development of electric nets provided a very useful tool for studying the behaviour of tsetse in relation to sampling methods and for testing trap efficiency in the absence of a human observer.

Swynnerton (1933) first experimented with electric traps, but at that time the means for providing a source of electric power made them impractical. It was not until 1974 that Vale (1974a) followed up the idea, followed by Rogers and Smith (1977) who described a new electrical circuit, which could function with small lightweight batteries. Subsequently electric screens or 'nets' were further developed and have proved invaluable in studying the responses of tsetse to traps, odour baits and animals.

Rogers and Randolph (1978a) compared electric trap and hand-net catches of *G. p. palpalis* and *G. tachinoides* in the Sudan vegetation zone of northern Nigeria. The latter species was not caught as efficiently as *G. p. palpalis* with electric backpacks as it was low-flying; a lower-level electric trap might have been better. The electric trap was more efficient than hand-nets for catching *G. p. palpalis.* Early trials indicated that approximately 95% of flies released near electric nets flew into them, suggesting that they were almost invisible to the flies. However, the efficiency of electric nets is not as high as originally thought, as a large proportion of flies are not killed when they hit the electric wires and about 15–20% appear to 'see' the electric net and avoid it (Griffiths and Brady, 1994).

Hargrove (1977) proposed relating all trap catches to the efficiency of electric trapping devices (Vale, 1974b). Using this method in Zimbabwe, Hargrove found that classical 'animal' traps caught only a small percentage of tsetse that approached them; and although the presence of ox odour increased the number of flies coming to the trap, it did not affect trap efficiency. Later, video recordings of tsetse behaviour around a net showed that the nets may be invisible only when approached head-on but not when approached obliquely (Packer and Brady, 1990). Thus, the best system had an efficiency of only 55%, not taking into account flies which avoided the net, and 46% when behavioural avoidance was taken into account, rather than Vale's figure of 94% efficiency. There were also significant differences in efficiency with fully charged batteries and partly charged batteries. Nonetheless, Packer and Brady concluded that the use of electric nets to sample tsetse populations had 'revolutionised our understanding of tsetse behaviour and ecology'.

Vale and Hargrove (1979) further developed their method of estimating trap efficiency using an incomplete ring of electrified nets placed around the trap or target. The distribution of catches in these systems was used to estimate the minimum efficiencies with which the trap captured flies that were initially attracted by odours derived from a herd of cattle hidden below ground. They derived two formulae for estimating the efficiency of a trap based on the assumptions that flies approached and departed from a trap only once and in random directions. Equations were, firstly:

$$\% \text{ Efficiency} = y \times 100/[y + (x/p)] \tag{6.1}$$

where y is the number of flies caught in the cage of the test trap, x is the total number caught on the inside of the screens and p is the proportion of the perimeter covered by the electric screens.

The second equation was based on the number of flies caught on the outside of the screens (z), in which case:

$$\% \text{ Efficiency} = y \times 100/z/(1/p - 1) \tag{6.2}$$

Kyorku *et al.* (1990) used alternate treatments each day to take account of site effects and therefore modified the equation to:

$$\% \text{ Efficiency} = y \times 100/(z/p) \tag{6.3}$$

where y is the number of flies caught in the cage of the trap without the screens and z and p are as defined above.

Ryan and Molyneux (1982) defined absolute trap efficiency as the number of flies n, expressed as a percentage, removed from a population of size *N*, per trapping unit time. As population size or growth rate is not usually known, only relative efficiency can be assessed by comparison with other traps. They described the design of experiments for comparison of trap efficiency. The efficiency of conventional non-odour baited traps is estimated to be less than about 25% and may be affected by site, season and geographical location (Jack, 1939; Vale, 1971; Hargrove, 1977).

6.3 REPELLENCY OF HUMANS

Although fly-rounds were used to sample tsetse populations for many years, particularly in anglophone African countries, a series of papers by Vale in the late 1960s and 1970s cast doubt on the validity of much of this work. Samples caught using hand-nets were strongly biased as humans are repellent to *G. m. morsitans* and *G. pallidipes* (Vale, 1969). Studies using electric nets showed that only 'desperately hungry' flies probed humans whereas less hungry flies probed an ox in the presence of humans (Vale, 1974b). Thus, only very hungry flies, or males looking for females with which to mate, would come to humans. This was the reason for the low proportion of females detected by fly-rounds in tsetse populations, even though approximately equal proportions of males and females emerge from puparia collected in the wild (Nash, 1930). Nash (1930) conjectured that female *G. morsitans* were probably as abundant as males within the 'true habitat' but were inactive between meals. Humans appear to be recognized by the upright appearance and odour.

Many researchers at the beginning of the 20th century were misled by catches showing a predominance of males in the sampled population and produced various explanations for this before the bias of sampling methods was realized. For example, the mortality in female flies fed experimentally on crocodiles was high compared with those fed on birds or mammals (Kleine, 1909). Therefore, Kleine assumed that the predominance of males in catches resulted from a high female mortality after feeding on certain hosts. Fiske (1913) reviewed reasons for the low proportion of females, and was aware that this could result from sampling bias although he presented evidence from which he concluded that 'a sufficiently large series of caught flies offer a very good index of the proportions actually prevailing in the locality'. He discounted Kleine's suggestion of higher mortality in females and proposed a hypothesis of migratory activity to explain differences in the sex ratio. He justified his hypothesis by stating that it would be odd if tsetse flies did not have some mechanism for dispersal in the way that some seeds and other insects had.

As early as the 1940s it was known that some species of tsetse, for example *G. pallidipes*, were reluctant to come to humans (Jack, 1941; Vanderplank, 1944). At the time, fly-rounds were considered to give the best index of a tsetse population that could be obtained, although it was recognized that there was considerable variability in the results obtained (Glasgow, 1960). Other sampling methods that were tested to overcome the repellency of humans included the use of bait animals, moving vehicles and traps (Jack, 1941).

Experiments with electric nets have shown that the presence of humans also influences the catch of some species of tsetse at bait oxen, and that female flies visit them more frequently than was previously thought (Vale and Hargrove, 1979). Hargrove (1991) showed that the presence of humans in a trapping system comparing various traps reduced catches of *morsitans* group tsetse significantly ($P < 0.05$).

6.4 FACTORS AFFECTING TSETSE POPULATION SAMPLES

Many factors affect tsetse population samples; amongst them, the concept of availability is discussed in Chapter 9 and the use of odour attractants in Chapter 10. Some other causes of variation in sample size and sampling biases are referred to here.

Total catches of tsetse by most sampling methods depend on activity as well as the population density. This activity is related to the feeding cycles of tsetse and was discussed in Chapter 8 (section 8.6). Adult males appear to be active for about 30 min of flight per day only, and females are probably active for a shorter period. The probability of capture is therefore low, and biased towards the small proportion of active flies. Man fly-rounds catch mainly hungry flies for this reason. Nonetheless, fly-round data can be useful in looking at long-term fluctuations in tsetse populations, as can trap catches.

Owaga and Challier (1985) studied the catch composition from biconical traps with rotating screen attachments, compared with standard, stationary biconical traps. A trap with screens rotating at 20 revolutions per minute caught the most representative sample in terms of age category of females; as the speed of rotation increased, fewer females were caught and few of those were old flies. They explained this in relation to the frequently observed attraction of male *morsitans* group tsetse to moving objects, in search of a mate. Stationary traps caught predominantly hungry flies.

Rather than trying to find a sampling technique that catches representative samples of tsetse, an alternative approach is simply to try to quantify the biases and correct for them. Availability of flies to different sampling methods is low soon after feeding, increasing slowly at first and more rapidly as flies enter the feeding phase.

Factors to be considered in examining samples of tsetse are:

- intrinsic variability;
- climatic and seasonal variations;
- siting of traps/sampling areas;

- nutritional status;
- reproductive status;
- sex and age composition.

6.4.1   Intrinsic variability

Potts (1930b) drew attention to some of the inaccuracies in assessing tsetse density from fly-rounds and suggested some ways of improving the technique using what he called 'ecological fly-rounds'. In experiments in which catches of tsetse from the same area, using either fly-rounds or Morris traps, were compared, there was no correlation between sample sizes from each method and catches were variable, casting doubt on the use of such data for estimating the level of trypanosomosis risk or tsetse challenge (see Chapter 14) to which livestock are subject (Smith and Rennison, 1961a,b). Glasgow (1960) studied the variability of fly-round catches and concluded that a 7.5 km fly-round carried out once a week could not detect less than a fivefold change in the mean catch and to detect a twofold change would require a 15 km fly-round carried out twice per week. 'Availability' varied in one place from day to day.

In addition to variations in trap catches due to seasonal changes in tsetse population sizes, trapping probability and the stochastic variation in the sex ratio, there are considerable day-to-day variations. Spatial variability has been ascribed to various site factors such as trap visibility and shading (Morris and Morris, 1949). Climatic factors (rainfall, relative humidity and temperature) are responsible for daily fluctuations in addition to the nutritional state of flies and relative movements of hosts and flies (Hargrove and Vale, 1978; Dransfield, 1984). Williams *et al.* (1990b) considered that among many factors contributing to the fluctuations in trap catches, the most important included:

- The density of flies in the vicinity of the trap – a function of birth and mortality rates, immigration and emigration.
- The activity of those flies – dependent on the physiological state of those flies and on the climate.
- The efficiency of the trap – this may depend on the nature of the habitat and climatic factors.

There are systematic changes in apparent density (trap catches) over long periods and related changes from one day to the next, but variation over periods of 3–4 days show little correlation. Williams *et al.* (1990b) set out to separate systematic changes from stochastic fluctuations by first trying to establish the intrinsic variation due to the stochastic nature of the sampling process. There is usually a functional relationship between the mean number of insects sampled and the variance of the number of insects sampled, described by Williams *et al.* (1990b) as a 'power law relationship'. If insects are randomly distributed over an area, the number of insects in a given area of a given size will follow a binomial distribution; in such a case a square root transformation will stabilize the variance of the numbers of flies caught. When there is a clumped distribution a logarithmic transformation will stabilize the variance in the observed numbers of counts. Using plots of these transformations can give an indication of the degree of clumping of a population.

The intrinsic variability of sampling data could be estimated from differences between the numbers of males and females if these were taken as two sample measures of the mean population (Williams *et al.*, 1990b). The sampling distribution of the trap catches at Nguruman, Kenya, followed a Poisson distribution and Williams *et al.* (1990b) therefore used a square-root transformation for stabilizing the variance, following which they estimated a minimum number of days over which sampling should be done to achieve a given level of accuracy in the mean monthly catch. At Nguruman, the catch on a single day gave an estimate of the monthly mean apparent density that was only accurate to within a factor of 3. If 50% accuracy were required with 95% confidence, trapping would have to be carried out over a period of at least 8 days.

6.4.2   Climatic and seasonal variation

Reasons for the considerable intrinsic variability in day-to-day trap catches are poorly understood, although changing meteorological conditions may be important causes. Williams *et al.* (1990a) examined the effects of humidity and temperature on day-to-day and seasonal variations in trap catches at Nguruman, Kenya. Seasonal variation was significantly correlated with maximum daily temperature; catches increased with temperature below a threshold of 34°C and decreased with temperature when it was above 34°C. Correlations between trap catches and relative humidity were weaker. Day-to-day variation was significantly greater than the intrinsic variation due to the stochastic nature of the sampling process and, for some traps, was correlated with temperature and humidity. Movements of flies in response to animal movement or to other climatic factors may explain some of the variability. Minimum temperature occurred at dawn and therefore would not be expected to correlate with trap catches.

There was no evidence of seasonal change in trap efficiency at Nguruman (Brightwell *et al.*, 1997).

6.4.3   Siting of sampling devices

The siting of traps or targets can significantly affect their efficacy and this has been studied in Zimbabwe. The amount of canopy cover can greatly affect catches, and should not exceed 50%. Sunny, open places are therefore better for catching species such as *G. pallidipes*. However, if flies are to be dissected for study of trypanosome infection rates, mortality may be high in such situations, requiring flies to be collected at frequent intervals. Visibility is an important factor, even when traps are used together with odour attractants, and catches are higher when the trap is in a position where visibility is good.

6.4.4 Nutritional and reproductive status

Most early studies of *morsitans* group tsetse populations were based on data for male flies, as females were not easily caught. Biconical and other traps have allowed analyses to be made using female flies.

Early investigations into hunger staging and sampling of tsetse suggested that males concentrated around hosts not only to feed, but also to find mates (Vanderplank, 1947a). Male flies caught in traps had lower fat reserves than those caught on a bait animal, whilst the fat reserves of females were higher (Jack, 1941). Jack concluded that *G. pallidipes* only came to a bait animal when it was hungry. The interpretation of these and other data was that flies caught in traps were looking for shade, or resting sites, whilst flies caught on bait animals were in search of a bloodmeal.

Bursell (1961a) investigated the behaviour of tsetse with regard to sampling bias, in particular the changes in behaviour or activity during the 'hunger-cycle'. A comparison of the fat content and residual bloodmeal of tsetse sampled in different ways revealed that fat content of flies in which the bloodmeal was in a late stage of digestion was high for a standard catch with hand-nets, low for resting and bait catches, and intermediate for a vehicle catch. The proportion of females caught showed the opposite trend – a low proportion of females caught with hand-nets and higher with a vehicle catch. These results were interpreted in terms of changes in behaviour of flies during the following four phases of the male hunger cycle:

1. Phase of inactivity.
2. Phase of activity characterized by sexual appetitive behaviour.
3. Phase of activity characterized by appetitive behaviour in relation to the feeding reaction indifferent to the stimulus of moving objects.
4. As for 3, except that moving objects constituted a stimulus for release of the feeding reaction.

Vale and Phelps (1978) investigated biases of different techniques used for sampling tsetse populations. They compared catches in electric nets at stationary sources of ox odour, hand-net catches at natural or artificial refuges and catches from mobile baits. Temperature, time of day and type of refuge or site affected the magnitude or composition of some of the catches but odour sources or refuges gave samples that roughly represented the sex and species composition of the wild populations. Mobile baits resulted in the least representative catches. Flies caught at natural or artificial refuges included a wide range of nutritional stages but electric nets at an odour source caught predominantly hungry flies. In a similar study in

Zimbabwe, Hargrove and Packer (1992) found that male *G. m. morsitans* and *G. pallidipes* caught in artificial refuges had 27–31% more fat than those caught in odour-baited epsilon traps (Hargrove and Langley, 1990); mean haematin levels were also higher. Arising from this study, a model for blood-meal metabolism was developed, which accounted for 79% of the variance of the *G. pallidipes* field data.

Rogers (1984) examined sampling biases in some detail and subsequently studied sample biases in relation to the nutritional status of captured tsetse. As trap catches are representative of the feeding portion of the population (Rogers and Randolph, 1986a; Randolph and Rogers, 1986) it was proposed that nutritional data alone, which are simpler to collect than detailed reproductive data, could be used to estimate the relative sampling efficiency of traps for females over the pregnancy cycle. However, this was disputed by Langley and Stafford (1990), on the grounds that only the hungriest of flies attracted to a trap enter it. Randolph *et al.* (1991a) collected both nutritional and reproductive data in a study at Nguruman in Kenya to test their hypothesis. From analyses of those data they concluded that the apparent reliability of nutritional data for estimating female sampling biases resulted from the uniform and distinct condition of females on each day of the pregnancy cycle in the laboratory. Their previous analyses were based upon such data, whilst field flies showed more variability. The probabilities for assigning flies to a day of pregnancy fitted with field data despite much variability. They stated that if relative sampling efficiency of a trapping system for females needed to be estimated as accurately as possible, there was no alternative to ageing the flies by detailed ovarian dissection – for example, to obtain a baseline for further quantitative analysis of a tsetse population. On the other hand, nutritional data are sufficient to reveal general patterns of sampling biases for comparing trapping systems, different species of tsetse or different ecological conditions. Randolph *et al.* (1991a) concluded that the feeding interval for an individual fly might vary from 86 h to 56 h, in contrast with other observations (Challier, 1973; Snow and Tarimo, 1983) that three synchronized bloodmeals were taken at set points during the pregnancy cycle.

Discriminant analysis has been used for the estimation of sampling biases for female tsetse (Randolph and Rogers, 1986). It was suggested that this could be used for estimating total population size provided the absolute sampling efficiency on any one day of the pregnancy cycle could be established, perhaps by mark–release–recapture experiments (section 6.5). Discriminant analysis determines the probability of an item of unknown origin belonging to one of several groups, on the basis of certain measured characteristics of that item, and the previously determined characteristics of each group. Certain assumptions made are:

- That the distribution of characteristics is multivariate normal.
- That the variance of each characteristic, and its covariance with all other characteristics, changes little from group to group.

Studies on the sexual responses of *G. m. morsitans* and *G. pallidipes* around host animals, traps and targets in Zimbabwe showed that males of both species seemed to accumulate around a stationary bait. Traps attracted males with a range of hunger states but mainly hungry males of both species (Wall, 1989b). Groenendijk (1996) reported that target catches of *G. pallidipes* were biased towards female flies in the first 3 days of pregnancy.

Marking flies

The technique of marking flies has been employed for experiments to estimate population size as well as to investigate resting sites of tsetse. Knight and Southon (1963) described a novel method for marking haematophagous insects. This depended on the detection of a dye, tryptan blue, in the bloodmeal of such insects, which had previously been injected into an ox. Using a flutter valve to slow down the rate of administration and prevent adverse effects, 200 ml of a 2% aqueous solution of tryptan blue was administered intravenously. The dose of about 4 g was given over a period of about 20 min as considerable distress is caused if it is given more quickly. The dye could not be found in mosquitoes more than 2 days after feeding on an injected animal; in tsetse it could be detected 8 days after feeding on an animal injected with the dye the same day. Other possible uses of the technique would be to assess biting rates on cattle for assessing challenge.

Macleod and Donnelly (1958) described four methods of marking insects: individual marking with paints; mass powdering with dyes; radioactive labelling with ^{32}P; and a combination of the last two. They also briefly described a fifth method in which the emerging fly labels itself with fluorescent dust. The latter technique was used by Tibayrenc *et al.* (1971) for studying tsetse resting sites.

Turner (1980a) described an original method for mark–release–recapture experiments for investigating tsetse behaviour, based on the transfer of fluorescent pigment from marked males to unmarked females during copulation. Powder, applied between the coxae of the male, lodges in the scutellar groove of the female during mating. Powder persisted well on males before mating and on indirectly marked females. Individual males could mark several females and the marks on different males could be detected on females after multiple mating.

6.5   ESTIMATION OF POPULATION SIZE

6.5.1   Mark–release–recapture

The technique of marking, releasing and recapturing tsetse flies has been used to study the dynamics (density, dispersal, survival) of tsetse populations and to compare the behaviour of laboratory-reared flies with wild flies after release in the field. It is a technique that can be used for estimating absolute population size, but is difficult practically, requiring a relatively large number of flies to be caught as probabilities of recapture are low. Low sample sizes and low recapture probabilities result in high coefficients of variation and, sometimes, biologically impossible results. In addition, Rogers (1977) suggested that the wide-ranging and random pattern of movement of tsetse within their natural habitat might explain the low recapture rates that characterize mark–release–recapture experiments in large infested areas. Furthermore, whilst being subject to many other sources of error, the estimate of population size applies to a large and ill-defined area. A method of minimizing these errors and making population estimates more reliable is to pool data from several consecutive samples (Hargrove and Borland, 1994).

Jackson (1933) first used the method for estimating population sizes of tsetse, having adapted a technique used and described by Lincoln (1930) for estimating duck populations. Jackson (1953) later described how, by the use of nine colours applied two at a time, it was possible to mark upwards of 25,000 tsetse without duplication. He also adapted the technique in order to take into account such factors as death and birth rates and the effects of migrations.

The principle of the mark–release–recapture technique

The principle of the method is simple. A random sample is collected from the population to be measured. The individuals are marked so that they may be recognized again and are then released in such a way that they may be expected to distribute themselves randomly with respect to the rest of the (unmarked) individuals in the population. At a later date, another random sample is taken from the population and the numbers of marked and unmarked individuals in it are recorded. It is important that either the initial marking or the subsequent catching be done evenly over the area selected for study. The second sample will contain some marked and some unmarked individuals. Assuming the marks are not lost, and provided that: (i) the marked individuals redistribute themselves randomly with respect to the unmarked ones; (ii) marked flies are neither more nor less readily caught than the unmarked ones; and (iii) between the times of release and recapture there are no gains or losses by births, deaths or migration, the equation to estimate population size x (derived by Bailey, 1951) can be expressed as:

$$x = an/r \qquad (6.4)$$

where a is the number of marked flies, n is the number of recaptured marked and unmarked flies and r is the number of marked flies recaptured.

In nature, condition (iii) is never likely to be fulfilled, although gains and losses might approximately balance each other out over short periods. However, by suitably designing the experiment and making appropriate extensions to the fundamental equation, the method allows estimation of total population size in the area and of the rate at which it is changing. Moreover, the rate of change may be divided into its four components of births, deaths, immigration and emigration. Bailey (1951, 1952) thought Jackson's use of a curve fitted to standardized recapture rates was unsatisfactory. He described the use of a maximum likelihood method for estimating population size with more precision, and for determining the degree of variance associated with these estimates.

Flies tend to be recaptured in greater numbers on the first and second days after marking. In order to minimize variability of catches from day to day or week to week during the course of an experiment, Jackson (1948b) defined a corrected recapture rate, y_n:

$$y_n = 10^4 R_n / C_0 C_n \tag{6.5}$$

where C_0 is the catch in the week of release, R_n is the number of flies originally marked on day 0 (when the catch was C_0) that are recaptured on day n in a total catch of C_n flies. This corrected recapture rate, determined by Rogers (1977), has a variance:

$$\text{var}\,(y_n) = [10^8 R_n (C_n - R_n)] / C_0^2 C_n^3 \tag{6.6}$$

There is a logarithmic decline in the recapture rate with time, due to loss by death and emigration. In Rogers' experiment this rate was equal to 21% per day, estimated from the formula:

$$\ln N_r = \ln N_0 - pt \tag{6.7}$$

where p is the proportion of flies dying or emigrating from the marking site, N_r is the number remaining after time t and N_0 is the initial number of flies.

Rogers and Randolph (1986a) subsequently modified the formula for the recapture rate, describing the maximum likelihood estimate for the average corrected recapture rate on day n as (p_n), which is approximately defined as:

$$p_n = [(\sum_{i=1}^{k} R_{ni} / \sum_{i=1}^{k} C_{0i} C_{ni}) \times 10^4] \tag{6.8}$$

where there are k replicates. Thus, C_{ni} is the number of flies caught on day n of the ith replicate of the experiment.

Assuming Brownian diffusion from a central release point (for example, a village in a study area), the proportion (P_r) of released, marked flies still within a radius r at time t after release (and therefore available to a system of traps within the village) is given by:

$$P_r = 1 - \exp(-r^2/4wt) \tag{6.9}$$

where w is the diffusion coefficient (Southwood, 1978). Introducing mortality at the daily rate of m, the proportion of flies surviving within the radius r is given by P_{sr}, where:

$$P_{sr} = \exp(-mt) \times [1 - \exp(-r^2/4wt)] \tag{6.10}$$

Further adaptations and explanations of these models are given by Rogers and Randolph (1986a), who showed that activity of female tsetse sampled with biconical traps was related more to the interlarval period than to the shorter duration of feeding cycles as in males.

Leslie and Chitty (1951) described the derivation of maximum-likelihood equations for estimating the death rate in a population using the mark–release–recapture technique. The method for mark–release–recapture was based on that used by Jackson (1939, 1948b). It was assumed that catching and marking flies had no effect upon their behaviour or survival, that the sampling method was random and that the chances of a fly being recaptured were not affected by a previous capture. The theory described by Leslie and Chitty is summarized below as its principle has been the basis of more recent developments.

Suppose a population consisting of an unknown and variable number (N_t) of individuals is sampled at equidistant intervals of time $t = 0, 1, 2, ...,$ T, and that at each sampling R_t individuals are captured, marked and returned to the population. Among the R_t ($t = 0, 1, 2, ..., T$) individuals captured and examined at time t, let u_t = the number unmarked, and s_t = the number with one or more marks (i.e. the total number of recaptures), so that:

$$R_t = u_t + s_t \tag{6.11}$$

Let $r_{0, 1, 2, ..., t}$ = the number of individuals recaptured at time t, bearing the previous marks 0, 1, 2, ... ($t - 1$). Assume that over the sampling period $0 - T$ the death rate per unit time remains constant.

Let P = the constant survival factor, so that out of a population of individuals N_t alive at time t, PN_t are alive at time $t + 1$.

The death rate Q in the population as a whole over each interval of time is then:

$$Q = 1 - P \tag{6.12}$$

If we assume that this death rate is approximately the same for all age classes of the population then the expectation of life of young individuals from the age they enter the population at risk of capture is:

$$C_0 = -1/\log_e P \tag{6.13}$$

The estimation of P in Leslie and Chitty's method comes from distributions of recaptures and refers to the time between the original and penultimate sampling. No estimate can be made for survival during the last sampling period. Leslie (1952) later described the estimation of total numbers in a population.

As a result of these studies, carried out under conditions where immigration and emigration were thought to be minimal or absent, a method was derived for estimating populations based on probability of capture, the theory of which can be summarized briefly as follows.

The probability that a marked fly is captured on day i is proportional to:

$$Y_m/(N_m - X_m) \qquad\qquad (6.14)$$

where N_m is the number of marked flies captured.

Assuming no difference in the behaviour of marked flies, an equation can be derived for the probability of capturing an unmarked fly and, from this, an equation for an estimate of the number of unmarked flies in the population can be derived:

$$N_u = (Y_u/Y_m)(N_m KS_m) + X_u \qquad\qquad (6.15)$$

where N_u is the population of unmarked flies and Y_u is the number of unmarked flies captured. An estimate of the total population of flies at time i is obtained by adding the known number of marked flies at time i to the above estimate.

6.5.2   Removal trapping

Assuming a sufficient level of trapping is imposed in an area, catches within that area would be expected to decline with time. Removal trapping (Zippin, 1956; Southwood, 1978) and other methods have been used at Rekomitje in Zimbabwe for estimating sizes of tsetse populations (Phelps and Vale, 1978). Removal trapping has also been used to control and estimate population sizes of tsetse in Côte d'Ivoire and in Zambia (Ryan *et al.*, 1981). The biconical traps used had clear and dramatic trapping-out effects on the tsetse populations in both sites. Estimates of population size, although variable, were considered realistic, but an assumption (often made) that marked flies become completely mixed within a population was probably false. The impossibility of achieving equal probability of capture of all flies was the weakest point in the estimation of population density. Zippin's removal trapping technique, based on the maximum-likelihood method of estimation of population size, is said to work for an isolated situation where removal of individuals (tsetse) should result in progressively fewer individuals caught in successive attempts. It assumes that the population is stationary, that the probability of capture during a given trapping period is the same for all animals exposed to capture, and that this probability of capture is constant. Catches of tsetse can also be influenced by temperature as observed by Phelps and Vale (1978) in Zimbabwe. Population estimates were low compared with those resulting from the method described above, possibly due to decreased activity resulting from low temperatures; when corrected for, the estimates of the tsetse population from Zippin's method were similar to those obtained from the other methods. Hargrove (1981) pointed out that all the estimates could have been unreliable as the population was not isolated, and therefore Phelps and Vale's assumptions may have been invalid.

Hargrove (1981) described discrepancies between estimates of tsetse populations using mark–release–recapture and removal trapping techniques. He modelled a *G. m. morsitans* and *G. pallidipes* population on the basis of a first-order kinetic system (Fig. 6.4), estimating rate coefficients for recruitment, loss and trapping from published data using non-linear regression techniques.

A linear effect of temperature on catches of tsetse was assumed, based on the experiments of Phelps and Vale (1978). The model explained 92–98% of the variance of catches of mature flies and a similar model explained 92% of the variance for teneral female *G. pallidipes*.

Good predictions for catches of *G. pallidipes* were made, with errors of between 7 and 10%. As there were no correlations or patterns of the residuals with time or temperature, Hargrove (1981) felt it unnecessary to complicate the model by constructing components to explain them. Estimates of the mature population obtained were one-third to one-half of those obtained using mark–release–recapture and the estimate of the daily emergence of teneral *G. pallidipes* females was 1/28th of the previous estimate. Hargrove's model predicted considerable emigration and immigration, which Zippin's method ignores. He therefore considered that the agreement in estimates resulting from the methods used by Phelps and Vale (1978) was coincidence. Differential availability due to behavioural differences or greater mortality of marked flies were suggested as possible explanations for these discrepancies.

Clearly, the estimation of absolute population sizes of tsetse poses many difficulties, and attempts to do this and to model their population dynamics are very complex. Nonetheless, considerable progress has been made, particularly by Hargrove and Borland (1994) with populations of tsetse on

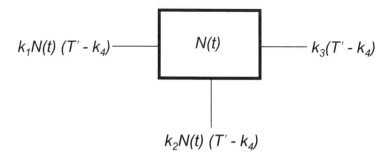

Fig. 6.4. Model of a first-order kinetic system for tsetse (from Hargrove, 1981). $N(t)$ is the number of flies present at time t, in days, after the start of the experiment; k_1, k_2 and k_3 are the coefficients of loss, trapping and recruitment rates, respectively; k_4 is the temperature at which these rates are all zero; T' is the daily maximum temperature.

an island in Lake Kariba, Zimbabwe. Population estimates and models of tsetse population dynamics should ultimately contribute to decision support systems for facilitating the sustainable control of tsetse populations.

Chapter 7
Ecology – Distribution and Habitats

In this chapter some general factors influencing tsetse distribution and abundance and some quantitative methods that have been used for their study are reviewed. The distribution of the *palpalis, morsitans* and *fusca* groups is then described.

Much early work on tsetse flies concentrated on their distribution and habitats, with the objective of determining priority areas for the control of the flies, and areas where people or domestic livestock were at risk. Early research was directed at identifying vegetation types with which tsetse are associated, and environmental factors that determine their distribution and abundance. As an example of apparent links with vegetation types, a report on tsetse in Nigeria stated that *Mimosa asperata* was 'practically an indication of the presence of *G. tachinoides*' (Pollard, 1912).

From the 1960s onwards, ecological studies became more directed to the population dynamics of the fly in relation to control, rather than to its habitats and distribution. Since the early 1990s, geographical information systems (GIS) and remotely sensed satellite data have been used as tools to examine environmental factors linked with distribution and abundance (e.g. Rogers *et al.*, 1996). This was emphasized by Hendrickx *et al.* (1997) and Hendrickx and Napala (1997) who discussed the prioritization of areas for trypanosomosis and tsetse control using GIS techniques to link habitat, vegetation and tsetse distribution.

7.1 TRADITIONAL IDEAS REGARDING TSETSE HABITAT

It may be useful to describe some of the traditional ideas regarding tsetse habitat, which were based on catches of tsetse on man fly-rounds. These ideas have changed or been modified following the discovery that at least some tsetse species are actually repelled by humans, and therefore the segment of the population being sampled by fly-rounds was biased and unrepresentative, comprising mainly hungry, male flies (e.g. Isherwood,

1957). The repellency of humans was shown to be a result of their upright-ness and odours of lactic acid released from the skin. It was observed that in open areas a few, mainly male flies needing a bloodmeal could be found, and these areas were termed 'feeding grounds'. In other more shaded woody areas flies were less hungry; these were considered to be the 'home' or true habitat of tsetse, and species such as *G. morsitans* concentrated in such areas of semi-evergreen woodland in the dry season.

These observations led to the following beliefs:

- The shape of certain trees was important to tsetse.
- They hunted by sight, and open areas facilitated host location.
- Removal of certain trees would make the habitat unsuitable and cause a reduction in density.

These old ideas changed largely as a result of work by Vale (1978) in Zimbabwe, following observations that discriminative bushclearing did not always successfully reduce tsetse populations (Bursell, 1966b; Pilson and Pilson, 1967). Investigations showed that a moving human bait attracted 'desperately hungry' flies unable to find anything else. In contrast, catches of tsetse attracted to a stationary ox resulted in large catches, comprising a high proportion of females. Therefore, the idea that in the dry season females concentrated in semi-evergreen woodland was wrong, and the concentration of males was less marked than formerly supposed. Eventually, by measuring physiological state, Bursell (1966b) showed that in the peak dry season, the leafless miombo and mopani woodlands (section 7.4) supported self-sustaining populations of tsetse prospering better than those in the leafy 'homes'.

According to Vale, this (i) invalidated the theory of tsetse control by discriminative bushclearing; (ii) suggested that much more of the habitat would need to be ground-sprayed than previously estimated; and (iii) showed that alternative sampling methods were required as the fly-round only sampled sexually appetitive males.

7.2 FACTORS AFFECTING DISTRIBUTION AND HABITATS

The general distribution of tsetse flies, determined principally by climate and influenced by altitude, vegetation and the presence of suitable host animals, has been known for a long time. However, more precise limits of distribu-tion, particularly in areas of low population density, were not (and still are not) well defined. Broadly, it is possible to correlate distribution of the three subgenera of tsetse flies fairly well with climatic and ecological parameters. Since the 1960s, when the efficiency of traps and odour attractants (see Chapter 10) was improved, more precise limits of tsetse distribution have been obtained from surveys using various sampling techniques. Recently, the use of GIS and remote sensing techniques has been discussed and experimented with as means of obtaining this sort of information.

According to Ford (1962), the southern limits of *Glossina* distribution in Africa lie north of a line drawn from Benguela, in Angola, to Durban, in South Africa. The northern limits are roughly a line from Dakar in Senegal across to Ethiopia and Mogadishu in Somalia on the east coast. The Semitic and Hamitic people in the north and the Bushmen and Hottentots in the south live outside the tsetse-infested areas. In 1906 Carter recorded the presence of *G. tachinoides* in south Yemen in the Arabian peninsula, but no other reports confirmed this. It was therefore thought that tsetse in the Arabian peninsula may have died out. However, *G. f. fuscipes* and *G. m. submorsitans* have since been detected in southwestern Saudi Arabia (Elsen *et al.*, 1990).

Among the principal studies on tsetse distribution, Ford (1963) reviewed distribution in relation to ecological requirements and historical events. Distribution maps were published by Ford and Katondo (1974), revised in 1977, and further updated by Katondo (1984). Moloo (1985, 1993a) published distribution tables in terms of presence or absence for each infested African country. Gouteux (1990) added to, and corrected, the distribution of tsetse in west and central Africa, particularly for *G. frezili*, first described in 1987 (Gouteux, 1987a). Distribution of the three subgenera of *Glossina* in Africa is shown in Fig. 7.1.

Some of the physiological parameters limiting tsetse survival and affecting their distribution and abundance have been estimated from field or laboratory studies. However, these characteristics vary according to species and sex; furthermore, it can be difficult to extend experimental data obtained under controlled conditions of temperature and humidity to the field where these conditions are not constant. Experimental observations indicate that the optimal temperatures for pupal fat consumption and development lie between 23 and 25°C (Bursell, 1960c; Phelps and Burrows, 1969a). The critical fat content for adult survival is approximately 0.5 mg, or 4.2% of the residual dry weight (Glasgow and Bursell, 1961; Glasgow, 1963).

7.2.1 Climatic factors determining tsetse distribution

It was recognized early on that climate was an important factor determining the distribution of tsetse, if not *the* factor controlling the basic pattern of distribution. The limit of distribution is closely correlated with the tropical savanna (summer rain) climate, which follows the 508 mm annual isohyet. Climate, though dependent on latitude, is modified by altitude and of course has a great effect on the vegetation, which is vital for providing shade and maintaining a suitable microclimate for tsetse as well as a habitat for their vertebrate hosts. As a generalization, the tropical rain forest (equatorial) climate controls the habitats of the *fusca* and *palpalis* groups, and the surrounding woodlands are the habitat of the *morsitans* group. Altitude influences tsetse distribution through its effect on climate, particularly temperature. In Ethiopia, 1600 m was considered the rough upper altitudinal

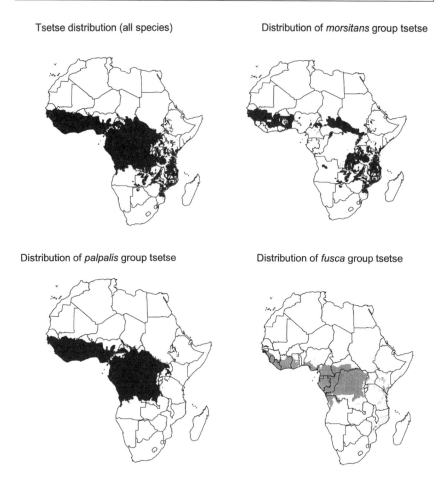

Tsetse distribution (all species)   Distribution of *morsitans* group tsetse

Distribution of *palpalis* group tsetse   Distribution of *fusca* group tsetse

Fig. 7.1. Distribution of tsetse in Africa (based on the distribution maps of Ford and Katondo, 1984; updated by Russ Kruska and Robin Reid, ILRI, Nairobi, 1998, unpublished maps).

limit to tsetse distribution (Langridge, 1976). Subsequently, however, *G. pallidipes* was found at 1700 m altitude, and '*G. m. ugandensis*' (*G. m. submorsitans*) at altitudes up to 2200 m (Tikubet and Gemetchu, 1984).

Ecological zones and their characteristics are summarized in Table 7.1. In West Africa the bioclimatic zones are roughly parallel to the equator. Bursell (1963a) suggested that the distribution of *G. morsitans* and *G. swynnertoni* might be limited by exhaustion of fat reserves during development under extreme climatic conditions. Another important factor limiting species distribution referred to by Bursell was the water balance of stages within the puparium; the degree of resistance to desiccation is closely linked with habitat.

Table 7.1. Summary of ecological zones and their characteristics (adapted from Jahnke, 1982).

Characteristic	Ecological zones				
	Arid	Semi-arid	Sub-humid	Humid	Highlands
Area ('000 km²)	8327	4050	4858	4137	990
Rainfall (mm)	< 500	500–1000	1000–1500	> 1500	Variable
Moisture index*	36	20	0–20	0	Variable
Growing days	< 90	90–180	180–240	> 240	Variable
Cattle (%)	21.3	30.8	22.2	6.0	19.7
Area of tsetse ('000 km²)	438	2036	3298	3741	195
Area of tsetse (%)	4.2	50.3	68.2	89.7	1.9
Predominant tsetse group	*Morsitans*	*Morsitans/ Palpalis*	*Palpalis/ Fusca*	*Fusca/Palpalis*	None

*Moisture index is the relationship between rainfall and evapotranspiration.

In contrast to some other insects, there is no period of diapause in the life cycles of tsetse; there is only a change in adult longevity and puparial duration, related to temperature. Thus, for example, in populations of *G. m. submorsitans* in West Africa, new individuals are being added to the population throughout the year. The northern limit of the distribution of *G. m. submorsitans* is determined largely by the survival rate among puparia, which have to develop during the dry season; the limit corresponds closely to the 762 mm isohyet, associated with a dry season of 6 or 7 months (Nash, 1948). In northern Nigeria, during the wet season and the first half of the dry season, numbers of *G. m. submorsitans* may increase for brief periods as births exceed deaths, and the innate capacity for increase (*r*) is positive; but as the dry season advances *r* becomes negative, and for the remainder of the year deaths exceed births and numbers decline (Nash, 1948). Nash described how the flies then begin to evacuate the drier parts of their range in the open savanna woodland, slowly becoming restricted to the islands of forest or the denser vegetation along rivers. By March, which is the hottest and driest month in Nigeria, tsetse are to be found only in these moister, shadier areas. At this time of the year flies enter adjacent woodlands to hunt only in the early mornings and evenings. When the rains begin again, *G. m. submorsitans* begins to disperse into the savanna woodland, and individuals travel as far as 13 km from places where they had spent the dry season. During the dry season, larvae are deposited almost entirely in thickets, but during the wet season they may be deposited under logs in open savanna. New generations breed in the newly invaded areas, and in due course the flies disperse far and wide through the savanna woodland. The period of multiplication and population expansion ends with the next dry season (Nash, 1948).

Climate and habitat

Many studies have been carried out to determine the effect of climate on *palpalis* group tsetse, particularly *G. palpalis*, at the extreme hot, dry northern limit of their range. Cumulative effects of a series of long rainy seasons or long dry seasons are thought to have been of importance in influencing advances and recessions of tsetse populations. Such advances were often preceded by several years with short dry seasons and prolonged rainy seasons. Similarly, several years of below average rainfall and long dry seasons could have an adverse effect, causing recessions. Nash (1948) observed a relationship between the northern dry distributional limits of *G. m. submorsitans* and the number of dry months of the year; no such relationship was found for the southern wet limits. Buxton's book, *The Natural History of Tsetse Flies*, reviewed the knowledge of the relationship between climatic factors and tsetse populations at that time in considerable detail (Buxton, 1955a). These relationships were particularly well studied in Nigeria where *palpalis* group species are now responsible for much of the transmission of trypanosomosis to both humans and cattle. As Rogers (1991) emphasized, climate cannot determine abundance directly, but does so only through its effect on birth, death, immigration and emigration rates.

7.2.2   Desiccation as a cause of mortality and limit to population growth

Desiccation of adult tsetse was at one time considered to be an important cause of death in natural tsetse populations although Bursell (1959a, 1960a) suggested that it was not the most important cause. Certainly, for the savanna (*morsitans*) tsetse species, Bursell's (1960a) studies of fat and water content demonstrated that this was not the case. However, tsetse have to take measures to avoid desiccation within their habitat, either resulting from evolutionary adaptation to avoid water loss, or by behavioural patterns that result in avoidance of desiccating conditions. Avoidance of desiccation is thus usually a result of the following:

- **Puparial adaptation** Bursell (1958a) conjectured that tsetse evolution is related to a transition from hygrophytic to xerophytic habitats. During this process tsetse acquired increasingly efficient water retention in the puparial stage leading to a relationship between the habitats of tsetse and the ability to retain water.
- **Utilization of niches within the habitat which provide suitable microclimates** The principal habitat of *G. palpalis* is the gallery forest, which, in savanna areas, provides a distinct environment formed by the canopy, and creepers, providing a microclimate that is more hospitable to the fly than the adjacent savanna.

Glossina palpalis is almost always found close to water, and in West Africa it has to remain near water in the late dry season in the extreme northern part of its range, unless the vegetational insulation is adequate. The water and vegetational screen of the gallery forest provides a level of humidity essential to the fly. Flies are in no danger of desiccation at 65% relative humidity, but at lower humidities, below 45%, desiccation becomes an important cause of mortality. Nash (1933a) demonstrated a negative correlation between saturation deficit and the apparent density of tsetse.

In northern Nigeria, when the mean relative humidity ranges between 43 and 57% in the late dry season, flies may suffer from desiccation and the longevity of *G. m. submorsitans* and *G. tachinoides* is then reduced (Nash, 1936). The dominant factor determining longevity appeared not to be humidity, although this was important, but maximum temperature, with which longevity was negatively correlated. *Glossina tachinoides* males appeared to live longer than *G. m. submorsitans* males under the conditions of that study. Nash thought it unlikely that flies of either species would live much more than 2.5–3 months under the most favourable conditions. The apparent density of *G. m. submorsitans* in Nigeria varied according to season, and could be correlated with evaporation rate (Nash, 1931). When evaporation rate increased, the apparent fly density decreased. There was a delayed effect (by 1 month) of high evaporation rate on tsetse apparent density – a delay which one would expect to occur biologically. Similarly, there was a lag of 1 month in the increase in apparent density following a decline in evaporation rate. Evaporation rate is closely correlated with wind, temperature and humidity and is generally inversely related to the density of *G. palpalis*. Evaporation rates increased considerably when vegetation had been burnt and a delayed drop in apparent density could therefore be expected, as occurred following bush fires in Nash's study in Nigeria. In a similar study in Tanzania, Nash (1933b) used fly-rounds to study the effect of climate on changes in apparent density of *G. m. morsitans*, measuring the degree of correlation between apparent density and each climatic factor. He also investigated the lags between changes in climate and effect on apparent density. There was a significant correlation between fly apparent density in one month and the evaporation rate of the same and the previous month.

Although, as Nash (1937) stated, 'climate is perhaps of paramount importance in the ecology of *G. tachinoides*', this species can withstand a much more severe climate than *G. palpalis*. Consequently, it is able to extend its range further north into the hotter, and drier, areas of West Africa, particularly along watercourses and into 'forest islands'. In these areas, the resting sites of tsetse are generally lower, and closer to water, providing a microclimate with a greatly reduced duration of unfavourable conditions. In the late dry season, *G. palpalis* is most often found only in those sections of thinly wooded streams where there are pools of water under well-defined banks.

The concentration of *G. tachinoides* and *G. m. submorsitans* within denser vegetation during the late dry season in northern Nigeria is believed to be a response to increasing temperature (Nash, 1936). The lower temperature limit for evacuation to occur was 39°C for both species. At normal temperatures, *G. morsitans* exhibits a positive reaction to light; if the temperature is raised sufficiently, the reaction of the flies is reversed and becomes strongly negative.

Whilst lacking reliable quantitative epidemiological evidence, researchers in West Africa described a theory of tsetse dispersion in relation to trypanosomosis risk to cattle. The theory was that after a succession of good rains *G. palpalis* tended to remain dispersed along streams and to travel widely. In years of moderate rainfall, when tsetse flies remained dispersed along hundreds of kilometres of stream, the risk to Fulani cattle must have been far greater than when tsetse were confined to the immediate vicinity of water-holes. After a season of abnormally heavy rainfall, or after a succession of years of good rainfall, prevalence of bovine trypanosomosis could increase until the water table was lowered. Flies would then concentrate and there would be little dispersal. This would result in reduced risk except for cattle visiting water-holes within a tsetse-infested habitat. However, Nash and Page (1953) doubted whether *G. palpalis* often travelled more than 3 km from its birthplace during its lifetime. On ecological grounds, attempts to control *G. palpalis* would best be timed to start at the outbreak of rains, when the population is lowest. *Glossina palpalis* does not breed far out in the open woodland in the heavy rains and puparia are safe from inundation except in years of quite exceptionally heavy rainfall. In such years, when inundation of downstream breeding grounds led to big reductions in fly density, the areas near sources of streams could be valuable survival foci.

In the early dry or cold season, the breeding grounds become concentrated to some extent into a much smaller area. In the 1950s and early 1960s there was considerable discussion about the cessation, or otherwise, of breeding by *G. palpalis* in West Africa during the rainy season. This was first referred to by Lloyd *et al.* (1924) as a cause of seasonal variations in trypanosome infection rates, which rose in the wet season in association with an increase in mean age of tsetse. At Ugbobigha, near the southern limit of *G. palpalis* distribution in Nigeria, Jordan (1962a) detected no wet season cessation of breeding, and suggested that this may only occur in areas of very heavy rainfall. Nonetheless, highest pregnancy rates at Ugbobigha were found during the months of heavy rainfall, lowest mean temperature, saturation deficit and evaporation rate. The pregnancy rate in *G. palpalis* was low compared with that at Kaduna, in the north of the country, and Jordan suggested that this might be due to factors causing abortion in the late egg and early larval stages. There was no evidence of a wet season cessation of breeding in northern Nigeria.

7.2.3   Effect of humans

Apart from more 'natural' factors, the effect of humans has had an important influence on the distribution of tsetse flies. Most African countries, particularly those in tsetse infested areas, have low human population densities. Nigeria, Africa's most densely populated country, has a population of 89 million (1992 census). (Other estimates put it at 100 million, equivalent to 108 people km^{-2}.) In addition, the country has been through a stage of rapid development with relative affluence as a result of oil revenues. Therefore, changes in human population density have probably had a more significant effect on tsetse populations than in other countries. Up to two-thirds of the potential *G. m. submorsitans* population of Nigeria may have been suppressed by human activity. This is partly because humans scare away or kill potential hosts, and partly because of destruction of the vegetation forming the flies' habitat, associated with agricultural development. The expansion of the road network in Nigeria has often been cited as an example of how humans have had an effect on the distribution and abundance of tsetse (Jordan, 1986). This network resulted in movement of local populations to newly established villages near commercial routes. Accompanying agricultural development may also have produced a drier climate and reduced shade, creating an unsuitable environment for tsetse (Popham, 1972). Generally, *G. m. submorsitans* was thought to occur in areas with human population densities ranging from 0 to $15\,km^{-2}$, occasionally in areas of $15–40\,km^{-2}$ but never when the population exceeds $40\,km^{-2}$ (Nash, 1948). In contrast to Nash's figures, *G. m. submorsitans* was found in 12 out of 19 districts in The Gambia with an average of more than 40 people km^{-2} (Rawlings *et al.*, 1993). Distribution of *G. m. submorsitans* in The Gambia was associated with the presence of warthogs, no flies being found where warthogs were absent. Climate and vegetation in The Gambia are mediated by the river running the length of the country, which may result in conditions under which tsetse can survive at human population densities which, in other localities, would result in their disappearance. Although human population densities obviously vary, most sub-Saharan African countries have densities below $40\,km^{-2}$. Rodhain (1926) had observed that a high level of human activity resulted in the decline of populations of *G. m. centralis* in the Democratic Republic of the Congo (formerly Zaire). In contrast, *palpalis* group flies, particularly *G. tachinoides*, are much less affected by human settlement, possibly because they are more able to adapt from a preference for feeding on wild mammals and reptiles to feeding mainly on humans and their domestic animals (Jordan, 1989a).

Omoogun *et al.* (1991) carried out a study in the derived savanna zone of Nigeria in an area where Fulani pastoralists had previously grazed their cattle for only part of the year. At the time of the study, cattle grazed throughout the year. They were unable to detect *G. m. submorsitans* and *G. longipalpis*, which previously occurred in the area and which are both important vectors of pathogenic trypanosomes to cattle. Increased human activity had destroyed much of the vegetation and depleted the wildlife population, which was believed to have caused the disappearance of the two *morsitans* group species. Two *palpalis* group tsetse, *G. p. palpalis* and *G. tachinoides*, were detected but were apparently less significant vectors as no trypanosome infections were detected in 200 flies dissected. The authors postulated that the same trend was occurring in other parts of the derived savanna and forest zones of West Africa. Biconical traps (Challier and Laveissière, 1973) and bait oxen were used in their survey and it is known that biconical traps are not efficient for sampling populations of *G. m. submorsitans*; thus these flies could have been present at a low density.

Similarly, in 1962 when large numbers of Tutsi refugees from Rwanda were settled in the Orichinga valley of Ankole district, Uganda, the dense tsetse-infested *Acacia* woodland was cleared after a few months, resulting in a reduction of the *G. m. morsitans* population (Ford, 1970a).

In contrast, Wilson (1958) reported advances in the distribution of *G. m. submorsitans,* mainly in the Guinea zone of northern Nigeria, but also in the Sudan zone when whole villages moved out of an area. Similar changes occurred as tsetse reinhabited areas occupied before the 1890s rinderpest epidemic. Following their reinvasion, the nomadic or recently settled people moved out, thus accelerating the advance. The occurrence of river blindness (onchocercosis, the helminth infection transmitted by *Simulium* spp.) and the need for a fallow period were also cited as reasons for the evacuation of the human population. In southwest Ethiopia, historical evidence suggests that an advance of tsetse (*G. m. submorsitans*) was responsible for a decrease in the area of cultivated land and associated settlement (Reid *et al.*, 1997c).

Tsetse flies may become associated with humans, adapting to peri-domestic habitats, with consequences for human sleeping sickness transmission. *Glossina tachinoides* can have a close association with human settlement. Long ago it was reported entering houses in a town along the river Benue in Nigeria, particularly in the wet season when it could even be found in office buildings (Pollard, 1912). It has also been commonly seen in villages following domestic pigs. *Glossina tachinoides* has adapted to man-made larviposition sites such as clumps of oil palm, cola nut and banana trees. Nash (1944a) briefly described a habitat for *G. tachinoides* in northern Nigeria created in streambeds where the palm *Raphia sudanica* grew and was exploited by humans for roofing poles. This provided a good habitat for the fly, giving rise to close human–fly contact. A consequence was a high incidence of sleeping sickness in nearby villages where people were engaged in the pole-cutting business. Licensing of pole cutting for the first 14 days of each quarter was introduced as a means of controlling this type of transmission.

The world's human population is expected to rise to over 8 billion by 2025 (McCalla, 1994). A result of this increasing population could be a diminution of tsetse distribution and abundance to levels at which trypanosomosis will cease to be a problem, and it has been argued that long-term research on ways of controlling the disease, or its vectors, is therefore unnecessary. The effect of a high human population in Nigeria on tsetse and trypanosomosis has already been described. Nigeria is a rather exceptional country, however, and it may be incorrect to assume that the same thing will happen in other African countries in the relatively short period of time to

2025. Using the observations of Nash (1948) and Rawlings *et al.* (1993), Reid *et al.* (1997) predicted changes in distribution of the *morsitans* and *fusca* groups over the next 50 years, concluding that significant areas of Africa will still have low human populations and there will still be large areas of tsetse habitat. Whilst the area infested by *G. morsitans* may diminish, *palpalis* group tsetse will still be important disease vectors.

The demographic effects of the HIV/AIDS epidemic remain unclear. It is likely that in some African countries human population growth rates will decline and, in some, populations may even fall.

7.2.4 Other influences on distribution – disease

Rinderpest

The biggest change in tsetse distribution in historical times occurred following the rinderpest epidemic in Africa in the 1880s and 1890s. One hypothesis is that the 1889 epizootic invaded Africa as a result of cattle imported by Italians into Eritrea, and swept through east and equatorial Africa due to the abundance of wild ungulates (Carmichael, 1938). Rinderpest is not confined to cattle but occurs in any wild ungulate. Buffalo (*Syncerus caffer*), eland (*Taurotragus oryx*), warthog and wild pig were among the chief victims as they are highly susceptible. Buffaloes were considered to be the main wild disseminators of the disease in Uganda whilst waterbuck were rarely infected, no natural case of infection being recorded. The epidemic caused a recession in tsetse distribution due to a reduction in numbers of hosts available. Some tsetse species, particularly *G. morsitans* subspecies, may still be recovering from this recession to reoccupy parts of their former range. The disappearance of tsetse from some areas was particularly marked, and well documented, in the Transvaal, Mozambique and Rhodesia. In northern Nigeria, a rinderpest epidemic occurred in 1886 and caused an estimated mortality of 80–90% of the total cattle population in that region (Putt *et al.*, 1980). Similarly, in Southern Rhodesia (now Zimbabwe) almost all the cattle were wiped out by rinderpest in 1896 (Cranefield, 1991). Following the rinderpest epizootic of 1917–1918 in Uganda, populations of *G. morsitans* either disappeared or were greatly diminished (Duke, 1919a; Carmichael, 1933). Some of the advances in tsetse distribution noted in recent years may be due to recoveries from the effects of that epidemic. One example is the advance in *G. m. submorsitans* distribution in parts of southwest Ethiopia, thought to represent reinvasion of areas occupied before the rinderpest epidemic. Alternatively, any advance into formerly occupied territory might be expected to have taken place much earlier, and recent advances may be due to unknown factors, possibly climate change.

Onchocercosis

Onchocercosis, an infection of the filarial parasite, *Onchocerca volvulus*, causes river blindness in humans and is common in some regions of Africa.

The 'savanna' form of the disease is transmitted by the black fly (*Simulium* spp.), whose larvae live in generally fast-flowing, oxygenated water and can be found in such situations throughout Africa. The disease can occur at a high prevalence in West Africa, causing evacuation or abandonment of many villages in infested areas. Advances of tsetse often followed these evacuations, following bush encroachment of abandoned farms. The World Health Organisation coordinates a control programme for onchocercosis which started in 1974, based on insecticidal spraying to eradicate the vector. The successful, large-scale control of *Simulium* and onchocercosis has significantly reduced this constraint to development. The disease has been eradicated from a number of countries in West Africa, including Burkina Faso, Mali, and Côte d'Ivoire, and control is being extended to countries with 'forest' onchocercosis. The epidemiology of onchocercosis and tsetse-transmitted trypanosomosis and their effects on human migration and settlement are closely linked, as the under-population and under-utilization of fertile land in river valleys results in suitable habitats for tsetse and increased prevalence of trypanosomosis. During the first half of the 20th century, trypanosomosis epidemics also played a role in the depopulation of river valleys in Burkina Faso (Remme and Zongo, 1989). In 1995 my colleagues and I interviewed people returning to areas from which the disease has been eradicated. Both sedentary and transhumant ethnic groups were settling in the same areas, keeping a mixture of trypanotolerant Baoulé and trypanosusceptible Zebu cattle and their crosses. Baoulé cattle were kept mainly by the sedentary ethnic group. As the area was infested by tsetse flies, trypanosomosis was a constraint to development. Tsetse control is now being carried out in some of these areas in order to allow the development of livestock production, whilst the increased settlement and cultivation is also contributing to a reduction in tsetse populations.

7.2.5   Migration and distribution of tsetse in wind fields

Molyneux *et al.* (1979a) suggested that tsetse could be distributed in wind fields; in West Africa this would occur along a southwest/northeast axis. They found that the distribution of human trypanosomosis foci was orientated in roughly parallel lines in a southwest/northeast direction. Wind-assisted long-distance migration of *Simulium* spp., which are smaller, lighter flies, can be important in the epidemiology of onchocercosis, but there is no evidence for migration of tsetse in this way. In 1913 Fiske hypothesized that female tsetse flies 'deliberately' migrated from favourable to less favourable habitats as a form of population control, replenishing populations in unfavourable habitats with excess flies from favourable ones. This would explain the low female : male ratio in population samples, although there is now abundant evidence that this was simply due to a sampling bias. There has been no subsequent evidence of migratory movements of this nature.

7.3 QUANTITATIVE METHODS FOR THE DETERMINATION OF DISTRIBUTION AND ABUNDANCE

Rogers and Randolph (1986b) took a quantitative approach to link tsetse distribution and abundance to climatic factors by producing climograms of mortality and reproductive capacity for some species of tsetse flies. From these, they were able to determine bioclimatic limits for those species. However, they observed that, whilst tsetse distribution should fall within those limits, non-tsetse areas would not always be outside them. This was because the occurrence of tsetse was also determined by other factors such as human impact on natural vegetation (Rogers and Randolph, 1986b). The distribution of any species could be described mathematically using a simple formula relating monthly fertility of a species to its monthly mortality resulting from abiotic, mainly climatic conditions (Rogers, 1979; Rogers and Randolph, 1986b). This formula is:

$$\sum_{n=12}^{n=1} \log f_n \geq \sum_{n=12}^{n=1} k_n \tag{7.1}$$

where f_n = fertility for each month, n, of the year and k_n = the corresponding monthly mortality. Near the boundary of a species habitat, numbers are usually low since, as conditions are suboptimal, the total numbers of births (the left-hand side of the equation) only slightly exceed the total mortality (the right-hand side of the equation). Improvement in climatic conditions for a species can result in an increase or change of its range. Rogers and Randolph (1986b) argued that at the edge of an animal's (tsetse fly's) distribution it is likely to be too scarce for biotic factors such as predators or competitors to have any marked effect.

Within the range of a species, the abundance will simply be a function of the difference between monthly fertility and monthly mortality, which can be described mathematically as:

$$\sum \log f_n - \sum k_n \tag{7.2}$$

The type of function affecting this formula will depend on other, biotic factors affecting mortality. Of the biotic factors that could affect mortality, desiccation was not considered to be very important, starvation being the major cause (Bursell, 1963b). If the density of a tsetse population is much below the carrying capacity of the habitat, the population will increase exponentially at the intrinsic rate of increase, r. The maximum intrinsic rate of increase for *Glossina* is estimated to be 2% per day (Brightwell *et al.*, 1997).

Meteorological and vegetation data obtained from satellites have been used to predict the distribution of tsetse in Côte d'Ivoire and Burkina Faso by discriminant analysis techniques. This allowed predictions with accuracies ranging from 67 to 100%, although abundance could be less accurately predicted (Rogers *et al.*, 1996).

7.4   DESCRIPTIONS OF HABITATS AND DISTRIBUTION

7.4.1   Vegetation types

Before describing the habitats of the three tsetse subgenera, a brief description of vegetation types is given. The distribution of vegetation types in sub-Saharan Africa is shown in Fig. 7.2 and vegetation profiles for the major habitats described are shown in Fig. 7.3.

Lowland rain forest
Lowland rain forest covers approximately 4.2 million km² of Africa, and consists of mostly evergreen and some semi-deciduous species. It occupies the Congo basin and the West African coastal belt except for the Dahomey gap, between Ghana and Benin. The Dahomey gap has a low rainfall/evaporation ratio compared with neighbouring areas in West Africa, and consequently has

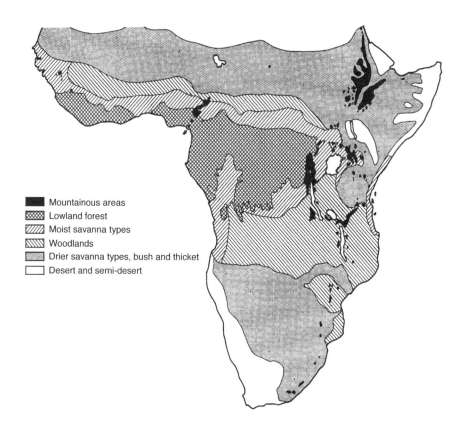

Mountainous areas
Lowland forest
Moist savanna types
Woodlands
Drier savanna types, bush and thicket
Desert and semi-desert

Fig. 7.2. Vegetation zones in Africa south of the Sahara (modified after Keay, 1960).

Advancing | Stable forest | Retreating

Diagram of forest profile, showing changes in ground cover.

Woodland, *Brachystegia, Terminalia* spp.

Thicket, *Commiphora, Combretum, Acacia, Teclea, Maba.*

Savanna, *Combretum, Acacia, Borassus.*

Acacia Savanna, *Acacia* spp.

Fig. 7.3. Sketch diagrams of vegetation profiles in tsetse habitats in sub-Saharan Africa. (a) Forest; (b) woodland; (c) thicket; (d) savanna; (e) *Acacia* savanna.

a different type of vegetation. The climate is characterized by high year-round rainfall and humidity and only minor fluctuations of temperature. This may be reflected in the stability of tsetse populations in areas of undisturbed forest. However, much forest has been felled, resulting in the disappearance of forest-dwelling species of tsetse.

Similar forests are also found along the East African coastal strip, but these are not part of the Congo system. The vegetation type results from the high rainfall and humidity of the coastal climate, but has been much altered by human activity, which has also had a significant impact on that region's tsetse populations.

Gallery forest

Gallery forests are bands of varying width of frequently dense vegetation and tall trees along the banks of rivers and streams. They allow tsetse (particularly of the *palpalis* group) to penetrate into dry savanna areas and are important habitats for the hosts of tsetse. Similarly, they provide a habitat from which forest species such as duiker (*Cephalophini*) and other antelopes can penetrate savanna areas. The bushbuck (*Tragelaphus scriptus*), which is one of the most important sources of bloodmeals for several tsetse species, is one of the commonest antelopes in this type of habitat. These antelopes emerge from the habitat in the evening on to the fringes of surrounding plains and thus live in close association with tsetse.

Savanna

Different types of savanna woodland arise according to the types of soil from which they are formed and the amounts of rainfall that they receive. Generally, the dry West African savannas are formed on granitic soils and are less fertile than those in East Africa, which are formed on fertile volcanic basaltic soils. This also partly explains the abundance of ungulate fauna in East Africa compared with West Africa.

MOIST SAVANNA WOODLAND. The moist savanna woodland (Guinea savanna) of Africa occupies about 10 million km^2 and together with the lowland rain forest accounts for most of the tsetse-infested area. Moist savanna woodland is the largest single vegetation type and consists of relatively open woodland with grassland. In the south, it is characterized by *Brachystegia* and *Julbernardia* species commonly known as miombo woodland, or by mopani woodland in which *Colophospermum mopane* is dominant. This may be associated with *Julbernardia globiflora*, *Combretum*, *Diospyros* and various species of *Terminalia*. In West Africa the moist savanna is known as doka woodland and is dominated by different species such as *Isoberlinia* and *Monotes*. *Mitragyna* also occurs in wetter areas around and within the drier savanna, depending on a higher water table – for example, in riverine flood basins. In addition to providing suitable climatic conditions for tsetse, these vegetation types provide habitats for the wild animals upon which the tsetse flies feed. Dominant tree species in the wetter Guinea savanna of West Africa include *Isoberlinia*, *Afzelia* and *Daniella* spp.

DERIVED SAVANNA. Derived savanna, described by Keay (1959), is considered to be intermediate between lowland rain forest and the southern Guinea savanna, and is formed largely through the effect of humans decreasing the tree cover (Sanford and Isichei, 1986). It often occurs in mosaic patches with some forest trees and oil-palms, with the grass *Andropogon tectorum* being common except in the north, towards the Guinea savanna.

Cutting forest edges creates ecotones where concentration of some vectors can occur, producing an edge effect (Coosemans and Mouchet, 1990). The wooded part (resting sites, reproduction sites) provides favourable microclimatic conditions for the vectors, while the cleared parts (feeding sites) allow an abundance of human hosts and domestic animals. *Glossina palpalis*, which inhabits this vegetation zone in West Africa, is rare in African tropical rain forests except where humans have partially destroyed the climax vegetation.

7.4.2   Distribution and habitats of the *morsitans* group

The distribution and abundance of *morsitans* group tsetse often corresponds with the distribution of wild animals (Simpson, 1918). An example mentioned previously is the distribution of *G. pallidipes* in the Lambwe valley, Kenya, which was positively correlated with the distribution of the preferred host, bushbuck. The important effects of seasonal variations in climate on distribution and abundance were described in section 7.2.1.

In the northern Guinea savanna zone of Nigeria the largest numbers of *G. m. submorsitans* could be caught in the early dry season (November–January) (Jordan, 1965a). As the climate became drier, the number of flies caught declined to a minimum at the end of the dry season or beginning of the rains (March–May). Most females were caught in the dry season (February) and fewest in the wet season. They rarely fed on the most frequently observed possible hosts (civet cats or duiker), but mainly on warthogs and humans, which were the next most commonly observed. In wet seasons of above average rainfall, males were evenly distributed in the habitat but were otherwise somewhat concentrated in areas of thicker vegetation. In the savanna, puparia were found in dry soil of forest islands and riverine vegetation in the dry season; wet season breeding sites were not discovered.

In Nigeria, in the wet season, *G. m. submorsitans* was widely dispersed in different woodland types. In the dry season, doka woodland was preferred in the upland, whilst densities decreased in the open 'orchard' woodlands and flies were mainly confined to islands of *Uapaca togoensis* vegetation (Mahood, 1962). Orchard, or parkland, in West Africa contains large numbers of the same economically useful tree species, such as *Butyrospermum* (butternut) and *Parkia*.

Similarly, at the Madina-Diassa ranch in Mali, *G. m. submorsitans* was evenly distributed throughout the savanna but concentrated in the dry season in riparian zones and thickets. The apparent density increased in the rainy season (Diallo, 1981). In Ghana, *G. m. submorsitans* was seldom found near rivers, being more abundant in savanna forest and open savanna country (Simpson, 1918).

Glossina longipalpis is less widely distributed than *G. m. submorsitans* and, in Nigeria, is found mostly in riverine fringing forest vegetation, similar to the habitat of *G. p. palpalis* (Davies, 1977). During the dry season in the northern part of its range, this species also concentrates inside fringing forests or forest islands.

Glossina pallidipes, regarded as a highly mobile species, is relatively large and its puparia have fairly impermeable membranes. This allows it to inhabit a wide climatic range, although in dry areas its habitat may be more typical of riverine species. In West Africa, the closely related *G. longipalpis* is the equivalent of *G. pallidipes* in terms of its ecology and its importance as a vector of pathogenic trypanosomes to livestock. It inhabits forest islands with plenty of shade, and moves to forest edges only during seasons of high rainfall (Page, 1959a). Thus, during seasons of high humidity *G. longipalpis* can temporarily invade upland savanna. In the southern Guinea vegetation zone of Nigeria, it becomes more concentrated around riverine vegetation and forest islands in the dry season (Baldry and Riordan, 1967). It is not found in Nigeria in areas with an annual rainfall of less than 1143 mm (Nash, 1948).

Glossina austeni is confined to the coastal zone of East Africa, from Somalia to South Africa, including the island of Zanzibar until the fly was eradicated from the island in 1997. In some places it has been found quite far inland; the first specimen described was apparently caught in fairly dense thicket vegetation in Jubaland, Sudan (Newstead, 1912) and it was recently found in eastern Zimbabwe, close to the border with Mozambique. It became more abundant along the coastal strip after Arab settlers left the area following abolition of the slave trade. Farms formerly belonging to Arabs became overgrown with bush, which formed a suitable habitat for *G. austeni*. Sacred forests, known as *Khayas*, of the coastal Mijikenda people also provided suitable habitats, and proved to be obstacles to control of the fly in the 1950s. Similar problems with tsetse control occur in parts of West Africa where sacred groves form the habitat of *palpalis* group flies and traditional beliefs make it difficult to apply current control methods.

Glossina austeni is a small fly, not easily caught in available traps, although a sticky trap designed for use on Zanzibar has been quite effective. The fly often occurs at low densities in a discontinuous distribution, yet it is an economically significant vector of pathogenic trypanosomes to domestic livestock as its habitat is associated with higher rainfall in coastal areas with intensive agricultural and livestock production. This species ranges close to its primary habitat in search of food, unless very hungry, and consequently domestic animals living close to an infestation may have a low prevalence of infection. In a study at Kilifi, Kenya, a high prevalence of infection only occurred in a herd of sentinel cattle when they were kraaled overnight within the dense vegetation forming the habitat of *G. austeni* (Paling *et al.*, 1987).

7.4.3   Distribution and habitats of the *palpalis* group

The distribution of *G. palpalis* in West Africa extends from the wet mangrove and rain forests of the coastal region northward into the drier savanna areas. They occur only in drainage systems leading to the Atlantic Ocean or the Mediterranean, but not those draining into the Indian Ocean. *Glossina palpalis* cannot tolerate a wide range of climatic conditions such as occur in the savanna belt, and is virtually restricted to the ecoclimate of the watercourses or islands of forest. Its northern limit of distribution therefore lies in the more arid regions bordering, and within, the Sahelian countries.

Human population and the distribution of wild animals are important factors affecting distribution of *palpalis* group tsetse. In Côte d'Ivoire and neighbouring countries of West Africa, *G. p. palpalis* is highly mobile but maintains a close association with villages (Baldry, 1980). Schwetz (1915) considered the three most important factors to be climate, vegetation and the distribution of water. By studying these three factors, Schwetz believed that one could predict the presence or absence of *G. palpalis*. Although this method of determining habitat suitability and predicting the presence of *G. palpalis* worked in some areas of the Congo, it did not work in others. Schwetz thought that this might have been due to adaptation of the fly to local conditions.

The habitat of *G. palpalis* is well-defined evergreen riverine forest, thick enough to form a closed canopy, with shelter at the sides provided by creepers that shield the interior of the gallery from hot, desiccating winds.

Recently, *G. tachinoides* appeared to have extended its range southwards and eastwards in Côte d'Ivoire and Togo (Kuzoe *et al.*, 1985; Mawuena and Itard, 1981). The new southern limit in Côte d'Ivoire was half a degree of latitude further south than had previously been reported; in Togo, *G. tachinoides* appeared to have advanced southwards to 2° further south than the previously known limit. This may simply reflect the inadequacy of surveys carried out in Togo prior to the study of Mawuena and Itard (1981) rather than a real southward advance. In Togo, the southern limit of distribution of *G. tachinoides* was thought to follow the northern edge of the dense humid forest zone, but the present southern limit follows the boundary of more intensive agricultural development (Hendrickx and Napala, 1997). Baldry (1966a) predicted that this southward extension would take place in Nigeria, in response to the opening up of the rain forest belt and adaptation of the fly to a peridomestic habitat. According to Baldry (1969a), *G. tachinoides* advanced into southern Nigeria from the north in the last few hundred years, partly due to human activities in the forest zone. The greatest concentrations of tsetse could be found in resulting peridomestic habitats (Kuzoe *et al.*, 1985).

In northern Nigeria, *G. tachinoides* is typically a riverine species inhabiting gallery forest vegetation with similar characteristics to the habitat of *G. palpalis*. Its range does, however, extend during the wet season into thicketed areas within suitable savanna woodland. Populations normally contract during the dry season when both species will concentrate around permanent water sources. In Nigeria, some *G. tachinoides* populations became so adapted to man-made habitats that Baldry (1969b) thought sub-speciation must have occurred.

Glossina fuscipes has poor 'waterproofing' and inadequate water reserves and is therefore confined to hygrophytic habitats. It is rarely found far from open water in lacustrine or riverine habitats. Jackson (1945) carried out experiments at Shinyanga, in Tanzania, in which he introduced *G. morsitans* and *G. f. fuscipes* into an area inhabited naturally by *G. swynnertoni*. The mean weight of *G. morsitans* was about the same as that of *G. swynnertoni*, allowing for difference in size between the two species. However, the weight of *G. morsitans* was lower than for that species in its natural environment. *Glossina morsitans* could produce a second generation in the environment of *G. swynnertoni* whereas *G. f. fuscipes* could not.

Clearly, the ecology of a tsetse species can alter according to differences in climatic factors, particularly at the limit of a species distribution.

The adaptation of certain tsetse species to human environments and advances in distribution are important as significant changes in the epidemiology of human and animal trypanosomosis are likely to occur.

7.4.4   Distribution and habitats of the *fusca* group

Jordan (1963) described the distribution of some *fusca* group tsetse in southern Nigeria and west Cameroon. Their habitat was always dependent upon forest vegetation, ranging from relatively dry forest islands and riverine forest in savanna to dense, humid, wet rain forest. The habitat of *G. tabaniformis* and *G. nigrofusca* in Nigeria is lowland rain forest; *G. medicorum* is found in forest outliers of derived savanna; and *Glossina fusca* occurs in both habitats (Jordan, 1962b). The rain forest habitat has a cooler and more humid climate than the forest outliers; the savanna is a more arid environment. The distribution of each of the species studied by Jordan (1963) was thought to be largely determined by climatic factors, of which rainfall and relative humidity are particularly important. *Glossina medicorum* was recorded only from the relatively dry northern part of the rain forest and forest islands or riverine forest in savanna in western Nigeria. *Glossina nashi* was found in dense, wet rain forest in west Cameroon. *Glossina tabaniformis* was found in forest reserves and wet rain forest. *Glossina haningtoni* was common in wet rain forest in west Cameroon and in western Nigeria. *Glossina fusca* inhabited a wider range of climatic conditions and habitat types than the other species, ranging from forest islands in savanna to wet rain forest. *Glossina nigrofusca* was rare but widespread in the two countries, occurring typically in wet rain forest. *Glossina tabaniformis* was most abundant during the middle of the rains and scarcest during the dry season. Large numbers were caught in areas of the rain forest where host animals were least disturbed by human activity.

Studies of the activity of *G. tabaniformis* showed that there were early morning and late afternoon peaks of activity throughout the wet and dry seasons; however, Jordan did not consider it to be a crepuscular species. Page (1959b) studied *G. tabaniformis* in the same area as Jordan and suggested that it was unlikely to be of economic importance as it had a low rate of trypanosome infection and was confined to the forest. Humans are likely to have an adverse effect on *fusca* group tsetse populations through forest clearing and hunting and Jordan (1962b) observed an inverse relationship between human population density and records of *fusca* group flies. However, high trypanosome infection rates were detected in *G. tabaniformis* captured on ranches in Gabon and the Democratic Republic of the Congo (formerly Zaire), where this species was an important vector of pathogenic trypanosome species to cattle (Leak *et al.*, 1990).

Two other *fusca* group species, *G. fusca* and *G. medicorum*, were likely to be of economic importance as vectors of trypanosomes to livestock, as their infection rates were high, and they dispersed into the savanna grassland (Nash, 1952; Page, 1959b). In contrast, *G. nigrofusca*, although having a high infection rate, was confined to the forest in Nigeria, whilst in Ghana it could be of importance as it occurred in the savanna. Generally, most species of the *fusca* group spend much of their time at rest on tree trunks (Nash and Davey, 1950; Nash, 1952).

In more recent studies, in the northern Guinea savanna zone of Burkina Faso, *G. medicorum* was readily caught in white biconical traps (Baldry and Molyneux, 1980). Its most northerly known location was between 9° 42' and 10° 1' N in the lower Komoe valley, where it inhabited places in which typical fringing forests were backed by extensive areas of *Guibourtia copallifera* forests with abundant wildlife. Activity was mainly during the periods of dawn and dusk, and the overall trypanosome infection rate of 29.4% consisted predominantly (64.3%) of *Trypanosoma vivax*-type infections.

Glossina brevipalpis and *G. longipennis* are anomalous members of the *fusca* group in terms of their distribution and ecology. Whereas most *fusca* group tsetse are found in West African forests, *G. brevipalpis* is found in East and southern Africa, where it requires dense thickets providing heavy shade. Consequently, it has a patchy, localized distribution (Swynnerton, 1936; Moloo *et al.*, 1980). Close to the Kagera valley, in southwest Ankole, a range of hills forms a barrier restricting *G. brevipalpis* to East Africa. In this valley, *G. brevipalpis* feeds on hippopotamuses (*Hippopotamus amphibius*) and occurs on both sides of the river.

In a very early study of the habits of *G. brevipalpis* in Nyasaland (now Malawi) this species occurred in limited areas, in contrast with the wide tracts of country infested by *G. m. morsitans* (Davey, 1910). It occurred in places with abundant shade provided by considerable growth of creepers and was most numerous under the largest trees. All localities where it was found were relatively low-lying, between 90 and 400 m above sea-level, and there was always water nearby. *Glossina brevipalpis* was hard to detect except in the evening and very early morning.

The Kagera river also forms the limit of distribution for *G. longipennis*, whose preferred host used to be the rhinoceros (*Diceros bicornis*). The rhinoceros has not been able to cross the Kagera river and is absent from Ankole; this may bear some relation to the distribution of *G. longipennis* (Swynnerton, 1936).

Although most *fusca* group species inhabit moist, evergreen forest habitats, *G. longipennis* lives in one of the driest types of vegetation inhabited by any tsetse fly. Its habitat consists of dry deciduous *Commiphora/Acacia* bush only 3–5 m high. A partial explanation for its ability to exist in such dry habitats is that its large puparium can retain abundant water reserves, enabling it to live in arid environments. *Glossina longipennis,* which is better adapted than any other tsetse to life in an arid environment (Glasgow, 1963), occurs mainly in East Africa, where it has a discontinuous distribution.

Glossina frezili is found in a geographically confined area of mangrove forest of Congo and Gabon. It is morphologically very similar to, and probably genetically close to, *G. medicorum* (Gouteux, 1987a).

7.5   LARVIPOSITION AND LARVIPOSITION SITES

7.5.1   General requirements and features of larviposition sites

Larviposition behaviour has largely been studied in the laboratory, due to the difficulties of studying wild flies. The female fly becomes very active a few hours before parturition, presumably whilst searching for a suitable larviposition site. Nash and Trewern (1972) observed a diurnal rhythm of larviposition in laboratory studies of *G. m. morsitans* and *G. austeni*. The rhythm was different for the two species. For *G. m. morsitans* there was a marked peak of larviposition between 10 and 11 hours, whilst with *G. austeni* there was a peak between 10 and 12 hours and a second peak at 14–15 hours. In a laboratory study of factors affecting breeding site selection by *G. palpalis*, black objects were strongly attractive and shaded soil was much more attractive than unshaded soil (Parker, 1956b). Addition of sand increased attraction (even though it made the site lighter in colour), as did a rough rather than a smooth surface. Thus, both tactile and visual responses seemed to play a role in breeding site location.

Lewis (1934) found that larvae of *G. tachinoides* and *G. m. submorsitans* were slow in crawling and burrowing and often could not burrow at all. They could penetrate coarse sand more readily than fine. He cited previous work of Fiske (1920) and Nash (1933b) indicating that the depth to which larvae burrow influences the effect of predators and parasites as well as the effect of heat from sunlight on the puparia. Finlayson (1967) described the behaviour of *G. morsitans* larvae from the moment of deposition to completion of puparium formation. Finlayson recognized six characteristic stages of behaviour between the time of deposition and pupariation: crawling (during which there is a strong photonegative response); circling and reversing; head retraction; 'barrelling', which Finlayson described as 'the assumption of the puparial shape'; hardening; and darkening. A review of larval behaviour in relation to metamorphosis was published by Zdárek and Denlinger (1993).

One of the first descriptions of tsetse puparia found in natural circumstances referred to an area of human sleeping sickness in Uganda (Bagshawe, 1908). Subsequently, Swynnerton (1923a) described the puparia of *G. swynnertoni,* which were found under thickets, under logs and under the shelter of overhanging rocks in Tanzania.

7.5.2 Larviposition sites of the *morsitans* group

In some situations and for some tsetse species there may be a change in larviposition sites between wet and dry seasons. This was suggested by Swynnerton (1923a) who, in early studies of tsetse larviposition sites, found puparia of *G. swynnertoni* under thickets, logs and the shelter of overhanging rocks. In the late wet season its puparia were detected in the largest numbers under logs, at tree bases, in thickets, under rocks and in rot holes of trees (Swynnerton, 1936). Swynnerton also claimed to have found numerous *G. pallidipes* puparia in thickets and under dry rocks. Other workers – for example, Potts (1950) – disagreed with the concept of wet season sites and thought that puparia were simply scattered so widely during the rainy season that searches at that time did not give good results. In subsequent studies no significant change to wet season larviposition sites was discovered (Burtt, 1952).

Harley (1954) studied the breeding sites of *G. m. morsitans* in Tanzania and investigated the possibility of a shift in breeding sites during different seasons of the year following previous observations by Nash (1939). Logs were by far the most important type of site under which puparia could be found. During the hot season larger sized logs were used as larviposition sites as well as rot holes, thickets and leaning trees; all those sites provided the most shade and the coolest temperatures. Harley (1954) concluded that changes in larviposition sites were associated with temperature; pregnant females were searching for cool places in hot weather and for warmer places in cold weather. There were significant differences in larviposition sites at various locations in Tanzania. These differences were especially marked between larviposition sites in woodland and in valleys, and were related to the amount of grass cover in the valleys. At one site woodlands were deserted in the hot season, when the trees were leafless, at which time puparia were found only in the more shaded thicketed valleys. Glasgow (1961b) also noted marked and abrupt changes in larviposition sites of *G. m. morsitans* in Tanzania, confirming Harley's observations. In August, during the early dry season, there was a very sudden change from the sites under logs to rot holes in trees. This change was reversed in November and December when the rains started.

Puparial shells of *G. m. morsitans* in the Ruizi valley, Uganda, could be found almost anywhere (Glover, reported by Lewis, 1947) but three vegetation zones favoured for larviposition were: (i) *Acacia* open woodland containing thicket elements such as *Rhus* sp.; (ii) *Olea* clump thicket with logs, stumps and rocks; and (iii) termite mounds with *Rhus*, *Grewia* and other thicket vegetation. Glover reported a 'definite correlation between the more important pupal sites and *Olea* clump-thicket relics'. By exploiting this knowledge, Moggridge (1949a) managed to significantly reduce the tsetse population in the Kamataiisi valley simply by removal of *Rhus* from termite mounds.

7.5.3   Larviposition sites of the *palpalis* group

The larviposition sites of *G. tachinoides* and *G. p. palpalis* differ. Johnson and Lloyd (1922) found that *G. tachinoides* had a preference for sandy sites, and out of over 8000 puparia, 7540 were found in coarse sand and 525 in fine sand. A total of 7970 puparia were collected in sites where the only shade was from high trees with no low growth. In contrast, *G. p. palpalis* were usually found in sites with low shade and an accumulation of vegetable debris on the ground surface. *Glossina tachinoides* at Nsukka in southern Nigeria can breed in peridomestic situations, such as soil beneath stacked coco-yam tubers, the base of fences of pig enclosures and in adjacent farmland (Baldry, 1970).

Palpalis group tsetse can be important vectors of human sleeping sickness, resulting partly from a close human/fly contact, which may be reflected in their peridomestic larviposition sites (Chapter 12, section 12.8).

In an endemic sleeping sickness area of the Liberian rain forest, puparia of *G. p. pallicera* were found in the leaf axils of oil-palms (*Elaeis guineensis*). This indicated that emerging tsetse might initially feed on humans and become vectors of human sleeping sickness (Kaminsky, 1984). The shade of oil-palms was also identified as a major breeding site for *G. p. gambiensis* in neighbouring Sierra Leone (Gordon and Davey, 1930).

After 518 h of searching for puparia of *G. palpalis* s.l. in the pre-forest zone (transitional forest savanna zone) of Côte d'Ivoire, 1909 puparia were found from 70 breeding sites (Sékétéli and Kuzoe, 1984). Puparia were found in both the dry and the rainy seasons. In locations with high pig density, a greater number of puparia were found either at the edge of the village or not further than 500 m from the houses, supporting observations that *G. palpalis* s.l. was peridomestic in those localities. Where the density of pigs was low, or where they were absent, peridomestic breeding was not observed. In such circumstances all puparia were found 500 m to 3 km away from the houses, in coffee or cocoa plantations and in woodlands.

Larviposition sites had the following essential characteristics:

1. The vegetation cover maintained adequate conditions of temperature, humidity and light intensity necessary for the pregnant female.
2. They provided resting sites for the female (trunks of fallen trees, roots of oil-palms or false date palms and low overhanging branches). Such sites also create a suitable microclimate at soil level for the larva and the puparium.
3. Soil with physicochemical qualities (texture, pH, water retention capacity, organic and mineral content composition) allowing the larva to burrow easily through the soil and allowing suitable conservation of the puparium until eclosion.

Peridomestic breeding sites of *G. f. fuscipes* in a sleeping sickness focus in Busoga, Uganda, were in coffee and banana plantations, and in *Lantana camara* thickets and around houses (Okoth, 1986). Although some reports imply recent adaptation of *palpalis* group tsetse to peridomestic habitats, it is likely that they may always have been closely associated with human habitation. In Liberia, *G. palpalis* s.l. occupied sites along rivers at bridges and fords where people drew water, or washed their clothes and pots, but not elsewhere, despite the presence of other potential wild hosts in the area (Morris, 1962a).

7.5.4   Larviposition sites of the *fusca* group

Fewer studies have been carried out on *fusca* group tsetse, mainly because of their lesser economic importance and possibly because larviposition is not concentrated in specialized sites in the relatively uniform environment of the rain forest. Preferred larviposition sites of *G. longipennis* in Kenya were logs, leaning tree trunks and the stumps of felled trees near shrubby or woody thickets (Lewis, 1942). They could also be found in open country many hundreds of metres from dense bush and in the shade of single trees in the savanna.

Chapter 8

Behavioural Ecology

Stephen (1986), in his book *Trypanosomosis, a Veterinary Perspective*, frequently criticized laboratory studies on trypanosomosis for using passaged and frozen stabilates of trypanosomes and rats or mice as hosts. Instead, he advocated the use of natural hosts in the field, and 'real' trypanosomes transmitted by tsetse flies. There has been similar criticism of laboratory studies of tsetse behaviour. For example, Brady (1979) quoted Bursell as having said: 'To embark upon research projects with laboratory reared tsetse is to study abnormal physiology and abnormal behaviour.' Brady disagreed with this view, stating that 'experience ... totally refutes this pessimism'. Brady (1979) supported his argument by summarizing a number of behavioural studies of tsetse, with ample references to laboratory studies. Despite Brady's view in favour of laboratory studies, it is necessary to be aware of the many potential differences in both behaviour and physiology of laboratory colonies of tsetse flies. Nonetheless, some behavioural aspects cannot practically be studied in the field, leaving no alternative to laboratory studies.

This chapter describes advances in the understanding of tsetse behaviour which have had a significant impact on the development and success of current control technologies.

8.1 MATING

The mating behaviour of tsetse was reviewed by Wall and Langley (1993) and the reader is referred to this for more detailed information. Tsetse flies can live at very low densities, maybe as low as 40 km^{-2} (Glasgow, 1963). It might be expected that mechanisms for the two sexes to meet would be required in such situations. Such mechanisms could involve visual and olfactory responses, and because of the possibilities of utilizing these mechanisms for the control of tsetse, the mating behaviour of tsetse flies has received much attention. A female tsetse fly needs to mate only once in her

104

lifetime but multiple mating occurs (Dame and Ford, 1968; Jaenson, 1980), though its frequency in nature is unknown. Potentially, this could have important implications for the sterile male tsetse control technique (Chapter 18, section 18.4); however, Dame and Ford (1968) doubted that it occurred sufficiently frequently to affect this method of population control.

Most female flies are successfully inseminated even at very low population densities (Teesdale, 1940). In the Lambwe valley, Kenya, where the tsetse population was reduced substantially by insecticidal spraying, the insemination rate did not fall below 72% and rose to 100% shortly after spraying (Turner and Brightwell, 1986). In Zimbabwe, mark–release–recapture experiments using flies released 12 hours after emergence, either at Rekomitje or on an island in Lake Kariba, showed that more than 90% of female *G. m. morsitans* were inseminated by age 4 days and of *G. pallidipes* by age 7 days (Hargrove, 1994).

Where habitats of different species overlap, cross-mating between species may occur. Jackson (1950a) observed cross-mating of *G. m. morsitans* with *G. swynnertoni* in Tanzania. Male hybrids are infertile.

The minimal mean duration of mating of virgin female *G. p. palpalis* ranges from 109 min, for 1-day-old flies, to 54 min for 10-day-old flies (Jordan, 1958). The average number of matings amongst 99 of these flies was 2.2 per female (Jordan, 1958). In an older group of flies the average was 1.4 matings. These experiments confirmed that mating of female *G. p. palpalis* is mostly confined to early life. After an age of 3 days the likelihood of a female fly being mated declines. Young flies may be prepared to re-mate but it is unlikely that females over 10 days old often do so. In laboratory studies, female flies do not seem to seek males, which appear to rely on visual attraction to a moving object to find mates (Dean *et al.*, 1968). The roles of visual and/or olfactory stimuli in eliciting mating responses have also been examined in the laboratory using decoys (Wall, 1989b). Young unfed males were significantly less likely to attempt copulation with a decoy female than were fed males, and copulatory attempts by *G. pallidipes* were significantly shorter than those of *G. m. morsitans*. For both species, the duration declined with increasing hunger (Wall, 1988). The male antennae sensed no volatile olfactory component of the female sex recognition pheromone, and attraction was apparently purely visual, though colour did not appear to be important in recognition.

8.2 THE 'FOLLOWING SWARM'

Over 80 years ago, Bruce (1915a,b) referred to swarms of *morsitans* group tsetse following moving objects such as cyclists, and speculated that the swarms, made up mainly of male flies, were formed in order to mate with females. The females in these swarms approached the moving object in order to feed, not to meet males. A large percentage of female flies in the swarm was considered to reflect a hungry fly population (Ford *et al.*, 1972). This hypothesis is supported by observations by Fiske (1920) who found that when wild host animals were removed from an island in Lake Victoria there was an increase in the proportion of female flies caught. Swarms consist principally of a few newly emerged flies, which land on the host and feed if given the chance, and a larger number of mature males that will not feed. Swynnerton (1936) therefore concluded that the non-feeding males were only there to intercept females coming to feed and mate with them; Bursell (1961a) described this as sexually appetitive behaviour.

Brady (1972a) described the visual responsiveness of tsetse to various factors contributing to 'following swarm' behaviour. He considered the 'following swarm' to be a unique behavioural system ensuring sexual encounters, in which a loose swarm of flies develops behind any large moving object. Brady's laboratory experiments showed that take-off, but not orientation towards a moving object, went through a daily cycle with morning and evening maxima and that this response increased linearly during a 5-day starvation period. He concluded that the four physiologically distinguishable phases of behaviour described in Chapter 6 (section 6.4.4) were unsatisfactory explanations of the responses of tsetse to hunger. Studies, using video cameras, of the behaviour of male tsetse flies when chasing and identifying potential mates showed that 'speed matching' might be a means of target identification for male flies (Brady, 1991).

8.3   VISION

The eyes of tsetse are basically similar to those of the other higher Diptera that have been investigated. In both sexes the eyes have a specialized zone of greater visual acuity in the forward-pointing region and are not significantly different, although the eyes of females have a narrower region of binocular overlap. This greater binocular overlap of males may serve as an 'early warning system' for the approach of potential females (Gibson and Young, 1991). The values of visual acuity are consistent with fast flight, visual detection of drift due to low wind speeds, mating chases and discrimination of cryptic hosts at high light intensities (Gibson and Young, 1991).

It has been suggested that zebras are protected from being fed on by biting flies, including tsetse, by their striped pattern (Waage, 1981). Gibson (1992) found that tsetse are less attracted to stripes than solid colours and appear to avoid horizontally striped objects. Zebras usually have a combination of vertical and horizontal stripes, but the horizontal stripes are on the lower part of the animal, which is where tsetse would normally feed. Studies attempting to investigate this are discussed further in Chapter 17.

8.4 FEEDING BEHAVIOUR

8.4.1 Host location

Tsetse recognize potential hosts by their visual and olfactory characteristics. These, and mechanical stimulation, will activate them and initiate host-oriented responses. Thus, the process by which a tsetse fly obtains a bloodmeal involves a series of behavioural responses. Commonly, the tsetse will first detect an odour plume and fly up it until it comes in sight of the host (source). Heat stimulation after landing on the host may then cause a probing response and subsequent feeding. This behaviour will obviously be affected by how hungry the fly is, which can be assessed from its nutritional state, in terms of fat and haematin content.

The strategy adopted by tsetse to reach a host after it detects an odour has been the subject of much study, and 'trap seeking' has been modelled by Williams (1994), who described three different strategies: 'point-and-shoot', 'careful-navigation' and 'intensive-search'. He conjectured that the choice of strategy would depend upon the amount of flight time (energy) available.

Approach to a stationary host appears to be by upwind flight, modulated by olfactory stimuli, flight speed being significantly reduced when a fly enters an odour plume (Warnes, 1990a). The final orientation towards the host is visual (Fig. 8.1). In laboratory and field studies of host location in Ghana, *G. medicorum* were not attracted to traps then available, nor to stationary objects, but were attracted to a slowly moving vehicle and responded to the smell of cows not visible to them (Chapman, 1961). They showed upwind orientation to odour in a wind tunnel. *Glossina medicorum* responds to moving objects up to 8–9 m away, whereas *G. morsitans* showed good response at distances up to 50 m. Attraction from a distance is in some cases mediated through chemoreceptors. Chapman concluded that as *G. medicorum* lives in a habitat with poor visibility, it probably locates its host initially through a response to host odour, followed by upwind orientation and finally, perhaps, visual approach.

Colvin and Gibson (1992), in a review of tsetse host-seeking behaviour, summarized previous studies (Brady, 1972a–c, 1975b) showing that both endogenous and exogenous factors influenced the host-seeking behaviour of tsetse.

Endogenous factors include a circadian rhythm of activity (Brady, 1972d), the level of starvation, age, sex and pregnancy status of the flies. Exogenous factors include temperature, vapour pressure deficit, and visual, mechanical and olfactory stimuli (Bursell, 1957b; Vale, 1974c; Huyton and Brady, 1975). Visual responsiveness of tsetse is related to their overall nutritional state and increases with the depletion of food reserves (Brady, 1975a). These changes in responsiveness are mediated by a circadian rhythm of activity and are summarized in Fig. 8.2; they are further described by Brady and Crump (1978).

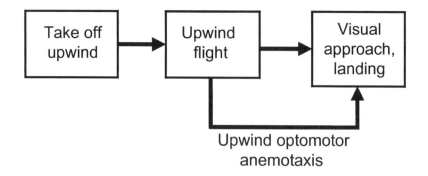

Fig. 8.1. Flight responses to odours: host location.

It might be expected that increasing wind speeds would straighten out odour plumes, making it easier for tsetse to navigate and increase the rate of arrival at the odour source. However, the relationship between wind speed and host location is complex, as odour plumes break up at high wind speeds (Brady *et al.*, 1995).

In a recent review (Willemse and Takken, 1994), the following four stages of host-locating behaviour of tsetse were described:

- **Ranging** – flying in search of a host in the absence of an external cue.
- **Activation** – change in behaviour caused by perception of external stimulus.
- **Orientation** – upwind anemotaxis in response to complex chemical and visual stimuli directing the insect to the host.
- **Landing**.

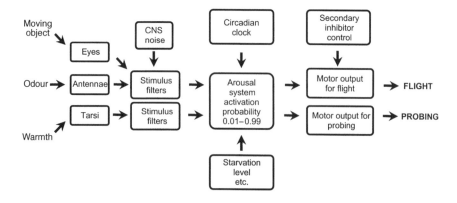

Fig. 8.2. Model for the central origin of the parallel, in-phase behavioural rhythmicity observed in tsetse flies (adapted from Brady, 1975b).

8.4.2 Host effects/preferences

Non-random feeding patterns, even where host and tsetse occupy the same habitat, may be explained by host responses to tsetse attack, such as tail flicking or skin rippling reactions. There is little evidence of long-range discrimination between hosts (Snow, 1980). As mentioned in section 8.3, the stripes of zebra are believed to protect them from feeding by tsetse. With humans, in addition to their upright habit, their unattractiveness to tsetse is an olfactory repellency due to lactic acid secretions in the skin. Hargrove (1976b) described experiments that showed that many more *G. pallidipes* landed and fed on an ox in the absence of humans than in their presence. The presence of humans not only repelled many tsetse completely but also inhibited the landing response of many flies which did approach the ox. This repellency was recently reconfirmed in Zimbabwe (Hargrove, 1991).

8.4.3 Feeding-related activity

Morsitans group tsetse are commonly most active early in the morning and/or late in the afternoon. Shown graphically, this appears as a roughly U-shaped pattern of activity during the hours of daylight. The U-shaped pattern of tsetse activity in nature was originally assumed to be solely due to environmental control. Further experiments and reanalysis of field data have now shown that control is by an endogenous circadian clock, mediated in the field by temperature (Brady and Crump, 1978; also reviewed by Brady, 1988). Seasonally cold temperatures eliminate or reduce the early morning peak in some areas. In a study of *G. pallidipes* activity in Zimbabwe in relation to estimation of trypanosomosis risk, Pilson and Leggate (1962a) found that the diurnal pattern of feeding activity was similar in all seasons, with a well-marked peak of activity at or shortly before sunset.

The studies by Brady and Crump (1978), which demonstrated the circadian rhythm of activity, showed the following:

- The U pattern occurs in nature independently of temperature.
- Biting activity is correlated positively with temperature in the morning and evening but negatively through the middle of the day.
- Light indirectly influences the daily rhythm by its effects at low intensity.
- Humidity has little influence on activity level.
- Roughly 80% of the amplitude of the U pattern in the field is due to endogenous circadian control and only 20% to direct control by temperature.

In West Africa, *G. tachinoides* are most active at temperatures between 27 and 29°C, and are inactive at temperatures below 16 or above 40°C (Nash, 1937).

8.4.4   Biting response

Heat is the prime stimulus leading to probing and subsequent feeding by tsetse. Contact of the tarsi with a substrate alone may also elicit a low level of probing response, provided the substrate is neither too smooth nor repellent. The antennae are the principal sites of thermoreceptors mediating probing, but some have also been reported from the tarsi of *G. morsitans* (Reinouts van Haga and Mitchell, 1975). Neither odour nor chemicals in contact with the tarsi are effective stimuli. Responsiveness increases as the fly becomes more hungry, but the fly becomes refractory as the final stages of starvation approach (Dethier, 1954).

8.4.5   Feeding process

Glasgow (1961a) observed the feeding process of *G. swynnertoni* in Tanzania, by catching flies in glass tubes whilst half-way through their feeds. Some flies alighted and probed unhesitatingly; others moved several times before probing. Hungry flies tend to settle with their heads upwards, which gives them a more stable position when gorged, and probably helps to avoid fouling themselves with excreta. A larger proportion of flies feed successfully in the shade than in the sun.

A tsetse fly alighting on a host still carries the labium enclosed between the palps. As it starts to probe at a suitable site, the labium moves from the palps, to an angle of 90° to the skin. Whilst the labella rests on the skin, the teeth on the inner surface of the labella are everted and penetrate it. At the same time, saliva is excreted from the hypopharynx. This saliva contains a powerful anticoagulant (Yorke and Macfie, 1924). In the normal type of feeding the teeth on the labellum then lacerate one or more capillaries, resulting in a haemorrhage which is rapidly sucked into the labrum. When the fly stops sucking, a small pool of blood forms around the damaged capillaries. This is known as 'pool-feeding'. If blood is not found the first time the fly penetrates the host it partially withdraws the labium and, bending it, makes a new penetration.

In a series of papers, Margalit *et al.* (1972) described the role of sense organs on the mouthparts of tsetse during the feeding process. Experimentally, these taste receptors were stimulated by ATP, which provided a strong stimulus for tsetse flies to feed through membranes. There is evidence for the presence of an ATP-sensitive cell, the LR7 sensilla, on the labium (Mitchell, 1976). The effect of ATP, ADP and AMP as stimulants producing a gorging response has been reported in all haematophagous insects in which that effect has been investigated, and has been exploited in the mass rearing of tsetse using membrane feeding.

Laboratory studies of feeding sites showed that tsetse are attracted to radiant heat and only fed where the pelt of experimental animals was less than 5 mm thick (Popham *et al.*, 1979). There was no evidence that host skin energy flux differentials were used to find sites from which blood was obtainable.

Unmated female tsetse take smaller meals than mated ones. This may be because the gut cannot physically be distended to the same extent, but could also be a hormone-mediated phenomenon related to their lower metabolic demands (Ejezie and Davey, 1982). There is some evidence for this, as female tsetse require a longer period to store sufficient tyrosine during the preovulation period than during other pregnancy cycles (Tobe and Davey, 1974a,b).

After feeding, tsetse flies buzz; this 'post-feed buzzing' causes a rise in thoracic temperature relative to the length of the buzz. The rise in temperature results in an increase in lift that the fly can produce. Buzzing flies also excrete excess water from the bloodmeal faster than non-buzzing ones. It has been suggested that this could have a strong selective advantage for the flies (Howe and Lehane, 1986).

8.5 BLOODMEAL COLLECTION AND IDENTIFICATION

The importance of determining the hosts of tsetse flies, for epidemiological studies and for control, led to attempts to identify bloodmeals many years ago. These were initially based on red cell morphology (diameter of erythrocytes, for example) but, due to the inaccuracies and low specificity of this method, more accurate serological techniques have been developed. The main difficulty in identifying the source of bloodmeals in arthropods arises from the small quantity of blood available to be tested, particularly after some digestion has taken place. Therefore, serological techniques that were developed had to be highly sensitive and adaptable to microtitration systems. The concept of serological diagnostic tests is that each vertebrate species possesses one or more plasma proteins with antigenic determinants unique to the species and so identification depends on the ability of antisera to recognize only the unique proteins of the host's blood.

The first two serological methods used involved precipitation and agglutination reactions, dependent on the production of specific antisera with high titres. These methods are limited by the specificity of the antisera and cross-reactions may occur with antisera of inadequate quality, particularly as many hosts of tsetse are closely related species. 'Cross-immunization' has been tried as a means of producing more specific antisera (Emslie and Steinberg, 1973).

8.5.1 Precipitin test

Nuttall, at the beginning of the century, found a correlation between antigenic relationships of animal sera and zoological classification of the species (Nuttall, 1904). He examined sera of 500 animals with nearly 30 different immune sera prepared from rabbits. Arising from this work, Weitz (1952) developed the precipitin test for bloodmeal identification. The test depends upon the interaction between a saline extract of the bloodmeal and a suitable antiserum, usually prepared from rabbits. Formation of a white precipitate of antibody and antigen indicates a positive reaction, and the absence of a precipitate shows that no antibody–antigen reaction has taken place. Weitz used a specially designed multiple serological dispenser for carrying out the test. Bloodmeal smears were collected by being expressed from the tsetse gut on to filter paper, which could be coated with a preservative such as sodium azide. An extract of the bloodmeal is made by cutting out the blood spot from the filter paper and placing it into a labelled test tube to which a quantity of normal saline is added. The volume of saline depends on the number of tests to be carried out and the quality of the blood smear. The use of as much saline as possible without diluting the extract beyond the sensitivity of the test is desirable. The tubes are kept at 4°C for 6–10 h before testing.

Symes and McMahon (1937) used the precipitin test to determine the feeding habits of *G. swynnertoni* and *G. f. fuscipes*. Sera were collected from a variety of wild animals in Kenya and Tanzania. Antisera were prepared from rabbits by injection of normal serum over a period of 5–6 days. Some cross-reactions occurred between hartebeest (*Alcephalus busephalus*) and roan antelopes (*Hippotragus equinus*), and one or two others, which they found inexplicable. Tsetse gut contents were dried for storage and a saline extract was subsequently made. Tests were carried out in capillary tubes along with control tubes. The inadequate specificity and sensitivity of the antisera limited the test's usefulness and showed its unreliability for differentiating between closely related animals. Nonetheless, the results obtained conformed fairly well to present ideas of the feeding habits of the tsetse species they studied.

Boreham (1972) described an adaptation of the precipitin test, which consisted of overlaying a sample of the bloodmeal extract on the specific antisera in a small capillary pipette. If homologous, the insect had fed on the test host and a white precipitate formed. It required antisera of high titre, specific for the animal or group. More reliable identification of related hosts could be achieved using a haemagglutination inhibition test (Weitz, 1970). The test required care in collection of bloodmeals, to prevent cross-contamination, and in preparation of antisera. Lack of adequate antisera has sometimes led to unreliable results being obtained. Furthermore, deterioration of the blood proteins due to digestion between the time at which the bloodmeal was taken and capture of the fly leads to difficulties in obtaining reliable results, particularly if good quality, sensitive and specific antisera are not used. Storage of the antisera is important to avoid cloudiness that might be confused with a positive reaction. Weitz (1952) described methods of storage as well as the test itself.

8.5.2 Inhibition test

The agglutination inhibition test is a more complex but more sensitive test requiring good quality bloodmeals in order to function adequately. It is based on red blood cells, coated with tannic acid, which are then capable of being sensitized with soluble antigens. The cells are thus coated with a selected serum protein antigen and will be agglutinated by an antiserum prepared against the protein with which they have been sensitized. Serum of the homologous species inhibits this agglutination (Weitz, 1956).

Lindquist *et al.* (1982) described methods of analysing bloodmeals from insects by haemagglutination inhibition and by enzyme immunoassay. They pointed out that most of the immunological methods with which it was possible to identify bloodmeals were compromised by extensive cross-reactions. In contrast, the indirect haemagglutination inhibition test showed excellent specificity and sensitivity, as did direct and indirect enzyme immunoassays.

The precipitin and inhibition tests were frequently used in conjunction; initial screening using the precipitin test determined the animal groups and then individual species were identified by the inhibition test.

8.5.3 Serological identification of reptile feeds of *Glossina*

Boreham and Gill (1972, 1973) described the serological identification of reptile feeds of *Glossina* using small amounts of antigen. To prepare antisera, equal parts of reptile sera and Freund's complete adjuvant were injected into the region of rabbit lymph nodes. The antisera obtained were tested against 24 different reptile sera. Results obtained in one investigation showed that 91% of feeds by *G. f. fuscipes* were derived from monitor lizards, 6% from snakes and 0.5% from tortoises.

8.5.4 Complement fixation test

A complement fixation test (Alton and Jones, 1967) was adapted by Staak *et al.* (1981) to provide specific identification from small samples of grade 4 quality bloodmeals (using the scale described by McLelland and Weitz, 1963).

Antisera were produced in rabbits by multiple intracutaneous and intramuscular injections of sera from different animal species emulsified in complete Freund's adjuvant, boosted 2 weeks later by intravenous injection of the same antigen on 3 consecutive days. Serum was harvested from killed rabbits about 2 weeks later when sufficiently high titres were obtained. Control antigens were obtained from 0.025 ml drops of heparinized blood dried on filter paper and kept at −20°C in sealed bags, in which they retained their antigenic activity for up to 12 months. Field samples were smeared from the guts of engorged tsetse on to filter papers. Even small drops from the foregut were sufficient as an antigen source. The samples were dried and sterilized by being dipped into acetone. One drop of control antigen was eluted with 7.5 ml of buffer and for field samples, according to quality, in the range of 0.5–3.0 ml. The complement fixation technique was adapted to a microtitration system. Antisera were box titrated against serial dilutions of homologous antigen for assessing the working dilution.

8.5.5   ELISA

Service *et al.* (1986) described a sandwich ELISA (enzyme-linked immunosorbent assay) developed to identify the host origin of bloodmeals in insects. Very small quantities of fresh blood (about 0.02 µl) could be detected. Rabbit antisera, obtained from rabbits immunized with IgG of the various host species, are used to coat the wells of PVC ELISA plates using a 10 µg ml^{-1} solution of the rabbit antispecies IgG kept overnight at 4°C. The wells are then washed out with PBS and dried at 37°C.

The double antibody sandwich ELISA consists of a series of incubation and washing steps. Firstly, a solution of the test bloodmeal is incubated in the coated plate. During this incubation any IgG in the bloodmeal will be captured by the homologous anti-IgG on the coated plate. The plate is then washed to remove unbound components of the bloodmeal before being incubated with peroxidase-labelled anti-species IgG conjugate. The homologous conjugate will attach to any IgG captured in the first step. Unbound conjugate is then washed off and enzyme substrate is added and incubated. This is initially colourless and will change to an orange-red colour wherever enzyme conjugate is present (i.e. in the positive wells). The colour change of positive reactions can be read visually or by measuring absorbance in a photometer.

Bloodmeal samples for testing are squashed on filter paper, numbered and dried. The samples are stored in plastic bags containing desiccant at 4°C or −18°C. To remove samples, the section of filter paper is cut out and placed in 1 ml PBS/Tween 20 for a period of at least 1 h. The procedure for analysing the bloodmeal is as follows:

- Place 0.1 ml of the filter paper eluate in each well to be tested. Positive and negative controls are used. Cover the plate and incubate in a damp chamber for 1 h and then empty, wash three times with PBS/Tween 20 and shake dry.
- Then add 0.1 ml of the diluted appropriate conjugate to each well and incubate the plate for 1 h in a humid chamber. It is then washed as before.
- Add 0.1 ml of the substrate solution to each well, cover the plate, place it in the dark and leave it for 20 min. One drop of 2.5 M HCl is then added to each well. The substrate was ortho-phenylenediamine (OPD), prepared 10 min before use by the addition of four OPD tablets to 12 ml of distilled water with 5 µl 30% H_2O_2 added immediately before use.

Results should be well coloured, and negative controls colourless. Sensitivity tests indicated that 0.02 µl of blood could be detected on a filter paper disc (Service *et al.*, 1986). Blood obtained from mosquitoes up to 26 h after feeding kept at 24°C could always be identified and even blood from mosquitoes 31 h post-ingestion could be detected without difficulty. No cross-reaction to anti-bovid sera was observed at low readings. Field kits of coated microplates and conjugates for cattle and human blood were prepared, together with all the associated ELISA reagents, and were tested in Zambia with satisfactory results.

Rurangirwa *et al.* (1986) described the preparation of specific antisera against sera of 46 species of vertebrates for use in an ELISA test for identifying bloodmeals from tsetse. They claimed that bloodmeals could be identified in 100% of fed flies at 20 h post-feeding but only 67.5% and 50% at 40 and 74 h post-feeding, respectively. Mixed bloodmeals from closely related species could be identified.

Reliability of these serological tests depends largely on the specificity of the antisera. The antisera used by Rurangirwa *et al.* were prepared by repeated immunization of animals with 1 ml of whole serum mixed with 1 ml of Freund's complete adjuvant and later in incomplete adjuvant at intervals of 2 weeks. Reimmunization was repeated until potent antiserum was obtained, as determined by agar gel immunodiffusion. One hundred microlitres of substrate were added to each well and left at room temperature for 1 h in the dark. Colour development was read visually or with a micro-ELISA autoreader. Any optical density reading of 0.100 and above was regarded as positive. The test was considered to be more sensitive than other methods, such as the agglutination test, quantitative precipitation, haemagglutination or immunofluorescent assays. As little as 58 µl of residual blood components could be identified.

8.5.6   Direct ELISA

Bloodmeals of *Anopheles* mosquitoes have been identified in Kenya using a direct ELISA assay (Beier *et al.*, 1988). This test made use of commercially available reagents allowing tests for seven hosts to be carried out in 4.5 h. Bloodmeals could be detected and identified up to 32 h after feeding in dried mosquitoes and after 23 h for frozen mosquitoes. In the direct ELISA, the bloodmeal sample is incubated directly in the microtitre plate well. A host-specific antibody–enzyme conjugate is used to detect homologous IgG in the bloodmeal sample. The test differs from the sandwich technique in that the indirect method uses an antiserum to capture a specific IgG and the direct ELISA uses the antibody–enzyme conjugate alone to bind host-specific IgG in the bloodmeal. The indirect ELISA is technically more difficult because an antiserum has to be produced for each host to be tested but is an appropriate method when information is required about a wide range of hosts. The direct method is potentially useful when it is the rates of human feeding that need to be known (for example, in the case of mosquitoes). Similarly, when estimating trypanosomosis risk to cattle or humans, it is the rate of feeding from those hosts which needs to be determined. Antibody–enzyme conjugates are now commercially available for a number of domestic hosts.

8.5.7   New developments

A latex agglutination technique was tested in the 1970s for other blood-sucking insects. The development of improved latex particles in recent years means that this type of test may potentially be developed as an extremely useful tool for identifying tsetse bloodmeals. Very recently a new method of bloodmeal analysis has been developed solely to distinguish human from non-human feeds, based on electrophoresis of superoxide dismutase (Diallo *et al.*, 1997). This test was designed for epidemiological studies of human sleeping sickness, for which the proportion of feeds from humans is an important parameter for risk estimation.

A micro-ELISA developed for identifying mosquito bloodmeals allows identification to the generic level with 58 µg of ingested blood serum (Burkot *et al.*, 1981). Bloodmeals of *Simulium* spp. have been identified using a similar, modified indirect ELISA (Hunter and Bayly, 1991), and an 'enhanced ELISA' for identification of *Culicoides* bloodmeals (Blackwell *et al.*, 1995).

8.5.8   General

Bloodmeals should be collected with great care to prevent contamination between samples. In order to protect against the risk of introducing pathogens with bloodmeal samples, they can be sterilized on filter papers with diethyl ether, acetone or chloroform or by physical treatment with ultraviolet light or heat at 60°C (Boreham, 1976).

There are large differences between the rate of digestion of captive, laboratory-bred flies and wild flies (Weitz and Buxton, 1953). In wild *G. swynnertoni* only 28% of bloodmeals could be identified after 3 days, compared with 90–100% of bloodmeals from captive flies 3 days after the bloodmeal was given. This may explain why greater sensitivity has been reported from laboratory evaluation of some tests, than is found when testing bloodmeals obtained from wild flies. At present (1998) the only services for the identification of tsetse bloodmeals are provided by the Robert Von Ostertag Institute in Germany, formerly under Professor Staak and at CIRDES in Burkina Faso using the technology transferred from Staak's Institute.

8.6   NATURAL HOSTS

Tsetse may originally have been reptile feeders living in forests, later becoming adapted to feed on mammals. It has been suggested that warthogs enabled *Glossina* to leave the primeval forest habitat, in which they may have originated, and enter the savanna ecosystems in which the *morsitans* group subsequently evolved as a separate subgenus (Ford, 1970b). Tsetse of this group then started to feed upon some of the antelope species inhabiting the savanna woodlands. Generally, one finds that antelopes *not* fed upon by tsetse are grazing species that naturally favour open, grassy places and those that *are* fed on are browsers. Table 8.1 shows the ecological niches of common hosts. There is a close correspondence between these and the habitats of tsetse, suggesting that overlapping habitats are among the main factors determining feeding patterns, which may be mediated to a lesser extent by behavioural characteristics of the host.

Sometimes the results of bloodmeal analysis can be misleading as variations in hosts fed upon may occur over very short distances. In a study carried out with sentinel cattle at Kilifi, Kenya, cattle had to be herded within the tsetse-infested bush before trypanosome infections could be detected (Paling *et al.*, 1987), suggesting that even over a distance of 1–200 m, feeding on these cattle would be significantly less frequent. Most analyses of large numbers of bloodmeals have been carried out on samples from areas with large wild animal populations, because that is where large numbers of recently fed flies can usually be obtained. Consequently, the data show what tsetse flies feed on when they have a wide choice. Tsetse flies are adaptable and will change to alternate hosts if their usual hosts become unavailable. An example is *G. longipennis*, which used to feed mainly on large animals – in particular, rhinoceros. As rhinoceros have become extremely rare, this tsetse species has adapted to feed on other large animals.

Areas in which domestic livestock are kept, and with denser human populations, generally have low tsetse densities and low populations of large wild animal. In such areas it is harder to obtain large numbers of bloodmeals for analysis and there are, therefore, far fewer data for these situations. Nevertheless, it would appear that in such areas domestic animals, particularly cattle, are fed on to a considerable extent.

Studies on the feeding habits of tsetse had already been reported by 1912 when Kinghorn and Yorke (1912a–c) published their findings from the Luangwa valley in present-day Zambia; most of their observations still hold today. Differences in the likelihood of wild animals being fed upon were partially due to differences in their habitats. For example, kudu (*Tragelaphus strepsiceros*), bushbuck and, to a lesser extent, waterbuck are usually found in thick cover from which they infrequently emerge; thus, they are constantly exposed to tsetse fly bites.

Similarly, Weir and Davison (1965) described the activity of various wild animals at water-holes during dry weather, highlighting the relationship between the important hosts of tsetse and their activity. The animals that are infrequently fed upon, such as impala (*Aepyceros melampus*), puku (*Kobus*

Table 8.1. Ecological niches of some of the commoner hosts of tsetse.

Family/Tribe	Common name	Species	Habitat/distribution	Remarks
Suidae	Bushpig	Potamochoerus larvatus	Wide range of forested and wooded habitats, preference for valley bottoms with dense vegetation	Bushpigs have diverged into true forest- and savanna-adapted species. Home ranges up to 10 km²
	Red river hog	Potamochoerus porcus	Main rain forest belt and in galleries with permanent water – valley bottoms. Rarely outside rain forest. Marked preference for river courses and swamp forest margins	
	Giant forest hog	Hylochoerus meinertzhageni	Localized distribution. Forest grassland mosaics wide range, including swamp forests, galleries, wooded savannas and thickets	
	Common warthog	Phacochoerus africanus	Widespread. Can live in arid or open areas. Home territory about 4 km²	Uses natural or self-dug burrows that are important sites for tsetse larviposition/resting
Tragelaphini	Bushbuck	Tragelaphus scriptus	Water dependent, requiring thick cover (e.g. small thickets)	
	Greater kudu	Tragelaphus strepsiceros	Thickets and evergreen forests along watercourses, deciduous woodlands	
	Eland	Taurotragus oryx	Woodland and woodland savanna	Home ranges 1400–1500 km²
Cephalophini	Bush duiker Forest duikers	Sylvicapra grimmia Cephalophus (17 species)	Savannas and woodlands of East Africa Widespread; some species inhabit riverine forests	Territorial. Absent from forests Some species are diurnal, with activity coinciding with that of tsetse. Territorial
Neotragini (dwarf antelopes)	Suni	Neotragus moschatus	Coastal thicket and forests with thick undergrowth	
Reduncini	Kob	Kobus kob	Low-lying flats close to permanent water	
Bovini	Buffalo	Syncerus caffer	Forest subspecies: grassy glades, watercourses and waterlogged river basins. Savanna subspecies: forest mosaics and savannas with patches of thicket, reeds or forest	Depend upon low-level browse in forests

vardoni) and wildebeest (*Connochaetes taurinus*), are usually found in open country, frequently remaining for the greater part of the day on wide, bare plains where the flies are less abundant than in the bush. In the Lambwe valley, Kenya, *G. pallidipes* were more attracted to a calf than a sheep or a goat in the absence of wild hosts (England and Baldry, 1972). They were more attracted to a human or to a Langridge tsetse trap (Langridge, 1975) than a sheep or a goat, although the trap attracted more flies than the non-Bovidae baits. The Langridge trap, no longer commonly in use, caught a more representative proportion of female flies than man fly-rounds did.

Factors determined as being most likely to affect host selection by tsetse (Weitz and Glasgow, 1956) included:

- Availability of hosts.
- Tolerance of the host.
- Digestion of blood of various animals (suitability in terms of effect on survival and reproduction).

Availability of hosts

Despite previous observations, indicating that tsetse flies do not simply feed on the most abundant host, Onyiah (1980) argued that host selection *might* be due principally to availability of hosts and not based on true preference, as tsetse are capable of adapting to new hosts in the absence of their usual ones. Field studies show that tsetse flies clearly do show definite selection and do not necessarily feed mainly on the most common host in an area. In contrast, tsetse feeding patterns often bear little relationship to host numbers (Lamprey *et al.*, 1962). In an area of Tanzania, although the mammalian fauna comprised only 3, 0.2 and 0.02% of warthog, rhinoceros and buffalo, respectively, these species provided 77, 2 and 14% of tsetse bloodmeals (Lamprey *et al.*, 1962; Boreham, 1979). Animals such as impala, comprising 70% of the fauna, provided only 1% of the bloodmeals. A more or less simultaneous survey of tsetse feeding habits and host abundance in the Serengeti area revealed significant differences in feeding habits, which corresponded well to host availability (Moloo *et al.*, 1971). Similarly, the natural hosts of *G. pallidipes*, *G. f. fuscipes* and *G. brevipalpis* in southeastern Uganda showed only small monthly variation, thought to be associated with availability of hosts (Moloo, 1980). Impala, gazelles, zebra, waterbuck and wildebeest, common in many tsetse-infested areas of Africa, are rarely fed upon by tsetse.

In the Lambwe valley, Kenya, where bushbuck were the preferred hosts of *G. pallidipes*, there was a positive correlation between the distribution of the two species. Although buffalo and roan antelope were also selected for food, neither was continually available to the tsetse fly population, and their presence or absence had no detectable influence on the distribution of the tsetse population (Allsopp *et al.*, 1972). In the presence of many wild animals neither humans nor cattle were frequently fed on by tsetse flies (England and Baldry, 1972).

Tolerance of the host

Complacency under tsetse attack might contribute to observed host preference (Ford, 1971). It has been observed that monkeys (*Cercopithecus* spp.) are able to kill nearly all tsetse flies that attempt to feed from them. Skin-rippling of antelope such as impala make it difficult for tsetse to feed successfully.

Suitability of blood from different host species

At one time it was thought that the blood of certain animals might be unsuitable for tsetse, and that they would not survive when fed on those hosts. This proved to be incorrect, although performance of flies in colonies can be affected by the species of host providing the blood. Lloyd (1914) thought that reptilian blood was unsuitable for *G. morsitans* as a continued diet. His observations must be treated with caution as his work was carried out on caged flies under conditions that were not ideal. More recently, Moloo *et al.* (1988) fed tsetse on blood from a number of different potential hosts. No significant difference in survival, reproductive performance, mean weight of the bloodmeal or in protein patterns of tsetse haemolymph were found between flies fed on different hosts. Thus, it was concluded that host selection by tsetse is probably not based on any aspect of nutritional value of the blood.

8.6.1   Natural hosts of the *morsitans* group

Wild hosts

As already pointed out, it is not always the commonest species of potential host that are fed upon. For example, results of a study in Tanzania using the precipitin and haemagglutination inhibition tests showed that half the 378 feeds of *G. m. morsitans* were from warthog or bushpig (*Potamochoerus larvatus*) although they only constituted about 10% of the available ungulate population (Weitz and Jackson, 1955). Hartebeest, although occasionally a source of bloodmeals for *G. m. submorsitans*, do not appear to be fed on by any other tsetse. At a site in Nigeria where the most common potential hosts were duiker or civet, these were rarely fed on by *G. m. submorsitans*, which seldom feeds on duiker (Weitz, 1963) but fed mainly on the next most commonly observed potential hosts: warthogs, followed by humans (Jordan, 1965c).

Studies in Uganda (Moloo *et al.*, 1980) and Kenya (Weitz, 1960) have shown the importance of Bovidae as hosts of *G. pallidipes*. In Uganda, bushbuck provided 44% of feeds, with buffalo an important alternative (20.8%). Eland, buffalo and unidentified bovids were also the most important hosts of *G. pallidipes* at Kiboko in Kenya (Weitz, 1960). In the Ruma area of the Lambwe valley in Kenya, *G. brevipalpis* took more bloodmeals from suids than *G. pallidipes*, although bovids (excluding cattle) were the most important source of bloodmeals for both species (Glasgow *et al.*, 1958). Bushbuck were the most important bovid host.

In Nigeria, Page (1959a) found that *G. longipalpis* fed predominantly on bushbuck (53%) followed by unidentified bovids (20%) and red river hog (*Potamochoerus porcus*) (18%). (In this study no serum had been available for testing for the red river hog, but it was assumed to account for 'unidentified suidae' as it was the only possible suid host in the area.) Page explained that some of the blood smears were of poor quality and could be identified no further than Bovidae, which accounts for the high percentage of unidentified bovids. These analyses were carried out using the precipitin test. Both Page (1959a) and Morris (1934) thought bushbuck and possibly duiker were the main hosts for *G. longipalpis*, not other smaller antelopes, even though they were abundant in its habitat. Sometimes wild animals can be highly infected with trypanosomes although they do not appear to be fed upon by tsetse. For example, in West Africa, the kob (*Kobus kob*), was highly infected with *T. b. brucei* in northern Côte d'Ivoire (30%) but no bloodmeals of *G. m. submorsitans* or *G. longipalpis* were detected as coming from that species (Küpper *et al.*, 1983).

Wild pigs, particularly warthogs, are clearly important hosts for *morsitans* group tsetse and their responses to this host were recently investigated. The feeding efficiency of tsetse visiting a single warthog could be as low as 12–18% (Torr, 1994). Between 26 and 31% of tsetse landed on the head region of live adult warthogs. Torr suggested that this concentration was a visual response to the dark patch produced by the pre-orbital glands of a mature warthog and was also a density-dependent response to changes in the grooming responses of warthogs.

At a site in Zambia, where *G. morsitans* bloodmeals came mainly from suids, followed by bovids and primates, most of the suid feeds were from warthogs, whilst puku and duiker were the most favoured bovids (Okiwelu, 1977b). Strangely, since suids are not reservoir hosts of *T. vivax*-type infections, only *T. vivax*-type infections were seen in the flies. A significant correlation was detected between infection rates in the flies and mean monthly temperatures.

Weitz and Jackson (1955) summarized feeding habits of tsetse in East Africa determined from the precipitin and inhibition tests as follows. Suidae formed about half the food supply of *G. morsitans* and *G. swynnertoni*, and nearly half the feeds of *G. austeni*. The remainder originated from ruminants, particularly bushbuck. Animals 'always bitten' by *G. morsitans* and *G. swynnertoni* were warthog and rhinoceros. Animals 'commonly bitten' were roan antelope, reedbuck (*Redunca* spp.), buffalo, kudu, bushpig, bushbuck, elephant and giraffe (*Giraffa camelopardalis*); animals 'rarely bitten' were eland, duiker, waterbuck, impala, baboon (*Papio* spp.), monkey, dogs and cats. Finally, animals 'never bitten' were hartebeest, topi, zebra and wildebeest (Weitz and Glasgow, 1956). Reptiles were important hosts for *G. f. fuscipes*.

Domestic livestock and humans

In an area of Zimbabwe, donkeys were one of the commonest host species and were the source of 86.4% of *G. m. morsitans* bloodmeals (Boyt *et al.*, 1972), despite not previously being considered as favoured hosts. In the study of tsetse feeding habits in the Lambwe valley referred to earlier, there was an inexplicably high incidence of pathogenic trypanosomes in cattle, even though tsetse were apparently not feeding on them. Although mechanical transmission was a possibility, it was thought more likely that the cattle were infected before having been brought into the area. Alternatively, it is possible that with a large population of tsetse, even a small percentage of feeds on cattle would result in a high incidence of infection. The location of bloodmeal collection sites could also have influenced these results. In Zimbabwe, Pilson *et al.* (1978) found that cattle were subjected to over three times the challenge by male *G. m. morsitans* than either sheep or goats although the results of the experiment were complicated by the confounding effect of human repellency to tsetse. In a related study, even in the presence of wild animals, cattle and donkeys were freely fed upon by *G. m. morsitans* and *G. pallidipes* (Boyt *et al.*, 1978). Cattle were more favoured hosts than donkeys or sheep.

8.6.2   Natural hosts of the *palpalis* group

Wild hosts

The *palpalis* group, generally inhabiting riverine or lacustrine vegetation, have a close ecological association with reptiles such as crocodiles and monitor lizards, which are important hosts; for example, Mohamed-Ahmed and Odulaja (1997) found that 73–98% of *G. f. fuscipes* feeds were from monitor lizards near Lake Victoria, Kenya. They are, however, opportunistic feeders and many mammals (including humans) and reptiles which enter their habitat, for water or any other reason, may be fed upon (Laveissière *et al.*, 1985a). One mammal, the hippopotamus, closely associated with water, may also be a significant host for *palpalis* group flies in some areas. Fuller (1975, 1978) showed a correlation between the apparent density of *G. fuscipes* and the abundance of hippopotamus in the Ghibe river in southwest Ethiopia.

In a study of *G. p. gambiensis* in gallery forests near Bobo-Dioulasso, Burkina Faso, 54.6% of feeds were from reptiles (crocodiles or monitor lizards), 26.1% from humans, and the remainder mainly from Bovidae, including bushbuck (Challier, 1973). These analyses were carried out using the precipitin test. Elsewhere, sitatunga (*Tragelaphus spekei*) were favoured hosts.

Nash (1941) suggested that bats could be an important source of feeds for *G. tachinoides*, based on observations that, under laboratory conditions, bats provided complacent hosts for tsetse, which fed on them readily. However, there is no field evidence to support this. In the Comoé national park in Côte d'Ivoire, 37% of feeds of *G. tachinoides* were from ruminants,

34% from hippopotamus and 19% from monitor lizard (Küpper *et al.*, 1990); 57% of the ruminant feeds were from bushbuck, which accounted for 21% of the total feeds. There were changes in the feeding pattern during the rainy season, during which the proportion of feeds from hippopotamus increased, whilst that from reptiles decreased.

Glossina pallicera, which inhabits the southern rain forest and coastal zones of West Africa, feeds mainly on Bovidae and birds (Gouteux *et al.*, 1982a). A study of *G. tachinoides* ecology in the moist savanna zone of West Africa showed that it fed mainly on wild mammals, whilst a quarter of the feeds were taken from reptiles (Laveissière and Boreham, 1976). A much higher proportion of feeds came from reptiles in this zone than for the same species in the dry savanna. In the cold season, 54% of *G. tachinoides* feeds were from birds and reptiles, and feeds from hippopotamus were recorded for the first time.

Domestic livestock and humans

In peridomestic situations in the Guinean forest savanna mosaic region of Côte d'Ivoire, 98% of bloodmeals from *G. palpalis* were from domestic pigs (Dagnogo *et al.*, 1985a); the remainder were from either humans or sheep. All *G. tachinoides* feeds were from Suidae. The fed flies were obtained for analysis from biconical trap catches made in the late afternoon, a short time after the flies were presumed to have fed. However, traps catch predominantly hungry flies and only 3.2% of the flies caught had recently fed. Because of their ability to maintain themselves on a reptile, human or domestic animal diet the *palpalis* group tsetse may not have been seriously affected by the rinderpest pandemic. In the forest zone of Côte d'Ivoire the feeding habits of *G. palpalis* varied with the availability of wild hosts, activities of humans in the environment, and the biotope in which the fly lived (Laveissière *et al.*, 1985a). The importance of interpreting results on feeding habits carefully with regard to the biotope from which they were collected was shown in this study. In village areas almost 100% of feeds were from domestic pigs, whereas in plantations *G. palpalis* took 40–50% of its feeds from humans. In human trypanosomosis foci outside the edges of villages, they fed equally on humans and antelopes, especially bushbuck.

Glossina tachinoides can survive in close association with humans in the virtual absence of wild mammals and reptiles (Jordan *et al.*, 1962). It will also readily feed on cattle but feeds have rarely been identified from domestic sheep and goats even when these were common in the fly habitats. Despite the few feeds apparently taken from sheep and goats by *G. palpalis* and *G. tachinoides*, those animals can be severely affected by trypanosomosis under certain circumstances. At various locations in northern Côte d'Ivoire, although trypanosome prevalence was significantly lower in sheep than in cattle, there was a significant correlation between tsetse challenge from *G. palpalis* and *G. tachinoides* and trypanosome prevalence in sheep (Leak *et al.*, 1988).

At Nsukka, in Nigeria, *G. tachinoides* was intimately associated with domestic pigs which accounted for 64% of identified bloodmeals, whilst less than 7.1% were from cattle (Baldry, 1964). In the dry season, the percentage of feeds from pigs was even higher (94.5%). Infections with *T. vivax*-type trypanosomes appeared to be commonest in the flies, which is surprising as pigs are considered refractory to infection with that species and would therefore not be good reservoir hosts. However, Baldry suggested that some of the infections could be old *T. congolense* infections in which the gut infections had died. Despite widespread availability, few meals were from humans.

In Nigeria and southern Cameroon, *G. palpalis* fed largely on humans and reptiles or on other wild animals when the former were not available (Jordan *et al.*, 1961). In the north of Nigeria, where *G. tachinoides* habitat is restricted to gallery forest, it feeds readily on whatever host is available. In the derived savanna or forest mosaics of the south it loses its opportunistic feeding habits, especially its readiness to feed on humans, and becomes a highly specific feeder on either domestic pigs or cattle. Thus, there is a gradual transition from northern riverine ecology to southern non-riverine ecology and a change from opportunistic feeding behaviour, with many meals from humans to specific non-human hosts (Baldry, 1970). Baldry (1969b) suggested that these differences could be related to genetically different populations. Weitz (1963) had earlier proposed the hypothesis that feeding habits of different tsetse species might be genetically determined.

8.6.3   Natural hosts of the *fusca* group

Wild hosts

Except for *G. brevipalpis* and *G. longipennis*, which are anomalous members of the group, these tsetse, whose main habitat is forest, can be expected to feed on forest-inhabiting species of animal such as forest pigs. The latter are a major host of many *fusca* group species. There are generally fewer vertebrate species in forests because of a deficiency of nutrients which are bound up in trees (Rogers, 1988c). *Glossina tabaniformis,* which feeds mainly on Suidae, occurs in rain forest, whilst *G. medicorum* occurs in small remnants of rain forest surrounded by grassy savanna. In Nigeria, *G. tabaniformis, G. nigrofusca, G. fusca* and *G. medicorum* fed mainly on bushbuck and the red river hog (Jordan *et al.*, 1958). Porcupine was the second most important host of *G. tabaniformis.*

Studies on *G. vanhoofi* in the Congo, near Lake Kivu, showed that it fed on wild pigs, *Potamochoerus* and *Hylochaerus* spp. These may have been the only hosts for *G. vanhoofi* in that area (Van den Berghe and Zaghi, 1963). Thirty-one per cent of feeds of *G. brevipalpis* were from hippopotamus.

Glossina longipennis is nearly always found in habitats where rhinoceros used to occur. This was its main host both in Tanzania (Weitz and Glasgow, 1956) and at Kiboko in Kenya, where they accounted for 74%

of feeds (Weitz *et al.*, 1960). The next most important host at Kiboko was buffalo, which provided the remaining 16% of feeds. Rhinoceros thus provided the 'staple diet' although elephant and buffalo were fed upon when available. More recently, rhinoceros have become extremely rare and this tsetse now depends on other large wild animals.

Domestic livestock and humans

Fusca group tsetse are not implicated as vectors of human trypanosomosis as they rarely feed on humans, nor are they commonly troublesome vectors of cattle trypanosomosis. Most *fusca* species are confined to densely shaded bush and thus do not come into close contact with cattle. Nonetheless, *G. tabaniformis* can be important vectors of trypanosomes to cattle in some circumstances (Leak *et al.*, 1990). In Kenya, where *G. longipennis* has a widespread distribution, it feeds readily on cattle and is a vector of pathogenic trypanosome species (Kyorku *et al.*, 1990). *Glossina fusca*, normally considered a zoophilic species, has been reported to attack humans occasionally (Page, 1959b; Yvoré, 1963).

8.6.4 General

Tables of the natural hosts of tsetse were compiled by Moloo (1993a) using all available data from published and unpublished sources between 1953 and 1991. These data are summarized in Table 8.2. Moloo (1993a) concluded that feeding habits could vary for a particular species within a region as well as from one region to another, depending on host availability, although probable feeding habits for a particular species may be predicted for a particular area with a reasonable degree of certainty.

Trypanosome infections in birds have occasionally been reported; for example, Molyneux described a new species, *Trypanosoma everetti*, from the black-rumped waxbill *Estrilda t. troglodytes* (Molyneux, 1973a), and also described experimental infection of tsetse flies with *T. bouffardi* which infect weaver birds (Ploceidae) in northern Nigeria (Molyneux, 1973b).

8.7 FREQUENCY OF FEEDING – HUNGER CYCLES

The frequency of feeding by tsetse is an important epidemiological parameter but precise estimates still seem elusive. This is partly because catches of tsetse are rarely entirely representative of a population, either in terms of sex or the stage in the feeding cycle of the fly. The feeding cycle, or hunger cycle, is the period between two consecutive bloodmeals.

As most traps catch hungry flies and provide biased samples of a tsetse population, Jackson (1930) classified 'hunger stages' of tsetse so that such sampling biases and stages in the feeding cycle could be quantified to some extent. His classification of the hunger stages of tsetse consisted of six grades:

Table 8.2. Summary of the main hosts of tsetse, determined from analysis of bloodmeals (from data compiled by Moloo, 1993).

Host	Morsitans group	Palpalis group	Fusca group
Domestic animals including man			
Man	3.6	13.00 (3)*	0.46
Cattle	3.56	4.20	2.38
Sheep and goats	0.26	0.40	0.20
Donkey	0.02	0.05	0.02
Dogs	0.15	0.26	0.02
Domestic pig	0.03	14.00 (2)*	0.17
Wild hosts			
Bushpig	5.60	2.20	28.0 (1)*
Warthog	31.60 (1)*	1.65	0.37
Total suidae	45.36	18.50	33.1
Bushbuck	10.18 (2)*	25.12 (1)*	11.2 (2)*
Buffalo	7.50 (3)*	0.90	7.59 (3)*
Kudu	5.37		
Total bovidae	43.22	45.50	28.1
Monitor lizard	0.036	6.02	0.05
Hippopotamus		4.04	15.84 (2)*
Rhinoceros			14.20 (3)*

* Numbers in parentheses show the ranking in order of importance, excluding rhinoceros and hippopotamus, which are from anomalous species (*G. brevipalpis* and *G. longipennis*).

Grade 1 Abdomen distended with red or black blood visible from outside.
Grade 2 Abdomen distended in both directions, but without black blood visible from outside.
Grade 3 Abdomen not distended or wrinkled, usually white in appearance and not concave below.
Grade 4 Abdomen beginning to wrinkle, sometimes distended by a gas bubble at the front end. Frequently straw-coloured on the ventral side.
Grade 5 Abdomen definitely concave or flattened, sometimes wafer-like.
Grade 6 Young (soft) flies.

Jackson later amalgamated grade 2 with grade 3, as the former was rare, and reclassified the hunger stages as follows:

Stage I Gorged.
Stage II Replete.
Stage III Intermediate.
Stage IV Hungry.
Stage V Young flies, which have not yet had their first meal.

Jackson (1933) described in detail how these stages could be distinguished. Due to the complications of pregnancy, female flies cannot be so easily classified using this method without dissection.

The hunger cycle, or feeding cycle, of *G. m. centralis* in the mid-dry season in Tanzania was estimated to be 3–4 days; for *G. swynnertoni* it was slightly shorter, at 2.5–3 days (Jackson, 1954a). This was thought to explain why *G. swynnertoni* was able to live in more arid conditions than *G. m. centralis* whilst in association with a generally higher density of host animals. In an earlier paper, however, Jackson (1933) had suggested that the hunger cycle of *G. swynnertoni* was about 5 days. Figure 8.3 illustrates the behavioural patterns of tsetse in relation to hunger stages.

Jackson summarized his observations on tsetse ecology in relation to hunger, emphasizing the behavioural patterns associated with each stage. For example, hungry flies usually attacked head-upwards; replete flies, if attacking, tended to adopt a head-down attitude. Most replete flies preferred to settle on the ground, or on objects other than humans. Hungry flies were the most persistent followers of moving objects. Jackson (1933) analysed the movements of marked flies and concluded that their activity supported the 'feeding-ground' theory. Considerable independent movement of flies occurred, apart from passive transport on host animals or humans. Some flies under such conditions travelled over 3 km in 3 h. The hunger cycle, or period required for a gorged fly to become hungry, varied with the time of year; drier conditions shortened the cycle. Relative humidity and saturation deficit were most closely associated with changes in hunger of wild flies. When flies were hungrier a greater proportion was found in feeding grounds and these flies tended to follow their hosts more persistently than replete flies. *Glossina fuscipes*, unlike the other species in his study, apparently became hungrier during the wet season.

Doubt was cast on Jackson's method of classification of hunger stages following a study in Uganda to compare catches of *G. pallidipes* caught either from cattle or from Morris traps (Smith and Rennison, 1961a,b). Unfortunately, no conclusions could be drawn because the recorders differed in their assessment of hunger stages. Consequently, it was believed that the subjective nature of the hunger stage classification would limit its usefulness in field studies. In place of Jackson's classification of hunger stages it is now considered that, rather than four physiologically distinguishable phases of behaviour, there is a more gradual, indistinct change (Brady, 1972a).

Responsiveness to mobile visual stimuli increases more or less linearly after a bloodmeal and Brady concluded that feeding behaviour is likely to be opportunistic. Feeding intervals are therefore likely to be influenced strongly by exogenous factors such as temperature or host-animal density. Tsetse behaviour thus appears to be the outcome of continuously variable responsiveness to visual and olfactory host stimuli, and not a succession of behavioural entities.

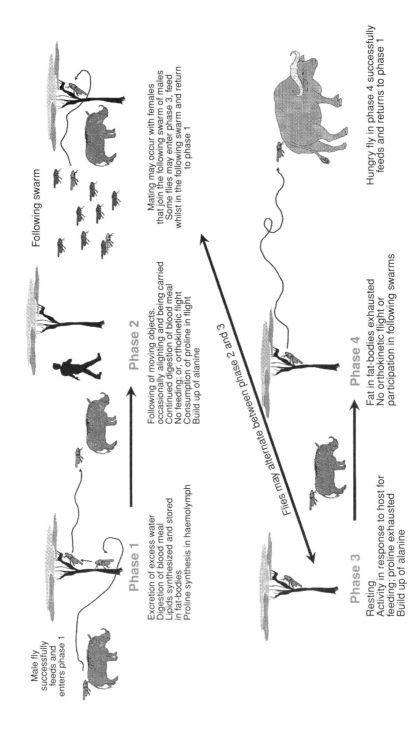

Following swarm

Mating may occur with females
that join the following swarm of males
Some flies may enter phase 3, feed
whilst in the following swarm and return
to phase 1

Hungry fly in phase 4 successfully
feeds and returns to phase 1

Phase 2

Following of moving objects,
occasionally alighting and being carried
Continued digestion of blood meal
No feeding; or, orthokinetic flight
Consumption of proline in flight
Build up of alanine

Flies may alternate between phase 2 and 3

Phase 1

Excretion of excess water
Digestion of blood meal
Lipids synthesized and stored
in fat-bodies
Proline synthesis in haemolymph

Phase 4

Fat in fat-bodies exhausted
No orthokinetic flight or
participation in following swarms

Phase 3

Resting
Activity in response to host for
feeding; proline exhausted
Build up of alanine

Male fly
successfully
feeds and
enters phase 1

Fig. 8.3. Hunger phases of tsetse flies (estimates of the duration of the hunger cycle vary between 2 to 5 days).

8.7.1   Nutritional state, fat/haematin concentrations and behaviour

The feeding interval is one of the most important parameters in trypano-somosis epidemiology. Consequently, some intriguing experiments have been carried out to investigate this rather controversial issue. The estimation of fat and haematin in tsetse flies has provided a means of determining the nutritional state of flies upon which the following discussion is based. These analytical methods have been used by a number of workers to study the categories of flies coming to various sampling devices and as a tool used to estimate the feeding cycle.

As described in Chapter 4, the digestion of a bloodmeal by tsetse flies involves the elimination of water, amino acids, salts and haematin. Haematin is a significant by-product of digestion, while energy reserves synthesized from the bloodmeal are stored as lipids in the fat-bodies. The behaviour of a tsetse fly in relation to stimuli from potential hosts will obviously be affected by how hungry the fly is, which may be assessed from its nutritional state. These changes in behaviour are important as they determine the trappability, or availability, to sampling techniques of a tsetse population. The probability of a tsetse successfully feeding on a particular visit to a host was estimated to be 50% when the host is an ox or kudu, and only 2–19% for other common hosts such as buffalo, bushbuck, bushpig or warthog (Hargrove, 1976b; Vale, 1977a,b).

There are clear differences between mature and immature tsetse and between the sexes in the way a bloodmeal is metabolized and used. These differences have been investigated through field experiments examining the variability in levels of fat, residual bloodmeals and residual dry weights in tsetse. Rogers and Randolph (1978b) produced scattergrams of these types of data and proposed hypotheses to explain the variations. Similar patterns of changes in residual dry weight (RDW) and fat level occurred for four species of tsetse but there were different patterns for the two sexes. A descriptive model was developed to show priority pathways for products of successive bloodmeals during the lifetime of adult tsetse. Tenerals of both sexes initially increased thoracic musculature (RDW) as a priority, without excessive production of fat. Males then used bloodmeals exclusively for the production of fat for storage as an energy source for flight, whilst females gradually increased fat and RDW, probably corresponding to development of a larva *in utero*. In each cycle of pregnancy, females accumulated a reserve of fat, apparently needed at the end of each cycle, when respiratory losses exceeded energy supply from bloodmeals. Analyses of changes in haematin concentrations indicated that females fed once during the beginning of each cycle, once when the larva *in utero* was approaching the moult from second to third instar, and finally during the middle of the third instar. This third and final meal of the reproductive cycle could be small and repeated.

This model was evaluated by Van Etten (1982a) in a comparison of two allopatric populations of *G. pallidipes* in Kenya that differed in physiological condition. The males coming from the population in better physiological condition appeared to use meals for increase in RDW as well as for fat storage, whilst those from the population under greater physiological stress only used meals for energy. Females from the area in which physiological condition was better apparently fed more frequently than those from the other area. Van Etten (1982a) concluded that although the two populations of *G. pallidipes* showed enzyme polymorphism, this was not a suitable technique for population characterization, as it did not appear to be linked to phenotypic variation.

In studies at Nguruman, in Kenya, Randolph *et al.* (1991a) combined analyses of ovarian age and nutritional status to relate feeding behaviour, activity and trappability of wild female *G. pallidipes* to their pregnancy cycle. Fat levels increased most rapidly during the egg *in utero* stage, while crude RDW (CRDW) increased significantly only during the larval stages. This culminated in a 4 mg increase in CRDW during the last day of the third larval instar. The haematin content indicated that females fed at approximately 3-day intervals and that they may feed on any day of the pregnancy cycle. The estimated time of feeding during the day corresponded with the observed time of peak activity, both of which were earlier in the day later in the pregnancy cycle. The rate of fat usage revealed significantly greater flight activity on day 5 of the cycle than on other days, agreeing with the high trappability on this day.

Despite such an analytical approach, there have been disagreements on the ways in which data from such studies have been interpreted (e.g. Langley and Stafford, 1990). Some of the controversy arises over different opinions of feeding strategies. One view is that because of risks of mortality associated with feeding (Randolph *et al.*, 1992) tsetse flies do not attempt to feed at intervals more frequent than about 3 days. Langley and others believe that there is a balance between hunger and host availability and flies will feed at shorter intervals of 2–2.5 days if hosts can be found. Baylis and Nambiro (1993a) estimated a feeding interval of 42–60 h for *G. pallidipes* in Kenya from the nutritional status of flies caught by electric nets when approaching, or leaving an ox. A differential equation model, based on fat haematin analysis, predicted a mean feeding interval of 54–65 h for the same species in Zimbabwe (Hargrove and Packer, 1993). Those authors showed that analysis of haematin content gave estimates of feeding frequency that varied according to the sampling method used. For example, tsetse caught in artificial refuges had about six times as much haematin and up to 32% more fat than flies caught in odour-baited traps, although haematin-specific fat levels did not differ significantly between methods. Supported by results of their study, Hargrove and Packer argued that *G. pallidipes* exhibits greater spontaneous activity than *G. m. morsitans* and would therefore be expected to utilize its bloodmeal more rapidly than the latter species. *Glossina pallidipes* would, therefore, start feeding earlier in the trophic cycle and would be more likely to visit stationary hosts for feeding. They also noted the behavioural difference that *G. m. morsitans* tends to feed on the flanks and back of hosts, while *G. pallidipes* predominantly feeds off the hocks, although, as Baylis (1996) pointed out, the site of feeding varies according to the host species. Thus, *G. pallidipes* feeds on the legs of cattle but on the back and face of warthogs.

In West Africa, Challier (1973) was unable to determine a feeding interval for *G. p. gambiensis* as males appeared to feed whenever they had an opportunity during their active phase. Females fed at more regular intervals, although, according to Challier, the number of feeds was low, only two or three meals being taken during an ovarian cycle of about 10 days.

Hargrove and Williams (1995) analysed feeding behaviour to clarify work described by Randolph *et al.* (1992) based on a simple model of a feeding hazard (f), which increases with the feeding probability (p):

$$f = f_0 + kp \qquad (8.1)$$

where f_0 is the constant mortality associated with feeding resulting from the presence of the bloodmeal which will increase vulnerability to predation, and k is a factor that measures the rate at which the feeding hazard increases with the feeding probability.

Three models were developed to assess the effects of different feeding strategies of tsetse on mortality and reproduction, but when assessing reproduction Hargrove and Williams (1995) took puparial and adult survival into account rather than just puparial production. This was in response to a criticism of Randolph *et al.*'s model in which it was assumed that puparial production alone was a measure of reproductive fitness. They pointed out that the survival of a fly depends on the size of the puparium from which it emerges, which depends on the amount of blood taken by its mother. The three models were:

Model I Prolonged non-feeding phase after each meal, followed by feeding at a constant rate; a constant probability of dying as a consequence of feeding.

Model II Feeding rate increases linearly after each meal.

Model III Feeding rate increases exponentially after each meal.

The first model could be considered superior to the second, provided feeding probabilities and hazards were assumed to be independent, as it allowed a similar peak productivity despite a higher feeding mortality. However, Hargrove and Williams (1995) did not believe this comparison to be valid, as it depends to what extent f and p are independent. They concluded that models II and III provided a better description of feeding than model I. Studies on defensive behaviour of cattle and tsetse feeding success in Kenya also supported models II and III, as hungrier flies were less deterred by defensive behaviour of the host (Baylis, 1996). Bloodmeal size is correlated with the state of a fly's fat reserves and therefore it is likely that there is some sort of feedback mechanism linked to a fly's fat reserves that affects the feeding process (Loder, 1997).

Colvin and Gibson (1992) cited other laboratory observations (Bursell and Taylor, 1980; Langley and Stafford, 1990), which showed that, to achieve maximal reproductive performance, female tsetse need to feed four or five times per interlarval period. This was not in agreement with other estimates of only two or three feeds per pregnancy cycle (Randolph and Rogers, 1986).

Wall and Langley (1991) thought a hunger cycle of 38 h would more accurately reflect observations that had led Bursell and Taylor (1980) to suggest a 3-day cycle. Further studies of reproductive performance of laboratory-reared tsetse flies provided evidence to support arguments concerning feeding intervals (Gaston and Randolph, 1993). The argument of Langley *et al.* (as put by Gaston and Randolph) was that flies reared in the laboratory required an opportunity to feed every day otherwise their reproductive performance deteriorated. A consequence of the low reproductive rate of tsetse in nature is that a population can sustain only low levels of mortality before it will start to decline. Clearly, where tsetse populations do not decline either mortality must be below this level or reproductive performance must be at an adequate level to maintain the population. It may be argued that wild tsetse populations must feed frequently enough to avoid the deterioration in reproductive performance observed in laboratory-reared flies when they are not offered a bloodmeal at least once every 2 days. Gaston and Randolph disputed this, pointing out that wild flies take larger bloodmeals than laboratory flies and digest them more efficiently. They referred to Bursell's (1963b) comments that flies in the laboratory behave abnormally and have abnormal physiology and that it is therefore dangerous to draw conclusions from such observations without supporting evidence from the field. Some estimates of feeding intervals are summarized in Table 8.3.

Hargrove and Packer (1992) concluded that feeding intervals had been overestimated in some studies, because a proportion of tsetse with high fat and haematin content will feed off mobile hosts early in the trophic cycle, and these flies have not been taken into account.

Feeding cycles, activity and trappability

The spontaneous activity and visual responsiveness of starved tsetse increase exponentially for about 5 days after feeding until a pre-moribund decline sets in. Bursell (1961a,b) stated that a feeding response starts when fat content reaches about 24%. Experimentally, abdominal weight and total body weight both change in parallel with behaviour and correlate better with it than with starvation time. Thus, mean levels of a fly's behavioural thresholds are modulated based on information about its overall nutritional state (Brady, 1975a). Supporting this hypothesis, Torr (1988a) suggested that the initial activation of resting flies is primarily mediated through endogenous, rather than host, stimuli.

Table 8.3. Mean estimates of tsetse feeding intervals.

Tsetse species	Feeding interval (h)	Type of study	Reference
G. pallidipes	72	Physiological studies	Randolph *et al.*, 1990
G. pallidipes	42–60	Nutritional status, Kenya	Baylis and Nambiro, 1993a
G. pallidipes	54–65	Fat/haematin analysis, differential equation model, Zimbabwe	Hargrove and Packer, 1993
G. m. morsitans	38		Langley and Wall, 1990
G. m. morsitans	96 (dry), 168–192 (wet)		Jackson, 1933
G. morsitans	72–96 (mid dry season)		Jackson, 1954a
G. swynnertoni	96–120		Jackson, 1933
G. swynnertoni	60–72		Jackson, 1954a
G. swynnertoni	84–108		Glasgow, 1961a
G. f. fuscipes	96	Mark–release–recapture	Rogers, 1977
G. p. gambiensis	80–120 (approximate range)		Challier, 1973

A later approach used to determine feeding cycles was based on the cycles of tsetse activity, which in males are related to feeding, whilst in females they are more strongly affected by the pregnancy cycle (Randolph and Rogers, 1986). Male *G. m. morsitans* appear to be active for about 30 min per day and females for less than that (Brady, 1972c; Randolph and Rogers, 1978, 1981; Bursell and Taylor, 1980). Estimates of sampling biases for female flies may be more accurate when based on analysis of reproductive condition – the stage in the pregnancy cycle of each captured female, as judged by ovarian dissection. Randolph and Rogers assumed that in the field, there must be equal numbers of females from each stage of the pregnancy cycle. Therefore, distribution of sampled flies along an axis representing days of the pregnancy cycle would provide a measure of the sampling bias. Complications arise because ovarian dissections must be done on freshly collected flies and they are difficult and time-consuming. They therefore suggested an alternative method using discriminant analysis based on the nutritional status of captured flies. Each fly is assigned to its probable position in the pregnancy cycle based on its fat and haematin content and residual dry weight. This method has the advantage of allowing flies to be killed and preserved in the field for later analysis in the laboratory.

Tsetse obtained from the Bristol University Tsetse Laboratory were used by Randolph and Rogers (1986) to establish the time course of changes in fat, haematin and CRDW reserves of the flies during the pregnancy cycle. These vary together with the maturation of each larva and can therefore be used to assign flies to one of nine groups for the 9 days of the pregnancy cycle. When using the results obtained from colony flies to calibrate the data from field-caught flies, the following assumptions were made:

- The pattern and timing of changes in the measured variables were the same in each population and between the two species of tsetse involved.
- The periodicity of feeding by flies offered food daily in the laboratory reflected that seen in the field.
- The relative timing of changes in the measured variable was unaffected by the length of the interlarval period (which is affected by temperature).

A possibility was discussed for improving the method if the changing nutritional characteristics of field flies were known more accurately during the pregnancy cycle.

The implications of hunger in relation to trappability are clear, and have been studied for *G. pallidipes* (Langley and Wall, 1990). Not all flies approaching a trap will enter it, as has been demonstrated using electric nets; some will approach and then depart. Flies caught around a trap, but not entering it, have significantly lower mean fat and haematin levels than those caught in a trap (Langley and Wall, 1990). Thus, it seems that flies directly entering the traps are mainly hungry flies actively looking for a bloodmeal, whilst other flies, approaching but not entering the trap, are perhaps only opportunistically looking for a bloodmeal.

The implications of this work for sampling tsetse populations are that feeding intervals estimated from trap catches are likely to represent the maximal intervals between feeds rather than the average feeding intervals as they are based on the hungriest flies in the population. The authors also felt that haematin concentration would not reliably estimate the time lapse since the previous bloodmeal. Furthermore, relationships between fat and haematin content were poor indicators of rates of fat consumption. Fat content was a more reliable indicator. These findings would substantially weaken conclusions of the work of Rogers and Randolph, which relied on the relationship between fat and haematin content to estimate rates of utilization of fat reserves and hence flight activity. Rogers and Randolph strongly dispute the conclusions of Langley and Wall (D.J. Rogers, Oxford, personal communication). Experiments with laboratory colonies of tsetse showed that for both *G. pallidipes* and *G. m. morsitans* there is a positive correlation between feeding frequency and adult survival, fecundity and puparial weight. The minimum feeding frequency that did not cause a decline in reproductive performance was four per inter-larval period for *G. m. morsitans* and five for *G. pallidipes*; six feeds was the optimum for each species (Langley and Stafford, 1990).

Some aspects of nutritional state in relation to trappability are discussed further in Chapter 6.

8.8   RESTING BEHAVIOUR

8.8.1   General features of resting sites

As the flight duration of tsetse flies is short, most of their time is spent inactive at resting sites. These sites provide shelter from high or low temperatures and from predators. They may also provide vantage points from which to seek hosts. The type of resting site can, therefore, change according to the time of day, and hunger stage of the fly. Recently engorged tsetse, which are unable to fly well, rest low down on tree trunks, and the height of the resting site above ground level increases in relation to the hunger stage of the fly (Nash, 1952). In Nash's study *G. fusca* and *G. medicorum* rested with their heads downwards; however, no hunger staging was carried out. The majority of recently gorged flies rest with their heads upwards, a position that would allow them to excrete water from the bloodmeal without soiling themselves. Hungry flies are more commonly found resting horizontally with the dorsal surface downwards.

In the past, much work was carried out to determine the resting sites of different tsetse species because of their significance for tsetse control operations in which resting sites are sprayed with a residual insecticide (described in Chapter 15). Early work indicated that, in addition to a number of other factors, the choice of resting sites varied with the habitat and tsetse species. Apart from normal visual observation, experimental techniques have included the use of sticky substances (e.g. Tanglefoot®), radioactive tracers, reflecting paints or glass spheres, and fluorescent powders or paints. The use of fluorescent powders to mark flies, which are later detected in their resting sites using UV lights, has been described by Tibayrenc *et al.* (1971). Table 8.4 summarizes the published reports of resting sites for the three tsetse subgenera.

8.8.2   Daytime resting sites

Daytime resting sites of morsitans *group tsetse*
The resting sites of inactive adult *G. pallidipes* in a given vegetation type can vary with season and/or time of day (Pilson and Leggate, 1962b). Pilson and Leggate observed that 89% of resting flies detected were at heights of 0–3 m, but whereas in November 49% of flies were at heights between 0 and 1 m, only 10–19% rested at this level at other seasons. The choice of more deeply shaded resting sites in November is thought to be due to the reversal of a normal positive reaction to light at temperatures over 35°C.

During the wet season in Tanzania, *G. swynnertoni* rests on the undersides of horizontal branches of trees at heights of between 1 and 5 m (Isherwood, 1957). At dusk, the flies moved to leaves between 0.5 and 3 m above the ground (Southon, 1959). These observations were confirmed by

Table 8.4. Resting sites of tsetse.

(a) Night-time

Group	Height			Sites			
	Range	Dry season Average	Wet season Average	Leaves (upper)	Small twigs	Creepers	Fencing of pig enclosures (oil and coconut palm fronds)
Palpalis	0–18	1.52	3.78	+++	++	++	++ (G. tachinoides)
	0–6.8	1.3	2.6	85%	+	+	

Note: females rested higher than males and both rested higher on moonlit nights.

Group	Height	Sites				
		Boles of trees (dusk)	Horizontal branches (underside)	Leaves	Small twigs	Creepers
Morsitans	15 cm–7 m	+++	++	+++	+	++

(b) Daytime

Group	Height		Sites				
	Dry season	Wet season	Underside of branches	Boles of trees	Rot holes	Base of tree trunks	Fallen logs
Morsitans							
G. pallidipes	0–3 m (89%)		+++	+++ (at high temp.)	++		
G. swynnertoni		1–5 m	+++				
G. m. submorsitans	0–4 m		+++ (horizontal)				
G. m. morsitans	0–4 m		+++	+++	++	++ (high temps)	++
Palpalis							
G. tachinoides			+++	+++		+++	
G. p. gambiensis			+++	+++		++	
G. palpalis	0–2.5 m					Creepers	
Fusca			Horizontal branches		Creepers	Bases of trunks	
G. longipalpis	20–240 cm		+++		+++	+++	
G. fusca	20–240 cm		+++		+++	+++	
G. medicorum			+++		+++	+++	

Level of site importance: + = low; ++ = moderate; +++ = high.

Chadwick (1964a), who determined that the undersides of branches were the most important sites and would therefore be suitable sites for insecticide application to control that species. He later tested that hypothesis by experimentally treating the resting sites of *G. swynnertoni* with endosulphan or dieldrin and achieved eradication in an area of 91 km² (Chadwick *et al.*, 1964).

At Rekomitje, not far from Chirundu, in Zimbabwe, for most of the year *G. pallidipes* rests, after feeding, on the undersides of branches of small trees and shrubs in riverine forests at heights of between 1 and 3 m. In the hot season, however, these sites are only occupied in the mornings and late afternoons. Towards midday, when the air temperature rises to 35°C and over, branch sites are evacuated. Flies can then be found at rest only on the boles of large trees, where they squeeze into the fissures of the bark, at heights generally less than 0.3 m from the ground. Otherwise, flies hide in rot holes, often quite high up, in big tree trunks. Feeding activity of *G. morsitans* in the same area was restricted to temperatures between 18°C and about 32°C (Pilson and Pilson, 1967).

Glossina tachinoides, *G. m. submorsitans* in Niger and *G. m. morsitans* in Mozambique conform to the general pattern of *Glossina* resting behaviour in that they rest on the woody parts of the vegetation during the day and on the foliage at night. Observations on resting sites of *G. m. submorsitans* in Nigeria confirmed conclusions of Nash (1952) that most engorged flies rested on tree trunks at an average height of 1 m above ground whilst others rested on the underside of horizontal branches at an average height of at least 2 m. In Niger, *G. m. submorsitans* and *G. m. morsitans* both preferred low horizontal branches as their main day-resting sites, while *G. tachinoides* rested mainly on the bases of tree trunks (Turner, 1980b). *Glossina m. morsitans* exhibited a three-phase pattern of resting behaviour: on foliage at night; low down on tree trunks for a short period after dawn; and on branches higher up for the remainder of the day. No relationship could be found between temperature or relative humidity on the choice and vertical distribution of night resting sites.

The physical shape of trees rather than tree species or texture of bark may be a controlling factor in the selection of resting sites (Mahood, 1962). In Nigeria, the most 'popular' resting sites of *G. m. submorsitans* were the undersurfaces of horizontal branches, 2–4 m above ground level. Straight, clean tree boles were not widely used by true resting flies although following flies would rest on them for short periods. A high percentage of resting flies caught from branches were males.

A later study in Nigeria showed that vertical (or near vertical) trunks and the underside of horizontal and inclined branches were the most favoured resting sites of *G. m. submorsitans* (MacLennan and Cooke, 1972). As it became hotter, flies tended to rest more frequently on tree trunks; as temperatures approached 90°C, they tended to choose lower resting places. The resting sites were also related to relative humidity. At dusk, the flies moved upward towards the tops of the trees leaving branches and the trunk deserted.

Okiwelu (1976) used a sticky substance, Tanglefoot® to investigate the seasonal resting sites of *G. m. morsitans* in Zambia. All resting flies were found below 4 m, and there was an inverse relationship between numbers of resting flies and the height of the sites. The predominant resting sites were boles, fallen logs, branches and tree canopies.

Daytime resting sites of palpalis group tsetse

In the pre-forest zone of Côte d'Ivoire, diurnal resting sites of *G. palpalis*, *G. pallicera* and *G. nigrofusca* in areas of human sleeping sickness were mainly ligneous, such as branches, rather than foliage (Gouteux *et al.*, 1984). The flies were uniformly distributed under dense forested cover but were at a lower level in more open places. *Glossina palpalis* was more frequently found near forest edges than the other two species. These forest edges, along which the tsetse flew, provided temporary resting sites during periods of activity.

Tanglefoot® was also used to investigate the resting sites of *G. p. gambiensis* in gallery forests of the northern Guinea savanna of Mali, where five resting site types were found (Okiwelu, 1981). These were the same as in the Zambian study: boles, fallen logs, undergrowth, branches and tree canopies. In any season, more than 80% of the flies were resting on fallen logs, with no significant difference between numbers on the top or the side of the logs. In the southern Guinea savanna zone of Nigeria, diurnal dry season resting sites of *G. palpalis* were mostly the undersides of stems or branches of climbing plants; exposed roots shaded by an overhanging river bank were also used (Abdurrahim, 1971). Over 90% of the resting flies were found at heights of 0.3–2.1 m. In Niger, the microclimatic conditions of resting sites for *G. tachinoides* were more favourable than ambient conditions during the middle of the day (Turner, 1980c).

A radioactive iron isotope has been used for detecting daytime resting sites of *G. p. gambiensis* in Burkina Faso, using a scintillometer to observe resting flies (Bois *et al.*, 1977). Resting sites were on branches, creepers, twigs and roots and there was a tendency to select sites on vegetation with a diameter of less than 10 cm, mostly 1–2 cm diameter. There was also a marked preference for sites within 0.5 m of water. No flies were observed more than 3.5 m from the edge of a stream. The sites were near the ground, mainly less than 30 cm from the ground. The undersides of creepers (lianas, *Acacia pennata*) were also the main daytime resting sites of *G. p. palpalis* in a focus of human sleeping sickness in Côte d'Ivoire, together with coffee bushes. Over 80% of resting flies were found on these sites regardless of season (Sékétéli and Kuzoe, 1994). Again, resting sites were low, not more than 2.5 m above ground level, and mostly below 50 cm; the diameters of the resting sites conformed fairly closely to those observed by Bois *et al.* (1977) for *G. p. gambiensis* in Burkina Faso. Most were less than 3 cm diameter.

Daytime resting sites of fusca *group tsetse*

There have been fewer studies of resting sites of *fusca* species. *Glossina longipalpis* and *G. fusca* both rested most frequently on the underside of stems and branches with a 1060 mm diameter during the daytime at Oke-Ako, Nigeria (Omoogun, 1985). Whether resting vertically or obliquely, *G. fusca* always rested with its head pointing downwards. More than 95% of both species rested between 20 and 240 cm above the ground. Preferred resting sites included shrubs, young saplings, climbers and, during high temperatures, at the bases of tree trunks. *Glossina fusca* rested in similar sites in the Central African Republic (Yvoré, 1962). These were, generally, on the trunks of trees or on vertical lianas, more rarely on low branches. In all cases they rested in a position parallel to the axis of their support, and with the head downward with the exception of mostly engorged flies. A preference for resting on branches of trunks of small size that were mostly at a height of between 60 and 160 cm was recorded. Other studies of resting sites of *fusca* group tsetse (*G. medicorum* and *G. fusca*) were carried out to determine where these species (which were not attracted to humans) could be caught (Nash, 1952). All the flies collected were from the trunks of saplings and stems of creepers, and all had their heads pointing downwards. Gorged flies were commonly caught from these resting sites.

8.8.3   Night-time resting sites

At dusk, tsetse flies generally move from daytime resting sites, on the trunks and large branches of trees, to spend the night on leaves or small twigs, possibly to avoid predators (McDonald, 1960). The choice of nocturnal resting sites can be strongly influenced by seasonal effects on the physical condition of the vegetation, such as leaf-fall.

The reverse migration seems to take place about one hour after sunrise. Brady (1987) suggested that the dusk migration might take place in response to light intensity falling below a critical level. The migration takes place within a few minutes but may occur at different temperatures at different seasons, implying that light intensity is the likely signal. Switching off the lights in laboratory colonies also stimulates a burst of activity, which Brady concluded represented the tsetse's natural upward migration at dusk.

Dean *et al.* (1969a) used radioactively labelled [tantalum[182]] flies to monitor tsetse activity and resting behaviour and observed that individual straight-line displacements occurred on average every 3 h. These movements were generally small and erratic in direction. The majority of flies rested between 2 and 3 m above the ground.

Techniques to investigate night resting sites have included the use of UV fluorescent and reflecting paints and marking flies with small reflecting glass spheres.

Morsitans *group*

Using UV fluorescent and reflecting paints to locate tsetse at night, Jewell (1958) observed *G. swynnertoni* resting on the boles of trees at dusk and on the underside of horizontal branches. After dark they moved to leaves. Southon (1958) also investigated the night resting sites of *G. swynnertoni*, in Tanzania, by marking them with small reflecting glass spheres. He obtained essentially the same results as Jewell; flies moved at dusk from diurnal resting sites on the undersides of branches to perch on leaves at night and returned to branch resting sites at dawn. McDonald (1960) adapted the Jewell (1956) method to detect nocturnal resting sites of *G. p. palpalis* and *G. m. submorsitans* in northern Nigeria prior to an experimental tsetse control trial. He used a luminescent pigment mixed with gum arabic glue and a small amount of detergent. Despite the apparently unpleasant nature of this mixture, painted flies survived as well in the laboratory as unpainted controls. In the field, more than 50% of *G. palpalis* were observed resting at night within 0.3 m of the ground on leaves and small twigs, but never on tree trunks or branches. *Glossina m. submorsitans* rested on similar sites – twigs, leaves and small creepers – from 0.3 to 5 m above the ground but mostly above 2.3 m.

Robinson (1965) marked *G. morsitans* in Zambia with reflective patches, detectable by torchlight at night, to study its night resting sites. He only detected 82 out of 507 marked flies, 22% of which were resting higher than 4 m above the ground. Nearly 60% of flies were resting on leaves, which, together with small twigs, accounted for 77% of resting sites. The maximum resting height was 6 m above ground level.

Palpalis *group*

Night resting sites of *G. p. palpalis* in a riverine swamp in northern Nigeria were detected with a UV light after marking the flies with UV reflecting paint (Scholz *et al.*, 1976). Most flies were resting on leaves, mainly on the upper surface, and on small twigs and creepers. The average resting height was highest in the mid-rains (3.78 m) and lowest in the dry season (1.52 m). The range was from ground level up to 18 m and females rested slightly higher than males. Both sexes tended to rest higher when the moon shone. These experiments may have given biased results as UV light itself can attract flies, whilst UV reflecting surfaces repel them (Green and Cosens, 1983; Green, 1989). The night resting sites of *G. tachinoides* in Nigeria, detected by torch-light, included dried fronds of oil and coconut palms used for fencing pig enclosures, mostly less than 60 cm above ground level (Baldry, 1970). Also in Nigeria, *G. tachinoides* were marked with UV reflecting paint (Spielberger and Barwinek, 1978) in a study which obtained almost identical results to those of Scholz *et al.* (1976), who studied *G. p. palpalis* at the same site in the north of the country. Almost 85% of the flies rested on leaves, mostly on the upper sides, whilst the remainder rested on small twigs and creepers. Resting sites ranged from ground level up to 6.8 m, with the average being 2.6 m in the mid-rains and 1.3 m in the early rains. Females rested slightly higher than males and both sexes rested higher on moonlit nights.

In Burkina Faso, *G. p. gambiensis* rested at night on extremities of vegetation close to the ground in the lowest part of the gallery forest, near to water (Challier, 1973). Challier suggested that it was attracted to such sites by the increased concentration of CO_2 in the atmosphere.

8.9 LANDING BEHAVIOUR

An understanding of the landing behaviour of tsetse can be used to increase the efficiency of control, particularly with respect to the use of targets (Chapter 16). Tsetse may be attracted to an object by visual or olfactory stimuli, but what activates a landing response once they have reached the source of the stimulus?

Landing behaviour is difficult to study in the field and, consequently, most information has come from laboratory studies. In one such study, *G. m. morsitans* appeared to avoid horizontal features in choice of landing site, and showed a preference for circular rather than square shapes, and square rather than triangular shapes (Doku and Brady, 1989). In contrast, horizontal resting positions are preferred in nature. These anomalous findings could not be explained satisfactorily. In other laboratory studies of landing responses to black and white patterns, *G. m. morsitans* showed a ninefold greater preference for blackness versus whiteness (Brady and Shereni, 1988). Landings were twice as frequent on a vertical black stripe pattern than on a horizontal one. In contrast, a vertical cylindrical target caught fewer flies than a horizontal one in field studies (Vale, 1974b). The difference between Vale's observations (supported by Torr, 1987) was thought by Brady and Shereni to be due to flies in Vale's and Torr's studies being in a 'host-seeking' mode, whereas Brady and Shereni's observations were of flies looking for resting sites. In Brady and Shereni's experiment, *G. morsitans* showed a marked preference for landing near the edge of the black on all targets and was strongly biased to the black side of an interface.

Black, blue and red targets resulted in strong landing behaviour in male *G. m. morsitans* in the laboratory, but not white or yellow targets (Green, 1993). Carbon dioxide increased the landing response of the same species on blue or black targets but had no effect with the white or yellow ones. UV light has a more complicated effect: whilst UV reflectivity reduces attraction to targets it can, under certain circumstances, increase landing behaviour.

8.10 ACTIVITY

Tsetse show a diurnal pattern of activity. This pattern may differ between species and according to the sex, hunger, pregnancy and nutritional state of different components of the population, as well as being affected by other mediating factors. A fairly large number of studies have been carried out on the influence of various factors, such as light, temperature, odours, wind direction and speed, physiological state and colour, on the activity of tsetse. Techniques such as radiolabelling, video recording and the use of wind tunnels have been employed to investigate factors determining what causes a fly to become active and the way in which it responds to a stimulus. The activity of *G. pallidipes* caught at hourly intervals in biconical traps in the Ghibe valley, southwest Ethiopia, is shown in Fig. 8.4. There was no significant seasonal difference in the pattern of activity, although at one site (Ghibe) peak activity was at 1800 h whereas at a second site (Tolley) peak activity was slightly earlier. This may have been due to cooler temperatures at the slightly higher altitude at Tolley (Leak, 1998, unpublished data).

8.10.1   Effect of light on activity

The effect of light intensity is closely linked with circadian and diurnal activity. Laboratory studies of the phototactic responses of *G. m. morsitans* to red and green light have demonstrated a strong attraction to short wavelengths, especially UV light (Green, 1984). There is a secondary peak of attractiveness with the red light wavelength that is not found in other insects, for which the secondary attractiveness peak occurs with green light. This suggested the existence of red receptors in the eyes of tsetse flies (not

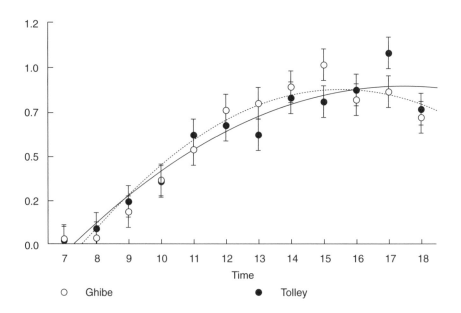

Fig. 8.4. Diurnal activity of *G. pallidipes* at Ghibe, southwest Ethiopia.

previously observed in Diptera), in addition to green receptors. Tsetse appeared to possess an unusual method for discriminating between red and green light, independently of intensity (Green, 1984).

The normal reaction of tsetse to light is a positive attraction, but if the temperature is raised sufficiently the reaction becomes strongly photo-negative. This has been shown for *G. morsitans* (Jack and Williams, 1937) and for *G. palpalis* (Mellanby, 1936).

Vanderplank (1941) found that *G. pallidipes* seemed to be more active on days following moonless nights than on days following moonlit nights. This suggested that flies were more active on moonlit nights than on dark nights and were therefore less active the following day. Both sexes of *G. pallidipes* may be active throughout the night on moonlit nights (Chorley and Hopkins, 1942). Some activity also occurred on a very dark night. *Glossina longipennis* is active at dusk and in the dark. Its activity patterns were studied at Lake Jipe in Kenya using man fly-rounds (Power, 1964) and more recently at Nguruman, also in Kenya, using traps (Randolph *et al.*, 1991b). Tsetse caught in the study at Lake Jipe were 93% non-teneral, hungry males; the females do not come readily to humans. An enormous increase in activity occurred at about sunset each day. The rise in activity at sunrise was more variable and less marked (Power, 1964). At Nguruman, trap catches were highest at the end of dusk, between 1900 and 1930 h, and this was associated with an increase in numbers of hungry flies, estimated to have fed more than 60 h previously (Randolph *et al.*, 1991b). *G. longipennis* shows limited activity between dawn and sunrise but during the day it is virtually inactive (Neave, 1912; Lewis, 1942).

8.10.2 Effect of season on activity

Harley (1965) studied the activity cycles of *G. pallidipes*, *G. f. fuscipes* and *G. brevipalpis* from catches made from bait oxen in Busoga district, Uganda. There were no large monthly or seasonal changes in the activity pattern. Male *G. pallidipes*, the most abundant species, showed a gradual increase in activity from early morning until evening followed by a rapid fall. Male activity started just before dawn with a peak in the second hour before sunset. There was little indication of an evening peak in activity of females, however. Peak female activity more closely coincided with maximum temperature than did male activity. Very little activity took place during the hours of darkness. Near the Kenyan coast, *G. austeni* is active throughout daylight hours with significant peaks of activity between 0900 and 1000 h and between 1400 and 1700 h (Owaga *et al.*, 1993). Its activity thus conforms to that of *G. pallidipes*, *G. fuscipes* and *G. morsitans* observed by other workers (Pilson and Pilson, 1967; Crump and Brady, 1979; Rowcliffe and Finlayson, 1982; Mwangelwa *et al.*, 1990). Owaga *et al.* (1993) thought that any night activity, such as that observed by Moggridge (1948) at the Kenyan coast, would be just occasional.

Harley (1965) found that the activity of *G. fuscipes* is similar for both sexes, with a rapid increase from early morning to a maximum in the middle of the day and a fairly rapid fall in the evening. In the shade, peak activity occurred earlier in the day than in open places. No tsetse flies were caught during the hours of darkness. The activity of *G. brevipalpis* was quite different from the previous two species: flies could be caught at any hour of the day or night and relatively large numbers were caught in short periods close to sunset and sunrise. There were more marked differences between catches from shady and open areas. In the open, there was a peak in activity of male flies during the hour after sunset and a smaller peak in activity during the hour before sunrise. Females showed a similar pattern but both the evening and morning peaks were bigger. Fewer flies were caught during the middle of the day. Yvoré (1962) noted the nocturnal activity of *G. fusca*.

In southern Nigeria, diurnal activity of *G. tabaniformis* showed an early-morning peak followed by a decline in fly numbers after about 0900 h; the flies were least active during the middle of the day (Jordan, 1962b). There was a second activity peak in the late afternoon and early evening, which was particularly marked in the dry season. During the wet season there was an inverse relationship between fly activity and both temperature and saturation deficit. In the same area, *G. fusca* was nearly equally active throughout the day in the wet season, with a suggestion of peak activity just before dark, but in the severe conditions of the dry season activity was negligible between about 1200 and 1700 h (Jordan, 1962b).

8.10.3   Effect of olfactory and visual stimuli on activity

Certain odours will stimulate activity in tsetse and will influence the direction of flight. The direction of flight of tsetse taking off in the presence of certain wind-borne odours shows a significant upwind shift (Bursell, 1987). Turns are steered in relation to wind direction if odours are present. Upwind flight in an odour plume is regularly preceded by a standing turn into the wind before taking off in upwind flight. This suggests that wind direction is assessed and flight direction determined before take-off. The type of flight behaviour exhibited in odour plumes following activation is described in more detail in Chapter 10.

8.10.4   Effect of climate (weather) on activity

Activity of *G. pallidipes*, *G. austeni* and *G. brevipalpis* has been studied at the Kenya coast in relation to climate (Moggridge, 1949b) and the findings contrasted with those of Potts (1940) at Shinyanga in Tanzania. Potts had found a strong positive correlation between catch and evaporation rate; whilst on the Kenya coast, where activity was confined to the early hours of the day, catch and saturation deficit were negatively correlated (Moggridge,

1949b). Laboratory experiments on the effect of humidity on activity have shown that hungry *G. m. morsitans* were more active in dry air than wet, and the intensity of this reaction increased with increasing desiccation (Bursell, 1957b). The reaction stopped at low light intensities, mediated by the compound eyes. The spiracular filters were identified as the sense organs responsible for the kinetic reaction to humidity. Vanderplank (1941) also studied activity of *G. pallidipes* in relation to climate in Tanzania, based on catches from bait oxen. Activity increased with temperature up to 30°C and decreased over 30°C. A similar effect was found with humidity, with activity decreasing over 25 mbar humidity. He found that the fly was active after dusk. Similarly, in Ghana and Nigeria, *Glossina longipalpis* are inactive until temperatures reach 23°C (Morris, 1934; Page, 1959a).

Activity patterns may differ for the same species of tsetse in different localities. This was observed for *G. pallidipes* in different areas of Kenya (Van Etten, 1982b). In the Nguruman area, peak activity occurred in the late afternoon, especially for males, while for the same species at Mwalewa, at the coast, there was a major activity peak in the morning, a smaller afternoon peak, and a midday depression of activity. Activity peaks in the early morning and late afternoon in the two areas were correlated with temperature. Reviewing published data, Brady (1974) found that correlation of activity with temperature in the field was not as good as generally supposed. Low temperature only affected the endogenous rhythm slightly, affecting activity level rather than pattern. High temperatures (above 33°C) caused the normal activity pattern of tsetse to break down, resulting in no late afternoon or evening peak of activity, this being replaced by a peak at dawn (Hargrove and Brady, 1992). Conversely, low temperature (below 24°C) resulted in the replacement of the dawn peak by a peak at dusk.

8.10.5 Spontaneous activity and circadian rhythms

In addition to environmentally stimulated activity, tsetse flies exhibit spontaneous activity in laboratory experiments (Brady, 1970). This activity appears to result from an underlying endogenous circadian rhythm demonstrable in constant darkness but obscured in constant light. The amount of activity increases roughly exponentially during the course of starvation, with long intervals of inactivity. Activity occurs intensely for brief periods of less than one minute. Explanations relating physiology and proline metabolism were suggested, although there were some anomalies, which could not be explained in this way (Chapter 4).

8.10.6 Activity of pregnant flies

In laboratory studies, the circadian pattern of activity of tsetse is markedly changed by larviposition. Twenty-four hours before larviposition, pregnant *G. m. morsitans* are inactive; activity peaks in the afternoon 2–3 h before and during parturition (Brady and Gibson, 1983). After larviposition, activity falls to normal levels. Forty-eight hours before larviposition, the morning peak of activity is twice the normal level and over 70% of pregnant females will feed. Several species of tsetse, including *G. m. morsitans*, are bimodally active in a 24 h light cycle, but this pattern of activity is not retained under constant darkness or constant light.

Other laboratory studies have shown that pregnant tsetse become less active as pregnancy progresses but activity on the day of larviposition rises significantly, although visual responsiveness does not (Abdel Karim and Brady, 1984). As pregnancy progresses, there is an increasing tendency for tsetse to rest in darker sites (Rowcliffe and Finlayson, 1981, 1982). This photonegative response increases with temperature, thus indicating a search for cooler, shaded areas. The effect of such behaviour would be to prevent deposition of larvae in an unsuitable environment. Conversely, a sudden drop in temperature experienced by a fly may be the immediate stimulus to deposit a fully mature larva. This would occur when she passes through leaves, which often cover larviposition sites, to the bare soil, where the temperature is usually 1–3°C cooler. Temperature, therefore, seems to be important in the timing of parturition and a sudden reduction of temperature may be of particular importance. There may also be a circadian rhythm of parturition (Zdárek *et al.*, 1992) as *G. m. centralis* kept under a 12:12 h light:dark photo-regime, at constant temperature, showed a rhythm of parturition occurring in the late afternoon. Although the rhythm dampened in response to continuous light, the authors thought that its persistence for some time after the switch was evidence of a circadian rhythm.

Chapter 9

Population Dynamics

Tsetse population densities vary seasonally, primarily as a result of climatic factors, which cause varying mortality and influence the rate of development of pre-adult stages. In addition, the rate at which reserves of fat, proline and water can be replenished from bloodmeals is a major determinant of tsetse population dynamics. Real changes in population densities have not been easy to detect, due to deficiencies in sampling techniques. These sampling techniques measure the apparent (or relative) density of a population, which may reflect the actual population size. Such measures of apparent density also depend upon the 'availability' of the flies to the sampling technique – the concept of availability is described below.

Tsetse populations often seem to have fairly stable densities compared with other insect species, although these may exhibit long-term, as well as smaller, seasonal fluctuations. The characteristic level of density may vary widely from as low as 40 individuals per km^2 according to Glasgow (1963), to tens of thousands per km^2 (for example at Nguruman, or the Lambwe valley in Kenya). In contrast with malaria vector populations, which vary greatly in space and time (Smith *et al.*, 1995), tsetse density is much less variable and these variations may account for about 50% of the variability of apparent density, depending upon the sampling technique used.

An understanding of tsetse population dynamics is essential for assessing tsetse control interventions and for understanding the epidemiology of human and animal trypanosomosis. Knowledge of tsetse population dynamics can usefully explain the events following an insecticidal tsetse control programme, as for example in two areas in Côte d'Ivoire where tsetse control was carried out (Randolph and Rogers, 1984a). In an area of low natural mortality rates there was, naturally enough, a high tsetse population (*G. p. palpalis*). Insecticide spraying killed a large proportion of the resident population, but because of the population's high resilience, there was a rapid recovery, supplemented by reinvasion from neighbouring areas. At a site with a high natural mortality rate and low tsetse population, the numbers were largely supported by immigration under

147

natural conditions, and insecticide spraying therefore only affected a small proportion of the local resident population. Because of the low resilience at this site, a repeated additional mortality would cause the population to decline.

9.1 AVAILABILITY

The concept of 'availability' was used in early studies of tsetse distribution and population dynamics, as a means of assessing the way in which catches reflected the abundance of a tsetse population. Fly-rounds (Chapter 6, section 6.1.1) were commonly used to determine the standard availability of tsetse, which was estimated as the proportion of the fly population appearing to a catching party when 10,000 yards (9144 m) of linear catches were carried out in each square mile (2.6 km²) of fly belt (Jackson, 1954b). Similarly, Glasgow (1963) defined availability as the relationship between fly-round catch and the true density of the population, whilst Ford *et al.* (1959) stated that a fly-round catch reflected the product of population density and availability. Often, availability was defined as the proportion of the total male tsetse population caught on a standard fly-round traverse. Mark–release–recapture experiments provided data for determining the proportion of the population which was caught (Chapter 6, section 6.5). This was considered extremely valuable in allowing fly-round data to be related to population studies. For example, Jackson (1954b) gave the standard availability of *G. morsitans* in the *Brachystegia/Isoberlinia* wood-land at Kakoma, Tanzania, as 13%, a figure which he later amended to 10%. In practice, there is no 'standard availability' for each tsetse species; it will vary according to location and circumstances, as was shown shortly afterwards by Harley (1958).

Harley (1958) pointed out that estimates of availability were influenced a number of factors, including activity, physiological state (nutrition), pregnancy/sex, season, sampling method, weather, time of day, visibility and vegetation types. The dynamic interactions between these factors, deter-mining changing availability of tsetse, are shown in Fig. 9.1.

In an attempt to understand how physiological state could influence the availability of tsetse, Randolph and Rogers (1981) examined the fat haematin content of *G. m. centralis*. This varied according to the sampling method used, and previously it had been shown that nutritional state could also differ between vegetation types (Bursell, 1966b). In Uganda, availability of *G. pallidipes* is greatest towards the end of the day (Curson, 1924; Fuller and Mossop, 1929; Smith and Rennison, 1961a,b).

One of the problems in assessing availability, and, indeed, of studying tsetse population dynamics, is that sampling methods do not catch a completely representative sample. This was apparent from the high percentage of males caught on fly-rounds, which did not reflect the true sex

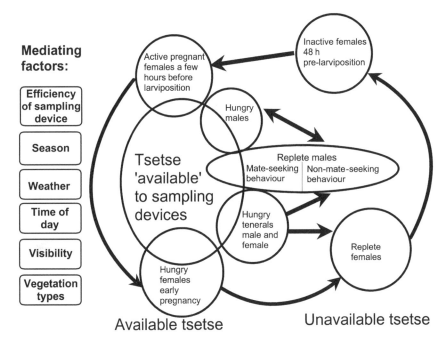

Mediating factors:

- Efficiency of sampling device
- Season
- Weather
- Time of day
- Visibility
- Vegetation types

Active pregnant females a few hours before larviposition

Inactive females 48 h pre-larviposition

Hungry males

Tsetse 'available' to sampling devices

Replete males
| Mate-seeking behaviour | Non-mate-seeking behaviour |

Hungry tenerals male and female

Replete females

Hungry females early pregnancy

Available tsetse

Unavailable tsetse

Fig. 9.1. Flow diagram describing the availability of tsetse.

ratio of the population being sampled, and from comparisons between different sampling methods that caught different sub-samples of the population. For example, the larger catches of female tsetse in Morris traps, compared with catches from bait oxen and standard fly-rounds, indicated to Smith and Rennison (1961a,b) that females were not as active between feeds as had previously been thought. Morris traps were used in the hope that they would provide better estimates of tsetse challenge than fly-round data, although Smith and Rennison concluded that, because of each method's particular sampling bias, none was particularly appropriate.

There is some seasonality in the emergence of new (teneral) flies from their puparia, determined by the temperature conditions during the puparial period. In West Africa, the range of the puparial period for *G. tachinoides* over a year varies from 25 to 58 days for males and from 22 to 55 days for females, and is determined by climatic variations (Laveissière *et al.*, 1984a,b). Delayed emergence in the cold season caused unhatched puparia to accumulate and a sudden population increase resulted when temperatures rose. This resulted in a high percentage of young flies in the population at the end of the cold season and beginning of the hot season, followed by an increased population growth rate to compensate for the effects of the cold season. Relative humidity never fell below a level that would cause puparial mortality. Temperatures, however, did fall below the critical limit in the cold season and rose to lethal levels in the hot season in sand. During the hot season tsetse adapted to the climatic conditions by depositing their larvae throughout the gallery forest, not only in the sandy riverbanks.

Despite the low reproductive rate of tsetse, their residual fertility must interact with a density-dependent mortality somewhere in the life cycle if there is to be any characteristic level of abundance (Rogers, 1985a, 1988a). Total catches of flies by various sampling methods depend on activity as much as the population density of the flies being sampled. In an investigation of the physiological correlates of availability of male *G. m. centralis* to different sampling methods, Randolph and Rogers (1981) estimated a daily flying time of 32 min per day from fat haematin curves. With such low levels of activity, it is likely that samples are biased towards a small, currently active segment of the total population. Hand-net catching seems to suffer from a 'saturation effect'. Comparisons of capture methods using electric nets showed that hand-net catches, as a percentage of electric-net catches, decrease as the numbers of flies arriving at the trap increase. Availability is low soon after feeding, then increases slowly, and subsequently more rapidly as the flies enter the feeding phase. Females are relatively inactive for the 48 h period preceding larviposition except for a burst of activity for a few hours before larviposition.

9.2   DISPERSAL WITHIN A HABITAT

A number of studies have been conducted on the movement, or dispersal, of tsetse, as the distances that they can travel affect the success of tsetse control schemes. The effectiveness of traps and targets currently used for control depends on activity of tsetse and their patterns of dispersal. In the days when barriers of bushclearing to prevent reinvasion were in use, the distance a fly could travel was critical in determining the width for effective clearings. In Zambia, a bushclearing 1 km wide was the standard for a 'holding line'; in Uganda, much wider clearings – up to 8 km wide – were used (Wooff, 1968).

Although movement within a habitat appears to be random and is assumed to consist of fairly constant step lengths (Rogers, 1977), it is affected by natural factors that tend to limit movement to within the habitat and may reduce its randomness. These factors include humidity, availability of shade, host density and odour plumes.

Early investigations of dispersal attempted to correlate the distance flies moved with natural factors such as bush cover and environmental conditions. At Shinyanga, Tanzania, Jackson (1946) introduced *G. morsitans* to an area where they did not naturally occur and in which they were isolated. Males dispersed in any one direction to a mean perpendicular distance of 0.3 km per week, increasing up to 0.84 km in 4 weeks. *Glossina morsitans*, which lived 6 weeks after marking, dispersed for an average of only about 2 km in that time (Jackson, 1948b), whilst the average for all flies was only about 0.8 km. At Shinyanga, *G. morsitans* took about 3 weeks to travel 0.8 km in any one direction. Some estimates of dispersal rates are summarized in Table 9.1.

Moggridge (1936a) also conducted a series of experiments to determine the effectiveness of bushclearing barriers to prevent or reduce dispersal of *G. swynnertoni* at Shinyanga. He pointed out that environmental factors would affect the efficacy of clearings, which may be effective at a given width in some situations and not in others. Moggridge found that a moving object in an open area up to 300 m away from the bush could attract *G. swynnertoni* in the dry season, provided visibility was unobstructed by long grass. *Glossina swynnertoni* would follow wild animals, cattle or even humans for distances up to 1440 m from the edge of their habitat into cleared areas. The distance they would venture into cleared areas was greater during the wet season than in the dry season, partly because *G. swynnertoni* feeds less frequently during the wet season. Moggridge speculated that in the dry season they would feed quickly on any host that they were following before departing again to suitable resting sites.

9.2.1   Theoretical models and analysis of dispersal

In an attempt to model the movement of a population of *G. f. fuscipes,* Rogers (1977) carried out a single mark–release–recapture experiment to estimate the feeding cycle, losses from the marking area, distribution of the population within the habitat, and the daily movement of flies. The feeding interval for male flies was about 4 days and fly distribution was related to the amount of light reaching any particular area, this being determined by

Table 9.1. Estimates of dispersal rates of tsetse.

Tsetse species	Random movement (m day^{-1})	Linear advance	Reference
G. morsitans		334 m week^{-1}	Jackson, 1946
G. m. morsitans	283		Jackson, 1948
males	357		Jackson, 1944
	146		Jackson, 1933
	320		Ford, 1960
G. m. submorsitans	411		Davies and Blasdale, 1960
G. swynnertoni males	539		Jackson, 1948
G. pallidipes	175		Brightwell *et al.*, 1992
G. fuscipes	338		Rogers, 1977
G. fuscipes	200	20m day^{-1} (7 km year^{-1})	Williams *et al.*, 1992

vegetational cover. Rogers gave two methods for investigating the outcome of two-dimensional random movement appropriate for tsetse fly studies. The first method was based on a formula to predict the mean distance, d, moved away from the starting point by *G. morsitans*, assuming a constant step length, s (the distance moved per unit time), and a variable number of steps, x:

$$d \cong sx^{\frac{1}{2}} \tag{9.1}$$

A logarithmic transformation ($\log d \cong \log s + 0.5 \log x$) of that equation showed a linear relationship of movement over time with a predicted slope of 0.5 and an intercept on the ordinate of the log step length. This model provided an approximation to the random movement of *G. f. fuscipes*. A more accurate definition of the daily displacement was derived by Hargrove (in Bursell and Taylor, 1980) as:

$$d = 0.9\ sx^{\frac{1}{2}} \tag{9.2}$$

The second method was based on modelled predictions of a series of random movements away from a release point. These predictions involved probability distributions for directions moved, and distances covered per step, but could be simplified by assuming that tsetse only fly for a few minutes per day and have a relatively constant step length. From his model Rogers concluded that male *G. f. fuscipes* moved about 338 m day^{-1} on average (probably further when hungry, and less far after feeding). His analyses supported the idea of random movement.

Other models have estimated rates of advance of tsetse at 20 m day^{-1}, or 7 km year^{-1}, based on a root-mean-square displacement of 200 m day^{-1} and a population growth rate of 1% per day. These estimates correspond well with field estimates (Williams *et al.*, 1992).

A Gaussian diffusion model with an exponential mortality term (Williams *et al.*, 1992) can describe dispersal of insects. The rate of diffusion (dispersal) will be defined by the root-mean-square displacement in 1 day (λ). If this is high, tsetse will disperse quickly into the vicinity of traps and there will be a rapid reduction of the population. The rate of dispersal of a tsetse population is perhaps a less critical parameter for barriers in tsetse control schemes using impregnated traps and targets where the range of attraction (a) of a trap barrier is more important. Williams *et al.* (1992) investigated the relationship between these parameters.

In Burkina Faso, the linear dispersion of *G. p. gambiensis* and *G. tachinoides* in a gallery forest was studied using marked flies, which were recaptured with biconical traps deployed along the gallery forest (Cuisance *et al.*, 1985). The distances travelled varied according to the season and females tended to spread out more widely than males. Maximum distances travelled were up to 25 km and the median ranged from 0.6 to 2.4 km. Humidity was the main factor resulting in seasonal dispersal of *G. pallidipes* at Nguruman, in Kenya, but availability of shade and host density were also important (Brightwell *et al.*, 1992). Dispersal up to at least 3.5 km from thicket areas took place after rain, and could be accounted for by random diffusion with an average daily displacement of about 175 m day^{-1}. The same factors were important for *G. longipennis*, which had a greater range of dispersal. Flies navigating upwind in odour plumes appear to respond to gaps in thick vegetation rather than flying over the top or through it (Paynter and Brady, 1992).

9.3 POPULATION REGULATION

9.3.1 Ageing of tsetse flies

As the determination of the age of tsetse flies is crucial for studies of population dynamics, methods of ageing flies are outlined here.

Techniques for ageing vector populations such as tsetse flies or mosquitoes have been developed for epidemiological and population dynamics studies, and for monitoring control operations. The age structure of a population enables survival or mortality rates to be calculated, and these are essential for estimating the rates of changes in population size. These can then be related to seasonal climatic fluctuations or to additional mortality imposed on a population as a result of vector control.

The age structure of a tsetse population can be useful for evaluating and regulating tsetse control using sequential aerial application of ultra-low volumes of non-residual insecticides (Chapter 15). For aerial spraying, the timing of the spray sequence depends on the duration of the puparial period in order to ensure that newly emerged flies are sprayed shortly after emergence, before they have had an opportunity to mate and deposit their first larvae. Ageing of tsetse in a spray block before the end of the sequential spraying is usually carried out in order to ensure that any flies detected are newly emerged ones. If older flies are detected this may be due to either a gap of unsprayed habitat between swaths (possibly resulting from navigational error), or ineffective insecticide or reinvasion from neighbouring unsprayed areas. Davies (1978) used ovarian ageing techniques to evaluate aerial spraying against *G. m. centralis* in Botswana, but a difficulty that he encountered was the small number of flies that survived spraying or which emerged shortly after spraying, and could easily be missed. A further difficulty he reported was the time-consuming nature of the technique.

For monitoring trap or target control schemes, the declining age of a population will enable the additional mortality imposed on the population to be estimated, and the time to extinction, or adequate control, of the population to be predicted. This is not quite as simple as it sounds as fluctuations due to seasonal effects are usually superimposed on the effects of the control technique and are usually difficult to disentangle.

Three methods have been developed to estimate the age of tsetse: wing-fray analysis of the rate of wear of the wings; the rate and stage of development of the ovaries of the female fly; and the rate of accumulation of fluorescent pigments (pteridines) in the heads of tsetse. All these rates of change or stages of development may be calibrated to the age of adult flies.

The age of puparia can also be assessed by dissection, using a method described by Bursell (1959b), though this technique has seldom been used. Bursell developed it for his physiological studies on water balance of tsetse.

Wing-fray

Jackson (1946) was the first to describe the method of ageing *G. m. centralis* from the progressive fraying of the trailing edge of the wing with age as a reasonable indication of mean population age. Several workers have subsequently demonstrated a correlation between ovarian age and wing-fray (Saunders, 1962; Harley, 1967a,b; Challier, 1973; Ryan *et al.*, 1980; Leak and Rowlands, 1997). Ryan *et al.* (1980) concluded that the rate of wing-fray with age varied between species. Furthermore, the wings of females frayed more quickly than those of males for all the species investigated and there was evidence of seasonal variation in the rate of wing-fray. In contrast, at Shinyanga, no significant seasonal differences were observed in the rate of wing-fray of *G. m. centralis* (Jackson, 1950b). Despite these drawbacks the method is a convenient, if crude, field method for ageing males and there are few practical alternatives.

Ovarian ageing

Russian entomologists first showed that female *Anopheles* mosquitoes could be aged accurately by determining the number of follicular relics present in the follicular tubes after successive ovulations. Subsequently, Saunders (1960, 1962) and Challier (1965) adapted the method for ageing tsetse flies. The technique is based upon the changes occurring in the ovaries during successive gonotrophic cycles. Each ovary of *Glossina* contains two polytrophic ovarioles, each of which contains a single egg follicle at any one time. The single eggs are produced alternately from right and left ovaries. No follicles in either of the ovaries are ever at the same stage of development together. At the time the adult female fly emerges, the right ovary is larger than the left and produces the first egg in the sequence. The two ovarioles in the right ovary are designated A and B; those in the left ovary, C and D. Age determination depends primarily upon evidence that an ovariole has ovulated. This is shown either by an open expanded sac in the follicular tube or by a small follicular relic (corpus luteum) which remains on the posterior end of the follicular tube after the regression of the open sac. The number and position of these structures, combined with the relative sizes of the four egg follicles and the state of the egg or larva in the uterus, are then used for determination of the age of the fly. Serial follicular relics, which are seen with mosquitoes, are not found in tsetse. The method is straightforward only up to the end of the fourth ovulation. After this time the ovarian development is not only temperature dependent but also dependent upon the nutritional state of the female (Saunders, 1972); starved pregnant females take 30–40% longer for the interlarval period. Saunders (1962) showed that nulliparous females (category 0) could be divided into 0A (less than 5 days old) and 0B (more than 4 days old) according to whether the first egg follicle was less than or greater than 0.6 mm in length. This work was first done for *G. morsitans* and then applied to *G. pallidipes*, *G. palpalis* and *G. brevipalpis*. A detailed method of dissection of female *G. tachinoides* for ovarian ageing was described by Itard (1966b). He noted the presence or absence of sperm in the spermathecae and the volume occupied, the presence or absence of an egg or larva in the uterus and the instar of any larva present. In this way, the age of a female could be reliably estimated up to age 80 days. There are slight differences between tsetse species in egg follicle lengths related to species size (measured from wing vein length). When using size of the first egg follicle to categorize ovarian age into 0A or 0B, there could be significant underestimation of the proportion of females of *G. pallidipes* and *G. brevipalpis* in the 0A category (less than 5 days old) (Wall, 1990).

Pteridine ageing

In a review of methods for ageing insects, Tyndale-Biscoe (1984) included the description of a method for ageing *Stomoxys calcitrans* from the accumulation with age of pteridines (fluorescent eye pigments) first described by Mail *et al.* (1983). The technique was subsequently adapted for ageing adult male and female tsetse (Lehane and Mail, 1985). This method is simple and allows large samples to be processed quickly. Standard curves for the regression of pteridine fluorescence showed that the technique was reliable for *G. m. morsitans*, *G. p. palpalis* and *G. tachinoides,* but not for *G. austeni* (Langley *et al.,* 1988b). Comparing the method with ovarian dissection, Langley *et al.* (1988b) found that the variances increased with increasing age with both methods, resulting in poorer correlations for flies between 44 and 71 days old.

Pteridines were believed to accumulate linearly with age, in laboratory conditions, up to at least 63 days in males and 140 days in females. However, interpretation of results can be difficult, as pteridine content varies with fly size and its rate of accumulation increases with increasing temperature between 16 and 27°C. Fly size can vary seasonally in a predictable fashion but because of the large variation in pteridine content, and as its rate of accumulation is not linear in all species, there is no clear relationship with physiological age as was initially thought (McIntyre and Gooding, 1995). There is evidence that pteridine accumulation is under genetic control (McIntyre and Gooding, 1996). The pteridine ageing method is not suitable for accurately ageing very young flies because of the variability of the eclosion levels of different batches of tsetse flies. Previously, a mathematical correction of the data was performed before age determination, by adjusting the frequency distribution of age to bring the youngest group to zero. A technique for ageing such very young flies was developed based on their high levels of abdominal fluorescence (Msangi and Lehane, 1991). Pteridine fluorescence has also been used for age determination of the screw-worm fly, *Chrysomya bezziana* (Wall *et al.*, 1990).

Other ageing methods

Two other experimental methods of ageing tsetse have recently been described. In one, cuticular growth layers of the thoracic phragma (the dorsal partition between the thorax and the abdomen), corresponding to the age in days of adult tsetse, was assessed (Schlein, 1979). The technique, which was first developed for ageing mosquito populations, does not appear to have been tested with wild populations of tsetse. The second method, developed for ageing *G. tachinoides*, measures the level of grey coloration in the wings (de La Rocque *et al.*, 1996).

9.4   DENSITY-DEPENDENT AND DENSITY-INDEPENDENT MORTALITY

Environmental (density-independent) mortality factors, often correlated with rainfall or temperature, have a clear and important effect on tsetse populations. This led to a short-lived theory that population densities of *G. morsitans* were maintained by climatic change (Jackson, 1937). Long-term records (Glasgow and Welch, 1962; Jordan, 1964a) show that tsetse numbers fluctuate very little about their characteristic population levels compared with other insects. Density-dependent processes are thought to maintain population levels, and only one of the four demographic processes (birth, death, emigration and immigration) needs to depend on density for regulation to occur, although more than one of them may actually do so (Rogers and Randolph, 1985).

Rogers (1979) used Moran curves to estimate density-independent mortality of tsetse populations. Moran curves are plots of the logarithm of population apparent density at one time point against the logarithm of population density at a previous time point. The log transformation transforms the phase of a population's exponential growth to a linear form from which density-independent mortality can be estimated. Although Moran curves only apply to populations (such as laboratory colonies) where only density-dependent mortalities affect them, Rogers (1979) described a method from which density-independent mortality could be estimated. Rogers then examined the relationships between estimates of mean monthly density-independent mortality and climatic variables, particularly saturation deficit, for a number of sites. There were strong correlations with mean monthly saturation deficit and less strong correlations with mean monthly temperature. From these data Rogers (1979) was able to construct climograms for *G. m. submorsitans* in Nigeria and thus define environmental optima for this species, and in a similar way for *G. m. morsitans* in Zambia.

Very few studies have been carried out to examine natural density-dependent population control as such factors were thought to be less important, due to the low reproductive rate of tsetse. Residual fertility of tsetse must interact with a density-dependent mortality at some stage in the life cycle if there is to be any stable level of population, even if that stable level is affected by long-term changes in climatic cycles (Rogers, 1985a).

9.4.1   Predators and parasites

There has been some debate about the significance of density-dependent predation of adults and puparia of tsetse flies. Rogers (1974) formerly argued that it was likely to be important whilst Hargrove (1988, 1990) and Jordan (1986) were less convinced of its significance. Based on the levels of parasitism of puparia collected in the field, both Hargrove (1990) and Taylor (1979) assumed that puparial mortality was negligible and therefore adult mortality must be high. They ignored losses due to predation. Few attempts have been made to estimate mortality due to predators and parasites. In one early study, mortality due to parasitism of tsetse larvae by wasps was estimated as high as 25% over the larval/puparial period (Chorley, 1929). A later attempt to estimate puparial predation of *G. f. fuscipes* in Uganda (Rogers, 1974) was criticized, as it was thought that the methodology could influence the level of predation (Jordan, 1986). The experiment showed that predation (by various ant species, especially those in the genus *Pheidole*) was density dependent and was significant above a threshold density of four puparia per linear metre. There was no such density-dependent predation by invertebrates. The predators evidently responded to increasing tsetse densities by concentrating their search. Because of the criticism, Rogers and Randolph (1990) repeated similar experiments with *G. pallidipes* at Nguruman in Kenya. In contrast to the previous findings on puparial predation, and to data obtained by Dransfield and Brightwell (1989) at the same site, this time they found no evidence of density dependence. Explanations put forward were that natural puparial densities were too low for it to occur, or possibly adult predation rates were too high for density dependence to be detected. In a study at Nguruman, in Kenya, the loss rate between puparial and teneral adult stages was density dependent and was possibly due to puparial predation (Dransfield and Brightwell, 1989). The density-dependent loss rate between 3- to 8-day-old flies and 9- to 17-day-old flies probably resulted from starvation or emigration after competition at the host.

The occurrence, or otherwise, of density-dependent mortality was referred to in a deterministic mathematical model for tsetse control by sterile male release (Simpson, 1958). For the model, the important assumption was made that no density-dependent mortality occurred. Although no density-dependent factors affecting tsetse populations were known at the time, the possibility that they might exist was noted.

Robber flies of the family Asilidae and *Bembex* spp. wasps have been recorded as predators of *G. morsitans* adults, and a beetle, *Melyris pallidiventris*, was detected in Tanzania whose larvae were predators of laboratory-bred tsetse puparia (Nash, 1933b). Tsetse have been found in nests of hymenopteran wasps, *Sphex* spp. and *Synagris* spp., in the Democratic Republic of the Congo (Bouvier, 1936); and the spider, *Hersilia setifrons*, which wraps its prey in a 'shroud' of silk, has been observed preying on tsetse. By searching for the spider's conspicuous silk bags of prey, rough estimates of predation were made (Southon, 1959). Predation rates of 17% per week were estimated but Glasgow (1963) considered this figure to be too high, representing about half of the weekly natural mortality. Further details of predators and parasites are given in Chapter 18 and summarized in Table 18.1.

Seasonal variations in puparial mortality of *G. p. gambiensis* were detected during the dry season in West Africa, due to desiccation. There was also a seasonal abundance of predators and parasites of puparia at the beginning of the rains and at the time of the heavy rains (Challier, 1973). Challier therefore believed that puparial mortality was likely to play an important role in population dynamics of *G. p. gambiensis*. In an experiment to evaluate deltamethrin as an insecticide for tsetse control, about 90% of moribund flies were predated (mainly by ants) within 2.5 h (Holloway, 1990). Generally, at the edge of a species' (tsetse fly's) distribution the animal is likely to be too scarce for biotic factors such as predators or competitors to have any marked effect.

9.5 SURVIVAL AND MORTALITY RATES

Epidemiological models for vector-transmitted diseases are particularly sensitive to changes in tsetse mortality rates (Rogers, 1988a,b, 1990). Their calculation is therefore of interest both for measuring the importance of factors determining the distribution and abundance of populations and for understanding the epidemiology of disease transmission. Some of the methods employed to estimate mortality of tsetse populations are described below.

9.5.1 Saunders's probabilistic method

Saunders (1967) showed how a survivorship curve could be obtained from the probable age composition of a population sample by ovarian ageing and described how this can be used to calculate daily survival rates. A number of assumptions had to be made and their validity was discussed in that paper. One of the main assumptions was that the sampling was unbiased.

Unfortunately, all sampling methods have some bias. For estimating potential reproductive capacity, Saunders relied on data for age-specific fertility rates from tsetse colonies as such information was not available for wild tsetse populations. It was assumed that if a tsetse population was in a stationary, stable state, the age structure could be used for estimating survival rates. Saunders also assumed a logarithmic adult death rate, and a survivorship curve for a tsetse population was calculated by constructing a straight line through the logarithms of the number of flies in each ovarian age category, using $\log 10_n$ to eliminate a negative index. A curve of the survivorship values, l_x, was then calculated with the zero of the abscissa starting at the time of larviposition. Teneral flies were excluded so the logarithmic adult death rate started after the first bloodmeal, assumed to be taken 2 days after the emergence of the adult fly. A 10% mortality between emergence and time of the first bloodmeal was also assumed. Estimates of age-specific fertility rate (m_x) are required in order to calculate the intrinsic rate of increase (r), and it also has to be assumed that the probability of a female producing a daughter at each birth is 0.5. The estimate for m_x obtained in this way would be the maximum as it does not account for the various reproductive losses which take place in the wild. Colony data may give a minimum figure unless performing optimally, so the actual curve might lie somewhere in between.

9.5.2 Van Sickle's method – the Euler–Lotka equation

Van Sickle (1988) claimed that several published estimates of the rate of increase of tsetse populations based on sampled ovarian age distributions were invalid because of the assumption that $r = 0$. Estimates considered incorrect were those of Saunders (1967), Taylor (1979), Ryan (1981) and Allsopp (1985a). Some of those methods are described below. Van Sickle and Phelps (1988) suggested an alternative method based on the Euler–Lotka equation, but, although the method was valid, its sensitivity to sampling errors rendered it impractical. This method (described in section 9.6.2) was later used by Williams *et al.* (1990a).

Ryan (1981) summarized previous studies on tsetse population growth rates and showed relationships between population parameters and temperature for several tsetse species. He discussed these relationships in terms of their relevance to tsetse control.

9.5.3 Challier and Turner's method

Challier and Turner (1985) described a method of calculating survival rates in tsetse populations based on the geometric mean survival per ovarian cycle of an age-graded sample and the duration of the interlarval period. This method involved refining the Saunders method (section 9.5.1) by estimating a daily survival rate and simplifying the mathematics from which it was derived.

The survivorship curve is used in the form:

$$\log y = a + bx \tag{9.3}$$

where y is the number of females in each ovarian category x, a is the log of the number of flies on day 1, and b is the log of the survival rate per ovarian cycle. If the duration of each ovarian cycle is λ days, then the daily survival rate, Φ, is given by:

$$\Phi = \text{antilog } b/\lambda. \tag{9.4}$$

The duration of the ovarian cycle is temperature dependent and can be calculated from a formula given by Glasgow (1963). Challier and Turner gave an example using data for *G. p. gambiensis* from Burkina Faso for which the equation:

$$\log y = 2.54551 - 0.1192x \tag{9.5}$$

was obtained for the survivorship curve. The interlarval period was between 9 and 10 days and the daily survival rate was therefore between antilog $-0.1192/9 = 0.970$ and antilog $-0.1192/10 = 0.973$.

A simplified calculation of the survival rate per ovarian cycle is derived from the geometric mean formula:

$$\Phi = \text{antilog } \{\log [4 + 4n] \ldots + \log [7 + 4n]\} - \{\log [0 + (4 + 4n)]$$
$$\ldots + \log [3 + (7 + 4n)]\}/16 \, \lambda]. \tag{9.6}$$

Challier and Turner (1985) suggested that an alternative to the geometric mean formula would be an arithmetic mean, which would be easier to calculate. The arithmetic mean formula is as follows:

$$\text{Arithmetic mean survival rate/4 cycles}$$
$$= (4 + 4n/63 + 5 + 4n/44 + 6 + 4n/33 + 7 + 4n/17)/4 \tag{9.7}$$

$$\text{Daily survival rate } (\phi)$$
$$= \text{antilog } [\log (\text{arithmetic mean survival rate})/4 \text{ cycles}]/4\lambda \tag{9.8}$$

9.5.4   Climate and mortality method (Rogers *et al.*)

In Nigeria, Moran curves were used to estimate density-independent mortality rates from long-term fly-round data. Estimates varied seasonally and were most closely correlated with saturation deficit but also depended on temperature (Rogers and Randolph, 1986b). The difference between the maximum rate of increase of a tsetse population and the observed increase provides a measure of mortality over the period concerned. Enclosing areas of equal mortality with contours allowed bioclimatic optima to be identified, which corresponded to those calculated experimentally by Nash (1937). An alternative approach was to try to estimate current mortality rates from the age structure of the population rather than from changes in total sample sizes. When estimated in this way, the mean monthly mortalities of *G. p. palpalis*, caught from biconical traps in Côte d'Ivoire over an 18-month period, were, again, significantly correlated with mean temperature and saturation deficit (Rogers *et al.*, 1984). In contrast, Gouteux and Laveissière (1982) found a stronger correlation over a shorter period with relative humidity rather than with temperature.

Rogers and Randolph (1991) found a significant correlation between data from meteorological satellites (normalized difference vegetation index, NDVI; and saturation deficit) and mortality rates of tsetse estimated using Moran curve techniques. In this way, satellite data could be used to predict both mortality rates and abundance of tsetse over large areas of Africa to produce maps of high risk areas of disease transmission.

Mortality increases both with increasing saturation deficit and with increasing maximum temperature. The decrease in tsetse populations throughout the dry season is due to a general decrease in longevity associated with rising saturation deficit coupled with a decrease in the breeding rate associated with the cold dry season. In the late dry season dangerously high maximum temperatures, leading to a minimal tsetse population, further reduce longevity.

9.5.5 Effect of sampling bias on estimates of mortality

Sampling biases can affect the accuracy of mortality estimates, as the sample age composition has to be representative of the population age structure. Challier and Turner (1985) discussed the errors in estimating the ovarian age category 0 in particular, resulting from sampling bias. They suggested that if category 0 was omitted from the calculations together with the (4 + 4n), a more realistic estimation of survival rate would be obtained. Underlying assumptions are that the mortality rate in each age category is constant, and that the population is relatively stable. These conditions may not be met where large seasonal (climatic) differences occur. In Zimbabwe, catches from odour-baited traps were biased in that they caught more old flies (*G. pallidipes* and *G. m. morsitans*); therefore, mortalities calculated from age distributions of the trap samples would be underestimates (Hargrove, 1991).

According to Challier and Turner (1985), the Saunders method would overestimate the survival rate with large sample sizes and underestimate it with small sample sizes. Knipling (1979, in Hargrove, 1988) stated that tsetse populations could not possibly survive unless the adults have a very low death rate. Hargrove (1981) and Rogers *et al.* (1984) estimated an approximate daily death rate of about 3.5% for a non-decreasing population. Theoretically, if one could measure a total population size accurately, the actual increase from one month to the next could be compared with its potential increase and the difference taken as the level of mortality suffered (Rogers and Randolph, 1986b).

9.5.6   Size and mortality

Physical factors such as size or weight of adult tsetse can give an indirect measure of mortality of a tsetse population as they are closely correlated with the nutritional state of the flies. A fly population under nutritional stress will produce smaller puparia and adults, which are less likely to survive. In addition, the fraying of wings, which increases with age, may decrease a fly's ability to obtain a bloodmeal or avoid predators. Thus, wing-fray might impose a physiological limitation on the lifespan of tsetse and contribute to population regulation (Allsopp, 1985b).

After desiccation was ruled out as a major cause of mortality in tsetse populations, other potential causes for population decreases in the hot season were investigated. Starvation is the most likely cause of mortality, and is closely related to the size of flies. Larger flies have greater reserves of fat and are less likely to die from starvation than smaller ones. Thus, tsetse populations are characterized by marked, and regular, fluctuations in mean size and fat content symptomatic of variations in nutritional stress. The link between size and mortality can be used as a means of estimating mortality rates. Selection for size occurs in natural tsetse populations, smaller, teneral flies being much more likely to suffer mortality (Glasgow, 1961c). Gooding and Hollebone (1976) estimated the heritability of adult weight in *G. m. morsitans* and concluded that only 9–16% of the weight variation was due to additive genetic factors whilst environmental factors, particularly maternal nutrition, were the most important, accounting for most of the remaining variation. There is a high correlation between blood consumption of the mother and puparial weight, with bloodmeal size accounting for 80% of the variance.

Size-dependent mortality was first shown to occur in *G. morsitans* (Jackson, 1948a) and was further demonstrated in *G. pallidipes* at Nguruman, Kenya (Dransfield *et al.*, 1989). According to Jackson (1952), seasonal variations occurred in the size of tsetse, estimated from wing-vein length, and this estimate of size was correlated with saturation deficit 2 months previously. At Nguruman, seasonal changes in size, also measured using wing vein length, were associated with a fourfold change in population mortality rates (Dransfield *et al.*, 1989). Changes in the size of nulliparous females and wing-fray category 1 males were correlated with the relative humidity 2 months before capture. Size was thus correlated with density-independent mortality acting on the parent population 2 months previously. However, only about 50% of the variability in mortality rates could be explained by variation in wing-vein length, making it an unsatisfactory predictor (Dransfield *et al.*, 1989).

Vegetation cover, distribution of wild hosts and temperature experienced by the puparia during development, as well as variations in maternal nutritional stress, can also indirectly affect size of the emerging adult fly. Thus, size could be used as an index of the intensity of density-independent mortality on a parent population. Phelps and Clarke (1974) reported seasonal elimination of some size classes (estimated from measurements of thoracic area) of male *G. m. morsitans* in Zimbabwe; smaller flies were selected against for about 7 months of the year, and this selection was temperature dependent. Small individuals of both sexes appear to be selected against in nature (Glasgow, 1963; Phelps and Clarke, 1974). Large individuals may also be selected against (Glasgow, 1961c) but, as small size reflects a decreased availability of nutrients for energy, the selection against small flies is likely to be of most significance.

9.5.7 Feeding-related mortality

A model of feeding-related mortality (Randolph *et al.*, 1992) was referred to in Chapter 7. This model balances the effects of death by starvation, after a period set at 6 days without successful feeding, with mortality specifically associated with feeding, in order to determine the optimum theoretical feeding strategy for tsetse. The optimum strategy is determined as that which allows the maximum reproductive output, and the model shows that the best strategy is to make no attempt to feed for 3–4 days after the previous meal, followed by attempts to feed that are likely to have a high probability of success.

In the model, life-time fertility (F) – the number of female puparia produced by a single female tsetse – is given by:

$$F = 0.5e^{-17d}/(1 - e^{-9d}) \qquad (9.9)$$

where d is the daily adult mortality rate, made up of daily background mortality, incorporating all mortalities not associated with the feeding process, a starvation mortality, and a feeding mortality assumed to operate only when the flies find and feed on a host. The average daily starvation mortality, s, is give by an approximation:

$$s \approx e^{-r(m-n)}/m \qquad (9.10)$$

where n is the time after their last meal that flies start to look for their next meal; if they have failed to obtain a meal by day m, they die of starvation. Days $n - m$ thus represent the feeding phase; r is the rate at which they find and feed on hosts. A proportion, u, of flies will not have fed by day m. The equivalent daily mortality due to feeding, g, is:

$$g \approx (1 - e^{-r(m-n)}p/i \qquad (9.11)$$

where p is the proportion of flies that do find a host ($1 - u$) which suffer a feeding mortality. This mortality operates over the mean feeding interval, i, of the flies that successfully find a host.

9.6 POPULATION GROWTH RATES

The growth rate of any animal population is determined by the difference between the birth and death rates. Survival rates affect longevity, population density and thus vectorial capacity. Knipling (1979) estimated that tsetse, with a lower reproductive potential than any other insect, would have a maximum growth rate of only tenfold per year; however, other workers (Rogers, 1979; Hargrove, 1988) calculated a much higher potential growth rate per annum. Annual growth rates of laboratory colonies of 400–550-fold have been reported (Jordan and Curtis, 1972), whilst Turner and Brightwell (1986) estimated a 200-fold annual growth rate for *G. pallidipes* in the Lambwe valley, Kenya.

Taylor (1979) used seasonal age measurements from ovarian ageing of a wild tsetse population to construct a life table for *G. m. morsitans* in Zimbabwe. This life table was used to estimate the intrinsic rate of natural increase r_m, his estimate of which was 0.01204 (Table 9.2).

Allsopp (1985a) estimated the natural rate of increase (r_s) of a population of *G. m. centralis* in Botswana. The value of r_s varied from −0.0099 to 0.0022, with a mean of −0.0053, reflecting a declining density during the study period (Table 9.2). The cold dry season was least favourable for the fly and would therefore be the period of choice for insecticide application for tsetse control. Allsopp calculated the rate of increase from entire age distributions of the population. Larval and puparial mortalities had to be estimated, as no reliable information was available for those age groups. Tsetse were sampled on fly-rounds using hand-nets; females were aged physiologically as well as by wing-fray category and their reproductive condition was noted. Tables of the monthly age composition were then constructed. Wing-fray and ovarian age for each monthly sample were closely correlated. Survivorship values (I_x) were calculated from the monthly age distributions, assuming female age-specific fecundity (m_x) to be a constant 0.5 (the maximum possible). Age intervals (x) were taken as 10 days for larval development and age groups 1–7. Age group 0 was taken as 15 days. The duration of each age group (P) was calculated each month according to the formula of Glasgow (1963) based on mean monthly temperature (Chapter 3, section 3.4.2). Life tables were calculated from the above and r_s was calculated according to a method described by Southwood (1978). Monthly estimates of the rate of increase were correlated with temperature and relative humidity.

Table 9.2. Tsetse population growth rates.

Rate	Tsetse species	Estimate	Reference
R_m	*G. m. morsitans*	0.01204	Taylor, 1979
R_s	*G. m. centralis*	−0.0099–0.0022 (−0.0053)	Allsopp, 1985a

R_m is the intrinsic rate of natural increase. It is the instantaneous growth rate coefficient expressed when the population is growing in a constraint-free environment and age structure has become stable. R_s is the capacity for increase; it is an approximation for R_m.

Dransfield *et al.* (1990) used similar methods to monitor the control of tsetse using odour-baited traps at Nguruman in Kenya. They estimated mortality rates due to trapping at 4–5% per day, based on age distributions and numbers of flies killed in the traps. Female mortality rates were estimated from the relative numbers in age categories 1–7+ after first pooling the numbers in age categories 4+ to 7+ and then using a maximum likelihood estimation (Dransfield *et al.*, 1986b). The abortion rate in *G. pallidipes* at Nguruman was estimated at between 1 and 5% (Dransfield and Brightwell, 1989)

9.6.1 Hargrove's model

Hargrove (1988) estimated the rate of growth of tsetse populations by calculating the dominant eigenvalues of appropriate Leslie matrices, which was the first use of this technique for modelling tsetse population growth. The theory is relatively simple and relates population growth rate to the factors determining birth and death rates, the difference between which determines the growth rate. His model was based on the four variables of pre-adult and adult survival probability, interlarval period and puparial duration. These four factors were used with Leslie matrices to estimate population growth from birth and death rates. Hargrove's model used data taken from the literature for stage-specific mortalities and fecundity. The log of the growth rate ($-R$) varied approximately linearly with adult and pre-adult death rate and linearly also with the log of fecundity and of puparial duration. From his model, Hargrove concluded that if one could impose, and sustain, an added mortality of 4% per day on any female tsetse population then it must become extinct regardless of the strength of density-dependent processes. He considered that in most field situations an added 2–3% mortality only would be required if it was sustained. Many insecticidal control techniques give a higher but unsustained additional mortality. Odour-baited targets give a much smaller increased death rate but this is sustained over a much longer period. A rule of thumb was drawn from Hargrove's model, that a change of 1% in the mortality of a tsetse population changes the growth rate by about one order of magnitude. This agreed with other estimates (Langley and Weidhaas, 1986) and with unpublished data of Vale. A long puparial period cushions a tsetse population against high adult mortality at the cost of allowing only low growth rates when adult mortalities are low.

9.6.2 Williams's model

Williams *et al.* (1990a,c) believed mortality rates to be central to studies of tsetse population dynamics and the development of tsetse control programmes, and discussed the estimation of mortality from life-table data. For a population at equilibrium, with a stable age distribution, the age-specific mortalities may be estimated directly from the number of individuals in each age class, but a correction must be applied when the population is growing or declining. They used the Euler–Lotka equation, which relates age-specific mortality and fecundity to the overall growth rate of the population, to study the loss rate of *G. pallidipes* as a function of puparial mortality, adult mortality and mortalities applied to each age class separately. They produced a simulation model to quantify and set limits on the precision of estimates of mortalities when the mortalities themselves are changing (Williams *et al.*, 1990a). However, the assumptions that the population is changing at a constant rate and that the age distribution is stable are not generally true.

The Euler–Lotka equation relates the growth rate of a population (r) to the age-dependent mortalities, $m(\tau)$, and fecundities, $\beta(\tau)$. Williams *et al.* (1990c) use the survivorship, $s(\tau)$, which is equal to $-m(\tau)$. The following integral form of the Euler–Lotka equation was used:

$$\int_0^\infty \beta(\tau) \exp\left\{\int_0^\tau s(\tau')d\tau' - r\tau\right\}d\tau = 1 \qquad (9.12)$$

Provided $\beta(\tau)$ and $s(\tau)$ are known, the equation can be solved for the growth rate of the population (r). This provided a 'convenient starting point for the analysis of life-table data' (Williams *et al.*, 1990a) and a further equation was derived allowing the relationship between growth rate, adult mortality and puparial mortality of tsetse flies to be determined.

9.7   FLY MOVEMENT AND DISTRIBUTION WITHIN HABITATS

Fly movement is important, at least in the short-term regulation of fly numbers. This is especially so for *G. pallidipes,* which is a particularly mobile tsetse species. Immigration of *G. palpalis* into villages in Côte d'Ivoire can also be a density-dependent mechanism of population regulation (Rogers and Randolph, 1984a,b).

In the pre-forest zone at Vavoua, in Côte d'Ivoire, meteorological conditions were optimal for tsetse only in forest patches and cocoa and coffee plantations (Gouteux and Buckland, 1984). In these areas, breeding occurred continually, and both teneral and male flies were present in trap catches in proportion to their abundance in the total population. Such sites, however, are not ideal for feeding so *G. palpalis* ventures from them to villages with large domestic pig populations.

Gouteux and Laveissière (1982), studying the ecology of tsetse in relation to human sleeping sickness in the pre-forest zone of Côte d'Ivoire, used the Shannon Diversity Index (Shannon and Weaver, 1948) for estimating the diversity of populations in trap catches or of biotypes frequented by tsetse. The Shannon index has the formula:

$$I_{Sb} = -\sum_{i=1}^{n} p_i \log_2 p_i \; (\log_2) \qquad (9.13)$$

where n represents the number of species (or of biotopes) and p_i the relative frequency of each species (or biotope). This formula may also be expressed as:

$$H' = -\ni p_i.\ln p_i \qquad (9.14)$$

where H' = diversity and p_i = the proportion of individuals in the ith species.

Glossina palpalis occurred at high mean apparent densities, showing only small seasonal variations, in peridomestic biotopes around village water points, and was the only species that penetrated the centre of villages. In contrast, *G. pallicera*, which was more abundant in the cocoa plantations, occurred at increasing mean apparent densities with increasing distance from villages. *Glossina nigrofusca* had a similar distribution to *G. pallicera* but at lower densities. Male *G. palpalis* dominated in the more shaded biotopes, such as forests and water points, whilst the reverse was the case in more open areas, such as the edges of villages, possibly due to differing phototactic behavioural responses. Both sexes of *G. pallicera* were distributed more homogeneously in all biotopes. Calculation of survival rates based on age structure showed that important seasonal variations in distribution of *G. palpalis* took place, according to changing climatic conditions. Thus, females were found in shady areas such as plantations, forests and, above all, water points in the dry season. Survival rates were greatest during the rainy season and lowest during the dry season.

An alternative form of the formula, used by Dagnogo *et al.* (1997), was:

$$I_{Sb} = 3.322 \; [\log (Q - 1)]/Q \; \text{Sum} \; q_i \log q_i \qquad (9.15)$$

where Q = the number of the sample and q_i = the number of the species i. Using this formula, Dagnogo *et al.* examined the diversity of three tsetse species in three different biotopes of the forest belt of Côte d'Ivoire (gallery forest, plantations and a village) to determine if there was any change in species composition in relation to changing land use patterns. Two species, *G. nigrofusca* and *G. pallicera*, were found to be quickly disappearing from plantations and gallery forest. Deforestation increased the predominance of *G. palpalis*, which could be of significance in the transmission of human sleeping sickness.

Chapter 10

Odour Attractants

One of the most important developments in the 1970s and 1980s was the identification and synthesis of odour attractants for some species of tsetse, which have greatly increased the efficiency of traps and targets, making tsetse control by these means a feasible alternative to the use of insecticides. The development of odour attractants is reviewed in this chapter.

The possibility of using odour attractants for sampling or controlling tsetse has been recognized for a long time. For example, 85 years ago a trap was made for use with a bait consisting of a central wick soaked in a solution of human or animal skin secretions (Balfour, 1913). Further development of odour baits for attracting tsetse took place in the 1940s, using skin washings and urine of various animals. It was observed that cow urine was attractive to tsetse in Uganda as early as 1948 (Chorley, 1948). Chorley used the 'scent' of cattle dung and urine to attract *G. pallidipes*, but trap designs were then rather inefficient and the attractiveness of odour baits was insufficient for any useful developments to arise. In other experiments, a lanolin product from sheep's wool and a substance from pig skin scrapings and hair sprayed on to vegetation and stones were examined. These increased catches of *G. pallidipes* from fly-rounds using a black cloth screen carried on the route along which the attractant substances had been sprayed (Langridge, 1960). Lanolin almost doubled catches of *G. pallidipes* whilst the pig extract almost quadrupled the catch.

An indication of the increase in knowledge of tsetse odour attractants in recent years can be obtained from the small amount of information on this subject in *The African Trypanosomiases* (Mulligan, 1970) in comparison with current literature. Much of this increase in knowledge has arisen from improved techniques of analysis and detecting components of odours which stimulate tsetse. These include the use of wind tunnels to investigate responses of tsetse in odours, development of electric nets to make more accurately quantified estimates of attraction, electroantennograms to measure electrical responses of antennae to odours, and high pressure liquid chromatography to isolate and identify chemicals in substances such as animal urine or skin washings.

Recently it has been suggested that, with potent odour baits, simply placing a number of baits scattered throughout an area – without any target, trap or insecticides – might control tsetse. The tsetse flies would respond to the baits, flying around and using up their energy reserves without finding a host, and would die of starvation. This technique, termed the Cheshire trap, is based on studies of the energy reserves of tsetse and the estimated short time that tsetse can afford to spend in daily flight (D.J. Rogers and S.E. Randolph, Nairobi, 1988, personal communication). Clearly, this method would be unlikely to work in areas of adequate host density.

Odour attractants can be classed into three types:

- Those associated with animal breath, such as acetone, octenol and CO_2.
- Those associated with urine, such as the phenols.
- Those associated with skin secretions, such as sebum.

The latter attractants have proved less promising than products from breath and urine.

Research into odour baits was stimulated after it was shown that a fly's ability to detect a host depended more upon odours than upon visual attraction. At about the same time, in the 1970s, the biconical trap was being developed. Some of the substances previously investigated, such as urine and skin washings, were re-tested, with the objective of isolating their active fractions and synthesizing attractants for use with traps and targets.

Experiments in Zimbabwe, with tons of live oxen in a pit, provided one of the greatest stimuli to the development of odour attractants (Vale, 1974d; Hargrove and Vale, 1978). One experiment included bait of 37 cattle, 22 goats, 43 sheep, a donkey and a buffalo (11,500 kg of stock) in the pit. Hargrove and Vale found that, with the attraction of this odour, large catches of tsetse could be maintained for extended periods. There was a linear relationship between the log numbers of tsetse caught and the log weight of the livestock bait. This indicated that either the population density was very high, or the area into which the odour passed was extremely large, or that recruitment into the sampling area was very rapid. The latter explanation seemed the most likely, supporting the feasibility of using a few thinly spread, but highly effective, baits for tsetse control. This optimism has been borne out in view of the progress in trap/odour bait technology for tsetse control (Chapter 16).

Odours from ox, donkey, goat, sheep, buffalo, bushbuck and bushpig were attractive to *G. m. morsitans* in Zimbabwe but human odour was repellent. The head region of the animals was the most attractive, but apparently not only due to the CO_2 emitted in breath.

10.1   ODOURS IN HOST BREATH

10.1.1   Carbon dioxide

In one of the early observations on the effect of CO_2 on tsetse (Bax, 1937), exhaust fumes from a lorry, presumably containing concentrations of CO_2, seemed to stimulate *G. swynnertoni* 100 m downwind, whilst a small black screen at about 35 m resulted in a response equivalent to that from six oxen.

In later experiments in Uganda, catches of *G. pallidipes* in Morris traps were significantly increased when the traps were baited with CO_2 (Rennison and Robertson, 1960). Solid CO_2 was used for the experiment, and for obvious logistical reasons this was not recommended for widespread use. Carbon dioxide appears to be the only odour substance in ox breath which increases the landing responses of tsetse to targets in the laboratory (Green, 1993). As CO_2 concentrations fall rapidly to background levels after dispersing from a potential host, it is likely to be only a close-range attractant (Torr, 1990).

10.1.2   Acetone

Acetone, also a component of ox breath, was identified as a relatively cheap and practical tsetse attractant. A mixture of CO_2 and acetone released at a visual target in Zimbabwe resulted in a fivefold increase in catches of *G. m. morsitans* and *G. pallidipes*, using an electric net to kill the attracted flies. Also in Zimbabwe, it was shown that tsetse which fly upwind to the end of an odour plume and find no visual target continue upwind for a few metres and then return downwind for about 8 m (Vale, 1984). The maximum distance from which tsetse responded to CO_2 and acetone (released at concentrations of 0.5 g min^{-1} and 4 l min^{-1}, respectively) was estimated to be 45 m. In West Africa, when CO_2 was released at a dosage of 0.5 to 20 l min^{-1} at the base of biconical traps, catches of *G. tachinoides* increased by a factor of up to 3.2 (Galey *et al.*, 1986).

Butanone can be used as an alternative to acetone; it has a slower rate of evaporation and may therefore be more economical. It was successfully substituted for acetone during a tsetse control trial with targets in Zambia (Willemse, 1991).

10.1.3   1-Octen-3-ol (octenol)

Octenol is another attractant that has been isolated from ox breath. Experimentally, it elicits high responses by some species of tsetse to electroantennograms, increases upwind flight in wind tunnels and results in increased attractiveness of ox odour and mixtures of carbon dioxide and acetone. It is, however, repellent at high doses (Vale and Hall, 1985).

Octenol has been isolated from many natural sources, mainly plants and fungi, and has been identified in volatiles from clover and alfalfa (Hall *et al.*, 1984).

Catches of very young female tsetse and relatively recently fed males are increased when octenol is used. As these two groups of flies are usually under-represented in trap catches, the addition of octenol may provide a more representative sample of the tsetse population (Randolph *et al.*, 1991a).

10.2 ODOURS IN HOST SKIN SECRETIONS

Skin washings from pigs, concentrated by evaporation with petroleum spirit, were used as an attractant for use with 'scented traps' to catch *G. f. fuscipes* and *G. pallidipes* in Uganda (Persoons, 1966). Catches of *G. pallidipes* were significantly higher than in unbaited traps, but catches of *G. f. fuscipes* only increased when fresh, untreated pig washings were used. There was only a short residual effect of the attractants, which limited their usefulness. The potential of skin secretions, from oxen, was reassessed in Zimbabwe (Warnes, 1990b). The sebum increased catches of *G. m. morsitans* and *G. pallidipes* in F3 traps by 80 and 29%, respectively, but the results were variable. In this experiment screens impregnated with sebum were sited adjacent to the traps. Active components of sebum were in the phenolic fraction, in contrast with those of urine, which are in the acidic fraction. In laboratory studies of behavioural responses of tsetse to sebum, Warnes (1995) observed that following tarsal contact with sebum the duration of contact with the target was reduced, but flies tended to return to the treated target more frequently. This was in contrast with video observations by Packer and Warnes (1991) but these may have been confused by the presence of other flies (Warnes, 1995).

10.3 ODOURS IN HOST URINE

In a comparison of buffalo and cow urine as attractants, cow urine increased biconical trap catches of *G. pallidipes* by 1.8 times whilst catches using buffalo urine increased by 9.6 times compared with unbaited control traps (Owaga, 1985). Saini (1986) showed that the antennae of *G. m. morsitans* responded to buffalo urine and fresh urine was as attractive as 4- or 8-day-old urine. Acetone, when added to the two odour treatments, increased catches by 2.8 times. Owaga (1984, 1985) found that older urine was more effective than fresh urine.

In Zimbabwe, the responses of tsetse to urine and host odour residues collected on bedding sacks were investigated (Vale and Flint, 1986). Catches of *G. pallidipes* and *G. m. morsitans* were roughly doubled by the presence of two bushpig bedding sacks or jars containing 200 ml of urine from ox or buffalo. The urine attracted *G. pallidipes* to the traps, following which the residues appeared to increase the trap-entering responses. The residues lasted at least 2 months but urine poured on the ground quickly became ineffective. Octenol did not affect the catches at the concentrations used.

No components of ox urine have been identified that significantly affect landing responses and there may be a dose-dependent repellent effect reducing, rather than increasing, landing responses to targets (Green, 1993). Two litres of urine were repellent to *G. pallidipes* and *G. morsitans* in Zimbabwe (Vale and Flint, 1986).

The components of cattle and buffalo urine attractive to tsetse are shown in Table 10.1. The fractions of buffalo urine most attractive to tsetse are phenolic, consisting of seven simple phenols including phenol itself (Hassanali *et al.*, 1986). Female *G. m. morsitans* were more responsive than males to chemicals isolated from buffalo urine, and responded to a wider range of chemicals. Of seven phenolic compounds identified from buffalo urine, 4-methyl-phenol and 3-*n*-propylphenol seemed to be the most attractive and appeared to act synergistically (Owaga *et al.*, 1988). Components of cattle urine (also phenolic components) were identified by Bursell *et al.* (1988) whilst Saini and Dransfield (1987) identified chemicals to which male *G. m. morsitans* responded in electroantennogram experiments. Electroantennogram responses can vary between species as well as with age and hunger state (Den Otter *et al.*, 1991).

Phenolic compounds derived from cattle urine are also attractive to *G. tachinoides*; 4-methyl-phenol and octenol in a 3:1 ratio remain attractive for a long period, increasing biconical trap catches by 2.5 times for up to 10 weeks. Cow urine alone is more effective in attracting *G. tachinoides* than fractions of urine, and the phenolic fractions were closest in their attractive power to pure urine (Filledier and Mérot, 1989a,b). The latter workers specifically used fractions isolated from Baoulé rather than zebu cattle, as it has been suggested that Baoulé cattle owe part of their trypanotolerance to a lower attractivity to tsetse flies. Previously, Baoulé urine had appeared more attractive to tsetse than zebu urine (Filledier *et al.*, 1988). The attractivity of

Table 10.1. Attractive components of host odours.

Cattle urine	Buffalo urine	Other chemicals
4-Methyl-phenol	4-Methyl-phenol	Acetone
Ethyl-phenol	3-*n*-Propyl-phenol	Butanone
Propyl-phenol	Phenol	Formaldehyde
		Methyl-ethyl-ketone
		Methyl-vinyl-ketone
		1–Octen-3-ol
		Pentanal

host odours from humans, domestic pig and a Baoulé cow to *G. tachinoides* was dependent upon the number of animals used (Mérot *et al.*, 1986). Cow urine was less effective than in other trials in attracting *G. tachinoides* in Ethiopia (Slingenbergh, 1988).

10.4   COMBINATIONS OF ATTRACTANTS

In Zimbabwe, catches of *G. pallidipes* and *G. m. morsitans* in traps baited with acetone and octenol were increased by the addition of a synthetic mixture of eight phenols found in cattle urine to a level equal to or greater than those with natural urine (Vale *et al.*, 1988). Addition of natural urine to the synthetic mixture did not increase catches further, indicating that the phenols account for essentially all the attractiveness of cattle urine. The naturally occurring components essential for attractiveness are 4-methyl-phenol and 3-*n*-propylphenol, whilst 2-methoxyphenol reduces attractiveness. In these experiments, mixtures of phenols that increased the attractiveness of traps to tsetse showed similar effects with targets but at a slightly reduced level (Vale *et al.*, 1988). Phenols have two effects: they increase attraction from a distance and enhance the trap-entering responses of tsetse that arrive near the traps. With respect to *G. pallidipes* there is a synergism between the two active phenols.

In Kenya, neither acetone nor cow urine increased catches of *G. longipennis* when used alone, but together they increased catches by four or five times. Acetone with *p*-cresol, 3-*n*-propylphenol and octenol gave a better index of increase than acetone and urine when used with electric nets and targets. Age composition of the catch was not affected (Kyorku *et al.*, 1990). A possible explanation for the lower attractivity of odours for *G. longipennis* could be that the mean temperature is lower when the fly is active at dawn and at dusk. A higher temperature would result in a higher release rate of the attractants.

Tests of odour attractants for *G. tachinoides* in Côte d'Ivoire, showed only a slight increase in attraction with acetone and octenol, whilst CO_2 had a much lower attractivity than for *morsitans* group flies in Zimbabwe (Küpper *et al.*, 1991). Phenol and indol mixtures gave dose–response relationships. It was concluded that odour was probably less important in host finding for *G. tachinoides* than for savanna species, although the urine of bushbuck, a widely used host, significantly increased catches (Späth, 1997).

Experimental work has suggested that acetone and octenol appear to increase the responsiveness of tsetse to visual cues (Torr, 1990).

Acetone and octenol are effective but generally less potent as attractants for *G. m. submorsitans* (Politzar and Mérot, 1984) and *G. palpalis* (Cheke and Garms, 1988) than for *G. m. morsitans* or *G. pallidipes*. The fact that octenol and acetone improve the catches of *G. m. submorsitans* is of particular interest as this tsetse species is difficult to catch in unbaited traps and the use of attractants may now increase the possibility of controlling this species using trap or target technology. Table 10.2 lists some of the field trials demonstrating the effectiveness of attractants.

10.5   BEHAVIOUR OF TSETSE IN RESPONSE TO ODOURS

The complex responses of tsetse to host odours, dynamics of odour plumes and how insects respond to them were reviewed by Brady *et al.* (1989) and Murlis *et al.* (1992). Odours arrive in tsetse habitats from all directions, only 30% with a bias towards the true source direction. Tsetse either 'range' until they arrive by chance in a host odour plume, or else the odour activates resting flies, which then take off roughly upwind. In flight they usually turn upwind upon entering a plume, and turn back when they lose contact (Brady *et al.*, 1990). In Zimbabwe, the maximum range of odour attraction for *G. morsitans* and *G. pallidipes* is about 90 m (Vale, 1977a). A summary of behavioural responses of tsetse to odours is given in Table 10.3.

Video recordings have proved useful in studies of tsetse flight behaviour in odour plumes, and were used by Gibson and colleagues in Zimbabwe. In one of these studies, most flies leaving odour plumes turned sharply (95°) without regard to wind direction (Gibson and Brady, 1988), supporting Bursell's (1984a) hypothesis that tsetse attracted to an invisible odour source respond weakly, if at all, to wind direction while in flight. Fewer flies entering a plume cross-wind turned, and those that did generally made smaller turns that were biased upwind. There was clearly in-flight sensitivity to wind direction as more flies entering an odour plume turned upwind rather than downwind. When flies lost contact with the odour they made a sharp turn, the extent of which varied considerably and was uncorrelated with wind direction. These turns may serve to arrest the upwind progress of the fly (Gibson and Brady, 1985).

Wind-tunnel experiments show that tsetse navigate up host odour plumes in flight by responding to the visual flow fields, due to their movement over the ground (optomotor anemotaxis), even in weak winds blowing at a fraction of their ground speed (Colvin *et al.*, 1989). However,

Table 10.2. Field trials demonstrating the effectiveness of synthetic odours.

Odour	Species	Country	Reference
Octenol	*G. morsitans*	Zimbabwe	Vale and Hall, 1985
"	*G. pallidipes*	Zimbabwe	
"	Tabanidae	Zimbabwe	French and Kline, 1989
Acetone + octenol	*G. m. submorsitans*	Burkina Faso	Politzar and Mérot, 1984
	G. palpalis	Liberia	Cheke and Garms, 1988
Acetone + 4-methyl-phenol	Tabanidae	Burkina Faso	Amsler *et al.*, 1994a

Table 10.3. Behavioural responses to odours.

Stimulus	Response
Entering a host odour plume	Turning upwind, upwind flight, reduced speed
Carbon dioxide	Activation, upwind anemotaxis
CO_2 + acetone or octenol	Elicit upwind flight
Acetone or octenol alone	Increased visual responsiveness
Ox sebum	Increased flight activity
Unknown odour	Elicit landing response
Human odour (lactic acid)	Repellent
Host body washings	Landing and probing

Kennedy (1983, in Torr, 1988b) doubted that tsetse employed optomotor-steered anemotaxis but instead landed frequently and took off upwind when stimulated by odour. Other wind-tunnel experiments have shown that substances from body washings of host animals at close range (up to 20 cm from the source) resulted in cessation of flight, alighting and probing responses by male *G. m. morsitans* (Saini *et al.*, 1993). Active substances in the phenolic fraction were most effective in eliciting a probing response.

Electrocuting nets used to study tsetse behaviour in odour plumes show that they generally fly upwind, with 50–60% flying within 35° of due upwind and more than 80% flying at < 50 cm above ground level (Torr, 1988b). Flight is not perfectly straight, nor precisely upwind, but at some angle to the wind direction. When flies lose contact with the odour plume they carry out a reverse turn within about 2 m and if they regain contact with the plume they again turn upwind. In those experiments, the propensity to land in the vicinity of the odour source was greater for *G. pallidipes* than for *G. m. morsitans*, greater for immature than mature flies and greater for males than for females.

10.5.1 Responses to odours and visual targets

Bursell (1984a) observed that *G. pallidipes* tended to overshoot an odour source in the absence of a visual target, and thought that the orientation of tsetse to host odours may involve a step-wise approach. Further video studies showed that when no visual target was present, flights in odour were strongly biased upwind and, in the absence of odour, strongly biased downwind (Gibson *et al.*, 1991). With a target present, between 16 and 40% of the upwind approaching flies responded visually by circling the target as they passed it. They then responded orthokinetically by slowing down as they passed the target, thus confirming Bursell's observations. When an odour source (containing a mixture of octenol, propyl- and ethylphenols in the proportions 4 : 1 : 8) was 5 m upwind of a 1 m square black target, flight tracks of tsetse arriving at or leaving the target were significantly biased upwind, although only marginally when only acetone was present. Thus, an upwind anemotactic response was elicited by the $4:1:8$ mixture, similar to that from CO_2, whilst acetone only elicited a weak response and may potentiate visual responses (Brady and Griffiths, 1993).

Among olfactory and visual stimuli which cause resting *G. m. morsitans* and *G. pallidipes* to become active and take off are 100% ox odour, 0.08% CO_2 or a visual stimulus from a 0.75×0.75 m black target about 5 m from the refuge of the resting flies (take-off response stimulated by the target was low) (Torr, 1988a). Stimuli of 25% ox odour, 0.8% CO_2, acetone and octenol did not cause take-off.

Tsetse flying upwind in host odour plumes may respond to visual features whilst in flight. In behavioural experiments, 45% of flies could be diverted to square targets and 30% to a black oblong target but no significant diversion occurred towards a bark-coloured, vertical oblong target (Torr, 1989). Black was better than blue, which was better than red, which was better than yellow in terms of the degree of diversion from upwind flight.

Male *G. pallidipes* of all ages are equally attracted to both odour-baited stationary or mobile traps, but nulliparous females are not attracted to a stationary trap (Langley *et al.*, 1990b).

10.6   GENERAL FIELD OBSERVATIONS

Table 10.4 summarizes indices of attraction for various odour attractants that have been tested. A detailed description of technical aspects of odour dispensers was given by Torr *et al.* (1997).

In West Africa, Späth and Küpper (1991) found that most *G. tachinoides* alighted on the lower blue part of unbaited biconical traps but were equally distributed all over the blue part of odour-baited traps. Only 18% of males and 9.3% of females approaching a biconical trap entered it, although efficiency increased up to 90% when odours were used and numbers attracted to the vicinity of the trap increased by as much as 200%. The hungrier the flies were, the more they were attracted to the trap. Optimal combinations of odour attractants for West African tsetse, *G. longipalpis*, *G. medicorum* and *G. tachinoides*, were described by Späth (1995b). In his experiments, 3-methyl-phenol and 4-methyl-phenol from the phenolic fraction of ox urine showed the highest attractiveness for those tsetse species.

Catches in devices baited with live oxen are significantly higher than when the same devices are baited with synthetic compounds. This indicates that some component has still not been identified; thus, there is still a potential for increasing, and possibly doubling, the efficacy of odour-baited targets or traps. Experiments conducted in Zimbabwe suggest that the unknown substance has a high volatility and a low molecular weight

(Warnes *et al.*, 1995). Some of the odour attractants referred to also increase trap catches of Tabanidae and probably other biting flies.

Table 10.4. Indices of catch increases from odour attractants.*

Location	Tsetse species	Odour attractants	Index of increase	Reference
Zimbabwe	*G. m. morsitans*	Carbon dioxide 2.5–15 l min^{-1}	× 6	Vale, 1980
		Acetone 0.3–300 g l^{-1}	× 6	Vale, 1980
		Octenol	× 3	Hall *et al.*, 1984
	G. pallidipes	Carbon dioxide 2.5–15 l min^{-1}	× 6	Hall *et al.*, 1984
		Acetone 0.3–300 g l^{-1}	× 6	Hall *et al.*, 1984
		Acetone, phenol mix	× 20	Vale and Hall, 1985
		Octenol	× 3	Hall *et al.*, 1984
Kenya	*G. pallidipes*	Acetone, phenol mix	× 6–8	Baylis and Nambiro, 1993b
		Buffalo urine	× 9.6	Owaga, 1985
		Cow urine	× 1.8	Owaga, 1985
	G. longipennis	Acetone + cow urine	× 4–5	Kyorku *et al.*, 1990
Ethiopia	*G. pallidipes*	Cow urine + acetone	× 2–3	S.G.A. Leak, unpublished report, ILRI, 1995
Somalia	*G. pallidipes*	Acetone, phenol mix	× 3–4	Torr *et al.*, 1989
		Phenol mix	× 1.6	Torr *et al.*, 1989
Côte d'Ivoire	*G. tachinoides*	3 : 1 4-Methyl-phenol + octenol	× 2.5	Filledier and Mérot, 1989
		Phenolic fraction of bushbuck urine	× 1.8	Späth, 1997
		Monitor lizard skin washings	× 1.34	Späth, 1997
		Warthog skin washings	× 1.46	Späth, 1997
	G. longipalpis	Warthog urine	× 1.58	Späth, 1997
		Domestic pig urine	× 1.91	Späth, 1997
		Bushbuck urine	× 2.51	Späth, 1997

* Different trapping systems and rates of dispensing odours were used to obtain these indices of increase; thus, the figures provide only a rough guide. Details of experimental methods are given in the references.

Part II

Epidemiology

Chapter 11

Host–Parasite Interactions

The interactions between trypanosomes and their vertebrate and invertebrate hosts are of considerable importance in the epidemiology of trypanosomosis. These interactions and relationships are examined in this chapter. In the first two sections, the evolutionary origins and development of trypanosomes in tsetse are described. Sections 11.3 and 11.4 describe the factors influencing trypanosome infection rates in tsetse and mechanisms affecting their establishment and maturation. The final two sections describe the effect of trypanosome infections on tsetse and the relationships between tsetse, trypanosomes and vertebrate hosts.

11.1 ORIGINS OF THE HOST–PARASITE RELATIONSHIP

It has been suggested that pathogenic African trypanosomes were first transmitted to mammalian hosts by leeches and then found their way to *Glossina*, which already carried trypanosomes in the hindgut (the 'posterior station') but had the midgut and mouthparts (the 'anterior station') free (Baker, 1963). Hoare (1970) suggested that trypanosomes of mammals evolved recently from the Stercoraria (the group of trypanosomes, such as *T. grayi*, which are transmitted by contamination of the host's oral membranes, usually with the vector's faeces, close to the site of a bite). An alternative theory was that the anterior-station trypanosomes of mammals evolved recently from aquatic reptiles rather than from leeches (Woo, 1970), although perhaps the leeches were the vectors of the trypanosomes to aquatic reptiles. The basis for believing that trypanosomes infecting mammals had evolved recently was the high degree of pathogenicity exhibited in many of these hosts (Woo, 1970).

Only tsetse are known to be capable of transmitting African trypanosomes cyclically, whilst biting insects may transmit them mechanically and are responsible for mechanical transmission of some trypanosome species in South America. *Trypanosoma cruzi*, the causative organism of Chagas' disease of humans in South America, is transmitted by blood-sucking bugs (e.g. *Rhodnius* spp.). There have been reports of trypanosomes being transmitted by ticks and, experimentally, a midge (*Culicoides nubeculosus*) has transmitted *T. bakeri* to parrots (*Psittacula roseata*) (Miltgen and Landau, 1982). Hippoboscids (*Melophagus* spp.) are capable of transmitting trypanosomes (*T. melophagium*). The transmission of *T. b. gambiense* by *Mansonia uniformis* mosquitoes was suspected and investigated in the Congo (Heckenroth and Blanchard, 1913) but there is no evidence of trypanosomes responsible for human sleeping sickness being transmitted by vectors other than tsetse flies.

The classification of trypanosomes of economic and epidemiological importance is shown in Fig. 11.1.

11.2 DEVELOPMENT OF TRYPANOSOMES IN TSETSE

During development in tsetse, trypanosomes change from a non-Krebs cycle respiratory pathway, used in the mammalian host, to a Krebs cycle pathway. This change has been studied mainly in *Trypanozoon* but also occurs with *Nannomonas* and *Duttonella* trypanosomes, and is associated with their extensive development of mitochondria in *Glossina*. This transformation is necessitated by a change from the oxygen-rich environment of the mammalian host's bloodstream to the oxygen-deficient environment in tsetse. Both bloodstream stages of trypanosomes, and the vector midgut developmental stages, are free swimming, but on reaching the tsetse salivary glands *Trypanozoon* mesocyclic trypanosomes attach to the gland microvilli and multiply as attached epimastigotes. Attachment is maintained during the differentiation of the metacyclic trypomastigotes, with the latter becoming free again for injection with the saliva. The length of time taken for trypanosomes to complete development in tsetse is quite variable and also depends upon the maintenance temperature of adult tsetse. Some experimental observations on temperature-dependent maturation times are summarized in Table 11.1.

Duttonella *subgenus:* Trypanosoma vivax

A comprehensive review of *T. vivax* (Gardiner, 1989) is referred to for a full account of its transmission by tsetse.

Trypanosoma vivax has the simplest life cycle in tsetse, normally developing in the proboscis, although infections can sometimes also be detected in the cibarium and the anterior gut (Jefferies *et al.*, 1987; Moloo and Gray, 1989). Moloo and Gray (1989) found trypomastigotes, pre-epimastigotes and epimastigotes in the region of the cibarium and the oesophagus of tsetse dissected 1–48 h after an infected feed. The cibarium is not usually thought to be a site of development for *T. vivax*, although Bruce *et al.* (1911) referred to this possibility. There was some evidence that

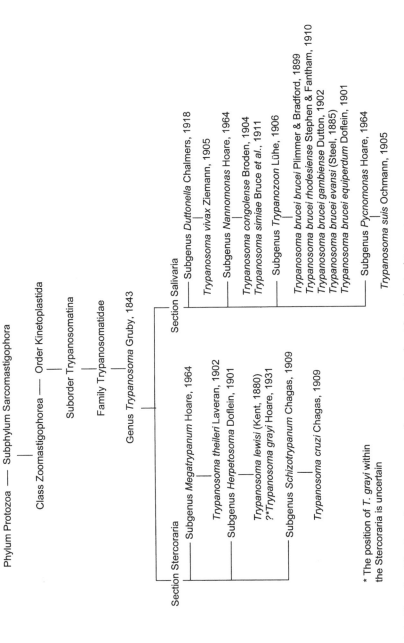

Fig. 11.1. The classification of trypanosomes of economic and epidemiological importance.

Table 11.1. Maturation times of salivarian trypanosomes in tsetse.

Trypanosome species	Conditions	Time taken to maturation (days)	Reference
T. vivax	*G. palpalis* puparia 26°C adults 29°C	5	Desowitz and Fairbairn, 1955
	adults 22°C	12–13	
T. congolense		19–53	Hoare, 1970
	23–24°C	15–20+	Harley and Wilson, 1968
		5.5 ± 4.6	Dale *et al.*, 1995
T. simiae	*G. morsitans* 28.3°C *G. brevipalpis*	20	Bruce *et al.*, 1913; Peel and Chardome, 1954
T. b. rhodesiense and other *Trypanozoon* species	*G. m. centralis* 28°C *G. m. centralis* 30°C	23 12 17–45 18 ± 5.22	Fairbairn and Culwick, 1950 Fairbairn and Culwick, 1950 Hoare, 1970 Dale *et al.*, 1995

cyclical development of *T. vivax* initially occurs in the cibarium/oesophageal region, where parasites migrate to the food canal of the proboscis.

Development is completed to the infective metatrypanosome stage in the hypopharynx and is a temperature-dependent process. According to Desowitz and Fairbairn (1955), *T. vivax* completed its developmental cycle in *G. p. palpalis* in 5 days when the puparia had been kept at 26°C and the adults were then kept at 29°C. In adults kept at 22°C the cycle was completed in 12–13 days.

Developing trypanosomes become attached to the interior of the tsetse proboscis (Vickerman, 1973). Electron microscopy of epimastigote clusters of *T. vivax* in the proboscis of *G. fuscipes* showed that some trypanosomes were attached directly to the lining of the labrum by their flagellae, mainly aligned along the long axis of the proboscis. Other trypanosomes were attached indirectly via other trypanosomes, their flagellae being fixed to those of the trypanosomes directly attached. Long zonular hemidesmosomes attach the flagellae to the proboscis wall and series of closely set macular desmosomes link the flagellar membranes of adjacent trypanosomes. Vickerman concluded that these complexes served to maintain anchorage of the trypanosomes during the tsetse blood feeding. Gardiner (1989) suggested that the increased flagellar adhesiveness of late bloodstream forms might be an adaptation to help attachment when ingested by tsetse.

A significant feature of *T. vivax* differentiation in tsetse is the loss of the surface coat. In contrast to the situation with *T. b. brucei*, infection rates of 100% could be achieved with *T. vivax*, by repeatedly feeding flies on infected hosts. Flies do not have to be teneral in order for *T. vivax* infections to become established. Mature *T. vivax* infections in the proboscis can persist for up to 58 days, but may eventually disappear (Soltys and Woo, 1977).

Nannomonas *subgenus:* Trypanosoma congolense

Trypanosoma congolense develops in the midgut and the proboscis of tsetse, and takes longer to development than *T. vivax.* The parasite is therefore exposed to anti-trypanosomal factors in the fly gut. In laboratory experiments, male *G. m. centralis* were significantly more susceptible to infection with a strain of *T. congolense* than females, but no significant differences were detected between male and female *G. pallidipes* (Moloo *et al.,* 1992a). Male *G. p. palpalis* were also more susceptible to infection with a *T. congolense* stock from Uganda, suggesting that maturation of *T. congolense* may also be controlled by a recessive gene on the X chromosome. Experiments using *T. congolense* resulted in higher immature infection rates in female *G. m. morsitans* than in males but higher mature infection rates in males (Mwangelwa *et al.,* 1987). Mwangelwa *et al.* (1987) speculated that the lower infection rates in females might result from differing rates of digestion between male and female flies. This could result from a greater concentration of protease in the midguts of females resulting in a stronger barrier to infection. Initially, Mihok *et al.* (1994a) found no correlation between protease activity and trypanosome infection rates to support that hypothesis, although trypsin-like proteases were found in the posterior midguts of *G. m. centralis* and *G. m. morsitans.* However, in later studies, protease levels were weakly correlated with trypanosome infection rate in *G. m. morsitans* (Mihok *et al.,* 1995a).

Wide variations in the duration of the developmental cycle of *T. congolense* have been reported. These range from 19 to 53 days (Hoare, 1970) and from 15 to 'well over 20 days' at 23–24°C (Harley and Wilson, 1968). More recently, a stock of *T. congolense* was shown to have a mean maturation time of 15.5 days with a standard deviation of 4.6 days (Dale *et al.,* 1995).

Some trypanosomes never mature; the percentage of trypanosome infections successfully maturing will vary between species, and is influenced by factors within populations of tsetse including anti-trypanosomal factors, lectins and rickettsia-like organisms in the tsetse midgut (see section 11.4). One of the barriers to infection in tsetse with *T. congolense* is the migration of trypanosomes forward from the posterior segment of the midgut. This is also the case with *T. b. brucei* (Dipeolu and Adam, 1974). Penetration of the proventriculus and invasion of the mouthparts appeared to present no obstacle. *Trypanosoma congolense* that had been repeatedly syringe-passaged appeared to lose their ability to invade the hypopharynx of tsetse.

In 50–60% of *G. m. morsitans* and *G. austeni,* infections of *T. congolense* could persist in the midgut for up to 3 days after the infective feed (Dipeolu, 1975). However, in only a few flies did mature infections become established. The maintenance temperature of *G. m. morsitans* did not affect their susceptibility to *T. congolense,* when raised from 26 to 31°C. The duration of development was longer in *G. austeni* than in *G. m. morsitans,* and the concentration of trypanosomes in the infective feed affected the initial infection rate but not the number of established infections. Syringe passage of *T. congolense* appeared to reduce the infectivity of the trypanosome in tsetse (Dipeolu, 1975).

Epimastigote attachment of *T. congolense* in the tsetse proboscis appears to be a requirement for its metacyclic development, as differentiation does not occur *in vitro* if attachment is prevented by shaking culture bottles (Hendry and Vickerman, 1988).

Nannomonas *subgenus:* Trypanosoma simiae

Trypanosoma simiae is a *Nannomonas* trypanosome with a similar cycle of development to *T. congolense*, with which it may be confused when detected in tsetse flies. All the literature on *T. simiae* up to 1965 was reviewed in a monograph on pig trypanosomosis by Stephen (1966a). The cycle of development of *T. simiae* takes about 20 days in *G. morsitans* or *G. brevipalpis* at 28°C (Bruce *et al.*, 1913; Peel and Chardome, 1954). The susceptibility of *G. m. submorsitans* and *G. tachinoides* to infection with *T. simiae* was compared by Agu (1984a), using 592 *G. tachinoides* and 348 *G. m. submorsitans* in attempts to transmit *T. simiae* to pigs. *Glossina m. submorsitans* was very good at transmitting *T. simiae* infection to pigs but *G. tachinoides* was very poor (Agu, 1984a).

Pigs are particularly susceptible to *T. simiae* and suffer an acute fulminating infection, resulting in death a few days after trypanosomes are observed in peripheral blood. Sheep and goats show variable susceptibility and cattle are refractory to infection (Stephen, 1966a).

At the Kenya coast, Janssen and Wijers (1974) observed a less acute disease in pigs than normal, and reported a correlation between virulence and the species of tsetse transmitting *T. simiae*. *Glossina brevipalpis* transmitted virulent strains, *G. austeni* transmitted moderately virulent strains and *G. pallidipes* transmitted only chronic strains of the trypanosome. They found that *T. simiae* strains could be made more or less virulent by cyclically passaging them through either *G. pallidipes* or *G. brevipalpis*. The statement made by Janssen and Wijers that *G. pallidipes* is less susceptible than *G. brevipalpis* to infection with *T. simiae* was found to be invalid by Moloo *et al.* (1994b).

Trypanosoma simiae is not pathogenic to warthogs, which are probably the main wild reservoir hosts of the parasite (Stephen, 1966a).

Trypanozoon *subgenus:* Trypanosoma brucei *group*

The *Trypanozoon* subgenus trypanosomes include *T. b. brucei*, which infects wild and domestic animals, and *T. b. rhodesiense* and *T. b. gambiense*, which infect humans. The cyclical development of *Trypanozoon* species in the tsetse fly takes place in the midgut and culminates in the metacyclic trypomastigote form in the vector's salivary glands. The metacyclic stage is infective for the mammalian host and, unlike the other developmental stages in the tsetse fly, these metacyclic trypanosomes possess a surface coat like that of bloodstream forms.

Trypanosoma b. rhodesiense took about 23 days to develop in *G. m. centralis* maintained at 28°C, and 12 days at 30°C (Fairbairn and Culwick, 1950). The range for the duration given by Hoare (1970) is from 17 to 45 days or longer. The maturation time for *Trypanozoon* stocks is variable; in recent experiments one stock took a mean of 18 days with a standard deviation of 5.22 days (Dale *et al.*, 1995). Maturation was a sex-dependent phenomenon in *Trypanozoon* but not in *T. congolense*.

The means by which trypanosomes migrate from the midgut to the salivary glands has been a controversial subject. In 1972 *T. brucei* was found in the haemocoeles of tsetse (Mshelbwala, 1972), suggesting an alternative route to the accepted view at that time. It was subsequently found that trypanosomes could penetrate the anterior midgut cells of tsetse (Ellis and Evans, 1976). This is discussed further in section 11.5.1.

Trypanosoma grayi

Trypanosoma grayi is a stercorarian trypanosome, parasitic in crocodiles. It develops in the hindgut of tsetse flies and is transmitted contaminatively via their excreta, from which the parasite may enter the circulation of its vertebrate host through wounds or abrasions. The significance of *T. grayi* in the epidemiology of trypanosomosis of domestic livestock is that it may wrongly be classed as an immature *Nannomonas* or *Trypanozoon* infection when detected in the guts of dissected tsetse. When occurring simultaneously with *T. vivax* in the fly, the infection may be wrongly identified as a mature *T. congolense* type. This could result in overestimates of the level of tsetse challenge or trypanosomosis risk. It is a large trypanosome, with a free flagellum in the vertebrate host, whilst in the fly this is comparatively small. *Trypanosoma grayi* is distinguished from other trypanosomes in the gut of tsetse by the invariable presence of crithidial forms. It has a large macro-nucleus, oval or compressed, situated not far from the posterior end of the organism, and a large, usually rod-shaped kinetoplast. Lloyd and Johnson (1924) found infections in 70 flies, usually in the hind part of the midgut. They detected *T. grayi* in the midguts of *G. tachinoides* and *G. p. palpalis* but never in *G. m. submorsitans*. It appeared to be a blood parasite of one of the common hosts of riverine tsetse and was found in 2.8% of *G. p. palpalis* and 24% of *G. tachinoides* when those flies had fed on non-mammalian blood. Experimentally, Hoare (1929) was unable to infect monitor lizards with the parasite and he did not detect them in wild monitor lizards at Entebbe, Uganda. All the randomly selected crocodiles examined by Hoare (1931) were infected with *T. grayi* and he therefore suspected that most adult crocodiles would be infected by the parasite, although his sample size was small.

Hoare obtained very high experimental infection rates (61%) with the parasite in *G. palpalis*; wild flies had a mean infection rate of 11.2%. There was a developmental stage in the midgut of tsetse in which the trypanosomes multiplied and diminished in size. Between the 13th and 34th days of infection, in about 50% of the flies all the parasites had migrated to the hindgut, where the infective stages develop.

Little work has been carried out on this trypanosome species other than Hoare's descriptions of the parasite's life cycle and confirmation that the parasite attaches itself to the tsetse hindgut by hemidesmosomes (Molyneux, 1977). Molyneux speculated that heavy infections with *T. grayi* in female *Glossina* could disrupt physiological processes related to reproduction.

Trypanosoma theileri

Trypanosoma theileri, of the subgenus *Megatrypanum*, is a cosmopolitan parasite of domestic cattle. It is, as the subgeneric name implies, a large trypanosome, and is normally considered to be non-pathogenic. It does not undergo cyclical development in tsetse flies, but it is of interest and it is important to be aware of this parasite so as to avoid misidentification. There have been instances of it previously being misidentified as *T. evansi*, and probably as *T. vivax* despite its large size. Parasites seen in the peripheral blood of hosts show a wide variation in body length from 10 to 130 µm. It is very easily grown in culture. It is naturally transmitted contaminatively in the faeces of infected Tabanidae (and possibly by Hippoboscidae) (Noller, 1925; Wallace, 1962; Bose and Heister, 1993). These parasites may then infect cattle through contamination of the oral mucosa. Highest incidence appears to occur in adult animals. It is likely that infected tabanids lose their infection through secretion of metacyclic stages as they occur entirely in the hindgut (Bose and Heister, 1993). The parasite probably multiplies in Tabanidae and may live in them for prolonged periods (Wallace, 1962). Studies on *T. theileri* were reviewed by Wells (1972) and the life cycle in Tabanidae was described fully for the first time by Bose and Heister (1993).

Other trypanosome infections

Infections with other trypanosome species have been detected in *Ixodes ricinus* ticks in Austria (Rehacek *et al.*, 1974) and Britain (Mungomba *et al.*, 1987); the latter may have been *T. melophagium*. These authors tabulated the species of ticks in which trypanosomes have been detected around the world, including identifications of *T. theileri* from ticks in Africa.

11.3 FACTORS INFLUENCING TRYPANOSOME INFECTION RATES IN TSETSE

Trypanosome infection rates are generally quite low in tsetse and many investigations have been carried out to determine the reasons for this. Le Ray (1989) referred to the intrinsic vectorial capacity of tsetse, by which he meant the intrinsic capability of a fly to develop a metacyclic infection. At that time very little was known about factors determining this; it is perhaps surprising that our understanding has grown significantly even in the short time since that remark was made. Welburn and Maudlin (1992) and others have studied this aspect in some detail, particularly in relation to lectins and rickettsia-like organisms. These investigations are summarized in the following sections.

Jordan (1974), reviewing work carried out on trypanosome infection rates in tsetse, showed that low infection rates were usual for the three salivarian trypanosome subgenera, especially for *T. brucei* group infections. The infection rates were thought to be mainly determined by the biology of the parasite and by the identity of the host animals from which the flies obtained most of their bloodmeals. *Trypanosoma vivax*, having the simplest life cycle, was expected to have the highest infection rates, which it usually did; whilst *T. b. brucei* was the most complicated. In 1930 Lloyd, carrying out experiments on factors influencing trypanosome infection rates in tsetse, concluded that 'the factors which influence the rate of infection in tsetse by trypanosomes are so many that it is impossible to estimate with any degree of accuracy the transmissibility of any particular strain' (Lloyd, 1930). Despite that conclusion, many workers have since carried out experiments designed to elucidate those factors.

Jordan (1974) and Molyneux (1976, 1980a) listed some of the factors influencing the potential of trypanosomes to develop in tsetse flies and these are summarized in Table 11.2.

These and other factors affecting the establishment of trypanosome infections in tsetse are discussed below.

Table 11.2. Factors influencing trypanosome infection rates in tsetse.

Endogenous factors associated with tsetse	Ecological factors	Parasite and host factors
Tsetse species	Climatic factors	Parasite numbers available to tsetse
Sex	Availability of infected hosts	
Age at infective feed		Parasite species and its infectivity to tsetse
Age structure of tsetse population	Hosts available for subsequent feeds	Immune state of host
Genetic differences/ variation between and within species		Subspecies: strains
		Susceptibility
Behaviour (i.e. host preference)		Intercurrent infection
		Behaviour and attractiveness to tsetse
Concurrent infections (viruses, bacteria, fungi)		
Interactions between lectins and RLOs		
Physiological/biochemical state		

11.3.1   Sex and infection

In natural field situations female tsetse usually have higher infection rates than males; however, this is to some extent because females live longer than males, and therefore have a greater likelihood of feeding on an infected host and picking up and maturing an infection. When age is taken into account, infection rates of males and females are more or less equal. In the laboratory, transmission experiments have produced some conflicting and thus confusing results but, on balance, male tsetse generally have higher mature infections with both *Nannomonas* and *Trypanozoon* trypanosomes. For example, in laboratory experiments male *G. p. palpalis* are more susceptible to infection with *T. congolense* than females and show a higher proportion of proboscis (mature) infections (Distelmans *et al.*, 1982). Similar experiments have sometimes shown differences between tsetse species as well as between mature and immature trypanosome infections.

Because *T. vivax* develops only in the proboscis, it is less exposed to anti-trypanosomal factors in the gut of the fly than other trypanosome species and differences in their activity between male and female flies do not appear to play a significant role in determining infection rates with this species. Significant differences in infection rates in male and female flies have not been observed.

Early observations suggested that there were differences between sexes of tsetse in infection rates with *Trypanozoon* trypanosomes, but some of these observations were contradictory and perhaps unreliable. For example, in laboratory experiments, female *G. f. fuscipes* appeared to be more susceptible to infection with *T. b. gambiense* or with *T. b. rhodesiense* than were male flies (Duke, 1933f). In contrast, Burtt (1946a) and Ashcroft (1959a) found that male *G. morsitans* had a higher infection rate with *T. b. rhodesiense* than females. Strangely, there was a higher rate of mortality in females than in males in Burtt's experiments and his findings have to be viewed with caution. Ashcroft (1959a) suggested that his conclusions may not be applicable to natural conditions. Harley (1971) noted that male *G. pallidipes* and *G. fuscipes* were more readily infected with *T. b. rhodesiense* than females. Van Hoof (1947) also noticed this for *T. b. gambiense*.

In more recent experiments more *Trypanozoon* midgut infections matured in male *G. m. morsitans* than in females except for one stock (Welburn *et al.*, 1995). For both sexes, significantly more human serum-sensitive *Trypanozoon* stocks (trypanosomes that are killed by incubation with human serum, and therefore considered to be *T. b. brucei*) matured than human serum-resistant (*T. b. rhodesiense*) stocks from an area of epidemic sleeping sickness in Uganda.

The question of whether sex of a fly influences infection rates has still not been definitively answered; Kazadi *et al.* (1991) found no significant difference in *G. p. gambiensis* infected in the laboratory with *T. b. brucei*. In an experiment in which *G. m. centralis*, *G. m. morsitans* and *G. pallidipes* were fed on blood containing D-(+)-glucosamine, total *T. b. brucei* infection rates were equal in males and females but twice as many infections matured in males (Mihok *et al.*, 1992a). Females had consistently 2–3 times as many parasites as males. The D-(+)-glucosamine enhanced infection rates by blocking midgut lectins that would normally kill procyclic trypanosomes in the midgut, but had no effect on maturation rates.

11.3.2 Trypanosome infection rates and temperature

The probability of a tsetse becoming infected when it feeds upon a host with trypanosomes in its blood may be influenced by the environment in which the fly's developmental stages were passed. In laboratory studies, infection rates are positively correlated with the maintenance temperature to which puparia and adult flies are subjected, and this also influences the duration of the development cycle within the tsetse (Hoare, 1970). Experimental data on the effect of temperature are summarized in Table 11.3. Kinghorn and Yorke (1912d) found that *T. b. rhodesiense* developed more readily in *G. morsitans* at higher temperatures; later, Burtt (1946b) showed that *G. morsitans* adults, bred from puparia incubated at a high temperature, are more readily infected with *T. b. rhodesiense* and transmitted the parasite more readily. The infection rate in flies emerging from puparia kept at 30°C was nearly three times greater than in flies from puparia kept at lower temperatures. Similar results were obtained by Ndegwa *et al.* (1992), who examined the effect of puparial incubation temperature on *T. congolense* infection rates in three species of tsetse.

Desowitz and Fairbairn (1955), investigating the influence of temperature on the length of the development cycle of *T. vivax* in *G. p. palpalis*, found that the trypanosome life cycle was completed in 5 days in flies emerging from puparia kept at 26°C and then maintained at a mean maximum temperature of 29°C. In flies that emerged under the same conditions, but that were maintained at a mean maximum temperature of 22°C, the cycle took 12–13 days. It was thought that the increased infection rate in flies emerging from puparia kept at higher temperatures resulted from the fact that they were more eager to feed on the day of emergence.

Although field studies have shown a similar effect of temperature on trypanosome infection rates, this effect is less clear, because in an uncontrolled situation many other factors play a role. In Nigeria, lowest infection rates for *T. vivax*-type infections occurred around January, the coldest months of the year. In Zambia, higher trypanosome infection rates occurred in *G. morsitans* in the hot season than in the cold season (Kinghorn *et al.*, 1913, in Nash and Page, 1953). In contrast, there was no significant difference in seasonal trypanosome infection rates in *G. m. submorsitans* in the northern Guinea zone of Nigeria (Jordan, 1965a).

Ford and Leggate (1961) found a positive correlation between trypanosome infection rates (all species combined) in tsetse and distance

Table 11.3. Laboratory studies on the effect of temperature on trypanosome development.

Tsetse species	Trypanosome species	Temperature regime		Observations	Reference
		Puparia	Adults		
G. palpalis	T. vivax	26°C	22°C	Development took 12–13 days	Desowitz and Fairbairn, 1955
G. palpalis	T. vivax	26°C	29°C	Development took 5 days	
G. morsitans	T. rhodesiense	26°C	30°C	Faster development and higher infection rates in adults kept at higher temperatures and in puparia	Kinghorn and Yorke, 1912d; Burtt, 1946b
G. m. centralis	T. congolense	29°C		High puparial mortality at 29°C	Ndegwa et al., 1992
G. brevipalpis		28 : 25°C (day : night)		Optimum temperature 25°C	
G. f. fuscipes		25°C		Higher infection rates at 29°C or 28 : 25°C day : night regime	
				Only midgut infection rates were higher for G. f. fuscipes, which is a poor vector of T. congolense	
G. palpalis	T. vivax	A range of adult and puparial maintenance temperatures from 23 to 30°C		Infection rate increased with increasing puparial incubation temperature but decreased with fly maintenance temperatures of 28–39°C. Survival of flies at 28–30°C was lower. Optimum temperatures were 28°C for puparia and 23°C for adults	Fairbairn and Watson, 1955

from the median (7°S) of the *Glossina* belt in Africa, although infection rates in *G. pallidipes* from Zululand did not fit this pattern. This positive correlation with latitude was associated with increasing mean annual temperature. The correlation with *vivax*-type infections was closer than that with *congolense*-type infections. Ford and Leggate (1961) suggested that the mean temperatures at which wild tsetse exist, possibly associated with the range of temperature experienced between seasons, may account for the general prevalence of infection. However, within the same neighbourhood variations may occur due to differences in local fauna providing the trypanosome reservoirs. This might help to explain seasonal periodicity of sleeping sickness epidemics, particularly in West Africa, which may have resulted from increased human–fly contact during the dry season when flies are restricted to riverine vegetation.

11.3.3   Age of fly at time of infective feed

The age of tsetse at the time of the infective feed can influence their trypanosome infection rates although, again, some contradictory observations have been made between field and laboratory studies. The effect of age was first shown by Van Hoof *et al.* (1937a), in experiments with *T. b. gambiense* in which highest infection rates were obtained in *G. f. fuscipes* fed when 0–1 day old on a host infected with *T. b. gambiense*. Salivary gland infections of *T. b. gambiense* were higher in flies that took their first feed from an infected mammal. This contrasted with results of earlier work with *G. f. fuscipes* (Duke, 1935a), but was confirmed by Wijers (1958a,b). Wijers obtained the highest infection rates when flies were induced to feed on the first day of life.

The infection rate was even higher in those flies that fed within 3 and 19 h after emergence. Wijers believed that Duke's conflicting results were due to his flies not having been fed on the infected host on the first day of life. The low infection rates found among tsetse from sleeping sickness areas in northern Nigeria were also believed to be due to the low proportion feeding on the first day of life (Wijers, 1958a,b). Flies most easily infected were tenerals that took the infective bloodmeal on the first day of life, as soon as possible after emergence.

Wijers (1958a) suggested the following conditions to increase the likelihood of *G. palpalis* becoming infected with metacyclic trypanosomes of *T. b. gambiense*:

- The fly must be willing to feed within 24 h of emergence.
- Its first host must be a human being.
- The human being must have sleeping sickness.
- The sleeping sickness case must be in the early stage, with a sufficient number of transmissible trypanosomes in the peripheral blood.
- The fly must then live for at least another 18 days to become infective.

It is now established that wild and domestic animal reservoirs of *T. b. gambiense* exist and therefore the chances may be higher of a fly feeding on a host infected with that species. Otherwise, the above conditions are most likely to be met in the West African situation during the late dry season when both flies and humans concentrate at the water-holes under conditions of close personal contact (described in Chapter 12). The flies are also most likely to feed early in life in hot dry conditions.

In contrast to Wijers's (1958a) observations with *T. b. gambiense*, age at first infective feed was less important in experiments on infectivity of *T. b. rhodesiense* to *G. morsitans* and *G. fuscipes* and there was little difference in susceptibility of flies between 2 and 11 days old (Harley, 1970). Thus, although younger flies are certainly more susceptible, Harley concluded that age was less critical for infections of *G. fuscipes* with *T. b. rhodesiense*. Other work has suggested that only a proportion of the fly population is active and feeds within 48 h of emergence (Jackson, 1946; Bursell, 1959b; Harley, 1966a). Adults fed on rats when less than 24 h old readily became infected, older flies less readily. However, in nature a small proportion of flies is thought to feed in the first 24 h after emergence.

Makumyaviri *et al.* (1984b) further examined the vectorial capacity of *G. m. morsitans* to infection with *T. b. brucei* and found a significant effect, on infection rates in the flies, of both sex and age at the time of the infective meal. The highest salivary gland infection rate was in males having taken the infective feed whilst tenerals, in the 32 h following eclosion. Thus, his findings were in agreement with those of Van Hoof *et al.* (1937a), Wijers (1958a) and Gingrich *et al.* (1982) for *T. b. brucei* and of Distelmans *et al.* (1982), who showed that only those *G. p. palpalis* fed within 1 day of emergence would develop mature *T. congolense* infections. Field data appear to show that prevalence of mature infections increases with age for both *Nannomonas* and *Duttonella* group trypanosomes for a variety of tsetse species (Woolhouse *et al.*, 1993, 1994; Leak and Rowlands, 1997). Lectins, described in section 11.5.2, play a major role in the greater susceptibility to infection of young flies compared with older ones; older flies have gut lectin levels 100–200 times higher than teneral flies (Jacobson and Doyle, 1996). Gibson and Ferris (1992) were able to superinfect *G. m. morsitans* with *T. brucei* or *T. congolense*, giving the first infective feed 2 days after emergence and subsequent feeds at intervals of about 72 h.

The susceptibility of *G. m. morsitans* to infection with *T. b. brucei* is also age dependent. In the laboratory, flies are susceptible when given an infected meal (of infected rat or rabbit blood fed through a membrane) 1–8 h after emergence (Otieno *et al.*, 1983). Males were more susceptible to infection than females, but numbers of trypanosomes ingested did not affect subsequent salivary gland infections, although there was an influence on midgut infections. In contrast, Page (1972) found that tsetse were more readily infected when an infected bloodmeal (from mice) contained a 'moderately high' number of stumpy forms (10,000–20,000 trypanosomes ml^{-1}) with a decrease above that level. The probability of a single trypanosome infecting a tsetse fly has been modelled by fitting a simple model to data on tsetse infection rates (Baker, 1991).

Molyneux and Stiles (1991) believed that the release of anti-trypanosomal factors (midgut trypanolysin and trypano-agglutinins) significantly affect the initial establishment of the ingested trypanosomes in the gut. Furthermore, as their release fluctuates with the normal digestive cycles, feeding patterns of tsetse would be important in determining susceptibility to infection. Repeated feeding at short intervals would render the fly refractory to infection by stimulating midgut agglutinin and lytic activities, whilst extended periods of starvation may render the flies more susceptible than usual, as agglutinin and lytic activity decreases with time after feeding. The lytic substance was found in *G. p. palpalis* but not in *G. p. gambiensis* (Stiles *et al.*, 1990), which could explain the lower trypanosome infection rates reported for *G. p. palpalis* in both the field (Ryan *et al.*, 1982b) and the laboratory (Moloo and Kutuza, 1988).

In the experiment referred to earlier by Mwangelwa *et al.* (1987), higher infection rates with *T. congolense* were obtained in starved 16 h or 2-day-old flies than in flies of 2 days or 7 days old which had been fed on a 'clean' bloodmeal before the infective one. In contrast, Gooding (1988b) was able to infect a significant proportion of post-teneral male *G. m. morsitans* and *G. m. centralis* with *T. b. brucei* without starving them before feeding them on infected rabbits. However, these tsetse were fed many times on the infected rabbits, the equivalent of which is likely to be a rare natural occurrence.

Jordan (1976) suggested that the principle that the older the fly, the more refractory it becomes to infection, could be extended to all *Glossina* species infected with *Trypanozoon* trypanosomes but concluded that flies of the *morsitans* and *palpalis* groups could be infected with *T. vivax* at any age. *Nannomonas* trypanosomes occupied an intermediate position. Similarly, when reviewing this topic, Maudlin (1991) concluded that both *T. b. brucei* and *Nannomonas* infection rates in tsetse were affected significantly by the age of the fly, but the case for *T. vivax* was still uncertain. Furthermore, the actual age is less critical than whether or not the fly has taken a previous, uninfected feed.

The greater susceptibility of teneral flies to midgut infection with *T. congolense* compared with non-teneral flies was confirmed by Welburn and Maudlin (1992), who showed that tsetse, once fed, remain relatively (but not absolutely) refractory to infection. Welburn and Maudlin suggested an explanation for this, involving the role of rickettsia-like organisms in potentiating teneral susceptibility to midgut infection. They concluded that the peritrophic membrane (discussed in section 11.5.1) does not act as a barrier preventing non-teneral flies from becoming infected.

Age structure of the tsetse population

Despite the controversy over the extent to which tsetse become infected with trypanosomes with increasing age, this clearly differs among trypanosome species. In contrast to experimental observations, field data suggest that the longer a fly lives, the more likely it is to become infected, despite the evidence that age and earliness of the infective feed are important (Ryan *et al.*, 1982b; Tarimo *et al.*, 1985; Leak and Rowlands, 1997). Thus, infection rates in natural populations may be expected to increase with age; and the longer an infected fly survives after becoming infective, the greater is its potential for transmission of trypanosomes. As described, this probably applies less, if at all, to *Trypanozoon* infections than to *Duttonella* or *Nannomonas* infections. Wild male tsetse can achieve a life span of almost 5 calendar months but this is probably unusual. Fraser and Duke (1912) found the greatest age for males was 168 days.

Harley (1967c) investigated the effect of sampling method on observed trypanosome infection rates. He used seven different sampling methods for simultaneously catching flies and found marked differences between the infection rates of the various samples. In the case of *G. pallidipes*, these differences were closely associated with differences in the mean age of the samples, demonstrating the importance of taking age into account if comparing the infection rates of tsetse caught by different means. Harley found that 80% of infections were found in flies more than 40–50 days old.

Ward and Bell (1972) found that *G. austeni* did not become non-infective as they grew older, and trypanosome transmission of *T. congolense* to mice from flies 100 days old was possible.

11.3.4   Influence of species of tsetse fly

Fly species differ in their capacity to transmit trypanosomes. *Morsitans* group flies, except for *G. austeni*, are good vectors of all trypanosome species. *Palpalis* group species appear to be poor vectors of most trypanosome species except certain stocks of West African *T. vivax*, although *G. f. fuscipes* can be an important vector of human infective trypanosomes (*T. b. rhodesiense* and *T. b. gambiense*). Many studies have shown that *palpalis* group flies are poor vectors of *T. congolense* (e.g. Godfrey, 1966; Harley and Wilson, 1968). From the few studies that have been conducted, forest species of the *fusca* group appear to be good vectors of *T. congolense* and *T. vivax* but poor vectors of *Trypanozoon* trypanosomes.

Variation between individual tsetse flies

Maudlin (1982) considered that, irrespective of age, species, sex, host or infecting organism, as 100% trypanosome infection rates cannot be achieved experimentally, other factors associated with individual fly genotype must determine susceptibility to infection. In breeding experiments, susceptibility to trypanosome infection of F_1 generation tsetse flies was inherited maternally

(Maudlin, 1982). The results could have been due to variation in the maternal environment but the effect persisted for many generations, indicating that this was not the case. As male parental phenotype did not affect the progeny, it appeared that susceptibility to *T. congolense* infection was an extra-chromosomally inherited character. However, the maternally inherited character applied only to the establishment of midgut infections; their maturation appeared to be associated primarily with trypanosome genotype. The character also only applied to teneral flies; flies having had an uninfected meal were like outbred flies. Maudlin concluded that maturation of *Trypanozoon* infections is probably controlled by a sex-linked recessive allele. It was later shown that rickettsia-like organisms (RLOs) were the extra-chromosomal factor (Maudlin, 1985).

11.4   MECHANISMS DETERMINING SUSCEPTIBILITY TO TRYPANOSOME INFECTION

For successful cyclic transmission of African trypanosomes, successive 'biological barriers' have to be overcome. These are associated with the ecology and physiology of the host(s) and vector(s), and the parasite's ability to adjust to the various substrates within those hosts and vectors (Lambrecht, 1980). It is only in the last few years that we have really begun to understand the nature of these barriers. The low transmission coefficients shown in Table 11.4 demonstrate the importance of barriers to trypanosome infection in tsetse and vertebrate hosts. These estimates for transmission of pathogenic trypanosome species from vector to host and from host to vector are drawn from studies by Rogers (1988a), Milligan and Baker (1988) and Baylis (1997).

Baylis (1997) discussed the possible reasons for the disparity between his estimates and those of Rogers and of Milligan and Baker, which he believed to be overestimates resulting from the methods used for their

Table 11.4. Transmission coefficients for pathogenic trypanosome species.

Trypanosome species	Transmission from infected vector to susceptible vertebrate	Transmission from infected vertebrate to vector
T. vivax*	0.29	0.177
T. vivax†	0.024	
T. vivax‡	0.20	
T. congolense*	0.46	0.025 (average)
T. congolense†	0.0084	
T. congolense‡	0.20	
T. brucei*	0.62	0.065 (first meal only)

* From Rogers, 1988; † from Baylis, 1997; ‡ from Milligan and Baker, 1988.

estimation. Rogers' estimates were derived from the numbers of infections in cattle or mice resulting from the injection of macerated whole proboscides of infected *G. pallidipes*, whilst those of Milligan and Baker were derived from numbers of trypanosomes or 'mouse-infective doses of trypanosomes extruded by *G. pallidipes* when feeding or probing'. Baylis (1997) stated: 'In general, in the absence of precise data on the number of trypanosomes required to infect cattle, estimates of transmission efficiencies based on laboratory studies of trypanosome extrusion rates must be considered extremely unreliable.'

Experimentally it is much easier to infect tsetse of any species with *T. vivax* than with other trypanosomes and very high infection rates may be obtained. This may be simply explained by the fact that this trypanosome has only to establish itself in the first organ reached – that is, the proboscis. By this reasoning it would be expected that a *T. congolense* infection would be more difficult to establish, as it has to survive in the tsetse gut, while *T. b. brucei* would show the lowest infection rates as it also has to migrate successfully to the salivary glands. This theory is roughly borne out by the experimental evidence, but some related questions have been raised:

- Why were repeated infective feeds necessary to establish a *T. vivax* infection in 'almost' all flies (Clarkson and McCabe, 1970)?
- Why were significantly lower infection rates recorded for *T. simiae* than for *T. congolense* in both species of fly used by Roberts and Gray (1972) when both these trypanosomes develop in the midgut?
- Why were the infection rates of *T. b. rhodesiense* in *G. f. fuscipes* (Harley, 1971) much higher than the infection rates for *T. congolense* in the same species of fly (Harley and Wilson, 1968)?
- Why were such variable infection rates obtained with four fly species all infected with the same trypanosome, *T. b. rhodesiense* (Harley, 1971)?

These questions demonstrate that whilst differences in trypanosome life cycle within the fly clearly have a role in determining infection rates, there are other factors involved which must, in part, reflect genetic differences in both tsetse and trypanosomes.

Godfrey (1966) was unable to transmit *T. congolense* through *G. p. palpalis*, although 5% of the flies had midgut infections, while Harley and Wilson (1968) obtained only a 2.9% infection rate with *T. congolense* in *G. f. fuscipes*. In the latter study, it was concluded that differing feeding habits were not the only cause of the lower infection rate in wild-caught *G. f. fuscipes* compared with *G. pallidipes* from the same area (Harley and Wilson, 1968). Laboratory experiments led Buxton (1955b) to conclude that either an initial infection disposes a fly to further infection, or the two infections commonly occur together in one host, or certain flies are readily infected by both trypanosome species.

Roberts and Gray (1972) compared *G. m. submorsitans* and *G. tachinoides* as vectors of *T. congolense*, *T. simiae* and *T. vivax*. Mature infection rates in *G. m. submorsitans* used to transmit *T. congolense*, *T. simiae* and *T. vivax* were 6.9%, 3.9% and 78.4%, respectively, and the corresponding infection rates in *G. tachinoides* were 2.3%, 0.6% and 73.7%. It was concluded that *G. m. submorsitans* is a better vector of trypanosomes of the *Nannomonas* subgenus but both species are equally efficient vectors of *T. vivax*.

One of the many factors influencing the infectivity of tsetse may be the number of trypanosomes ingested at the infective feed. However, there is little evidence for a difference in transmissibility from blood with scanty parasitaemia compared with blood with high parasitaemia. A single *T. congolense* trypanosome can be sufficient to infect *G. m. morsitans* experimentally (Maudlin and Welburn, 1989). Above a threshold level, which was low (fewer than seven trypanosomes per fly), infection rates were independent of the trypanosome dose taken in by the fly. Maturation of midgut infections is completely independent of trypanosome dose in infective feeds.

11.4.1   Numbers of trypanosomes extruded during feeding

The number of trypanosomes extruded during feeding is one of the factors that might influence transmission rates from tsetse to vertebrate hosts. Harley *et al.* (1966) estimated the number of infective *T. b. rhodesiense* extruded by *G. morsitans* during feeding on blood pools through membranes. The blood pools were then inoculated into mice to assess their infectivity. The number of trypanosomes extruded was very variable; feeds in which no infective trypanosomes appeared to be extruded were much more numerous than expected, in contrast with observations from probing on to warmed slides, in which trypanosomes were generally detected.

In similar studies, Southon and Cunningham (1966) also estimated numbers of infective trypanosomes of the *T. b. brucei* subgroup in *G. m. centralis*. They concluded that the numbers of infective trypanosomes extruded in their experiments could vary by a factor of ten. The mean number of infective trypanosomes discharged at each feed was 23,000, which seems a surprisingly large number but, although larger than the numbers reported from direct counting methods, it was within the range recorded by Fairbairn and Burtt (1946).

Otieno and Darji (1979), using infected *G. pallidipes* from the Lambwe valley, Kenya, found that when trypanosomes in saliva from 'probed' tsetse were examined, flies infected with *T. b. brucei* always secreted saliva heavily infected with trypanosomes. In contrast, flies infected with *T. vivax* secreted saliva containing very few trypanosomes and sometimes none for several days. It is generally considered that few metacyclic *T. vivax* are extruded by tsetse when feeding.

11.4.2   Host-related factors

Differences in transmissibility, infectivity, antigenicity, pathogenicity and drug sensitivity among species and strains of the African pathogenic trypanosomes are of epidemiological importance (Gray and Luckins, 1980). Experiments have been conducted involving the cyclical transmission of stocks of *T. congolense* from locations in East and West Africa to rabbits, calves and sheep by *G. morsitans*. These studies showed differences in the occurrence of local skin reactions in different host species, whilst serological tests and immunization and challenge experiments showed antigenic differences between trypanosome stocks from different areas. The number of antigenic types of *T. congolense* in nature appears to be large (Wilson *et al.*, 1973), whilst that of *T. vivax* is comparatively lower, which may explain a higher degree of age immunity to *T. vivax* compared with *T. congolense* in cattle. A 'reversed age immunity' to trypanosomosis has been observed in young animals, in which some apparent immunity to infection is shown. This may result from young animals having a higher proportion of γδ T cells, which is a phenomenon observed with the young of various animal species. The link between T-cell populations and apparent immunity is speculation at present. Experimentally, negligible *in vitro* proliferative responses were observed in γδ T cells isolated from trypanosusceptible Boran cattle compared with marked proliferation in trypanotolerant N'Dama cattle, when stimulated with trypanosome antigens, although neither cattle type showed significant T-cell recognition of trypanosome VSG (Flynn and Sileghem, 1994). Younger cattle, which have higher levels of γδ T cells, are less susceptible to *T. congolense* infection (Wellde *et al.*, 1981) and this appears to be unrelated to the level of maternal antibody (Dwinger *et al.*, 1992).

11.5   ESTABLISHMENT OF TRYPANOSOME INFECTIONS

One of the barriers to successful development of trypanosomes in tsetse that has been discussed for many years is the crossing of *Trypanozoon* and *Nannomonas* trypanosomes from the gut to the ectoperitrophic space where maturation takes place before passing to the mouthparts or salivary glands.

It is now recognized that two processes are involved for a fly to become infective: firstly, the establishment of the infection; and secondly, the process of maturation of that infection. It has been established that lectins play a role in determining refractoriness to infection by killing procyclic trypanosomes, whilst symbiotic bacteria are involved in determining susceptibility to infection by a process that results in inhibition of those midgut lectins and thus reduces their killing effect. The progress in determining the nature of these processes is outlined in the following sections.

Harmsen (1973) investigated the 'barrier' to the establishment of *T. b. brucei* in the gut of *G. pallidipes* in an attempt to explain the extremely low natural frequency of infection in tsetse salivary glands. He argued that as trypanosomes could become established in the gut of tsetse there must be some barrier to the later stages of cyclical development in the fly. The only factor of major importance that he felt had been demonstrated was the age of the fly at the time of the infective feed. Harley (1971) obtained infection rates as high as 24% in male *G. pallidipes* that had taken their infective feed 15 h after emergence or thereabouts. Furthermore, laboratory puparia incubated at above-average temperature also usually produced flies that could attain higher infection rates when taking the infective feed early in life. This indicated that the barrier appears in mature flies, but was not fully developed until 24 h after emergence. Flies with depleted fat reserves (resulting from high metabolism during the puparial period) are likely to feed when younger than other flies and are thus more likely to pick up infections.

Whatever the barrier mechanism was, Harmsen believed it would exert a major mortality pressure on trypanosomes at a time of selective importance, and would involve more than simply the trypanosome's ability to penetrate the fly's peritrophic membrane (an aspect which is discussed further in section 11.5.1). Harmsen hypothesized that the barrier was related to the time the blood stayed in the crop. In older flies, due to growth of the peritrophic membrane, most blood passes to the midgut and relatively little to the crop. The blood that does go to the crop is transferred within 20 min. In very young flies, blood is retained in the crop for 45–75 min, but 24 h after emergence this time is much reduced. There is probably no dehydration, no digestive enzyme activity and no clotting or coagulation of blood proteins in the crop, in contrast with what takes place in the midgut. This less harsh environment could, therefore, allow trypanosomes to adapt physiologically to the fly. The evidence pointed to destruction of non-transformed trypanosomes in the midgut (after leaving the crop) of mature tsetse flies being the main barrier to establishment of infections.

Le Ray (1989) concluded that susceptibility of tsetse to trypanosome infection depended on two distinct barriers controlling the colonization of midgut infections and migration to salivary glands, respectively.

11.5.1 Role of the peritrophic membrane

The peritrophic membrane, secreted by midgut epithelial cells in the annular pad of the proventriculus, consists of an annular sheet of chitin with associated protein. The membrane forms an open-ended tube in the gut of tsetse flies and other insects. Wigglesworth (1929) first demonstrated the involvement of the peritrophic membrane in the trypanosome life cycle. There has since been much discussion on its role in tsetse as a barrier to infection with *T. congolense* and *T. b. brucei*-complex trypanosomes. The membrane's ultrastructure was described by Moloo *et al.* (1970) who discussed its penetration by trypanosomes in relation to structure.

Two theories put forward to explain how *Trypanozoon* trypanosomes pass from the gut to the salivary glands are as follows:

- Trypanosomes bypass the peritrophic membrane by passing around the open end in the hindgut before migrating back up the midgut and multiplying in the ectoperitrophic space; they then pass to the cardia and penetrate the soft, immature peritrophic membrane where it is secreted.
- Trypanosomes actively penetrate the peritrophic membrane and gut epithelial cells in the midgut and enter the haemocoel.

The original descriptions of the life cycle of *T. b. brucei* group trypanosomes assumed that they did not at any time penetrate the body cavity of tsetse. However, in a number of later studies, trypanosomes were detected in the tsetse haemocoel (Table 11.5) and Mshelbwala (1972) suggested that trypanosomes migrated to the salivary glands directly through the haemocoel. Large, vacuolated trypanosomes (giant multi-nucleate cyst forms) of *Trypanozoon* have been found within tsetse midgut cells and contain flagellae or whole moving organisms within them. Ellis *et al.* (1982) suggested that they might arise from some type of fusing of individual trypanosomes and might provide a mechanism for genetic exchange by giving rise to new individuals. East *et al.* (1980) suggested that as salivarian trypanosomes had been found in the haemocoel of tsetse, haemocytes might have a role in the control of such infections. They therefore examined the haemocytes of tsetse in haemocoelic fluid collected from the head after staining with Giemsa stain. Five types of haemocyte were described.

Interestingly, malaria ookinetes appear to penetrate the mosquito midgut by digesting the peritrophic membrane by means of a chitinase that they produce (Huber *et al.*, 1991). Trypanosomes may penetrate the tsetse peritrophic membrane in a similar way (Table 11.6). This involvement of the peritrophic membrane was dismissed by Welburn and Maudlin (1992), but the earlier studies on the role of the peritrophic membrane are reviewed here.

Table 11.5. Trypanosome infections detected in the tsetse haemocoel.

Tsetse species	Trypanosome species	Reference
G. p. palpalis	*Trypanozoon*	Mshelbwala, 1972
G. m. submorsitans	*T. b. brucei*	Mshelbwala, 1972
G. m. morsitans	*T. b. brucei*	Otieno, 1973, 1976
G. tachinoides	*Trypanozoon*	Mshelbwala, 1972; Otieno, 1973, 1976
G. p. gambiensis	*T. brucei*	Foster, 1963, 1964

Table 11.6. Trypanosome infections detected penetrating the peritrophic membrane or in the ectoperitrophic space.

Tsetse species	Trypanosome species	Remarks	Reference
	T. b. brucei	Between annular pad of secretory epithelial cells and invaginated portion of foregut	Yorke et al., 1933
G. palpalis	T. b. rhodesiense	EM photomicrographs of trypanosomes penetrating the peritrophic membrane	Fairbairn, 1957
G. morsitans	T. congolense	Passing through peritrophic membrane	Gordon, 1957
	T. b. brucei	In association with peritrophic membrane and in ectoperitrophic space 30 min after infective feed	Freeman, 1973
G. m. morsitans	T. b. rhodesiense	Can penetrate midgut cells and the peritrophic membrane 7 days post-infection (EM photomicrographs)	Evans and Ellis, 1975; Ellis and Evans, 1976, 1977
G. m. morsitans	T. congolense	In epithelial cells of anterior midgut	Ladikpo and Seureau, 1988

Yorke *et al.* (1933) showed that *T. b. brucei* was present in the space between the annular pad of secretory epithelial cells and the invaginated portion of the foregut. They also concluded that the parasite passed through the peritrophic membrane in this region, where it was being secreted and where, presumably, it was still fluid, to reach the lumen of the proventriculus and oesophagus. Photomicrographs of *T. b. rhodesiense* penetrating the peritrophic membrane in this region of the proventriculus of *G. palpalis* subsequently provided evidence for this (Fairbairn, 1957). Fairbairn (1957) concluded that, towards the centre of the proventriculus, even the peritrophic membrane lining was still sufficiently fluid for trypanosomes to penetrate. At around the same time *T. congolense* was shown to pass through the peritrophic membrane of *G. morsitans* and Gordon (1957) thought the trypanosomes passed through the liquid portion of the membrane immediately adjacent to the proventricular press. Describing the development of the peritrophic membrane and its relation to infection with trypanosomes, Willett (1966) and Ward (1968) suggested that trypanosomes might be able to penetrate the soft part of the membrane to enter the ectoperitrophic space. This theory was rejected by Harmsen (1973), who found that the peritrophic membrane did not grow faster during the period of transfer of blood from the crop to the gut. Furthermore, blood did not pass beyond the peritrophic membrane during or shortly after feeding.

Trypanosoma b. brucei has been found in the lumen of the gut, and in association with the peritrophic membrane, as well as in the ectoperitrophic space of tsetse 30 min after ingestion of the infective bloodmeal (Freeman, 1973). This was believed to result from trypanosomes penetrating the freshly secreted peritrophic membrane at its anterior end. Maudlin (1991) queried the fluidity of the freshly secreted peritrophic membrane, referring to studies of *Stomoxys calcitrans* which suggested that its peritrophic membrane is impassable (Lehane, 1976). Trypanosomes were not detected in any flies killed later than 30 min after the infective feed in either the peritrophic membrane or the ectoperitrophic space. *Trypanosoma b. rhodesiense* can penetrate the midgut cells and the peritrophic membrane (Fairbairn, 1957; Evans and Ellis, 1975; Ellis and Evans, 1976, 1977). However, no evidence was found of penetration of the 'soft' portion of the fully formed peritrophic membrane (Ellis and Evans, 1977) or of the hindmost portion of the midgut, and trypanosomes observed in that region appeared to be dying. Long midgut trypomastigotes have no surface coat, and were believed to penetrate the fully formed peritrophic membrane in the central two-thirds of that region. These observations supported the theory of an active penetrative process by the trypanosome, rather than passive uptake.

In laboratory experiments, penetration of the peritrophic membrane of *G. m. morsitans* occurred in the midgut 7 days after infection (Ellis and Evans, 1977). By 21 days post-infection the trypanosomes reached the proboscis, and 28 days post-infection they were found as clumps of epimastigotes attached by hemidesmosomes to the labrum. Later, free forms could be found in the hypopharynx. (The term hemidesmosome is used to describe the junctions between trypanosome flagellae and microvilli of internal organs of tsetse to which trypanosomes become attached; Molyneux, 1980b.)

Parasite migration down the labrum to the hypopharynx tip might occur and could provide the route by which parasites reach the salivary glands. Although Bursell and Berridge (1962) doubted that trypanosomes could withstand the acidity (pH 5.8) in the hindgut, Maudlin (1991) found that procyclic forms can withstand pH 5.5 and therefore supported the classical theory that trypanosomes pass round the free end of the membrane in the hindgut.

The following research questions were suggested to investigate the possibility of trypanosomes reaching the salivary gland lumen from the haemocoelic fluid:

- What proportion of wild-caught *Glossina* have haemocoelic infections?
- What method can be used most suitably for the detection of haemo-coelic infections in the field?
- Do the subgenera *Duttonella* and *Nannomonas* occur in the haemocoel?
- At what stage in the life cycle of tsetse do haemocoelic infections occur; is there a correlation between fly age and the presence of such infections?
- How do trypanosomes survive in the haemocoel of *Glossina*; is there an immune response to their presence mediated either humorally or through haemocytes; if a response does occur, how does this affect the ability of the fly to become or remain infective? (Jordan, 1976).
- Does encapsulation or melanization of foreign materials occur in the *Glossina* haemocoel? (This has been reported for *Crithidia* infections in some insects; Schmittner and McGhee, 1970.)

The theory of a developmental stage of trypanosomes passing through the haemocoel was considered unlikely (East *et al.*, 1983; Maudlin, 1991) following the discovery of an anti-trypanosomal humoral factor in haemolymph which specifically reduced motility of culture forms of *T. b. brucei* (Croft *et al.*, 1982). This humoral factor is trypanocidal *in vitro* (East *et al.*, 1983), and is evidence of innate immune control of trypanosome development in the haemocoel. East *et al.* (1983) believed that trypanosomes were unlikely to survive in the haemolymph and that invasion of the haemocoel was unlikely to be of significance in their developmental cycle.

Experimentally, addition of D-(+)-glucosamine to bloodmeals of tsetse flies restores the teneral state with regard to their infectability, and results in significantly increased prevalence of midgut trypanosome infections (Welburn and Maudlin, 1992; Welburn *et al.*, 1994). This may be interpreted as evidence that the peritrophic membrane represents no physical barrier to infection in older flies, and that its developmental state could not be responsible for the greater susceptibility of teneral flies.

The role of the peritrophic membrane has been investigated for other parasites and their vectors. In addition to the chitinase produced by malaria ookinetes referred to earlier in this section, *Leishmania* parasites also produce a chitinase, which allows them to escape from the peritrophic membrane of their vectors and avoid being excreted (Schlein *et al.*, 1991). The peritrophic membrane also acts as a barrier to infection of *Simulium* with microfilariae of *Onchocerca volvulus*, most of which become trapped (Lewis, 1953). In contrast, apart from being a potential barrier to infection, the peritrophic membrane of *Phlebotomus* sandflies seems to promote survival of *Leishmania* parasites by protecting them from enzyme activity in the vector midgut until they have differentiated into more resistant stages (Pimenta *et al.*, 1997).

11.5.2   Role of lectins and rickettsia-like organisms

Lectins

Lectins are a group of proteins, or glycoproteins, with a range of properties, notably that of binding specific carbohydrates, in which capacity they therefore act as agglutinins. Endogenous lectins or lectin-like substances are believed to be involved in cell–cell recognition mechanisms. Lectin-binding phenomena play a role in the interactions between parasites and host cells, and are important in invertebrate immunity (Rudin *et al.*, 1989). Both cellular and non-cellular immune reactions may occur against pathogenic organisms infecting insects; and agglutinins, of which lectins are the commonest identified, are amongst the non-cellular factors that have been investigated.

Pereira *et al.* (1981) first suggested the involvement of lectins in transformation of trypanosomes in their vectors, following their demonstration in the crop, midgut and haemolymph of *Rhodnius prolixus*, the vector of *Trypanosoma cruzi* in South America. Pereira and co-workers carried out much of their work with *T. cruzi,* and in the previous year had reported the presence of lectin receptors in its developmental stages (Pereira *et al.*, 1980). The lectins of *R. prolixus* agglutinated epimastigotes of *T. cruzi* but not the trypomastigote stage. Binding of lectins to the surface of *T. rangeli* stages and to midgut structures of *Rhodnius prolixus* has also been reported (Rudin *et al.*, 1989). Sugar residues exposed to the interacting surfaces appeared to be important in enabling the trypanosomes to pass through the gut epithelium. Lectins show specificity for carbohydrates such as sugars, for carbohydrates of erythrocyte membranes and for protozoan parasites such as trypanosomes. Lectin-like agglutinins have also been identified from sandflies, *Phlebotomus papatasi*, in which they are responsible for agglutination of erythrocytes and *Leishmania* parasites (Wallbanks *et al.*, 1986).

Table 11.7 lists the studies in which lectins of *Glossina* spp. have been demonstrated or characterized.

Maudlin and Welburn (1987, 1988) suggested that protective lectins were responsible for the increasing resistance, with age, to infection with both *brucei* group and *congolense* group trypanosomes. Lectins were apparently absent or blocked in the unfed (teneral) fly, thus permitting infection; in older flies, lectin production was stimulated by the bloodmeal, preventing subsequent infection. There is, however, evidence that older flies and non-teneral flies can be infected, particularly if starved (Gingrich *et al.*, 1982; Makumyaviri *et al.*, 1984b). Welburn *et al.* (1989) suggested that starving the fly would lower the level of lectins in the tsetse gut, thus allowing infection to take place.

Agglutinins from the midgut of *G. austeni* can be effectively blocked by D-(+)-glucosamine *in vitro* (Ibrahim *et al.*, 1984). Production of D-(+)-glucosamine results from activity of endochitinases generated by the

Table 11.7. Demonstration and characterization of lectins in *Glossina* spp.

Tsetse species	Location	Remarks	Reference
G. m. morsitans G. austeni G. p. gambiensis G. tachinoides	Haemolymph	First demonstration	Croft *et al.*, 1982 Ibrahim *et al.*, 1984
G. m. morsitans	Haemolymph	Agglutination of human erythrocytes	Ingram and Molyneux, 1988
G. tachinoides G. p. gambiensis G. f. fuscipes	Haemolymph	Isolated and partially characterized	Ingram and Molyneux, 1990
G. p. palpalis G. p. gambiensis	Midgut	Trypanolysin Trypanoagglutinin	Stiles *et al.*, 1990
G. longipennis	Midgut lectin–trypsin complex		Osir *et al.*, 1995
G. morsitans	Bloodmeal-induced lectin	Properties described	Abubakar *et al.*, 1995
G. tachinoides	Midgut	Two lectin systems described	Grubhoffer *et al.*, 1994

presence of RLOs in tsetse midguts. Blocking midgut agglutinins by feeding tsetse D-(+)-glucosamine with an infective bloodmeal results in a significant increase in the midgut infection rates of *T. congolense* and *T. b. rhodesiense* (Maudlin and Welburn, 1987). Experimentally, over 90% of *G. m. morsitans* fed with D-(+)-glucosamine for the first 5 days were infected 21 days later with *T. congolense*; that is a remarkably high percentage. Susceptibility to trypanosomal infection in tsetse, therefore, appears to be mediated through midgut lectins responsible for killing of trypanosomes as they enter the fly midgut. In susceptible tsetse, it is assumed that the action of endochitinase produced by RLOs leads to an accumulation of glucosamine in the fly midgut, which blocks the lectin-mediated trypanocidal activity.

MATURATION OF TRYPANOSOME INFECTIONS IN TSETSE. Trypanosomes entering the midgut of a tsetse fly are exposed to secreted lectins, which can have two quite different results (Fig. 11.2). Firstly, they may cause trypanosome death and may prevent the establishment of infections depending on the characteristics of the trypanosome, or the fly, or an interaction of both. The trypanosomes appear to be killed by lysis as a result, possibly, of membrane damage caused by lectin action. Secondly, the process of maturation of *T. congolense* in tsetse is initiated by lectins secreted in the fly midgut (Maudlin and Welburn, 1988; Welburn and Maudlin, 1989). Those studies showed that an established midgut infection required at most 72 h exposure to midgut lectins to begin the process of maturation. When lectin activity in the midgut was inhibited, the trypanosomes remained as pro-cyclic forms. Maudlin and Welburn (1988) concluded that midgut lectins secreted by

Midgut

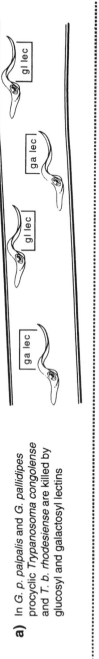

a) In *G. p. palpalis* and *G. pallidipes* procyclic *Trypanosoma congolense* and *T. b. rhodesiense* are killed by glucosyl and galactosyl lectins

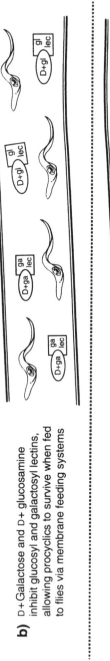

b) D+ Galactose and D+ glucosamine inhibit glucosyl and galactosyl lectins, allowing procyclics to survive when fed to flies via membrane feeding systems

c) In *G. m. morsitans* D+ glucosamine is generated by endochitinase produced by Rickettsia-like organisms (RLOs) in the tsetse midgut

$\overparen{\text{D+ga}}$ - D+ galactose
$\overparen{\text{D+gl}}$ - D+ glucosamine
◣ - Rickettsia-like organisms

ga lec - Galactosyl lectin ⎤ naturally occurring lectins. Both occur in *G. p. palpalis* and *G. pallidipes*
gl lec - Glucosyl lectin ⎦ Glucosyl lectins occur in *G. morsitans*.

Fig. 11.2. The role of lectins in trypanosome infection rates in tsetse flies.

tsetse were responsible for triggering maturation of pro-cyclic *T. congolense* and *T. b. brucei*. Furthermore, the maturation response of midgut trypanosomes to lectin stimulation varied between species and between stocks of the same trypanosome species, probably reflecting differences in numbers of lectin binding sites determined by trypanosome genotypes. Experiments with tsetse susceptible and refractory to infection suggested that susceptible flies did not secrete enough lectin (Welburn *et al.*, 1989). Mihok *et al.* (1992a) observed no effect of D-(+)-glucosamine fed *in vitro* to tsetse on their parasite loads, and following further experiments with a range of sugars concluded that the blocking of lectins was specific to D-(+)-glucosamine.

The activity of midgut lectins in tsetse may be affected by maternally inherited rickettsia-like organisms (RLOs) (Maudlin and Welburn, 1988). Lectin concentrations are higher in non-teneral than in teneral tsetse, which may explain why teneral flies are easier to infect with trypanosomes (Kaaya, 1989a). In addition to the glucosyl lectins, which are inhibited by D-(+)-glucosamine, Welburn *et al.* (1994) found a second, galactosyl lectin in the midgut of *G. p. palpalis*. This could explain the low trypanosome infection rates usually detected in *G. p. palpalis*, as glucosyl constituents, such as D-(+)-glucosamine, would not inhibit the galactosyl lectin. RLO infections in that tsetse species would have a less significant effect in potentiating trypanosome infections than, for example, in *G. m. morsitans*, which appears to lack the galactosyl lectin. In addition to lectin secreted into the midgut, a second lectin present in the haemolymph was essential to complete maturation (Welburn and Maudlin, 1990).

In mosquitoes, lectin binding is thought to be part of a multi-step process by which parasites recognize and attach to host cells prior to invasion by malaria parasites (Rudin and Hecker, 1989). It was suggested that interactions between tsetse lectins and parasite surface coats determine trypanosome transmissibility and may be partly responsible for the distribution of trypanosomosis in Africa within the limits determined by tsetse and cattle distribution (Welburn and Maudlin, 1990).

TRANSMISSION INDEX (MAUDLIN AND WELBURN). Maudlin and Welburn (1994) described a transmission index defined as the number of flies with mature infections divided by the number of flies with midgut infections, expressed as a percentage. This is clearly affected by the species of fly, as lectin and RLO contents differ between species; thus, *G. m. morsitans* gives a transmission index close to 100% with *T. congolense* whereas *G. p. palpalis* has a very low transmission index with the same stock.

Different stocks of trypanosomes also have different levels of transmissibility. *Trypanosoma b. rhodesiense* has a significantly lower transmission index than the non-human-infective *T. b. brucei*, which may partially explain the focal nature of human sleeping sickness compared with trypanosomosis of livestock (Welburn *et al.*, 1995).

Rickettsia-like organisms

Susceptibility of flies to infection with *T. brucei* and *T. congolense* is a maternally inherited character (Maudlin, 1982; Maudlin *et al.*, 1986). In studies of both laboratory-reared and wild flies, susceptibility appears to be associated with the presence of intracellular rickettsia-like organisms (RLOs) within the flies' ovaries and midgut cells (Maudlin and Dukes, 1985; Maudlin and Ellis, 1985). They are thought to act by producing chitinase in the larval and puparial stages of the fly. This leads to an accumulation of glucosamine in the midgut which blocks the anti-trypanosomal action of lectins on the first bloodmeal.

Rickettsia-like organisms (Fig. 11.3) were first reported from tsetse mycetomes by Reinhardt *et al.* (1972). Intracellular RLOs were subsequently described from midgut epithelium, cells associated with the fat-body and developing oocytes of *G. morsitans, G. fuscipes, G. brevipalpis* and *G. pallidipes* (Pinnock and Hess, 1974). Their discovery in developing oocytes led to the suggestion that they might be maternally passed from generation to generation via the egg cytoplasm or via the milk glands. RLOs from tsetse midguts show high levels of endochitinase activity, and tsetse with a high incidence of RLO infection have significantly greater chitinolytic activity and are more susceptible to trypanosome infection than tsetse with a low RLO incidence and which are refractory to midgut infection (Welburn *et al.*, 1993). Reinhardt *et al.* (1972) suggested that the link between rickettsia and susceptibility to trypanosome infection could provide a means of trypanosomosis control if antisera could be raised in domestic hosts against the rickettsia.

The role of RLOs was disputed by Moloo and Shaw (1989) who felt that, although they may play some part in the establishment of trypanosome infections in tsetse, they are not the only, or the most important, of such factors. This conclusion was reached following an experiment carried out on infections of *T. congolense* in *G. m. centralis* alongside electron microscopical examinations, which revealed RLOs within the mycetomes and the midgut epithelial cells of all teneral and non-teneral flies. The RLOs were more numerous in older flies. As RLOs were found in all uninfected or infected flies, susceptibility was not considered likely to be associated with these organisms. However, this conclusion has been refuted as various workers have shown that susceptibility to trypanosome infection is a quantitative character, dependent on RLO numbers (Maudlin *et al.*, 1986; Baker *et al.*, 1990; Welburn and Maudlin, 1991). Figure 11.4 shows the effect of increasing GlcNAc concentration in the infective bloodmeal of teneral male *G. m. morsitans* on infection rates with *T. b. rhodesiense*.

RLOs from tsetse have been cultivated *in vitro* by infecting a mosquito cell line with haemolymph from teneral tsetse (Welburn *et al.*, 1987). Maudlin *et al.* (1990) found a strong association between trypanosome (*T. congolense*) and RLO infection in wild *G. palpalis* in Liberia. However, infections of both organisms were at low levels, suggesting that the population under study was highly refractory to infection.

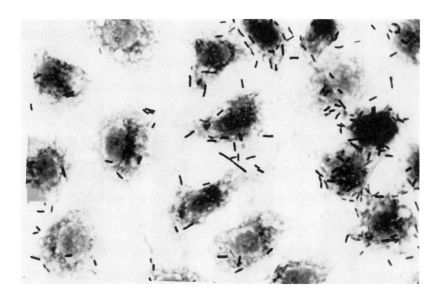

Fig. 11.3. Rickettsia-like organisms (RLOs) cultured *in vitro* in the cytoplasm of *Aedes* cells (from Maudlin and Welburn, 1988).

Flies carrying RLOs in the midgut were six times more likely to be infected with trypanosomes than flies without RLOs (Maudlin *et al.*, 1990) and their presence seemed to be homogeneous in the area studied. In a smaller sample of *G. pallicera* there seemed to be a similar but smaller correlation, but for *G. nigrofusca* there was little correlation, and most of the flies were RLO infected.

Moloo *et al.* (1987), comparing the susceptibility of different tsetse species for East and West African stocks of *T. vivax*, obtained mature infection rates in *G. m. centralis* and *G. brevipalpis*, East African stocks, of 61.1% and 75.3% for IL 2241 and 36.2% and 58.2% for IL 2337, respectively. In *G. austeni* and four *palpalis* group tsetse the infection rates for those two stocks were very low, ranging from 0% in *G. p. palpalis* to 1.8% in *G. austeni* for IL 2241 and from 0% in *G. f. fuscipes* to 5% in *G. tachinoides* for IL 2337. The West African stocks gave much higher infection rates for all species. Moloo *et al.* suggested that the differences could be due to biochemical characteristics of the attachment sites and efficiency of attachment. They discounted the possibility that RLOs played a role in determining susceptibility of tsetse to *T. vivax* infection, stating that, as development took place in the proboscis alone, they were not exposed to the RLOs in the midgut. Moloo referred to work of Maudlin *et al.* (1986), who also suggested that RLOs would not affect *T. vivax* infections in tsetse. Moloo *et al.* (1987) did, however, acknowledge that lectins could perhaps play a role in trypanosome differentiation and attachment, and subsequently observed that the maturation of *T. b. brucei* in tsetse is a

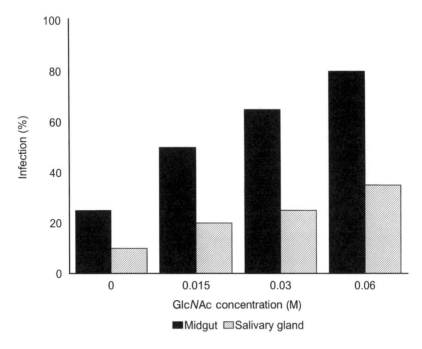

Fig. 11.4. Effect of increasing GlcNAc concentration in the infective bloodmeal of teneral male *G. m. morsitans* on midgut and salivary gland infection rates with *Trypanosoma b. rhodesiense* (from Welburn *et al.*, 1993).

complex interaction involving many interrelated physiological factors rather than simply an interaction between lectin and pro-cyclic trypanosomes in the midgut of non-teneral tsetse (Moloo *et al.*, 1994a).

The effect of temperature on trypanosome infection rates (section 11.3.2) might result from increased numbers of RLOs in the tsetse midgut when puparia are maintained at higher temperatures. This increase in numbers of RLOs then results in a higher susceptibility to trypanosome infection (Welburn and Maudlin, 1991).

A DNA probe for the detection of RLO in tsetse midgut has been developed from a fragment of the RLO genome representing a family of highly conserved repeats. It has been used for identifying RLO by dot-blot hybridization of homogenized midgut preparations (Welburn and Gibson, 1989). The usual method of detecting RLOs has been to stain them bright red with basic fuchsin, using a technique not suited to field use. Welburn and Gibson (1989) suggested that their dot-blot probe would provide a rapid and reliable method of assessing the potential susceptibility of a tsetse population to *T. b. brucei* and *T. congolense* infections. Maintaining the puparial stage of successive generations of tsetse at 3°C lower than normal reduces the number of RLOs carried by emerging flies (Welburn and

Maudlin, 1991). This reduces the susceptibility of these flies to midgut infection with *T. congolense* compared with control flies maintained at normal temperature, and provides evidence that the relationship between RLO and susceptibility is quantitative, teneral flies with heavier RLO infections being more susceptible to trypanosome infection. Welburn *et al.* (1993) proposed a model for susceptibility to trypanosome infection based on the generation of GlcNAc by RLO endochitinase activity in tsetse puparia inhibiting midgut lectin in teneral flies.

Virus-like particles

In addition to RLOs, virus-like particles (VLPs) are also found in the midguts, salivary glands or reproductive organs of tsetse. These particles have been described from the midgut epithelium of *G. f. fuscipes* (Jenni and Steiger, 1974a) and three kinds from *G. morsitans* (Jenni and Böhringer, 1976), one of which was the same as those found in salivary glands, discussed in section 11.7.5. They did not find VLPs in *G. austeni* or *G. brevipalpis*. Stiles *et al.* (1988) later reported virus-like particles in the nuclei of the midgut cells of *G. p. gambiensis* and suggested a possible correlation between VLP infection and the level of susceptibility to trypanosome infection. VLPs appear to occur within a small proportion of RLOs (Shaw and Moloo, 1993), and the presence of large numbers of VLPs results in a significant increase in the size and shape of infected RLOs. It may be through this effect that a correlation between VLP infection and the level of susceptibility to trypanosome infection occurs. In the study of Stiles *et al.* (1988), VLPs were apparently absent in flies refractory to infection with *T. b. gambiense* but present in susceptible flies.

11.6 GENETICS AND TRYPANOSOME INFECTION RATES IN TSETSE

Glossina species clearly differ in their response to infection; moreover, individuals and sexes of the same species respond differently, all of which suggests some involvement of the genotype. Jordan (1974) considered that any genetically controlled refractoriness to infection in tsetse would probably only be found in relation to *congolense* and *brucei* group trypanosomes and not in *T. vivax* infections, because of the relative ease with which *Glossina* are infected with this trypanosome. Heritability of vectorial capacity in haematophagous insects was reviewed by Gooding (1988b).

In many insect vectors, susceptibility to infection depends on the genotype of the insect and often on the inheritance of a single major gene (Wakelin, 1978). In addition to other factors, a genetic effect on the susceptibility of *Glossina* to infection has been inferred from work showing that infection rates were characteristic for particular tsetse populations. It is still not known whether there are genetic mechanisms that regulate the number of RLOs in tsetse midguts or whether environmental factors and random variation in numbers transmitted from a female to her offspring influence this. Maudlin (1982), who investigated the part played by the tsetse genome in determining a fly's susceptibility to infection with trypanosomes, suggested that a recessive gene on the X chromosome promotes the maturation of *Trypanozoon* infection. Further studies suggested that the gene product may be the haemolymph lectin previously shown to control trypanosome maturation (Welburn and Maudlin, 1990). Invasion of the tsetse salivary glands by *T. b. brucei* can also be affected by a sex-limited factor controlled by a single recessive gene on the X chromosome (Maudlin *et al.*, 1991; Moloo *et al.*, 1992a). This results in higher maturation rates in males, although in this study maturation of *T. congolense* was independent of sex.

Milligan *et al.* (1995) showed that the probability of maturation of *Trypanozoon* stocks that they examined decreased as the maturation time increased. They proposed a model in which it was assumed that product(s) from an X-linked gene(s) kills or prevents the establishment of a mature infection. These experiments showed that transmissibility of human-infective *T. b. rhodesiense* was lower than for *T. b. brucei*. Transmissibility of *T. b. gambiense* in the laboratory is close to zero (Dukes *et al.*, 1989).

The effect of the salmon eye mutation on heritability of vectorial capacity was referred to in Chapter 5. Mutant flies are more susceptible than wild-type flies to *T. congolense* and *T. b. brucei* infections.

In experiments on susceptibility to trypanosome infections in allopatric populations of *G. pallidipes* from Kenya, a population from Nguruman was significantly more susceptible to stocks of *T. congolense* from West and East Africa (Moloo, 1993b) and had higher infection rates with *T. vivax* (Moloo and Okumu, 1995). However, the *T. vivax* infection rate was only significantly higher for the West African stock.

11.6.1 'Sex' – genetic exchange between trypanosomes in tsetse

Since Tait (1980) and Ellis *et al.* (1982) referred to the possibility of genetic exchange in trypanosomes, this has been demonstrated between different strains of *T. brucei* occurring together as mixed infections in tsetse (Jenni *et al.*, 1986; Gibson *et al.*, 1995). It is thought that this non-obligatory genetic exchange takes place in the salivary glands and probably proceeds with fusion of diploid trypanosomes, followed by meiosis and fusion of a single pair of haploid nuclei, the rest being destroyed (Gibson *et al.*, 1995). How genetic exchange occurs is unclear, as mixtures of stocks are not observed in cattle, and may be rare in tsetse as it is claimed that the majority of *Trypanozoon* and *Nannomonas* infections are acquired in a single teneral feed (Welburn and Maudlin, 1992). However, novel diagnostic techniques have indicated that mixed infections in tsetse may not be uncommon (Masiga *et al.*, 1996a; Woolhouse *et al.*, 1996).

Experimentally, tsetse can become infected with double or triple mixed infections and can transmit these mixed infections to susceptible hosts

(Moloo *et al.*, 1982). Indeed, it can be comparatively easy to super-infect flies with a second trypanosome species or stock (Gibson and Ferris, 1992). This is contrary to expectations from previous studies (Van Hoof *et al.* 1937c; Wijers, 1958a; Distelmans *et al.*, 1982; Maudlin and Welburn, 1987). Mixed trypanosome infections can occur naturally in tsetse, including genetically different populations of the same species (*T. brucei*) (Johnson and Lloyd, 1923; Duke, 1936a; Clarke, 1969; Majiwa and Otieno, 1990; Stevens *et al.*, 1994). Thus, to develop a mixed infection, which is a prerequisite for trypanosome mating, flies do not necessarily have to pick up both parental trypanosomes on their first feed.

In a study of trypanosome infections in wild tsetse in Côte d'Ivoire, only 30% of 70 isolates tested by PCR from 124 infections isolated from tsetse were single infections. The remainder were mixtures of two or three of *T. congolense* savanna and forest types and *Trypanozoon* infections (McNamara *et al.*, 1995). The authors pointed out that if it is correct that successful midgut infections are only likely to establish at the first teneral bloodmeal (Maudlin and Welburn, 1987; Welburn and Maudlin, 1992) the three kinds of *T. congolense* must coexist in the same hosts, which is unlikely. They therefore concluded that tsetse could be infected sequentially with *T. congolense*. In a similar study using PCR to detect infections in *G. longipalpis* in Côte d'Ivoire, 16 mixed infections of the same two genotypes of *T. congolense* (forest and savanna) were detected out of a total of 50 infected flies (Solano *et al.*, 1995). Mixed trypanosome infections in tsetse, of either the same or different species, may not be uncommon, but further investigations are required, both for improving our understanding of the genetics of trypanosome populations and for evaluation of disease risk.

Studies of *T. brucei* stocks from a sleeping sickness area in Tororo District, Uganda, suggested that genetic exchange did not regularly occur and that clonal reproduction was the major genetic process in the epidemic (Stevens and Welburn, 1993). Tait *et al.* (1996) also believed that during an epidemic most tsetse flies would be infected with a single trypanosome stock and asexual reproduction would predominate.

11.7 EFFECTS OF TRYPANOSOME INFECTIONS ON TSETSE

Schaub (1994) reviewed findings regarding the effects of trypanosomes on tsetse flies and other insects. The main effects observed have been on feeding success and behaviour and on survival of infected flies.

11.7.1 Biochemical and physiological changes associated with trypanosome development in tsetse

Under favourable conditions, *T. b. brucei* undergoes a two-stage physiological transformation following ingestion by a tsetse. Firstly, development to pro-cyclic forms takes place (Newton *et al.*, 1973a,b); this probably starts in the crop (Harmsen, 1973) and is completed in the endoperitrophic space of the midgut lumen (Evans and Brown, 1972; Vickerman, 1974). Secondly, biochemical changes take place to allow proline to act as the substrate of oxidative metabolism in place of glucose (Evans and Brown, 1972). Glycoproteins in tsetse saliva almost disappear when metacyclic *T. b. brucei* are present in the salivary glands. It has been suggested that this may be associated with the formation of the glycoprotein surface coat when the trypanosomes differentiate to form metacyclics (Patel *et al.*, 1982).

11.7.2   Effect of trypanosome infections on feeding success and behaviour

Trypanosoma congolense parasites attach themselves to the cuticle of the tsetse proboscis by the flagellae, which form zonar hemidesmosomes. The flagellae are mostly attached parallel to the axis of the labrum and often pointed to its tip. Molyneux *et al.* (1979b) suggested a possible relationship between salivarian trypanosomes and *Glossina* mechanoreceptors on the labrum, following work done by Rice *et al.* (1973a,b) indicating that the function of the receptors is to ensure that the lumen of the labrum is not clogged by remnants of the bloodmeal. Molyneux *et al.* (1979) thought that there would be some impairment of the sensillae present in the labra of flies with salivarian trypanosome infections, which could result in more frequent probing. They observed a close relationship between the sensillae and *T. congolense* that colonized those areas. The trypanosomes were attached close to the bases of the sensillae in the labrum, and the tips of the sensillae were entangled in the rosettes of parasites. This was regarded as evidence that impairment of feeding would occur and that there would be more frequent probing, although the occurrence of more frequent probing was not confirmed.

It was later shown that trypanosomes become attached directly to the sensillae, which could impair their function (Thévenaz and Hecker, 1980). Both coated and uncoated forms were present in the labrum indicating that transformation to metacyclics may not be strictly limited to the hypopharynx. Uncoated forms were frequently attached to the cuticle of the common salivary duct and hypopharynx by hemidesmosomes. Thévenaz and Hecker thought this attachment could partly explain the low number of metacyclics deposited when flies probed on a warm glass slide.

Various workers have observed that trypanosome infections may affect feeding behaviour of tsetse. Thus, *T. b. brucei* infections in tsetse salivary glands cause them to probe more frequently, and feed more voraciously, than uninfected tsetse (Molyneux *et al.*, 1979b; Jenni *et al.*, 1980; Livesey *et al.*, 1980). The presence of clusters of epimastigotes of *T. congolense* in the labrum could possibly interfere in some way with the electrophysiological control of probing by tsetse flies (Evans and Ellis, 1983). This could have an important effect on the transmission potential of the fly. *Glossina m. morsitans* also probed more frequently when infected with *T. congolense* and required more time to engorge than uninfected flies (Roberts, 1981). In contrast, Moloo (1983) found that rosettes of trypanosomes in the flies' proboscides *did not* significantly affect feeding behaviour or probing. In a repeat of the first experiment, Moloo (1985) concluded that rosettes from *T. b. brucei*, or from *T. vivax* and *T. congolense* infections in the proboscis, caused no effect on feeding. This confirmed his view that the feeding behaviour of tsetse on mice (the same experimental animals used by Jenni *et al.*, 1980) was not impaired by trypanosome infections.

Because of the differences in behaviour of *Trypanozoon* infected flies observed by some workers, Livesey *et al.* (1980) claimed that the use of trypanosome infection rates as a measure of tsetse challenge would not be valid. Infected flies lived longer than non-infected ones and Livesey *et al.* suggested that challenge must be assessed in the field after accounting for variability in the behaviour of infected flies. Rosettes of *T. congolense* and *T. vivax* occur in close association with mechanoreceptors and evidence from scanning electron microscopy shows that the trypanosomes attach to the base of the mechanoreceptor sensilla or entwine around them (Molyneux and Jenni, 1981). The changed feeding behaviour of infected flies was thus thought to result from impaired function of the labral mechanoreceptors (Molyneux *et al.*, 1979b; Jenni *et al.*, 1980; Molyneux and Jenni, 1981).

11.7.3 Competition for energy metabolites

Mammalian bloodstream forms of *Trypanosoma* obtain energy by partial oxidation of glucose to pyruvate, while the insect form in the midgut uses partial oxidation of proline, which is the main source of energy for tsetse flight; thus, there is competition between the vector and parasite for this substrate. Bursell (1981) estimated the effects of trypanosome infections on the energy budget of tsetse and concluded that such infections could have a marked adverse effect on longevity. This conclusion conflicted with previous observations that infections with *T. b. rhodesiense* increased longevity of *G. m. morsitans* (Baker and Robertson, 1957; Livesey *et al.*, 1980). Bursell's explanation of this contradiction was that starvation was not a factor in the population dynamics of laboratory flies that did not have to expend energy on active flight searching for a host. Significant differences were observed between infected and non-infected *G. longipalpis* in Côte d'Ivoire (Ryan, 1984). Infected flies were more active and had lower fat reserves and a higher residual dry weight. Explanations suggested by Ryan were that reserves for flight energy may have been metabolized by trypanosomes and may have been at a lower level because of impaired feeding ability. Ryan's study did not appear to allow for age differences. Infected flies are generally older than non-infected ones, so some of the differences he observed may simply have been differences between older and younger flies and not related to trypanosome infections.

11.7.4 Altered feeding behaviour in other haematophagous vectors

Trypanosoma rangeli *and* Rhodnius *spp.*

Infections of *Trypanosoma rangeli* in triatomine bugs (*Rhodnius* spp.) have a marked effect on feeding behaviour, causing them to probe more frequently (Añez and East, 1984). In studies of the effects of *T. rangeli* infections in *Rhodnius prolixus*, infection caused the vector to pierce the skin of a host more often, and to draw less blood. This led the authors to conclude that salivary gland infection impairs the ability of *Rhodnius* to locate blood vessels by affecting salivary antihaemostatic properties. Apyrase activity of the infected salivary glands was also reduced (Garcia *et al.*, 1994). Haematophagous insects use the enzyme apyrase to inhibit platelet aggregation by hydrolysing ADP to ATP.

Leishmania *and* Lutzomyia *spp.*

In contrast to observations by Moloo (1983) that trypanosome infections in tsetse flies had no effect on probing responses, *Leishmania* infections cause more frequent probing in the sandfly, *Lutzomyia longipalpis*. At the same time, the infected hosts feed less well, yet still transmit the parasite (Killick-Kendrick *et al.*, 1977). The transmission of *Leishmania* by sandflies is an interesting example. It has been shown that infected sandflies maintained on sugar diets are good vectors whereas those maintained on blood are not. In a recent study (Schlein *et al.*, 1992), sandflies fed on sugar diets showed damage to the cardiac valve and other valve tissues, as a result of which the sandflies regurgitated their bloodmeal – containing parasites – into the host tissue. Sandflies fed on blood did not show this valve damage, apparently because blood or haemoglobin inhibited secretion of chitinolytic enzymes responsible for the valve damage. This suggests that *Leishmania* trans-mission, not previously fully elucidated, depends on sandflies with damaged alimentary tracts vomiting the parasites into their hosts whilst feeding! This fits in well with the observations of Killick-Kendrick *et al.* (1977) that sandflies that regurgitated their food would need to feed again and would consequently probe more frequently and this would result in increased parasite transmission.

Plasmodium *and* Anopheles *spp.*

Infected *Anopheles punculatus* vectors of malaria in Papua New Guinea feed more tenaciously than uninfected mosquitoes (Koella and Packer, 1996). This modified feeding behaviour will increase the chances of the malaria parasites' transmission. It was not clear whether the improved feeding of infected mosquitoes was due to longer feeding or more frequent attempts to feed until a full bloodmeal had been taken. The epidemiological importance of this observation in relation to the transmission dynamics of malaria was noted.

11.7.5   Effect of trypanosome infections on survival of tsetse

Initial observations suggested that *G. f. fuscipes* (classified at the time as *G. palpalis*) infected with trypanosomes survived as well as, if not better than, non-infected flies and suffered no adverse effects of infection (Duke, 1928). Similarly, tsetse infected with *Trypanozoon* parasites lived longer than uninfected ones (Baker and Robertson, 1957; Livesey *et al.*, 1980). Hecker and Moloo (1981) investigated the effects of *T. b. brucei* on the midgut cells of *G. morsitans* to determine if there might be a pathological effect that could influence the survival of infected tsetse. No difference between the relative organelle composition of infected and uninfected cells was observed, but there were indications of differences in cell volumes and differences in the presence of intracellular microorganisms in infected tsetse. They suggested that trypanosome infections could change the metabolism of tsetse and might lead to multiplication of the microorganisms and to tissue invasion. Whilst further studies were needed they cautiously concluded that cellular functions did not seem to be strongly impaired by either trypanosome infections or the microorganisms.

In contrast to early observations, *T. b. rhodesiense* is reported to reduce significantly the longevity of both sexes of *G. m. morsitans* 60 days after an infective feed (Maudlin and Milligan, in Schaub, 1994). Maudlin *et al.* (1998) demonstrated that the 'hazard' of dying increases with age more rapidly in *G. m. morsitans* exposed to *T. b. rhodesiense* infection than in uninfected flies and this effect is more marked in males than in females. This effect was statistically significant with *T. b. rhodesiense* infections but slight and not statistically significant with *T. congolense* infections. In a comparable study, experimental infections of a mosquito parasite, *Crithidia fasciculata*, produced severe infections of the haemocoel in five tsetse species, causing their death (Ibrahim and Molyneux, 1987). Similar work, in western Kenya, showed that *Plasmodium* infections did not reduce survival in populations of *Anopheles* mosquito malaria vectors (Chege and Beier, 1990). Makumi and Moloo (1991) detected no effect of *T. vivax* infections in *G. p. gambiensis* on feeding behaviour, longevity or reproductive performance of flies in laboratory experiments.

Glossina m. morsitans infected with *T. b. brucei* are more sensitive to topical applications of endosulfan and suffer a higher mortality from topical application of a natural pyrethrum extract than uninfected control flies (Golder *et al.* 1982, 1984). Infected pregnant female tsetse showed a less significantly higher mortality than males, concurring with the findings of other workers that pregnant tsetse are less susceptible to insecticides (Burnett, 1962; Hadaway, 1972; Fenn, 1992). Similarly, Nitcheman (1988) undertook an experiment using groups of trypanosome-infected and uninfected tsetse flies to compare their susceptibility to synthetic pyrethroids. In his control groups, longevity of infected flies was significantly lower than uninfected ones. No explanations for this observation were given. In a continuation of this work, the susceptibility to deltamethrin of female *G. m. morsitans* uninfected and infected with *T. congolense* was compared (Nitcheman, 1990). The results confirmed earlier observations that infected flies were significantly more susceptible to insecticide than uninfected ones. Uninfected control flies had a greater longevity than infected flies exposed to insecticide. Other studies also suggested that infected tsetse have a reduced longevity (Tarimo *et al.*, 1985).

Van den Bossche (1996) explained decreased susceptibility of infected tsetse to insecticides in relation to nutritional status. Flies with high fat reserves are more tolerant of deltamethrin as the liposoluble insecticide is diverted to those reserves, where it accumulates. Infected tsetse may be less tolerant of insecticides because trypanosome infections disrupt feeding activity, causing more frequent probing without an increase in feeding frequency whilst at the same time competing with the fly for metabolic substrates. This leads to a poorer nutritional status and lower levels of fat reserves to which insecticides can be diverted.

Pathological effects of Trypanozoon *infections in tsetse salivary glands*

Whitnall (1932) first reported the apparent pathological condition of infected salivary glands in tsetse. Burtt (1942) carried out investigations near Tanga, on the Tanzanian coast, into the ability of *G. brevipalpis* to transmit *T. b. rhodesiense* and found an unusually high infection rate (0.84%) in their salivary glands. No distinction was made between the subspecies of the *Trypanozoon* subgenus in this work, but it was thought most likely that they would be *T. b. brucei*. Salivary glands of the infected flies were chalky white when viewed by reflected light and a large proportion were brownish to nearly black under low power microscopy.

Burtt (1945) reported that hypertrophied salivary glands in *G. pallidipes* seemed to predispose them to trypanosome infection. Gouteux (1987b) reported enlarged salivary glands in *G. palpalis, G. pallicera* and *G. n. nigrofusca* in an area of human sleeping sickness in Côte d'Ivoire for the first time in that country. He did not find trypanosome infections in those flies, however. The glands were enlarged by four to six times and were chalky white. Virus-like particles (VLPs) had previously been described from normal salivary glands of tsetse (Jenni, 1973; Jenni and Steiger, 1974a,b). Virus-like rods were associated with hyperplastic salivary glands of *G. pallidipes* in Kenya (Jaenson, 1978), and there were very heavy infections of trypanosomes in some of the hyperplastic glands. This renewed interest in the question of whether or not the VLPs favoured development of trypanosome infections in infected flies. Jaenson's study also showed an association between enlarged salivary glands in male flies (due to infection with VLPs) and sterility. The testes of the infected flies were not functioning properly and the spermathecae were usually empty, raising the possibility of using the VLPs as a biological control agent. Jaenson commented that one way of transmitting the VLPs seemed to be from the adult female to its progeny; subsequently, an attempt was made to transmit VLPs causing salivary gland hypertrophy to tsetse (Odindo *et al.*, 1981). This was done either by haemocoelic microinjection or oral feeding with a pipette, in order to see if tsetse could pick up the infection during life or whether trans-ovarian and/or trans-ovum transmission was the most likely. The VLPs proved infective to the flies through both methods. The lifespan of tsetse with salivary gland hyperplasia was significantly shorter than flies with normal glands (Jaenson, 1986).

Salivary gland hyperplasia is due to a DNA virus infection and is associated with the presence of rickettsia-like organisms in the glands (Ellis and Maudlin, 1987; Minter-Goedbloed and Minter, 1989).

Clearly, trypanosome and virus infections of salivary glands might be expected to have an adverse effect on tsetse. Infections with *T. b. brucei* or DNA virus particles can cause a 'profound change' to the ultrastructure of *G. pallidipes* salivary glands, due to proliferation of cells, which gives rise to a stratified epithelium and gland enlargement (Kokwaro *et al.*, 1991). In that study, trypanosomes were found within the degenerating cytoplasm and lumen of the cell. During development of *T. brucei*, normal functioning of the tsetse salivary gland is impaired (Golder and Patel, 1980).

11.7.6   Effects of host trypanosome infections on tsetse feeding success

Following observations with other vector-transmitted diseases, for which vectors feed more frequently on infected than uninfected hosts, a study was carried out in Kenya which showed similar effects of trypanosome infections in cattle on the feeding success of *G. pallidipes* (Baylis and Nambiro, 1993c). Feeding success was 75% greater on cattle infected with *T. congolense* than on uninfected cattle. It was suggested that this difference might have been due to a higher tolerance of fly bites by infected animals, or a greater ease of feeding due to vasodilation and/or anaemia (Baylis and Nambiro, 1993c). The experiment was repeated using groups of oxen infected with *T. vivax* or *T. congolense* and with no infection. In this second experiment, oxen infected with *T. congolense* attracted up to 70% more *G. pallidipes* than uninfected oxen, or oxen infected with *T. vivax*. The feeding success of these flies was 60% greater on oxen infected with *T. congolense* than on other oxen (Baylis and Mbwabi, 1995a). The greater feeding success could result from vasodilation induced by *T. congolense*. Studies on the effect of packed red cell volume (PCV) on tsetse bloodmeal size have suggested that a reduction in PCV and increase in host body temperature act to increase the volume of blood consumed. These effects are symptoms of infection with pathogenic trypanosomes and might act to increase the possibility that trypanosomes are ingested by tsetse flies (Baylis and Mbwabi, 1995b).

11.8   INTERACTIONS WITH VERTEBRATE HOSTS

11.8.1   Relationships between trypanosome infection in wild animals and in tsetse

The numbers of infected tsetse and the species of trypanosome with which they are infected are related to the host species fed upon. The overall rate of infection with trypanosomes in tsetse flies has been positively correlated with the proportion of bloodmeals taken from bovids (Jordan, 1965b). Thus, generally, flies feeding preferentially on bovids will have higher infection rates than those feeding on other hosts (Jordan, 1965b; Harley, 1966b). In contrast, as the proportion of feeds taken from suids increases, the proportion of *T. vivax* infections in the tsetse declines. This is because suids are refractory to infection with *T. vivax*.

Data on infection rates, feeding habits and geographical distribution of *G. medicorum*, *G. p. gambiensis*, *G. longipalpis* and *G. m. submorsitans* in Côte d'Ivoire (Ryan *et al.*, 1986) have been compared with data obtained elsewhere (Ford and Leggate, 1961; Jordan, 1965b; Harley, 1966b; Clarke, 1969; Moloo *et al.*, 1971). There were significant correlations between the percentages of *vivax*-type infections and of Bovidae feeds. However, no relationship between the infection rate and the degree of latitude from the tsetse equator of 7°S, or between the percentage of *congolense*-type infections and the percentage of Suidae feeds was detected. High trypanosome infection rates often occur in *G. longipalpis*, which feeds largely on bushbuck (Jordan *et al.*, 1961); overall infection rates, varying between 21.5% (Page, 1959a), 35% (de Sequeira, 1935) and 24–29% (Morris, 1934), have been reported. Page observed that 82% of infections were of the *vivax* type whereas Morris found that 80% of infections were *congolense* type. Page did not explain this difference, but pointed out that the high infection rates suggested that *G. longipalpis* was of great economic importance as a vector of pathogenic trypanosomes to cattle and may have been responsible for the absence of zebu cattle in northern Benin.

Jordan (1964b), studying trypanosome infection rates of *G. m. submorsitans* in Nigeria, concluded that 'within the over-riding influence exerted by temperature through geographical latitude, infection rates are determined by the type of host that forms the principal source of food'. In some populations of *Glossina* the influence of the nature of the main food source could be sufficient to obscure the general effect of temperature in determining the level of infection. In a survey of trypanosome infections in 62 warthogs in The Gambia, the overall prevalence was 11% (Claxton *et al.*, 1992a). All trypanosome isolates examined proved to be *T. simiae*, causing infections that resulted in a prolonged period of low-level parasitaemia. Analysis of bloodmeals of *G. m. submorsitans* showed that up to 90% of its feeds were from warthogs, indicating their importance as maintenance hosts of the fly. In that study area, warthogs were not important reservoirs of trypanosomes pathogenic to cattle as no *T. congolense* infections were detected and suids are probably refractory to *T. vivax* infections. A recently named trypanosome species, *T. godfreyi*, of the *Nannomonas* group, the main host of which appears to be the warthog, was first identified in The Gambia (McNamara *et al.*, 1994).

Influence of host availability and feeding preference on infection rates
Despite the relationship demonstrated between overall trypanosome infection rates and types, host types and proportions of feeds from bovids or suids, the distribution of detected trypanosome infections in many wild animals bears little relationship to tsetse feeding patterns (Ashcroft, 1959b). Trypanosome infections have frequently been demonstrated in waterbuck and reedbuck, yet they are seldom bitten by tsetse. Their habits are such that contact with *Glossina* is infrequent. The bovids are generally less well adjusted to trypanosome infection than are suids.

An examination of the relationships between trypanosome infection rates and natural hosts of three tsetse species (Moloo *et al.*, 1971) showed that *vivax*-type infections originated from bovids while *congolense*-type infections came from bovids and bushpig. There was a positive correlation between trypanosome infection rates and increasing mean maximum temperatures.

The association of *G. tachinoides* and *G. palpalis* with domestic pigs is referred to in Chapter 8 and, as might be expected, there are high prevalences of infection with *Nannomonas* and *Trypanozoon* trypanosomes where this association occurs. Thus, in such an area of Nigeria, Killick-Kendrick and Godfrey (1963a) found a high prevalence of *T. b. brucei* and *T. congolense* infections in domestic pigs – 23% with *T. congolense* alone, 3% with *T. b. brucei* alone and 60% with both trypanosomes. They noted that previous experimental studies had shown that experimentally infected pigs had long-lasting parasitaemias with those trypanosome species and therefore could be important reservoirs (Mettam, 1940; Stewart, 1947). Further studies suggested that the rate of infection with *T. b. brucei* in *G. tachinoides* increased as the proportion of feeds from suids increased and the proportion from bovids decreased (Baldry, 1966c). The ingestion of pig blood seemed to enhance the infection rate with *T. b. gambiense* (Baldry, 1966c).

Ashcroft (1959b), discussing the feeding habits of tsetse in relation to their importance as reservoirs of trypanosomes, concluded that some animals, such as warthog, might be less important as reservoirs for trypanosomosis than might be expected from their importance as hosts. In contrast, other animals such as kudu, giraffe and reedbuck may be more important. He suggested that the number of tsetse carrying trypanosomes and the relative proportion of the different species of trypanosome may be closely related to the host animals on which tsetse feed. Not all infected bloodmeals eventually give rise to mature trypanosome infections in tsetse.

At the Kenyan coast, *G. pallidipes* had higher trypanosome infection rates in areas with mainly domestic animals than in areas with wild animals only (Tarimo *et al.*, 1984). This was attributed to the higher parasitaemias occurring in domestic animals compared with wild animals. Rogers and Boreham (1973) suggested that a smaller proportion of *T. congolense* infected bloodmeals establish themselves in tsetse (*G. swynnertoni*) than do *T. vivax* infected bloodmeals, and probably even a lower proportion of *T. b. brucei*.

In the Lambwe valley, Kenya, the overall rate of trypanosome infection in wild animals was only 16%, but 90% of bushbucks, which were the preferred host of *G. pallidipes*, were infected with trypanosomes (Allsopp, 1972). This demonstrates the importance of determining hosts utilized by tsetse as well as host availability. Bushbuck is a common host of *G. pallidipes*, which is the only vector of human sleeping sickness in the Lambwe valley. The association existing between bushbuck and *G. pallidipes* was considered extremely important in creating disease foci where humans could become infected through poaching activities. In two separate surveys, in early and late 1970, infection rates in *G. pallidipes* in the Lambwe valley were 20% and 30.9% respectively. In contrast to the high trypanosome infection rate in bushbuck reported by Allsopp (1970), in a study in Zimbabwe Knottenbelt (1974) found only one *T. congolense* infection out of 44 bushbuck examined and one mixed *T. congolense–T. b. brucei* infection from ten kudu. These parasites were detected by examination of wet smears and stained blood smears, which, as Knottenbelt was aware, are rather insensitive diagnostic techniques. As low parasitaemias would be expected in these natural hosts, it is probable that the actual trypanosome infection rate would have been higher. No apparent pathological effects were observed in the infected animals.

In laboratory experiments, the species of host on which *G. m. morsitans* was maintained after an infected feed could significantly affect the vector's subsequent infection rates with pathogenic trypanosomes (Moloo, 1981). It should be noted that some of the hosts used for this experiment were not natural hosts – for example, the lowest transmission rate was in mice.

11.8.2   Trypanosome infections in wild animals

Attempts have been made for many years to determine trypanosome infection rates in wild animals but, due to the difficulties in this type of investigation, data are still sparse. Results of such investigations are summarized in Table 11.8. One of the earliest attempts was that of Kinghorn and Yorke (1912b) in the Luangwa valley, Zambia.

In a survey for trypanosomes infecting wild ungulates in the Luangwa valley, out of 43 animals shot and examined, ten were infected with trypanosomes (Keymer, 1969). Similar results were obtained from a study in the same area 50 years earlier (Kinghorn and Yorke, 1912b). In both studies,

Table 11.8. Surveys for trypanosome infections in wild animals.

Location	Overall trypanosome prevalence (%)	General remarks	Reference
Luangwa valley, Zambia	28.0	16% with *T. b. rhodesiense*. Bushbuck and waterbuck were the most commonly infected species	Kinghorn and Yorke, 1912a,b
Luangwa valley, Zambia	23 (43 animals shot/examined)	Bushbuck and waterbuck were the most commonly infected species	Keymer, 1969
Luangwa valley, Zambia	14.5 (including 9 mixed infections)	*T. b. rhodesiense* detected in *Kobus ellipsiprymnus* and warthog	Dillmann and Townsend, 1979
Luangwa valley, Zambia		*T. b. rhodesiense* detected in zebra and impala	Mulla and Rickman, 1988
Botswana	15.7 (in buffalo, lechwe and reedbuck)	Highest prevalence in buffalo occurred at 2.5 years and then dropped	Drager and Mehlitz, 1978
Serengeti, Tanzania	7.5 infection in mammals		Bertram, 1973
Comoé National Park, Côte d'Ivoire		Highest trypanosome prevalence was in *Hippotragus*, buffalo and *Kobus defassa*	Komoin-Oka *et al.*, 1994
Busoga, Uganda		Trypanosome infections detected in bushbuck, Defassa waterbuck and elephant	Burridge *et al.*, 1970
Queen Elizabeth National Park, Uganda	11	Highest prevalence was in bushbuck	Mwambu and Woodford, 1972
Lambwe valley, Kenya	16	90% prevalence in bushbuck	Allsopp, 1972

examination of thin blood smears and animal inoculation were the methods used for identifying trypanosome infections. Keymer found 23% of ungulates to be infected, compared with 28% by Kinghorn and Yorke. The limited results from both studies showed that bushbuck and waterbuck were the most commonly infected host species and also demonstrated the correlation between fly infection rates and the proportions of feeds taken from bovids.

It was soon recognized that wild animals acting as reservoirs of trypanosomes generally tolerate infections or have a state of premunition. Carmichael (1934) investigated the nature of this tolerance with the aim of determining whether it arose through prolonged exposure to trypanosomes or if it was an innate characteristic that could be observed in animals taken from a tsetse-free area. Carmichael described the results of attempting to infect various animals captured in the wild in Uganda, some of which were resistant to infection (oribi, waterbuck) and others of which died (jackal and 'Ntalaganya' – this last animal is the black-fronted duiker, *Cephalopus caerulus melanrrheus*). He gave no conclusion regarding innate or acquired resistance.

Kinghorn and Yorke (1912a) also attempted to isolate trypanosomes from tsetse by feeding them on 'healthy' monkeys. In this way they claimed to have isolated three species of *Trypanosoma*, although only one of these species names would now be accepted. The three were *T. b. rhodesiense*, *T. pecorum* and *T. ignotum*. Sixteen per cent of wild animals were infected with *T. b. rhodesiense*.

By feeding *G. morsitans* on a reedbuck infected with *T. b. brucei*, Corson (1936) obtained high infection rates and suggested that the species of animal host may be a determining factor in the rate of infection of tsetse. He obtained much lower infection rates when monkeys provided the infective feed.

In the Busoga region of Uganda a survey of trypanosome infections in wild and domestic animals was carried out to investigate their possible role as reservoir hosts for *T. b. rhodesiense* in an area of endemic human trypanosomosis (Burridge *et al.*, 1970). Trypanosomes were detected in bushbuck, Defassa waterbuck and elephant. Also in Uganda, a survey of animals in the Queen Elizabeth national park revealed an infection rate of 11% in 72 animals shot and examined (Mwambu and Woodford, 1972). The highest infection rate was in bushbuck.

Ford (1971) showed that infection rates with *T. vivax* and *T. congolense* in *G. morsitans* and *G. tachinoides* were much greater when large wild animals were present in an area than when they were absent.

Garnham (1960) reported trypanosome infections in hippopotamus in Uganda; these were not identified definitively but were thought to be *congolense* type, possibly *T. simiae*. He referred to Kleine and Taute (1911) who found unidentified trypanosomes in hippopotamus in Tanzania. DNA probes have been used to identify *Trypanosoma simiae* in the white rhinoceros (*Ceratotherium simum*) and the dromedary camel (*Camelus dromedarius*) (Mihok *et al.*, 1994b).

Dillmann and Townsend (1979) surveyed trypanosome infections in 546 wild animals of 34 species in the Luangwa valley, Zambia, and found 79 infections, of which nine were mixed. Fourteen stocks of *Trypanozoon* were tested by the blood incubation infectivity test (Rickman and Robson, 1970a,b), three (two from *Kobus ellipsiprymnus* and one from *Phacochoerus aethiopicus*) retained their infectivity for rodents and were therefore considered to be human infective parasites. Human serum-resistant *Trypanozoon* species have also been detected in zebra and impala in the Luangwa valley, notwithstanding that tsetse rarely feed on either of those animals. The trypanosomes detected were most probably *T. b. rhodesiense* (Mulla and Rickman, 1988). Mulla and Rickman proposed that zebra and impala are natural hosts of *T. b. rhodesiense* in the Luangwa valley and may play a small role in the transmission of human sleeping sickness. Kinghorn (1925) had found that waterbuck and bushbuck were natural hosts of *T. b. rhodesiense* in the Luangwa valley, and that waterbuck had infection rates of between 17.8 and 25%.

In a serological and parasitological survey of the prevalence of trypanosomes in wildlife in northern Botswana (Drager and Mehlitz, 1978), the overall trypanosome prevalence in buffalo, lechwe and reedbuck was 15.7%; not surprisingly, there was a higher prevalence in areas with higher tsetse density. Maximum trypanosome prevalence in buffalo occurred at the age of 2.5 years and then dropped.

Epidemiological studies in a sleeping sickness focus of the Serengeti found a trypanosome prevalence in wild mammals of about 7.5% (Bertram, 1973).

In the Comoé national park in Côte d'Ivoire thin smears, antigen detection ELISA and the kit for *in vitro* isolation of trypanosomes (Truc *et al.*, 1992) were used for identifying trypanosome infections in wild animals (Komoin-Oka *et al.*, 1994). *Hippotragus*, buffalo and *Kobus defassa* were the most infected animals.

Modern diagnostic techniques have shown that a single host animal can be simultaneously infected with more than one distinct population of the same species of trypanosome (Scott, 1981).

Chapter 12

Epidemiology of Human Sleeping Sickness

In this chapter aspects of tsetse ecology relating to the transmission of trypanosomosis to humans are reviewed, as well as socio-economic and political factors that may be associated with epidemic outbreaks of disease in endemic areas. Other, sometimes controversial questions relating to the origins of the disease, whether *T. rhodesiense* arose from *T. gambiense* or not, and differences in the epidemiology of *rhodesiense* and *gambiense* sleeping sickness, are discussed in the light of observations during past disease epidemics.

No discussion of the epidemiology of human trypanosomosis would be complete without reference to Ford (1971). Ford argued against some earlier theories for epidemic human trypanosomosis and proposed new theories based on the contributions of civil disturbances, and of erroneous decisions (resulting from inadequate knowledge) made by the former colonial powers. Ford's study provides a wealth of information and a valuable insight into the epidemiology of this disease.

12.1 DISCOVERY OF CAUSE OF HUMAN SLEEPING SICKNESS AND MEANS OF TRANSMISSION

Trypanosomes were not known to infect humans until R.M. Forde detected them in the blood of a steamboat captain on the River Gambia in 1901. The infection was probably acquired in a region infested with biting flies, including tsetse, along the river. The captain of the steamboat, who had been employed in The Gambia for about 6 years at the time, was admitted to hospital in May 1901 and subsequently returned to England to recuperate. He went back to his job in The Gambia in December 1901.

Forde (1902a,b) described the clinical features of this case. He referred to 'small worm-like extremely active bodies' which he at first thought to be filaria, and associated sickness of the patient with their presence. The idea that sleeping sickness was caused by filarial worms was prevalent in East Africa also. Patrick Manson wrote a preliminary note to an article by Cook in the *Journal of Tropical Medicine* in 1901, stating that the possibility of the disease being caused by *Filaria (Mansonella) perstans* should be 'finally settled'. Cook (1901) had frequently found these filaria in the blood of sleeping sickness patients, but questioned how filaria could cause such differing diseases as elephantiasis and sleeping sickness.

In West Africa, at about the same time, Dutton visited Bathurst (The Gambia) where he was shown the blood of the patient referred to by Forde. He immediately identified the parasite as a trypanosome, which he described, and named it *Trypanosoma gambiense* (Dutton, 1902, in Baker, 1959).

It was Castellani (1903a) and Bruce (1908) who actually connected the parasite with the occurrence of the disease known already as sleeping sickness. Castellani, in 1902, examining a specimen of cerebrospinal fluid taken by lumbar puncture from a sleeping sickness patient, was surprised to observe a living *Trypanosoma* (Castellani, 1903a). This was the first time they had been found in a patient but he subsequently detected them in other patients from the sleeping sickness epidemic in Uganda. He was so surprised to find the trypanosomes that he 'thought it deserved a short publication'. At that time sleeping sickness was known, although not the cause, and 'Trypanosoma fever' was also known, although the two were not realized to be the same thing. Trypanosoma fever was not thought to be a fatal disease, but sleeping sickness was known to be fatal. It was only after Castellani's discovery that Bruce connected the two as the same disease, with trypanosomes being the causal organism. In his initial publication Castellani (1903b,c) had stated that: 'It must be clearly understood that these cases of trypanosome fever bear no resemblance in their clinical features to sleeping sickness.'

Bruce proposed that the trypanosome be called *Trypanosoma ugandense*, and *G. palpalis* (*G. fuscipes*) was implicated as its vector in Uganda at that time. An account describing Bruce and his life and the discovery of trypanosomes as the cause of sleeping sickness was published by Davies (1962a).

In 1903, Broden reported the first case of trypanosomosis in a European in the Congo. This and another case were later described in more detail (Broden, 1904). According to Broden, Brumpt had previously found a case of trypanosomosis in a European, believed to have been contracted on a boat in the Congo, at about the same time, but had not published it.

A few years later, in 1908, the polymorphic *T. b. rhodesiense* was found in humans in northern Rhodesia (Zambia) (Stephens and Fantham, 1910). The parasite was found in two Europeans who became infected in the Luangwa valley and both died after approximately 6 months. Stephens and Fantham thought the parasite was a different species as it was more virulent than *T. b. gambiense*. *Trypanosoma rhodesiense* was later found in Tanzania close to a fly-belt between Tabora and Kigoma in about 1926 and was believed to have been introduced from the south (Fairbairn, 1948). A period of uncertainty subsequently arose, as these two parasites and *T. b. brucei* in animals were morphologically identical. The relationship between the two trypanosomes and *T. b. brucei* and their origins has been discussed ever since. Figure 12.1 shows the suggested progress of sleeping sickness through the Congo and eastwards and southeastwards between 1897 and 1912.

As a curiosity, some alternative methods of human sleeping sickness transmission have been suggested. Soltys and Woo (1968) thought that leeches (*Placbdella rugosa*) could possibly be vectors of human sleeping sickness as they could be experimentally infected with *T. b. brucei* when fed on infected mice. They also believed that the trypanosomes underwent

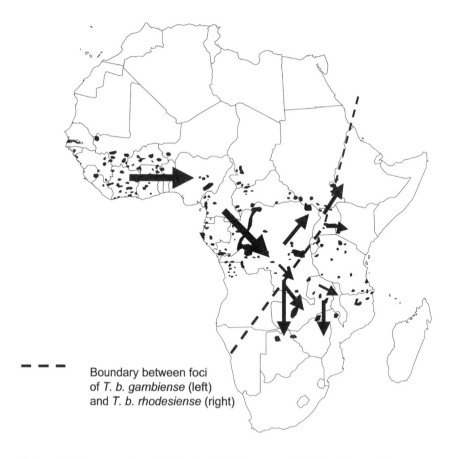

Boundary between foci
of *T. b. gambiense* (left)
and *T. b. rhodesiense* (right)

Fig. 12.1. Sleeping sickness foci and historical spread (1897–1912) from West to East Africa (adapted from Willett, 1965).

cyclical development in the leeches and suggested that further studies be carried out on their role in trypanosome transmission. Indeed, in recent experimental work *T. b. brucei* has been shown to be capable of reproducing inside the gut of a leech (*Hirudo medicinalis*) but experiments to transmit the parasite to mice were unsuccessful (Nehili *et al.*, 1994). Finally, it has been suggested that Maasai could become infected when drinking infected fresh cattle-blood (Soltys, 1971). This suggestion is not borne out by observations as sleeping sickness is uncommon among the Maasai people; furthermore, at least in recent times, the drinking of blood is a less common ceremonial activity amongst the Maasai.

12.2 EXTENT OF THE DISEASE: EPIDEMIOLOGICAL QUESTIONS RELATED TO ITS DISTRIBUTION

Human trypanosomosis is a major threat to human health in Africa, and occurs in distinct foci (Fig. 12.1). At present, there are about 200 disease foci in tsetse-infested areas in which the disease has occurred sporadically for very long periods. These foci, in 36 affected African countries, place 35–55 million people at risk (WHO, 1979) but only about 3 million of those people at risk are under surveillance and relatively few new cases are diagnosed annually (Knudsen and Slooff, 1992). Human sleeping sickness was largely under control in the 1960s, but its incidence has recently been increasing rapidly (Cattand and de Raadt, 1991). In 1994 it was estimated that there were as many as 150,000 cases in the Democratic Republic of the Congo (formerly Zaire), with prevalences as high as 70% in some villages (Cattand, 1994). It was recently estimated that approximately 300,000 people are infected (WHO, 1995), many of whom will die due to lack of effective treatment.

Clearly, the true number of people at risk or actually infected is very difficult to estimate with accuracy and it is accepted that the available figures are underestimates resulting from difficulties in surveillance and diagnosis (de Raadt, 1975; Kuzoe, 1989).

Human sleeping sickness has not recently appeared in such large epidemics as it did at the turn of the century, although Cattand (1994) stated that the disease was on the increase and should not be regarded as a disease of the past. This increase is associated with civil unrest and lack of effective administration in some countries of West and Central Africa. A recent estimate put the number of deaths from sleeping sickness in Africa at 55,000 per year (World Bank, 1993). It is characteristic of a recently evolved disease of humans to result in serious epidemics until it eventually becomes adapted to its new host and then persists, causing a less virulent disease which the host can survive. For parasitic diseases such as trypanosomosis, this would benefit both the host and parasite. This 'trypanotolerance' is marked in many species of wild animals and the trypanosomes that infect them.

At a workshop in Antwerp on modelling of trypanosomosis, although vector density was viewed as a factor of variable importance in estimating disease risk for humans, its measurement was considered necessary, as the ratio of vector to host numbers is a crucial variable in all present vector-borne disease models (Rogers and Muynck, 1988). De Raadt (1989), at the same meeting, thought that an epidemiological model might help to identify the correct parameters for answering questions determining the foundations of control operations.

Some questions specifically relating to surveillance and treatment for control of human trypanosomosis and requiring investigation were:

- Is there a level of vector density where a low endemic level of trypanosomosis can be assured?
- Is trypanosome prevalence, determined by passive surveillance (i.e. when sick people visit local dispensaries), a reliable guide to whether more resources should be devoted to, for example, active surveillance?
- What should be the interval between visits of mobile teams during such active surveillance?

There is little information concerning possible thresholds for transmission of human or animal trypanosomosis, although the threshold for human trypanosomosis could be much higher than for the animal disease (Rogers and Williams, 1993). Modelling data from a 'typical West African village situation', Rogers (1988a) concluded that human trypanosomosis could not be maintained below a density of about 2000 flies per person. The transmission threshold for *T. b. gambiense* was much higher than for *T. vivax* or *T. congolense*.

12.3 SOURCES OF CONFUSION

There are several confusing issues associated with accounts of human sleeping sickness. Firstly, and most straightforwardly, early accounts of the epidemic in Uganda refer to *G. palpalis* as the vector. This fly was subsequently reclassified as *G. f. fuscipes;* in the following account, this name is used to replace all references to *G. palpalis* in Uganda.

Other controversies relate to the identity of the parasites causing disease in humans, evolutionary origins of the parasites and theories concerning the geographical spread of the disease throughout Africa.

12.3.1 Identity of the parasite

Trypanosoma b. gambiense and *T. b. rhodesiense* are morphologically indistinguishable, and early diagnosis was made based on clinical signs of the disease in humans. Sleeping sickness caused by *T. b. rhodesiense* causes an acute disease, with death occurring after only a few months, whilst *T. b.*

gambiense causes a more chronic disease in which death may occur after several years. Diagnosis was supported by differing epidemiology and known geographical disease distribution. For example, *T. b. gambiense* generally occurred in West and Central Africa and was transmitted by *palpalis* group tsetse, whilst *T. b. rhodesiense* occurred in southern and East Africa and was transmitted by *morsitans* group flies. The two parasites were believed to overlap in Uganda (and possibly Tanzania and Sudan). The distinction between *T. b. rhodesiense* and *gambiense* in areas of overlap was further complicated as differences in virulence occur among both types of infection.

It has been argued that the original epidemic in Uganda was wrongly identified as being caused by *T. b. gambiense* instead of *T. b. rhodesiense* (Koerner *et al.*, 1995). If this is true, it probably applies to early diagnosis of *T. b. gambiense* in Tanzania also. Conversely, the current epidemic in southern Sudan is caused by *T. b. gambiense*, and the few early diagnoses of *T. b. rhodesiense* from that region are possibly incorrect. Similarly, the Rhodesian sleeping sickness epidemic in Ethiopia may have been wrongly diagnosed, as the nearest sleeping sickness focus is the one in southern Sudan (*T. b. gambiense*); Busoga, in Uganda, is the closest focus of *T. b. rhodesiense* sleeping sickness.

The parasites can now be identified by various techniques described in Chapter 13. However, we shall never know the true identity of the parasites causing early outbreaks in East Africa. The two parasite species are also morphologically indistinguishable from *T. b. brucei*, which does not infect humans, and until the late 1960s human infective trypanosomes could only be distinguished by inoculation and infection of human volunteers. For these reasons, much of the early research on fly transmission and reservoir hosts of specific parasites from those areas cannot be relied upon.

Controversy over the distinction between *T. b. gambiense* and *T. b. rhodesiense* continues despite current techniques for characterizing parasites, with some researchers still believing that *T. b. gambiense* and *T. b. rhodesiense* are simply strains of the same parasite.

12.3.2 Evolutionary origins of the parasites

Although it is generally accepted that sleeping sickness trypanosomes are derived from *T. b. brucei* (infecting wild and domestic animals), there are several theories about the evolution of species infecting humans. Firstly, the two types of sleeping sickness may be caused by one and the same parasite, and the disease is simply more acute when the trypanosome becomes adapted to certain species of wild animals, humans being incidental hosts (Ashcroft, 1963). Chronic disease occurs when the trypanosome is adapted to humans as its principal host. Duke (1935c) also thought that *T. b. rhodesiense* might 'metamorphose' into *T. b. gambiense* under certain circumstances.

Secondly, *T. b. rhodesiense* may have evolved as early as the late Miocene or Pliocene, when hominid types were exposed to certain strains of the *Trypanozoon* trypanosomes at the first stages of their occupation of the savanna. Subsequently, *T. b. gambiense* evolved from *T. b. rhodesiense* by mutation and subsequent adaptation when hominids invaded forest environments and became hosts of *palpalis* group tsetse (Lambrecht, 1964). In an interesting publication, Lambrecht also proposed theories for the differing virulence of the two trypanosomes.

Thirdly, the two subspecies evolved independently, from *T. b. brucei* or a common ancestral species (Baker, 1974). *Trypanosoma b. gambiense* probably evolved first, as there is evidence that it is better adapted to its human host and sleeping sickness in West Africa has been known for at least 600 years (Baker, 1974). In contrast, *T. b. rhodesiense* evolved relatively recently (perhaps within the last 100 years), in southeast Africa.

A fourth theory is that after *T. b. gambiense* spread or was introduced to savanna areas of southeast Africa, *T. b. rhodesiense* evolved from it and subsequently spread northwards (Willett, 1956). Ormerod (1960, 1961) shared this opinion, noting that a *gambiense* outbreak preceded every *rhodesiense* outbreak and resulted from a change in virulence of *T. b. gambiense* (Willett, 1956). Ormerod believed that the foci south of the Zambesi were the original source of *T. b. rhodesiense*, first suggested by Kinghorn (1925). Kinghorn observed that, although first diagnosed in Europeans in about 1909, local people maintained that the disease had been known 'as far back as their memories carry'. Furthermore, there were no signs of epidemic outbreaks that might have been expected had the disease been recently introduced.

Hoare (1962) disagreed with this theory, arguing that *T. b. brucei* was the ancestral form that gave rise first to *T. b. rhodesiense* and then to *T. b. gambiense* and that the evolution of the two diseases in humans proceeded from a purely animal infection to an anthropozoonosis (*rhodesiense* sleeping sickness), and then to a pure anthroponosis (*gambiense* sleeping sickness).

Following a review of the taxonomy of the sleeping sickness trypanosomes it was suggested that the two parasites should be classed as strains of *T. brucei* differing in terms of their epidemiology and clinical pattern (Ormerod, 1967).

Based on molecular characterization of the parasite, it is now considered most likely that *T. b. rhodesiense* had polyphyletic origins (Hide, 1997). Interestingly, Baker (1974) also considered this possibility but he thought it unlikely.

12.3.3   Geographical spread of sleeping sickness

Theories of the spread of sleeping sickness are outlined here, and further details are subsequently given in case studies of sleeping sickness epidemics.

Trypanosoma b. gambiense

A widely accepted theory, reviewed by Morris (1962b), is that *T. b. gambiense* sleeping sickness spread from West to Central Africa. In Morris's view the evidence was 'very convincing' that *T. b. gambiense* was introduced from West Africa into the East African Nile basin, along a route following the Ubangi and Uele rivers in the 1880s and 1890s. This spread resulted from big increases in trade and movement of people in the area. Morris pointed out the ecological similarity between areas of West Africa where sleeping sickness occurred and the area of the Albert Nile, referring to a continuous corridor of Sudano-Guinean climate and vegetation along which, he concluded, *T. b. gambiense* sleeping sickness spread.

Murray (1921) argued that *T. b. gambiense*, restricted to West Africa and the Congo until the late 1880s, spread throughout the Congo basin due to population movements associated with colonial military activities. The disease reached Uganda in 1896 and in the next few years spread around the shores of Lake Victoria and to Lake Tanganyika and parts of Zambia. Morris (1962b) also believed that West Africa and the lower Congo basins formed the original endemic foci from which *T. b. gambiense* was carried eastwards between 1885 and 1900.

Ford (1969), supported by Lyons (1992), thought it far more likely that the epidemic resulted from the prior spread of stress conditions such as famine, other diseases and social disruption (there have been few studies on effects of malnutrition on susceptibility to infection with trypanosomosis and there is some doubt about whether effects are synergistic or antagonistic; Targett, 1980).

Trypanosoma b. rhodesiense

It has been suggested that *T. b. rhodesiense* spread northwards either from Central Africa, through East Africa, or from the Zambezi valley in Zambia, where it evolved (Willett, 1956; Ormerod, 1960, 1961). This was based on first detection of the disease in Zambia, and its subsequent detection in western Tanzania in the early 1920s, Uganda and Kenya in about 1940 and the West Nile region of Uganda in the late 1950s and early 1960s. Ormerod (1961) described the spread of *T. b. rhodesiense* chronologically, from Zimbabwe to Tanzania and then to Uganda. It was believed that the disease increased in virulence the further north it progressed; thus, infections in Zambia were less acute than in Ethiopia, where *T. b. rhodesiense* sleeping sickness was only detected in 1967 (McConnell *et al.*, 1970). This is the northernmost extension of the known range of *T. b. rhodesiense* although there are unsubstantiated suggestions that the disease was present in the area much earlier (Langridge, 1976). Willett (1956) argued that *T. b. rhodesiense* originated in the Zambezi valley and spread to Malawi, Zimbabwe and Tanzania for the following reasons:

- Sleeping sickness reached the limits of *G. fuscipes* distribution at Lake Mweru, Lake Tanganyika and the Luapula river in 1907.

- No case had previously been documented in Malawi, Zambia or Zimbabwe.
- Trade routes existed between Lake Malawi and the mining areas of Zambia, the Congo and Lake Tanganyika.
- The first cases in Malawi in 1908 were imported and prevalence in Malawi and Zambia then increased from 1909 to 1911. The cases in Zambia occurred close to points where the trade routes crossed the Luangwa river.

An alternative explanation was that the cases arose in the Luangwa valley, from where they spread (Ormerod, 1961). Willett's only reason for rejecting that suggestion seems to be that the first case was found in the Luangwa only 4 years after it was found near Lake Malawi.

Van den Berghe and Lambrecht (1963) disagreed with Ormerod's opinion that *T. b. rhodesiense* sleeping sickness arose at the same time as its first identification in 1908 and suggested that it had simply not been reported before. They agreed that it had then spread northward, but thought this particular northward thrust may just have been one of a series, as other cases had been recorded in northerly places. They conjectured that the disease might have originated in West Africa and had then been spread southwards by the Bantu migrations. Ruppol and Kazyumba (1977) firmly considered that early foreign explorers certainly played a role in spreading sleeping sickness to previously uninfected areas; however, these theories have been strongly contested (Koerner *et al.*, 1995; Hide *et al.*, 1996).

Murray (1921), reviewing the introduction and spread of *T. b. rhodesiense* sleeping sickness in Malawi, disagreed with the conclusions of a Royal Society Commission that the disease was endemic in the country. In Murray's view, human sleeping sickness was first introduced to Malawi, either from the Lake Tanganyika region or from northeast Zambia, at the time of the discovery of *T. b. rhodesiense* in Zambia in 1907/8. Murray therefore believed that *T. b. rhodesiense* arose from *T. b. gambiense* in the Zambezi basin by some unexplained means, possibly involving its transmission through tsetse and wild animals. During this process the parasite became more virulent and adapted to *G. morsitans* as its vector, in place of *G. palpalis* or *G. fuscipes*.

Evidence against the spread of *T. b. rhodesiense* results from modern biochemical and molecular characterization, which has shown that strains of *T. rhodesiense* from Zambia and Kenya/Uganda probably have independent origins (Hide *et al.*, 1996).

Based on isoenzyme analysis and molecular characterization, *T. b. rhodesiense* appears to be a genetic variant of *T. b. brucei*, perhaps not justifying subspecific status, whilst *T. b. gambiense* does appear to justify such status (Hide, 1997). Hide concluded that the different subspecies probably originated from several independent sources and different strain groups were likely to be responsible for different sleeping sickness foci.

The two species of trypanosome were thought to overlap in Uganda, where the first recorded epidemic near Lake Victoria was thought to be due to *T. b. gambiense* and where, in the 1940s, *T. b. rhodesiense* was subsequently detected (Hide *et al.*, 1996). As both the acute and sub-acute forms of disease caused by *T. b. rhodesiense* occurred in Uganda, diagnosis was further complicated. Based on studies of molecular epidemiology, it is likely that both epidemics were caused by *T. b. rhodesiense* (Koerner *et al.*, 1995) although this was disputed by Gibson (1996). Currently, *T. b. gambiense* infections occur in northwest Uganda and *T. b. rhodesiense* in the southeast around Busoga. These epidemics are discussed further in this chapter.

Other anomalies in the beliefs regarding infection and transmission were evident in Nigeria where *T. b. rhodesiense* was thought to occur in an area where only *G. palpalis* was present (Lester, 1933). This was supported by evidence that *T. b. rhodesiense* could be transmitted by *G. fuscipes* in the laboratory (Duke, 1935c). Finally, Southon and Robertson (1961) proved the natural transmission of *T. b. rhodesiense* by *G. fuscipes* in southeast Uganda. At Tinde, in Tanzania, *T. b. gambiense* was transmitted by *G. morsitans* in the laboratory for 3 years. It thus seems that there is no fixed association between the tsetse groups and the trypanosome species causing sleeping sickness.

The first cases of human sleeping sickness were identified in Zimbabwe and Botswana in 1912 and 1935, respectively. This was regarded as evidence that the disease had not previously existed in the region and had only recently been introduced (Wyatt *et al.*, 1985). Thus, it was argued that *T. b. rhodesiense* was simply a virulent strain (of *T. b. gambiense*) introduced by travellers from the Democratic Republic of the Congo in the first decade of the century and became adapted to transmission by *morsitans* group tsetse.

12.4 EPIDEMIOLOGICAL DIFFERENCES BETWEEN *T. B. RHODESIENSE* AND *T. B. GAMBIENSE* SLEEPING SICKNESS

Human trypanosomoses, caused by *T. b. rhodesiense* and *T. b. gambiense,* differ in their epidemiology and in their virulence in mammalian hosts (Baker, 1974). Baker believed that *T. b. gambiense* did not occur with parasitaemias adequate to infect *Glossina* in most of the tsetse flies' hosts except for humans, and therefore depends on the human–fly–human cycle. In contrast, *T. b. rhodesiense* gives rise to higher parasitaemias and humans are adventitious hosts of an ungulate–fly–ungulate cycle. This view is probably incorrect as tsetse are able to become infected from hosts with parasitaemias undetectable by standard microscopical techniques.

Ashcroft (1963) proposed that the two types of sleeping sickness were caused by one and the same species of trypanosome. In his view, the disease was simply more acute when the trypanosome was adapted to certain species of wild animals, and humans were incidental hosts. Chronic disease occurred when the trypanosome was adapted to humans as its principal host.

'Gambian sleeping sickness' appeared to be a peridomestic disease in that the infection is acquired near the victim's home and all members of the family are equally at risk (Baker, 1974). In contrast, *rhodesiense* trypanosomosis was associated with areas where wild mammals exist that are fed on by the *morsitans* group 'game' tsetse. In consequence, individuals in high-risk 'bush' occupations were the group generally affected.

The Lambwe valley in Kenya is a well-studied example of a historical focus of human sleeping sickness in which isoenzyme analysis was recently used to study the genetic variation and phylogeny of trypanosomes. There are significant differences in zymodeme frequencies of *Trypanozoon* parasites over comparatively short geographical distances, suggesting that transmission of the parasite is somewhat localized. There is also a significant and stable association between the zymodeme and the mammalian hosts of the parasite, indicating that zymodemes are probably adapted to different host species (Cibulskis, 1992).

Isoenzyme analysis techniques have been used to identify two genetically distinct strains of *T. b. rhodesiense* associated with the acute and sub-acute forms of the disease from the geographical areas of Busoga in Uganda and Zambia. Human infective stocks of *T. brucei* collected in Uganda formed a clearly distinguishable population compared with other stocks circulating in the domestic cattle reservoir (Hide *et al.*, 1994). These data supported the occurrence of genetic exchange between trypanosomes from the cattle stocks, while an epidemic population structure involving limited genetic exchange was characteristic of the human infective stocks (Hide *et al.*, 1994). These workers also showed that in Tororo District, Uganda, the Busoga and Zambezi strain groups of *T. b. rhodesiense* must have very close origins and probably only represent genetic variants, rather than groups having differences of epidemiological significance.

Evidence that *T. rhodesiense* from the southern part of Africa was less virulent than that occurring further north has not been disputed, although the theory of virulence being associated with advances in distribution has. Recently, isoenzyme analysis has shown that two genetically distinct strains of parasite occur between Uganda and Zambia/Zimbabwe. In Uganda, both acute and semi-acute strains of the parasite occur together (Enyaru *et al.*, 1993).

The tolerance of *T. b. brucei* infection by domestic animals may be of great importance in the epidemiology of human trypanosomosis. Data presented by Hide *et al.* (1996) showed that, in areas of Uganda, between 21% and 33% of *T. brucei*-infected domestic animals (mainly cattle) could be infected with trypanosomes which were also infective to humans. This indicates their importance as reservoirs of *T. rhodesiense* infections, particularly where peridomestic wild animal populations are declining. Hide *et al.* (1996) estimated the probabilities of transmission of *T. rhodesiense* in southeastern Uganda shown in Fig. 12.2.

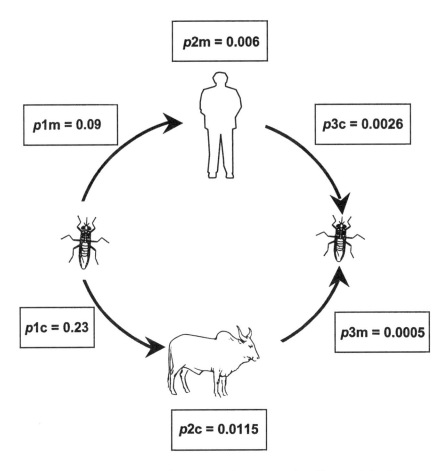

Fig. 12.2. The importance of domestic cattle as reservoirs of human-infective trypanosomes. Probabilities of transmission of *T. b. rhodesiense* calculated from data collected in Uganda. (From Hide *et al.*, 1996.) $p1$ = Probability of feeding on humans ($p1$m) or cattle ($p1$c); $p2$ = *T. b. rhodesiense* infection rates in humans ($p2$m) and cattle ($p2$c); $p3$ = the probability of a fly infected with *T. b. rhodesiense* is at least five times more likely to have picked up that infection from domestic cattle ($p3$c) than from an infected person ($p3$m).

12.5 BASIC REPRODUCTIVE RATE FOR TRYPANOSOMOSIS (R_0): ITS EPIDEMIOLOGICAL SIGNIFICANCE

The importance of the statistic for the basic rate of reproduction, R_0, in determining threshold levels was emphasized by Rogers and Williams (1993). R_0 is a measure of the potential rate of a parasite's transmission among hosts, and determines the number of new cases of a vector-borne disease that will arise in the future from a single initial case. This statistic is the main epidemiological parameter determining how resilient a protozoan parasite will be to control measures and is a significant factor in determining the level of vector reduction required for successful disease control. A successful evolutionary strategy of the parasite would be to maximize this value (Anderson and May, 1991). It is defined by the formula:

$$R_0 = a^2 mcb\, e^{-uT}/ur$$

where a is the biting rate of the vector; m is the ratio of vectors to hosts; c is the proportion of infective bites by the vector on uninfected hosts that give rise to infection in those hosts; b is the proportion of infected blood-meals that give rise to infections in uninfected vectors; u is the vector mortality rate; T is the incubation period of the parasite within the vector; and r is the rate of recovery of the host from infection.

The basic reproductive rate, R_0, for trypanosomosis is low, resulting in a disease with long periods of apparent very low endemicity, occasionally interrupted by epidemics. This is in contrast with malaria, which is a disease of high endemicity (Koerner *et al.*, 1995). This may have misled historians into believing that trypanosomosis 'spread'. What they observed (and misinterpreted) were intermittent epidemics in areas where the disease had always been endemic. Such epidemics appear to arise from various types of social upheaval, such as the economic activities associated with colonization in the Democratic Republic of the Congo (Lyons, 1992), consequences of the 'hut tax' in Uganda, and later, political instability during the Amin regime, also in Uganda. *Trypanosoma b. gambiense* has a long incubation period in humans before late stage clinical symptoms are apparent. This can vary from 2 to 10 years. An animal reservoir may not, therefore, be essential for its survival (although it has one), while *T. b. rhodesiense*, which kills its human host very rapidly, would undoubtedly require an animal reservoir. The misinterpretation of events and perceptions related to human epidemic sleeping sickness in the Congo, as seen by Lyons, is outlined in section 12.10.2.

Rogers (1988a) reported a threshold of around 0.26 flies per host for *T. vivax* transmission to animals and 150 flies per host for *Trypanozoon* transmission to humans. For trypanosomosis to persist, R_0 must be greater than 1.0; to be controlled, R_0 must be reduced to less than 1.0; and for eradication, it has to be kept below that level sufficiently long for the disease to disappear. This low level of R_0 may explain the highly focal nature of human trypanosomosis. For malaria, which has a very high value of R_0, it is estimated that R_0 would have to be reduced below 5 before any significant reduction in prevalence could be expected. However, illness or disease can be reduced before parasite prevalence, and for animal trypanosomosis there is a growing emphasis on control of disease rather than of infection (Dye, 1994a) except, perhaps, in areas where cattle act as reservoirs of human infective trypanosomes.

In relation to the dynamics of disease transmission and the basic reproduction rate, the question of whether control of the vector or detection and treatment of infected individuals would be the most appropriate strategy has been investigated. The traditional approach to control of human sleeping sickness in francophone Africa has been the latter, whilst in anglophone Africa control of the vector has usually been carried out in epidemic situations, in addition to treatment of infected individuals. Gouteux and Artzrouni (1996) compared the two strategies using a compartmental model, which indicated that in an endemic situation detection and treatment is more efficient than vector control, but when the disease is epidemic vector control becomes relatively more efficient than disease surveillance. The model showed that the probability of a tsetse bloodmeal being taken from humans was one of the most important parameters for determining control strategies. A possible criticism is that the model assumes that there is no animal reservoir for *T. b. gambiense*.

Welburn *et al.* (1995) speculated on the low transmissibility of *T. b. gambiense* and *T. b. rhodesiense* in relation to R_0 and the higher rate of maturation in male flies, pointing out that longevity of the fly was important as R_0 depends on the number of infective bites that will result from an initial feed on a primary case. If the mean lifetimes of male and female tsetse are about 23 days and 46 days, respectively, and the maturation time for human infective *T. brucei* subspecies is about 30 days, clearly the probability of transmission is small and is most likely to be due to female flies, despite their lower rate of trypanosome maturation. Welburn *et al.* (1995) concluded that the likelihood of genetic exchange occurring naturally between *T. b. rhodesiense* and *T. b. brucei*, or within *T. b. rhodesiense*, is small.

Trypanosoma b. rhodesiense is significantly less transmissible by tsetse than *T. b. rhodesiense*, providing further evidence of the divergence of these subspecies over time (Welburn *et al.*, 1995). In relation to the evolution of *T. rhodesiense* from *T. brucei*, the statistic R_0 has been discussed as an estimate for parasite fitness and trade-offs made between traits (Ebert and Herre, 1996). Welburn *et al.* (1995) suggested that the low transmissibility, close to zero, of *T. b. gambiense* by female tsetse in the laboratory is further evidence that this subspecies represents the final stages in an evolutionary process in which transmissibility through tsetse is traded for long-term infectivity to humans.

12.6   TSETSE RELATED ASPECTS OF EPIDEMIOLOGY OF HUMAN SLEEPING SICKNESS

Some aspects of tsetse ecology and research questions relating to the epidemiology of human sleeping sickness are listed below. These are discussed in the following sections in relation to examples of sleeping sickness epidemics. Factors thought to be responsible for the outbreak of epidemics are described, together with attempts that were made to control them.

- **Reservoir hosts.** Are there reservoirs of both *T. b. gambiense* and *T. b. rhodesiense*? If so, which are they?
- **The type of transmission cycle:**
  - human–fly–human contact – *T. b. gambiense*;
  - wild-host–fly–wild-host–human contact – humans as accidental hosts – *T. b. rhodesiense*.
- **Personal or impersonal contact between tsetse and humans.**
- **Adaptation of tsetse to peridomestic sites.** Association with domestic pigs or cattle which act as reservoirs of infection. Peridomestic breeding sites such as stacks of coco-yam tubers.
- **Sociological aspects affecting human–fly contact.**

12.7 RESERVOIR HOSTS, INFECTIONS IN WILD ANIMALS AND PROBLEMS OF DISTINGUISHING *T. B. RHODESIENSE* AND *T. B. GAMBIENSE*

Until recently, one of the difficulties in proving the existence of reservoir hosts for *T. b. rhodesiense* and *T. b. gambiense* arose from problems in distinguishing these parasites from *T. b. brucei*. Past methods used to demonstrate human infectivity of *Trypanozoon* stocks from wild or domestic animals included experimental infection of human volunteers (Denecke, 1941; Willett, 1956; Willett *et al.*, 1964; Onyango *et al.*, 1966), and the blood incubation infectivity test (BIIT) (Rickman and Robson, 1970a,b; Hawking, 1976a,b; Mehlitz, 1977). It is now possible to identify human infective forms of the parasites more reliably, using DNA hybridization, the polymerase chain reaction (PCR) (Noireau *et al.*, 1989) and biochemical methods, including the comparison of isoenzyme profiles of the trypanosomes (Godfrey, 1979; Truc *et al.*, 1991). The question of reservoir hosts has, therefore, been resolved to some extent and there is much more evidence for the existence of a range of wild and domestic animal hosts. It can still be concluded that they are likely to be less important in the epidemiology of *T. b. gambiense* than for *T. b. rhodesiense* sleeping sickness.

12.7.1 Reservoirs for *T. b. rhodesiense*

The main vectors of *T. b. rhodesiense* sleeping sickness are *G. morsitans* and its subspecies, *G. pallidipes* and *G. swynnertoni* (all from the *morsitans* group) and *G. fuscipes* of the *palpalis* group.

Initially, anecdotal evidence suggested that a wild animal reservoir was necessary for maintenance of the disease. For example, an area near the river Ugala, in Tabora district, Tanzania, was depopulated due to sleeping sickness in 1925–1926 and almost nobody lived there during the rainy season; yet when honey gatherers and Jackson's staff went there during the dry season some of them contracted sleeping sickness (Jackson, 1955). The wild animal population was suspected to maintain the disease as there was no possibility of infection from other infected people.

Many animals can be experimentally infected with *T. b. rhodesiense* and there is no doubt that the disease has an animal reservoir. Table 12.1 summarizes some of the natural and experimentally infected animals that may be reservoirs of the parasite. Fairbairn (1948) proposed that epidemics of *rhodesiense* sleeping sickness started in new areas after the arrival of a human being infected with *T. b. rhodesiense*. The role of wild animals was to act as reservoirs for maintaining the disease endemically following introduction.

Table 12.1. Reservoir hosts of *Trypanosoma b. rhodesiense*.

Experimental infections	References	Natural infections	References
Wild hosts			
Francolin	Corson, 1932a	Bushbuck	Heisch *et al.*, 1958; Kinghorn, 1925
Guinea-fowl	Corson, 1932a	Hyena	Geigy *et al.*, 1971, 1975; Awan, 1971, 1979; Onyango *et al.*, 1973
Muscovy ducks	Corson, 1932a		
Chimpanzee	Baker, 1962, 1968	Lion	Geigy *et al.*, 1971, 1975; Awan, 1971, 1979; Onyango *et al.*, 1973
		Hartebeest	Geigy *et al.*, 1971, 1975
		Warthog	Awan, 1971, 1979
		Zebra	Mulla and Rickman, 1988
		Impala	Mulla and Rickman, 1988
		Waterbuck	Geigy *et al.*, 1971, 1975; Awan, 1971, 1979; Kinghorn, 1925
Domestic hosts			
Cattle	Van Hoeve *et al.*, 1967	Cattle	Onyango *et al.*, 1966; Robson and Rickman, 1973; Gibson and Gashumba, 1983
Sheep	Fairbairn and Burtt, 1946; Van Hoeve *et al.*, 1967	Pigs?	Okuna *et al.*, 1986 (human infectivity not demonstrated)
		Dogs	Duke, 1914; Gibson and Gashumba, 1983

These lists are not exhaustive; other wild antelopes are presumed to be reservoirs of the parasite although not scientifically demonstrated. The methods used for identification of the trypanosome species include the card agglutination test (CATT), isoenzyme analysis, BIIT, DNA hybridization techniques and human inoculation.

Wild antelopes and hyena are well adapted to infections, which can persist for over 2 years without causing clinical disease and remain transmissible by tsetse, although there may be a tendency for the trypanosome to lose its infectivity to humans (Duke, 1937). Experimental infections have been made in less natural hosts such as chimpanzee (*Pan troglodytes*) with both *T. b. brucei* and *T. b. rhodesiense* (Baker, 1962, 1968); all three infected animals subsequently died. Previously, Corson (1932a) had shown that large species of birds, such as francolin, guinea-fowl and Muscovy duck, could be experimentally infected with *T. b. rhodesiense*.

Early attempts were made to isolate and identify *T. b. rhodesiense* from wild animals by Ashcroft (1958), Heisch *et al.* (1958) and Baker *et al.* (1968). Ashcroft (1958) and Baker *et al.* (1968) unsuccessfully attempted to show the existence of wild reservoirs of *T. b. rhodesiense* in Tanzania. In Ashcroft's experiments, a variety of wild animals were shot and blood taken from them was immediately inoculated into rats with the aim of isolating *T. b. rhodesiense*. When polymorphic trypanosomes were identified in Giemsa-stained smears, whole blood containing these strains was inoculated into human volunteers. No *T. b. rhodesiense* was detected, however, even though the samples were obtained from sleeping sickness foci.

Shortly after Ashcroft's unsuccessful attempt, Heisch *et al.* (1958) isolated *T. b. rhodesiense* in a bushbuck from a sleeping sickness focus on the shores of Lake Victoria in Kenya. Heisch *et al.* shot a variety of wild animals, and any strains of polymorphic trypanosomes detected were inoculated into human volunteers. The parasite isolated from a bushbuck was pathogenic to a European staff member. Heisch *et al.* (1958) suggested that one reason for the failure of Ashcroft's experiment may have been because too little blood was inoculated into his rats. On the nearby Sigulu island in Lake Victoria, waterbuck were thought to have been the reservoir of infection.

Using the BIIT test to investigate the prevalence of *T. b. rhodesiense* in host reservoirs in the Lambwe valley in Kenya, Robson and Rickman (1973) found that out of 18 positive identifications, 14 came from cattle. Cattle are undoubtedly significant reservoirs for the parasite in Uganda (Onyango *et al.*, 1966) and Kenya.

In an examination of 115 mammals in Musoma District, Tanzania, five trypanosome strains were obtained that showed a positive BIIT reaction, suggesting that they were *T. b. rhodesiense* (Geigy *et al.*, 1971). These strains came from hyena, lion, waterbuck and hartebeest. Further studies were carried out, using inoculation of human volunteers to confirm results of the BIIT to identify human-infective *T. b. rhodesiense* strains isolated from wild animals from the Serengeti area of Tanzania (Geigy *et al.*, 1975). *Trypanosoma b. rhodesiense* strains were isolated from lion, hyena and Coke's hartebeest.

During the 1980s sleeping sickness epidemic in Uganda, *T. brucei* infections were quite prevalent in domestic pigs on the northern shores of

Lake Victoria, but whether or not they were *T. b. rhodesiense* was not determined (Okuna *et al.*, 1986). Between 1983 and 1984, in the Busoga sleeping sickness focus of Uganda, low infection rates of between 1 and 2.4% were detected in *G. f. fuscipes*, with *T. brucei* group infections being the highest proportion (Okoth and Kapaata, 1986). The infection rates were lower than those detected during investigations in the early stage of the epidemic; for example, Rogers *et al.* (1972) found a 4.8% prevalence of *T. b. brucei* infection. In contrast, Kutuza and Okoth (1981) found only a 0.52% prevalence. In recent epidemiological studies of bovine trypano-somosis in Mukono District, on the northern lakeshore, infection rates with *T. brucei* in *G. f. fuscipes* of 3.8% were detected (unpublished data, LIRI/University of Berlin/ILRI collaborative research project). The latter infection rates and those of Rogers *et al.* (1972) are remarkably high for *T. brucei* infections in tsetse.

Role of reservoir hosts in localized distribution of T. b. rhodesiense infections in humans

As wild and domestic animals have been confirmed as reservoirs of *T. b. rhodesiense*, what explanation can be given for its localized distribution when compared with the distribution of vectors adjacent to sleeping sickness areas? Some of these animals are apparently better reservoirs of infection than others and Duke (1935b) suggested that *T. b. rhodesiense* might lose its power to infect humans after prolonged 'residence' in antelopes. Fairbairn and Burtt (1946) passaged an isolate of *T. b. rhodesiense* through sheep, various antelope species and other mammals; after passage by cyclically infected *G. morsitans* through these animals for 10.5 years, the trypanosome was still infective to humans. Maintenance in a single species of host without passage through tsetse resulted in a drop in infectivity to humans. Fairbairn and Burtt (1946) suggested that humans required a minimum infective dose of trypanosomes to become infected and that in wild animals the trypanosomes were 'diluted' with *T. b. brucei* to such a level as to be rarely infective for humans. Using a technique of inducing flies to probe on albumen-smeared glass slides, they estimated that the injection of 300–450 metacyclic *T. b. rhodesiense* were required for an average human to become infected. Later experiments to determine the minimum number of *T. b. rhodesiense* required to establish an infection in humans indicated that as few as ten infective parasites of that species might be required (Bailey and Boreham, 1969).

Another hypothesis for the localized disease incidence is that very low parasitaemias result from infections in some of the hosts, such as warthog, because of their resistance to *T. b. rhodesiense*. Consequently, the disease might not become endemic where those animals predominate. There was circumstantial evidence for this; for example, at Shinyanga, in Tanzania, where sleeping sickness was never diagnosed, warthogs were the main hosts of tsetse, whilst in Sakwa, where sleeping sickness occurred, bushbucks were the main hosts. Bushbucks have 'better' infections, i.e. higher parasitaemia, than warthogs. In a sleeping sickness focus in the Luangwa valley, Zambia, waterbuck were important hosts. The BIIT test was used in this area to investigate the potential reservoir hosts of *T. b. rhodesiense* (Awan, 1971, 1979). One stock was detected in a warthog, two from hyena, one from a waterbuck and one from a lion. This was the first time the trypanosome was positively identified in warthog.

Trypanosoma b. rhodesiense may be able to switch from human serum sensitivity to human serum resistance and may therefore exist in a region in both the human infective and non-infective forms, with domestic and wild animals acting as reservoirs for either form (De Greef *et al.*, 1989).

On several occasions symptomless human carriers of *T. b. rhodesiense* have been reported (Lamborn and Howat, 1936; Ross and Blair, 1956). In some cases trypanosomes were observed in the blood for several months with no clinical symptoms. Most infected people, however, are so ill that, according to Ashcroft (1959c), it is unlikely that they would have contact with tsetse and would therefore be unimportant as reservoirs of the disease. More recently, *T. b. rhodesiense* was diagnosed in humans and domestic and wild animals in the Luangwa valley using the card agglutination test for trypanosomosis (CATT), described in Chapter 13 (section 13.8.1) (Magnus *et al.*, 1978) and the antigen ELISA (Bajyana Songa *et al.*, 1991). Twenty-two per cent of humans in the area were, or had been, carrying subpatent trypanosome infections, suggesting that the *T. b. rhodesiense* reservoir in the region could be considerable. *Trypanosoma gambiense* can cause a similar asymptomatic infection in baboons (Kageruka *et al.*, 1991).

12.7.2 Reservoirs for *T. b. gambiense*

It has long been known that some wild animals can act as reservoirs for *T. b. gambiense*, although opinions still differ over their epidemiological importance (Khonde *et al.*, 1995). The possibility of a non-human reservoir was investigated in an attempt to explain the continuing presence of the disease in areas of very low endemicity in which it appeared impossible for a human reservoir to maintain the disease. The difficulty in experimentally infecting potential wild hosts of tsetse with *T. b. gambiense* suggests that the parasite is poorly adapted to animals and led people to believe that they were not important reservoirs. These views have subsequently changed to some extent, as difficulties in distinguishing the parasite from other *Trypanozoon* species have been overcome by modern techniques of diagnosis and characterization. The subject of *T. b. gambiense* reservoirs was reviewed by Mehlitz (1986).

Some of the early experiments on reservoirs of *T. b. gambiense*, particularly those carried out on parasites from the Ugandan epidemic at the beginning of the century, may be unreliable as there is doubt about whether that epidemic was really due to *T. b. gambiense* rather than *T. b. rhodesiense* (Koerner *et al.*, 1995). As early as 1915, Bruce carried out transmission experiments in which cattle were experimentally infected with *T. b. gambiense*. Laboratory-bred tsetse could pick up infections from these cattle and transmit the infection to others. Furthermore, infections of *T. b. gambiense* were found naturally infecting cattle and Bruce concluded that cattle could act as reservoirs for the parasite (Bruce, 1915b). In the 1940s, a natural *T. b. gambiense* infection was detected in a dog, and small domestic animals were suspected to serve as reservoirs (Van Hoof, 1947).

Both wild and domestic animals have been experimentally infected with *T. b. gambiense* (Table 12.2) and infectivity to humans can be retained when passaged via flies in goats, dogs and pigs for at least 18 months. Human infectivity could disappear after a longer period of maintenance in a given host (Van Hoof, 1947). It has been shown by xenodiagnosis that *T. b. gambiense* can persist in the blood of pigs for at least 70 days (Watson, 1962) and indigenous pig breeds are believed to act as ideal reservoirs for the parasite (Van Hoof *et al.*, 1937b). A polymorphic trypanosome of the subgenus *Trypanozoon,* and believed to be *T. b. gambiense,* has been isolated from domestic pigs in an area of human *T. b. gambiense* sleeping sickness in the Democratic Republic of the Congo (Kageruka *et al.*, 1977). Baboons can be experimentally infected with *T. b. gambiense* when infected by inoculation with a cloned parasite (although previously an innate serum factor which lysed *T. b. brucei* was reported) (Kageruka *et al.*, 1991). In a review of Van Hoof's work on *T. b. gambiense,* Fairbairn (1954) concluded that both wild and domestic animals could act as reservoirs for that species.

More recently, a wide range of tests (isoenzyme analysis, DNA hybridization techniques, the CATT test and resistance to human plasma) have been used to identify the parasite in wild animals, pigs, sheep, dogs and cattle from sleeping sickness foci in Congo, Côte d'Ivoire, Liberia and other parts of West Africa (Mehlitz, 1977; Gibson *et al.*, 1978; Mehlitz *et al.*, 1982; Zillmann *et al.*, 1984; Noireau *et al.*, 1986, 1989; Truc *et al.*, 1991). Dogs have been experimentally infected, although parasitaemias were low. Following characterization of *T. b. gambiense* in domestic and wild animals in Côte d'Ivoire, Mehlitz *et al.* (1982) suggested that human trypanosomosis was a zoonosis in that country.

These studies provided firm evidence of the potential of domestic and wild animals as reservoirs of human trypanosomosis, although the low parasitaemias and the low prevalence of infection commonly occurring in them have been cited as an argument against an animal reservoir (Noireau *et al.*, 1986). However, *G. m. centralis* can acquire trypanosome infections from cattle and goats with subpatent parasitaemia (Moloo *et al.*, 1986), and experimentally infected dogs can also be used to infect tsetse despite having low parasitaemias (Yesufu, 1971). Noireau's argument against these reservoirs may therefore be incorrect.

Table 12.2. Reservoir hosts of *Trypanosoma b. gambiense.*

Experimental infections	References	Natural infections	References
Wild hosts			
Duiker*	Van Hoof, 1947	Duiker*	Mehlitz, 1986
Waterbuck	Bruce *et al.*, 1911		
Bushbuck	Fraser and Duke, 1912		
Reedbuck	Van Hoof, 1947		
Baboons	Kageruka *et al.*, 1991		
Monkeys	Corson, 1938		
Domestic hosts			
Cattle	Bruce, 1915b; Moloo *et al.*, 1986	Cattle	Mehlitz, 1986
Goats	Van Hoof, 1947; Moloo *et al.*, 1986	Sheep	Noireau *et al.*, 1986; Scott *et al.*, 1983
Pigs	Van Hoof *et al.*, 1937b; Van Hoof, 1947	Pigs	Molyneux, 1980c; Noireau *et al.*, 1986; Gibson *et al.*, 1978; Mehlitz, 1977; Kageruka *et al.*, 1977; Zillmann *et al.*, 1984
Dogs	Van Hoof, 1947; Yesufu, 1971	Dogs	Denecke, 1941; Van Hoof, 1947; Zillmann *et al.*, 1984; Gibson *et al.*, 1978
Chickens	Mesnil and Blanchard, 1912	Chickens?	Zillmann and Mehlitz, 1979 (not proven to be *T. b. g.*)

* *Cephalopus dorsalis.*

T. b. gambiense is poorly adapted to wild animals and therefore occurs with low parasitaemias.

The methods used for identification of the trypanosome species include CATT, isoenzyme analysis, BIIT, DNA hybridization techniques.

12.8 HUMAN–FLY CONTACT

The belief that human–fly–human transmission was responsible for *T. b. gambiense* sleeping sickness formed the basis of control by 'sterilization' (treatment) of the human reservoir of infection. This was quite successful, indicating that transmission between humans and tsetse was indeed very important.

Yesufu (1971) suggested that *T. b. gambiense* sleeping sickness was most prevalent among people whose activities led them to frequent tsetse-infested streams and thickets; these were mostly farmers and hunters who kept dogs for companionship and hunting, and the dogs were implicated as potential reservoirs. A low prevalence of *T. b. gambiense* in domestic animals in the Congo led Noireau *et al.* (1986) to suggest that two cycles of human trypanosomosis may occur: firstly, a predominantly human-to-human cycle with trypanosomes of a low degree of virulence; and secondly, a minor cycle involving an animal reservoir. Whilst not disputing that an animal reservoir was possible or probable in West Africa, they believed that the epidemiology in the Congo differed in this respect.

12.8.1   Adaptation of tsetse to peridomestic sites and the effect of changing feeding habits on disease transmission

The distinction made by Baker (1974) between *T. b. gambiense* sleeping sickness as a peridomestic disease transmitted by *palpalis* group tsetse and *T. b. rhodesiense* as a disease associated with areas inhabited by wild animals and *morsitans* group tsetse is based on the habits and ecology of the vector species. In sleeping sickness foci in the humid savannas of West Africa, the main vectors, *G. palpalis* and *G. tachinoides,* are closely associated with domestic pigs, which are likely reservoirs of *T. b. gambiense* trypanosomes affecting humans. The ecology of tsetse in such peridomestic situations was described by Baldry (1980), who suggested that endemic transmission to humans could occur where pig densities were lower and flies therefore fed significantly more on humans, or away from villages with flies that had 'strayed' from the village transmission cycle. He also suggested that epidemic conditions could arise from changes in pig density resulting in altered tsetse feeding habits. Examples of peridomestic breeding sites of *G. tachinoides* described earlier (Baldry 1966b; 1970) were beneath stacked coco-yam tubers, at the bases of fences of pig enclosures, under *Lantana camara*, at the base of farmland fences and around derelict farm buildings. Such locations provided breeding sites with suitable levels of humidity throughout the year. Dried fronds of oil and coconut palms were the main night-time resting sites.

12.8.2   Personal contact or impersonal contact between humans and tsetse

The closeness of human–fly contact is particularly important for transmission of *T. b. gambiense* and is probably a major factor in determining the distribution of the disease. At one time it was thought that mechanical transmission must be responsible for epidemics, because of the low infection rates in *Glossina* and the difficulty in infecting them experimentally (Taylor, 1932a). Ecological situations in which close human–fly contact may occur have been described from several areas and its importance has been highlighted in Nigeria.

In Nigeria, the prevalence of sleeping sickness in an area may bear little relationship to the density of *palpalis* group tsetse present. Often, the incidence of disease may be high where the flies are scarce. Conversely, where tsetse are numerous, disease incidence is often low. Generally, when human habitation is close to a fly focus, disease incidence is higher. Sleeping sickness occurs mainly in the north of the country, despite the presence of one of the vectors, *G. p. palpalis*, throughout the country. This observation led to a comparison of human–fly contact in the north and south of Nigeria in an attempt to explain the distribution of the disease (Page and McDonald, 1959). The study was conducted at two sites: one at the West African Institute for Trypanosomiasis Research (WAITR) field station at Ugbobigha, considered to be typical of the humid forest areas of southern Nigeria, and the second near Kaduna in the north. Striking differences were found between the degree of human–fly contact at the two study sites, supporting the hypothesis that, although the vector was present in the south, contact was not sufficiently close for transmission to take place. In the late dry season near Kaduna, *G. p. palpalis* concentrated around permanent pools in streambeds, which provided the sole source of water at that time of year (Nash and Page, 1953). When people use these pools they become primary hosts of the flies and transmission of *T. b. gambiense* sleeping sickness can take place. Similar sites had previously been described in Nigeria (Nash, 1944b), where *G. p. palpalis* concentrated, in the dry season, near streambeds at which women queued for water. Tsetse flies were able to obtain their bloodmeals with little effort at these sites and if a person in the queue was infected, the parasite could easily be passed on to the flies and thus to other people.

The situation in Nigeria has been contrasted with trypanosomosis epidemiology in The Gambia, where *G. p. gambiensis* readily attacked canoe travellers along creeks of The Gambia river; but despite human–fly contact being close, it was impersonal (Nash, 1948; Mulligan, 1970). Endemic incidence of *T. b. gambiense* sleeping sickness was high around side creeks of the main Gambia river, where important foci of *G. p. gambiensis* existed in the dry season. In contrast, on the edge of the main river incidence was low, as human–fly contact only took place in the wet season. The type and the duration of contact determined incidence and intensity of infection. Impersonal contact gave a low level of infection, whilst personal contact gave high level infection.

In the Congo (Brazzaville), where *G. f. quanzensis* is a vector of sleeping sickness, mark–release–recapture experiments in villages along the Congo river suggested that each village formed an isolated focus with a low degree of dispersal of tsetse outside the village. Human–fly contact was high, favouring transmission, and this resulted in a mean trypanosomosis prevalence of 15%, ranging up to 40% in some places. Tsetse rarely flew from one village to another except when following people travelling in canoes (Eouzan *et al.*, 1981).

High intensity human–fly contact may occur with *G. palpalis* where human habitation is dispersed amongst plantations (Gouteux, 1985).

Consequently, Gouteux suggested that resettlement of scattered populations could be considered as a preventive measure against the disease. In a sleeping sickness focus in the Central African Republic, intensity of *T. b. gambiense* transmission was significantly correlated with the density of flies (Gouteux *et al.*, 1993), but, as in Nigeria and elsewhere, intensity of contact with humans is often inversely proportional to tsetse density (e.g. Nash, 1944b; Morris, 1952). In this area *G. f. fuscipes* fed more on humans than on pigs, compared with *G. p. palpalis* (Gouteux *et al.*, 1993).

Nash (1960) discussed the circumstances that might bring about close personal contact of tsetse with humans. He suggested that restriction of fly population movement would be an important cause and could result from ecological isolation, as in tsetse-infested sacred groves on the edges of villages, or from lack of natural hosts, where humans have destroyed wild animals close to villages. In the vicinity of villages the clearing of vegetation for cultivation would restrict tsetse to isolated areas of remaining habitat where increased seasonal contact could occur, particularly close to the hot, dry limits of tsetse distribution. All these factors apply particularly to *palpalis* group vectors of *T. b. gambiense* sleeping sickness in West Africa.

In Côte d'Ivoire, transmission of sleeping sickness occurs mainly at the boundaries of plantations and forest. The extended duration of human–fly contact increases the risk of contracting the disease, and the time spent in the plantation is a critical factor. The distribution of human trypanosomosis in the forest zone of Côte d'Ivoire is affected by social customs of 'closed' or 'open' ethnic groups (Hervouët and Laveissière, 1987). This is related to their methods of cultivation, which create vegetation types of differing suitability to tsetse populations and therefore affect levels of disease endemicity. Peridomestic transmission is more frequent in parts of Central Africa where tsetse are concentrated at the edges of villages, and where agricultural and domestic activities, and breeding of domestic pigs in the periphery of villages, are beneficial to tsetse populations. Familial infections are common in these foci, whilst in Côte d'Ivoire specific practices (cultivation techniques) are the factors affecting risk.

In *G. f. fuscipes* habitat in Sudan, watering sites for humans and their domestic livestock, mainly at wells, are major foci for the transmission of *T. b. gambiense* sleeping sickness. In the 1980s, infection rates of 11% were detected in people using these sites, compared with 4.5% in people using watering sites outside recognized *G. f. fuscipes* habitat (Snow, 1984).

Epidemiological studies of *T. b. rhodesiense* infections in humans in lower Kitete, northern Tanzania, showed that the disease was occupational in nature (Tarimo, 1980). Factors favouring infection were the herding of animals, *olpul* feasts for males, and firewood collection, water drawing and collection of building materials by women. Over 90% of all infections in the area were acquired by Maasai during *olpul* or firewood collection.

12.9 *TRYPANOSOMA BRUCEI GAMBIENSE* SLEEPING SICKNESS

In a series of reviews of the ecology of endemic sleeping sickness, Morris (1951, 1952, 1960, 1962b, 1963) argued that sleeping sickness in West Africa belonged, essentially, to the dry country of the savanna woodland zone, although the disease was assumed to have originated and spread from the forest country of the Gulf of Guinea far to the south. He quoted the earliest record of what is presumed to be sleeping sickness, from the upper Niger during the 14th century; Baron de Slane referred to this in his 1927 translation of the work of an Arab historian, Ibn Khaldoun. Morris suggested a relationship between the distribution of infection and the main, centuries-old trade routes used by caravans taking cola nuts from Ashanti to markets in the Niger and upper Volta region. Ibn Khaldoun, who wrote a history of north Africa, died in Cairo in 1406. Hoeppli and Lucasse (1964) also published translations of Ibn Khaldoun's description of the disease.

The disease was next reported on the Guinea coast in 1734 (Atkins, 1978). Several African names for the disease relate to the swollen lymph glands, which are a visible symptom. This came to be known as Winterbottom's sign after a description of the disease by Winterbottom in 1803. This sign was also recognized by slave traders, who did not wish to have slaves in whom it was visible.

The epidemiology of *T. b. gambiense* sleeping sickness was reviewed by Molyneux (1980c). *Trypanosoma b. gambiense* sleeping sickness is a slow disease and can take 2–10 years before the characteristic late-stage infection appears. Because of this long 'patent period' (the period between infection and appearance of clinical signs) the transmission dynamics that initiate an epidemic may have significantly changed by the time the epidemic is recognized. Mild or symptomless cases of *T. b. gambiense* sleeping sickness have been reported (Cooke *et al.*, 1937), but this may simply reflect the long period before late-stage symptoms appear.

Ashcroft (1959c) believed that *morsitans* group tsetse are unable to act as vectors of *T. b. gambiense* sleeping sickness and that distribution of the disease is largely coincidental with the distribution of *palpalis* group tsetse, halting in the Congo where the *palpalis* group flies give way to savanna tsetse. The main explanation for this was that *palpalis* group flies fed frequently on humans, whilst *morsitans* flies did not. *Glossina p. gambiensis, G. p. palpalis* and *G. tachinoides* are the main vectors in West Africa, whilst *G. f. fuscipes* was the vector in the Busoga epidemic in Uganda (but see section 12.11.4). Another hypothesis was that *T. b. gambiense* might transform to *T. b. rhodesiense* when passing through *morsitans* group flies.

12.10 *TRYPANOSOMA BRUCEI RHODESIENSE* SLEEPING SICKNESS

According to Ormerod (1961), the first recognized cases of *T. b. rhodesiense* occurred in present-day Zambia in 1908, and during the next 4 years there was an epidemic in Malawi. Human sleeping sickness had previously been reported from Zimbabwe in 1906 (Fleming, 1906, in Blair, 1939) and this was probably due to *T. b. rhodesiense*. In Ormerod's opinion, the epidemic spread northwards to central Tanzania in the 1920s and Uganda and Kenya in the 1940s. The disease became more acute with each extension, and the strains more lethal to experimental animals. In 1912, sleeping sickness was again reported from Zimbabwe, after the parasite was identified in an official of the British South Africa Company, based in Sebungwe district. He had never been north of the Zambezi, and Sebungwe district was said to be the only place at which he had come into contact with tsetse. The man in question became sick in March 1911 and subsequently died (Fleming, 1913). The disease remains sporadic and chronic in Zimbabwe with 'healthy carriers'. Ormerod suggested that *T. b. rhodesiense* parasites causing the Zambezi valley epidemic might have arisen from *T. b. brucei*, rather than from *T. b. gambiense*, as the site of the epidemic was some distance from any focus of the latter parasite.

Apted *et al.* (1963) compared the epidemiology of *rhodesiense* sleeping sickness in different parts of Africa; they confirmed that wild animals acted as reservoirs for the disease and described differences in epidemiology resulting from degrees of adaptation of the parasite to humans.

It is generally accepted that *T. b. rhodesiense* is mainly a parasite of the large African mammals and infects humans by accident due to a breakdown in the normal ecological relationships between wild hosts, humans and tsetse (Bursell, 1973b). Bursell thought this could also explain the virulence of the disease and why European hunters, constituting a small proportion of the population, had a high proportion of fatalities. This was due to the nature of their occupation, and travel by vehicles, which masked the unfavourable olfactory and visual stimuli of humans to tsetse.

12.11 EPIDEMIOLOGY OF SLEEPING SICKNESS IN WEST AFRICA

The following examples of disease outbreaks and their epidemiology illustrate the theories of the apparent spread of the disease from West, through Central to East Africa and the 'appearance' of *T. b. rhodesiense* outbreaks, which were thought to be spreading northwards, to Sudan and southwest Ethiopia. The initial virulence of new disease outbreaks and subsequent decrease in virulence may support some of the arguments put forward relating to the origins and spread of the disease. Convincing arguments have recently been made that the disease was always present in an endemic state, but went unrecorded until epidemics occurred, and the vector and parasite were identified. These epidemic outbreaks happened at the time of colonization, because of the resulting socio-political and economic disruption, which caused ecological imbalances, particularly in the Democratic Republic of the Congo and Uganda.

Scott (1957) suggested that epidemic trypanosomosis moved across West Africa from east to west during the first decades of the century. The earliest recorded occurrence of sleeping sickness was in the 14th century on the Upper Niger at latitude 5° W and the disease seemed to become widely endemic throughout West Africa during the 18th century. Epidemics occurred from Senegal to Burkina Faso between 1870 and 1900. At the time of the major epidemics in the Congo and central Africa, in the first decade of the 20th century, the situation in West Africa was much more stable, with mildly endemic levels. Exceptions were epidemics on the Senegal river and on the northern bend of the Black Volta river. In the 1920s other epidemics started in Cameroon, northern Nigeria, upper Côte d'Ivoire and Ghana. In 1926 an epidemic extended across northern Togo and Dahomey (Benin), and later northeast Ghana.

12.11.1   Côte d'Ivoire

In Côte d'Ivoire, studies have been conducted on the impact of human activities and behaviour on trypanosomosis epidemiology, particularly in relation to disease risk. Detailed studies on tsetse control using traps and targets with the participation of beneficiary communities have also been carried out in that country.

Glossina p. palpalis are widespread in all components of the ecosystem in sleeping sickness foci of Côte d'Ivoire, including forest, 'interstitial savannas', coffee plantations, tracks and villages (Challier and Gouteux, 1980). In these studies, most tsetse were caught at the forest edge whilst dispersal took place in open areas. Plantations provided resting, feeding and breeding sites, giving rise to close human–fly contact. Human behaviour is an important factor in determining the degree of contact with tsetse (Laveissière *et al.*, 1986b,c). Coffee plantations are more labour-intensive than cocoa plantations; thus, in the former, human–fly contact and infection risk are greater. The collective system of the Mossi people generated a continuous mixing between sick people, vectors and healthy people, resulting in a uniform distribution of disease among that ethnic group. Success of community participation in tsetse control in a sleeping sickness area depended on the rate of participation of 'planters', and their numbers at the village level compared with villagers with other occupations (Laveissière and Méda, 1992).

In West Africa, human trypanosomosis is frequently linked to peridomestic tsetse populations closely associated with domestic pigs. High densities of *G. tachinoides* are often found at villages and settlements where domestic pigs are kept (Kuzoe *et al.*, 1985). In some such foci, domestic pigs are important reservoirs of *T. b. gambiense* and close pig–fly contact, peridomestic feeding, breeding and resting behaviour may be important factors in maintaining transmission. Epidemic conditions may arise from changes in pig density and consequent changes in fly feeding behaviour (Baldry, 1980).

In one sleeping sickness area of Côte d'Ivoire, *G. palpalis* took almost 100% of its feeds from domestic pigs close to villages in the forest zone; whilst around plantations, 40–50% of feeds were from humans. The main human trypanosomosis foci were outside the village boundaries, where the fly fed equally on humans and antelopes, particularly the bushbuck (Laveissière *et al.*, 1985a).

12.11.2 Nigeria

Human sleeping sickness in Nigeria was predominantly a disease of the drier northern savannas and not of the southern forests, where tsetse were abundant (McLetchie, 1953). However, early records of the disease from Nigeria referred to its occurrence in slaves on the coast or in slaves who had been transported to America. These cases may have arisen from infections in slaves brought down to the coast from the interior, but the first recorded outbreak of any size was near the coast in Calabar province. Western Nigeria appeared to be free of the human disease (McLetchie, 1953), but was noted at Jebba, Ilorin province, in 1898 (Lester, 1933). The Fulani *Jihad* of the 19th century may have caused some spread of the disease although there is no conclusive evidence of this.

After 1900, there were major extensions of disease distribution resulting from socio-economic changes following the British occupation of Nigeria. The most important change was the free movement of infected humans from tsetse-free walled towns to infested bush villages (McLetchie, 1953). New villages and markets were constructed within the first 5 years of British administration and there was general movement back into regions depopulated by the slave trade (Duggan, 1962). Duggan (1962) considered there to be two main foci of the disease in Nigeria: in the Niger–Benue valley and the Lake Chad basin. *Glossina p. palpalis* and *G. tachinoides* occurred in the former focus, but only *G. tachinoides* in the latter. Sleeping sickness was relatively common in Benue province after 1900; and in 1912, following an epidemic of smallpox, the incidence increased. Macfie (1913a) suggested that, as sleeping sickness occurred sporadically throughout Nigeria, yet was not virulent or epidemic, it must have existed in the region for a very long time, allowing the indigenous people to acquire some degree of immunity. Even at this time (1912) the association of *G. tachinoides* with pigs in the southern part of the country had been noted.

There is little information on more recent incidence of sleeping sickness in Nigeria although it appears to have decreased and remains in only a few small persistent foci. The CATT test, used in an epidemiological survey of Bendel state in 1989, revealed that parts of the state were hitherto unreported endemic foci of sleeping sickness. Out of 670 individuals screened, 12.5% were positive and a further 6.7% had traces of antibody against *T. b. gambiense* (Edeghere *et al.*, 1989).

12.12   EPIDEMIOLOGY OF SLEEPING SICKNESS IN CENTRAL AFRICA

12.12.1   Democratic Republic of the Congo (formerly the Belgian Congo/Zaire)

There is little detailed historical information on human trypanosomosis in the Congo (Kinshasa) for the 1800s and the first few decades of the 1900s; consequently, the origins of the disease in the country are a topic of speculation and controversy. Locations of places cited in the text are shown in Fig. 12.3.

There are two conflicting theories regarding the spread of the disease in the present-day Democratic Republic of the Congo (Morris, 1962b; Ford, 1969; Lyons, 1992). One theory is that during this period there was an extension of the distribution of the disease, which probably began around 1885 with a first epidemic, followed by a second around 1920. This did not

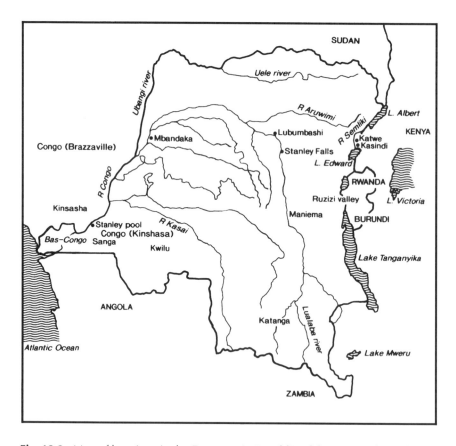

Fig. 12.3. Map of locations in the Democratic Republic of the Congo (formerly Zaire).

'stabilize' until 1945, by which time all west and central Africa were affected. The geographical origin of the first epidemic was thought to have been somewhere around the confluence of the rivers Congo and Ubangi, from where it rapidly spread northward, following the river Sanga, and westward, invading the present Republic of Congo (Brazzaville) (Burke, 1971). In 1887 the disease was endemic in the Stanley pool area, close to present-day Kinshasa, and by 1888 the disease reached Lake Albert on the border with Uganda. In the following year there was a severe epidemic in Busoga district of Uganda. In support of this theory it was suggested that the primary case(s) responsible for this epidemic were introduced via the caravans of Stanley who, in 1887, left his camp on the banks of the Lualaba to look for Emin Pasha. The first case further south in Katanga was diagnosed in 1906. The disease appeared to spread eastwards and was found in the Ruzizi valley and the shores of Lake Tanganyika around 1911 (Burke, 1971).

The British prime minister, Churchill, stated in Parliament in 1906 that the population of Uganda had been reduced from 6.5 million to 2.5 million, and there was an even less conservative estimate that the population of the Democratic Republic of the Congo was halved as a result of trypanosomosis (Burke, 1971). The epidemics resulted in the first International Congress on Trypanosomiasis held in London in 1907.

Alternative theories

The alternative theory, proposed by Ford (1971) and Lyons (1992), is that the disease had long been present in the region and was not introduced in recent times. Certainly the disease may have existed in the region for some time before, but no diagnosis was possible, nor had explorers witnessed it. Both Ford and Lyons suggested that social and ecological disruption during colonization resulted in epidemic outbreaks of disease. These disruptions included enforced population movements, shortages of food due to colonial forces travelling through the region and requiring food, and movements of people related to exploitation of natural resources – rubber, cocoa and coffee. The Arab slave traders also significantly disrupted human ecology in the region. The 'appearance' of sleeping sickness in the former Belgian and French Congo colonies coincided with the colonial occupation by French and Belgian people themselves. The theory that the disease simply happened to appear at the same time may be too much of a coincidence. Figure 12.4 shows a schematic framework of the way in which Lyons (1992) thought economic exploitation in the Democratic Republic of the Congo might have contributed to epidemics of sleeping sickness.

Responses to epidemics

In the Democratic Republic of the Congo, the approach to sleeping sickness control was historically one of 'sterilization' (treatment) of the human reservoir of trypanosomes rather than tsetse control. This was based on the

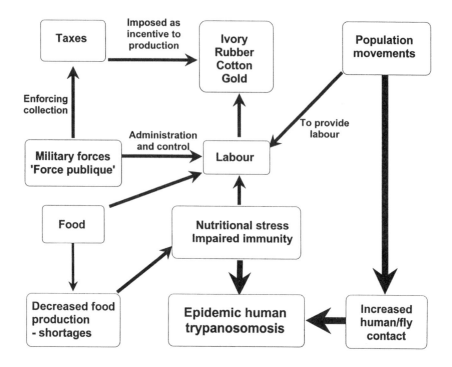

Fig. 12.4. Theory of economic exploitation of natural resources and epidemic human trypanosomosis.

premise that there was no significant animal reservoir of the disease. In 1906 a system for surveillance and treatment was established which remained the policy in most francophone countries. This consisted of:

- active surveillance or searching for suspected sick people;
- diagnosis by lumbar puncture;
- isolation of patients in quarantine;
- delimiting areas affected by the disease and establishment of medical diagnostic clinics;
- treatment with atoxyl, to which strychnine was added as an adjuvant;
- decreasing human/tsetse contact by bushclearing along river banks where there were points of access for people and boats or at livestock watering points.

These rules were made into law by the government in 1910, at which time powers were given to doctors to declare 'areas of infestation' – movements in and out of which could be officially controlled. Lyons (1992) viewed this control of movement as part of the colonial regulation, which caused further social disruption and, indeed, contributed to the epidemic.

Probably the first attempts at chemoprophylaxis were made by treatment of patients with injections of arseno-phenylglycine at 3-month intervals (Burke, 1971). Dr Lejeune, at Kiambi in Katanga, instituted a new method that became the basis of all subsequent prophylaxis (Lejeune, 1923). This consisted of using mobile teams of medical assistants to monitor the whole population of a region systematically, and establishing village centres of treatment at which people were methodically treated.

Atoxyl, which had been widely used for treating sleeping sickness, caused atrophy of the optic nerve and blindness in up to 30% of cases. Lower doses, which did not cause blindness, were not completely effective in killing all trypanosomes. Drug treatment improved from 1921 onwards, following experiments with Tryparsamide and Suramin (also known as Bayer 205). The drugs that have been used to treat human sleeping sickness are listed in Table 12.3.

In 1914 a new law introduced medical passports for people wanting to go further than 30 km from their chiefdom. Fishing licences were also introduced, along with other movement controls.

Progression of the disease in response to the control strategy

Unfortunately, few statistics on disease incidence are available for the period up to the early 1920s. Surveillance from 1918 into the 1920s revealed a declining prevalence of infection in Kwango, Ueles and Mayumbe districts, probably indicating that the control measures imposed were effective; by

Table 12.3. Drugs used for treatment or prophylaxis of human sleeping sickness in the Democratic Republic of the Congo.

Compound	Date of discovery (first use)	Remarks
Arsenic	1896	Highly toxic
Atoxyl	1905	Caused blindness at therapeutic doses
Antimony	1908	
Tartar emetic	1908	
Arsenophenylglycin	1908	
Suramin (Bayer 205, Antrypol)	1906	Early stage, both subspecies but mainly *T. b. rhodesiense*
Tryparsamide	1922	
Pentamidine	1937 (1944)	Early stage *T. b. gambiense* (*T. b. rhodesiense* has developed resistance)
Melarsen	1938 (1948)	
Melarsoprol (Mel B®)	1946 (1949)	Late stage
Eflornithine (DFMO)	1990	Ineffective against late stage *T. b. gambiense*

1930 more than 3 million people were being monitored annually, representing about 30% of the population (Burke, 1971). A better idea of the disease endemicity was possible after the end of the first world war and the incidence of new infections detected between 1926 and 1986 is shown in Table 12.4 and illustrated in Fig. 12.5.

Comprehensive data were available between 1926 and 1945 on numbers of sleeping sickness cases, and on the occurrence of drug resistance to arsenic compounds, which began to occur by 1930 (Van Hoof, 1947). Van Hoof believed that the more sensitive trypanosome strains were transmitted best by *Glossina*. At the same time, although initially they had accepted drug treatment, people began to show displeasure over the long series of injections to which they were subjected. In 1930, at the height of the epidemic, 33,502 new cases were detected. At about that time the Fondes Reine Elisabeth pour l'Assistance Médicale au Congo Belge (FOREAMI), was formed; it carried out control at several disease foci between 1931 and 1940

Table 12.4. Incidence of new cases of human sleeping sickness in Congo (Kinshasa), 1926–1997.

Year	New cases (%)*	Year	New cases (%)	Year	New cases (%)
1926	1.2	1947	0.1	1981	0.051
1927	0.95	1948	0.09	1982	0.057
1928	1.16	1949	0.068	1983	0.063
1929	1.12	1950	0.054	1984	No data
1930	1.2	1951	0.053	1985	0.072
1931	0.95	1952	0.044	1986	0.088
1932	0.75	1953	0.031	1987	0.107
1933	0.78	1954	0.022	1988	0.097
1934	0.63	1955	0.032	1989	0.096
1935	0.43	1956	0.027	1990	0.077
1936	0.36	1957	0.025	1991	0.058
1937	0.29	1958	0.022	1992	0.076
1938	0.27	1959	0.018	1993	0.114
1939	0.25	1960	0.010	1994	0.193
1940	0.24		No data	1995	0.182
1941	0.28	1976	0.038	1996	0.193
1942	0.26	1977	0.044	1997	0.22
1943	0.27	1978	0.058		
1944	0.27	1979	0.052		
1945	0.28	1980	0.048		

* Percentage of new cases amongst the population subject to monitoring.

Sources: Neujean (1963); Pépin *et al.* (1989). Data from 1976 to 1997 are preliminary data on the number of new cases provided courtesy of Dr P. Cattand, and incidence is calculated based on the estimated population at risk (10 million). These data therefore provide a rough indication of the disease situation for that period.

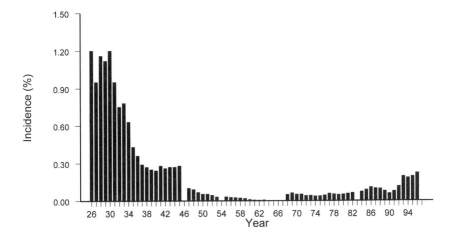

Fig. 12.5. Incidence of sleeping sickness in the Democratic Republic of the Congo, 1926–1997.

(Burke, 1971). During that period the incidence fell from 1.2% to 0.24%, whilst the population under surveillance increased gradually from 3 million to more than 5 million in 1940. The Second World War then disrupted sleeping sickness control, resulting in a reduction of the population under surveillance and an increased disease incidence.

Following the Second World War, a campaign of mass prophylaxis was introduced, using the new diamidines – propamidine and, later, pentamidine, which was better tolerated and was not affected by resistance to arsenic compounds (Burke, 1971). Pentamidine was useful for treating early cases but not for cases in which the central nervous system was involved. There was then a gradual decrease in disease until independence in 1960.

Active endemic foci remained in Kwango, Uele and Maniema and the problem of drug resistance increased (Van Hoof, 1947). It was therefore fortunate that at this time Freidheim (1949) discovered Melarsen and Melarsoprol, drugs that were capable of curing the disease (Burke, 1971).

As the incidence of disease was falling by 1959, people began to object more strongly to the continued monitoring and treatment. Nonetheless, at independence in 1960 about 6 million people were regularly monitored for human sleeping sickness. The incidence of infection continued to decrease just after independence, but so did the numbers monitored. At independence only one person in 10,000 was detected with sleeping sickness, representing 0.01% new cases; by 1997 this had risen to 0.22%.

Disease transmission

Since first being mapped by Dutton and Todd (1906), the distribution of trypanosomosis in the Congo basin has remained more or less the same

(Ford, 1969). The main geographical correlation is with the southern forest–savanna mosaic stretching from the lower Congo, between Mbandaka (Coquilhatville) and the northern Angola border, eastwards to beyond the Lualaba river, in a zone about 400 km wide. New epidemics in the 1970s and 1980s developed in exactly the same foci as the major epidemics of the early 1900s (Khonde *et al.*, 1995). This may have been a result of the ecological conditions at those sites. Khonde *et al.* found some evidence for acquisition of immunity after infection, which could explain the cyclical incidence of disease in sleeping sickness areas – a hypothesis suggested by Nash (1960), which merits further investigation.

12.13   EPIDEMIOLOGY OF SLEEPING SICKNESS IN SOUTHERN AFRICA

12.13.1   Malawi and Zambia

Kinghorn (1925) and others considered *T. b. rhodesiense* sleeping sickness to be an 'old' disease in the Luangwa and Zambezi valleys, and that the parasite originated there (Willett, 1956). The disease was virulent, and two Europeans who became infected in the Luangwa valley in 1908 both died in about 6 months. The virulence of the disease contrasted with the chronic nature of *T. b. gambiense* infections and was one of the main factors resulting in the identification of *T. b. rhodesiense* as a new species.

The first cases of the human disease in Malawi arose in patients who had either been travelling in areas in which the disease was found outside the country, or had been in areas where tsetse could have picked up infections from those first patients (Murray, 1921). The first case was detected near Nkata bay on Lake Malawi in an African who had accompanied a European to Tanzania and the Congo. He had returned home by steamer via Capetown and had been back in his home area, which was tsetse free, for 1 year without health problems. Nonetheless, trypanosomes were found in him and he appeared to be a 'healthy carrier' (Lamborn and Howat, 1936). This patient was moved to Dowa, 210 km south, in a *G. m. morsitans* infested area where, it is speculated, he was responsible for an epidemic of sleeping sickness that followed. Spread of the disease into Malawi resulted from population movements from Katanga and parts of Tanzania to the Lake Malawi region (Murray, 1921). Although the Malawi border was theoretically closed in 1907, in practice people were able to cross and enter northeastern Zambia. Despite surveillance of over 60,000 people for sleeping sickness, no case was detected, either before October 1908 or in previous years, in tsetse-infested areas of Malawi.

Buyst (1977), in a review of the epidemiology of sleeping sickness in the Luangwa valley (one of the main foci in Zambia), described the historical background as far as could be established. A high proportion of 'healthy carriers' characterized infections among the indigenous people of the Luangwa valley. It could be argued that this suggests that the disease had been established there for many years before its discovery. Buyst conjectured that haphazard contact between early humans and *fusca* or *morsitans* group tsetse led to resistance to infection with *T. brucei* strains from the savanna and forest, whilst frequent contact with *palpalis* group flies resulted in retention of susceptibility to riverine and lakeshore strains. Buyst suggested three main reasons for the occurrence of epidemic sleeping sickness in the Luangwa valley: firstly, what he called 'a collision of an expanding fly belt with the human habitat'; secondly, climatic stresses and a lack of wild animals on the northern edge of the Luangwa fly belt may have forced tsetse to feed more often on humans; and thirdly, wild animal movements may have resulted in expanding populations of tsetse during the rainy season, which then became hungry, accumulated near villages, and were heavily dependent upon humans for bloodmeals during the dry season.

The course of trypanosomosis in Zambians is usually sub-acute and thus differs from the acute course seen in Caucasians or in Africans from other parts of East Africa (Wyatt *et al.*, 1985). However, invasion of the central nervous system was very common by the time of diagnosis, and death rates were high unless early meticulous treatment was given.

Sociological aspects affecting human–fly contact and disease transmission
A study of risk factors associated with the acquisition of sleeping sickness in Zambia showed that birth outside the trypanosomosis endemic area did not increase risk (Wyatt *et al.*, 1985). Thus, although Buyst (1974, 1977) suggested that tribes that have lived for centuries in the tsetse fly-belts may have developed a partial resistance to *T. b. rhodesiense*, there was no evidence for this. Neither ethnic group nor main occupation affected risk, although fishing as an auxiliary occupation did increase risk of infection. Wyatt *et al.* (1985) suggested that members of the United Church of Zambia had twice the risk of other religious groups, perhaps because tsetse-infested areas had to be walked through to reach their scattered churches. Kinghorn (1925) also referred to the groups of people that appeared to be at risk in the Luangwa valley. Adolescents and adults were more at risk than young children; Kinghorn saw no cases in young children and rarely in people under the age of 15. The disease was more common in men than in women, and was associated with activities of men that took place in the bush, such as hunting, honey collecting and similar activities.

12.13.2   Botswana

Glossina m. centralis is the only species of tsetse present in Botswana and is found only in the N'gamiland and Chobe districts in the north of the country. In some years, human sleeping sickness reached epidemic proportions in those areas (Lambrecht, 1972). About 45,000 people are at risk near the Okavango swamps. The disease was first known as *Kotsela* – drowsiness – before sleeping sickness was described from Central Africa. It was first diagnosed microscopically in Botswana in 1934 and in 1939 the parasite was identified as *T. b. rhodesiense*. Examination of blood smears from 80 wild animals, shot to investigate the natural hosts of the parasite, showed that reedbuck were the most commonly infected species. Case histories revealed that most people became infected during hunting trips rather than from contact with tsetse in or near settlements. Thus, infection rates were higher in men than in women (71% compared with 25% in 1966; 86% compared with 14% in 1967). During an epidemic period between 1955 and 1965 there were recurrent increases during the dry season (Lambrecht, 1972).

12.14 EPIDEMIOLOGY OF SLEEPING SICKNESS IN EAST AFRICA

The question of whether the parasite causing the disease in southeast Uganda was *T. b. gambiense* or *T. b. rhodesiense* and associated theories of the spread of the disease across Africa have been discussed in section 12.3. These theories are further discussed below in the light of modern biochemical and molecular methods of characterizing trypanosomes. Figure 12.6 shows some of the locations in East Africa referred to in the text.

12.14.1   Uganda

One of the severest epidemics of sleeping sickness occurred in Uganda at the beginning of the 20th century. Possible causes were discussed in detail by Ford (1971). Up to the present, periodic epidemics have arisen – accompanied, as before, by periods of civil unrest and by epidemics of other diseases of humans and domestic livestock. Uganda has been the source of several important discoveries in the field of trypanosomosis and the origin of still controversial epidemiological questions regarding the occurrence of *T. b. gambiense* or *T. b. rhodesiense* as the parasite, and *G. pallidipes* or *G. f. fuscipes* (formerly called *G. palpalis*) as the vector. The elucidation of the epidemiology of the epidemics provides an interesting historical as well as scientific story.

Sleeping sickness in Busoga

THE 1900S EPIDEMIC. According to MacKichan (1944), the earliest record of sleeping sickness in eastern Uganda was in 1901, when workers at Namirembe hospital in Kampala noted cases of a mysterious disease similar to descriptions of sleeping sickness from West Africa. Ford (1971) believed that trypanosomosis had been in Busoga since 1896, based on a report by Christy (1903a). In 1902 over 20,000 people were reported to have died in Busoga alone, both on the mainland and on islands in Lake Victoria. Bruce and

Fig. 12.6. Map of the Ugandan and Kenyan shores of Lake Victoria showing locations referred to in the text.

Nabarro (1903) announced that the disease was caused by a trypanosome carried by *Glossina palpalis* (*G. f. fuscipes*). This caused great apprehension among Europeans, who realized that they were also at risk of contracting the disease. It was estimated that more than 200,000 people died in Buganda and Busoga between 1898 and 1906, and a further 5000 in 1907. Figure 12.7 shows the apparent progression of the epidemic in Buganda from 1900 to 1915, although some of the data, particularly for the early years, would not have been particularly reliable (Duke, 1919b). In 1905 the number of cases reached 100,000 and the following year it was decided to evacuate the population to a minimum distance of 3 km from the lakeshore. In Busoga, the evacuation took place to a distance up to as far as 24 km from the lake, corresponding to the limits of the disease during the 1940s epidemic.

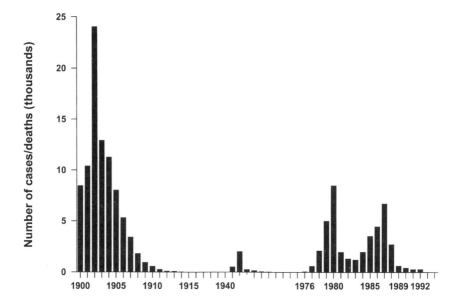

Fig. 12.7. Numbers of sleeping sickness deaths in Buganda Province, 1900–1915 and cases during the 1940–1946 and 1976–1989 epidemics.

Bruce and Nabarro (1903) reported that 'the Prime Minister of Uganda … considered the hut-tax system as the chief cause of the sowing of Sleeping Sickness all over the land'. Labourers were coming in thousands to Entebbe to work for the Government for 1 month instead of paying hut tax. They lived in huts built near the lake shore, about 1.6 km from Entebbe, along the Kampala road. Tsetse flies could be caught there that were capable of transmitting trypanosomes to monkeys experimentally (Bruce and Nabarro, 1903). The disease occurred along a 24 km wide strip of land on the northern shores of Lake Victoria. Figure 12.8 summarizes some of the indirect socio-economic factors contributing to the sleeping sickness epidemic.

The acting governor of the Uganda protectorate, Sir H. Hesketh Bell, reported in 1909 to the British government the current expert opinion that the sleeping sickness epidemic had originated from the Congo basin during Stanley's expedition in search of Emin Pasha. Stanley's porters and entourage had settled for some time in Busoga district. Alternatively, Emin Pasha's expedition could itself have introduced the disease into Uganda from the Congo (Bruce, 1915b). This epidemic gradually spread down both sides of Lake Victoria and invaded Tanzania (Fairbairn, 1948).

Duke (1919a) suggested that a dam constructed across the River Nile at Ripon falls, to raise the level of Lake Victoria, would reduce the density of tsetse flies inhabiting the shores of islands in the lake to such a level that the remainder could easily be controlled by other means.

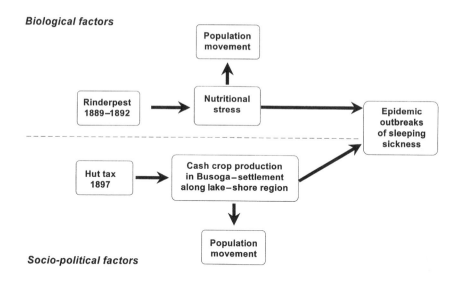

Fig. 12.8. Factors affecting epidemiology of epidemic sleeping sickness in Uganda.

T. B. GAMBIENSE OR *T. B. RHODESIENSE?* The causative organism of the 1901 epidemic was thought to be *T. b. gambiense* rather than *T. b. rhodesiense* (Bruce and Nabarro, 1903; Christy, 1903a). However, in the absence of modern techniques of identification, and as the parasites are morphologically indistinguishable, diagnosis was based solely on clinical symptoms of the disease. Furthermore, researchers at the time were unaware that there were two parasites causing sleeping sickness and *T. b. gambiense* had only just been identified in a human from The Gambia. In the later epidemic, described below, *T. b. rhodesiense* was identified as the causative organism and was thought to have displaced *T. b. gambiense*, which was no longer found (MacKichan, 1944). Results of DNA analysis led Koerner *et al.* (1995) to suggest that the sleeping sickness in the Busoga region of Uganda was always caused by *T. b. rhodesiense* and the first identification of *T. b. gambiense* was mistaken. Duke (1936b) stated that 'for many years now I have recognised the fallacy of assuming, without proof, that the lake-shore trypanosomes of Victoria Nyanza were of necessity *T. gambiense*'. Gibson (1996) questioned the assertion that the original epidemic was due to *T. b. rhodesiense*, as we can never be sure of the parasite's true identity.

THE 1940S EPIDEMIC. The 1940s sleeping sickness epidemic in Busoga dates from November 1940 (MacKichan, 1944). Up to the middle of 1943, 2432 cases were recorded with 274 deaths, the first case having been diagnosed in a schoolboy from Busoga whilst he was visiting Kampala. This epidemic was characterized by a rapid and virulent course of infection. In December 1941 the epidemic was shown to be caused by *T. b. rhodesiense*. This was the first identification of *T. b. rhodesiense* in Uganda. Duke (1944) had discussed the possible roles *G. f. fuscipes* and *G. pallidipes* as vectors of *rhodesiense* sleeping sickness and concluded that *G. pallidipes* was the vector of the disease in this epidemic, in contrast with the previous one. The role of *G. f. fuscipes* was not elucidated until much later, in 1960, when Southon and Robertson (1961) first detected what was believed to be *T. b. rhodesiense* in wild *G. palpalis* (*G. fuscipes*) in Busoga.

The epidemic area included the Lake Victoria shoreline from the source of the Nile at Jinja to the Kenyan border on the Sio river. The disease affected Buvuma island and crossed into Kenya. Inland, the road from Jinja to Busia bounded the affected area. Wild animals, including elephant and buffalo, were said to be plentiful at that time throughout the area, but no cattle were kept and goats were not numerous. At the time, it appeared that *T. b. gambiense* had been completely replaced by *T. b. rhodesiense* (Robertson and Baker, 1958).

There was considerable confusion over the identity of the trypanosome causing sleeping sickness in the region. Thus, although the epidemic was said to have started in November 1940, and was only in 1942 found to be caused by *T. b. rhodesiense*, a report of what was thought to be *T. rhodesiense* at Masindi in Uganda had been made a decade earlier (Duke, 1930). The parasite was later detected in humans by Duke (1930), who also referred to the detection of two infections with *T. b. rhodesiense* in humans from the Bahr el Gazal province of southern Sudan in 1922 and 1926. The latter infections may also have been misidentifications, as sleeping sickness in southern Sudan is caused by *T. b. gambiense*, and no *T. b. rhodesiense* has since been detected there (Gibson *et al.*, 1980). In 1926 Duke and Van Hoof detected what they thought was *T. b. rhodesiense* from a woman at Bambu near the western border of Uganda.

The most likely source of the 1940s epidemic was thought to be from along the west side of Lake Victoria in Tanzania. The persistence of the disease into the 1960s was associated with increased fishing activity and irregular settlement in the tsetse fly-belt of southeast Uganda (Robertson, 1963). Robertson noted that fishermen were well aware of the disease and could recognize the chancre (section 13.4.1), the swelling caused by trypanosomes at the site of an infective tsetse bite, even when they had little systemic illness. Between 1954 and 1960, 85% of cases were in males and 40% of these were fishermen, mainly from the eastern areas. There was a small reduction in the number of cases between July and October, when there was also a seasonal reduction in fishing.

THE 1976 EPIDEMIC. Sporadic cases of sleeping sickness continued to occur in Uganda, but in 1976 there was an epidemic rise in the number of infected people in Busoga district. Between 1976 and 1983, 10,414 male and 9560 female patients suffering from sleeping sickness were detected. The first cases were discovered in June 1971 north of Busesa, outside the zone previously regarded as the area infested by *G. f. fuscipes*, the only vector detected at that time (Matovu, 1982; Abaru, 1985). The inability of the control services to contain this outbreak adequately resulted in a larger epidemic in 1976. It was thought that the disease had spread from the old focus in Busoga, on the lakeshore (Gibson and Gashumba, 1983; Mbulamberi, 1989). The spread northwards took place in the early 1970s when, as a result of political turmoil in the country, the tsetse control services collapsed and, as agricultural practices changed, bush encroachment provided a suitable tsetse habitat. *Glossina f. fuscipes* spread, and could soon be found in areas of *Lantana* and overgrown coffee and banana plantations (Okoth and Kapaata, 1986). Regular close contact between humans and fly is assumed to have increased. During this epidemic, the miniature anion-exchange centrifugation technique for separating trypanosomes (Lanham and Godfrey, 1970) was adapted for diagnosing human trypanosomosis in patients with low parasitaemias in Iganga district (Gashumba, 1981). The modified technique has not been widely adopted elsewhere.

In the early 1980s aerial spraying of tsetse was undertaken and temporarily reduced the incidence of disease, which was thought to have peaked in 1982. However, by the late 1980s the number of cases reported had risen again. More recently the epidemic spread eastwards into Tororo district (Enyaru *et al.*, 1992), where the fly inhabited swamp and riverine vegetation (Maudlin *et al.*, 1990). The epidemic continued into the 1990s, although in 1992 the number of cases was low. In one of the largest tsetse-trapping operations in Africa, control was achieved using 13,000 pyramidal traps over an area of 39,000 km², combined with surveillance and treatment of both human and cattle populations (Lancien, 1991). The latter are believed to be important reservoir hosts of *T. b. rhodesiense* in epidemic areas, as are domestic pigs (Okuna *et al.*, 1986). *Trypanosoma b. brucei* was quite common in domestic pigs, some strains of which were resistant to human plasma in the BIIT test.

Lake Edward and Lake George endemic area

Almost 4000 km² of the area south of the Ruwenzori mountains have been designated as a sleeping sickness area since 1913. The Queen Elizabeth National Park, with large numbers of wild animals, was established in the area abandoned due to the disease. Sleeping sickness in that part of Uganda is probably an extension of the *T. b. gambiense* focus of the Semliki valley, in the present-day Democratic Republic of the Congo. That is a region which, according to tradition, was a focus of the disease since before the Belgians came to the area in 1896. The origin of the Semliki sleeping sickness focus was believed to have been the shores of Lake Edward, whence it spread via the Semliki river towards Lake Albert. A second probable source of infection was the seasonal migration of people from Semliki to the source of salt at Katwe (Van Hoof, 1947). This was an old-established trade in which salt was brought from Katwe and carried in canoes across Lake Edward. At the time of the epidemic there was a very high mortality from sleeping sickness among the canoe paddlers (Hodges, 1911). *Trypanosoma b. gambiense* was first diagnosed in a Congolese porter in 1908 and another case was detected near Lake Edward a year later. The earliest documented epidemic reached its height between 1915 and 1923, at which time resettlement was introduced to control the disease, and the whole population was relocated to higher, fly-free ground in 1924/25. An additional measure to control the outbreak was the establishment of an alternative market for salt at Kasindi. The latter proved unpopular and was of limited success.

Evacuation of the population and attempts to control the salt trade reduced disease incidence but a new epidemic occurred in Busongora in 1932. This was controlled by clearing tsetse-infested rivers and treating infected people. One of the rather bizarre-sounding developments for preventing the spread of sleeping sickness, particularly of the *T. b. gambiense* outbreak in the Albert Nile and West Nile regions, was the establishment of 'gland posts'. These were sited at strategic points on the boundaries of infected areas and no one was allowed to leave the area without a pass issued by a chief and countersigned by a gland palpator, certifying them free of sleeping sickness (Lester, 1939).

12.14.2   Kenya

The early documentation of sleeping sickness in the Nyanza region of Kenya started with a report by Christy (1903b), to the Sleeping Sickness Commission of the Royal Society. Following the discovery of the disease in Nyanza Province (Kavirondo), Christy concluded that the great Ugandan *T. b. gambiense* epidemic was spreading. It first appeared on the lakeshore at south Kavirondo (Central Nyanza) in October 1901, and at Kasagunga and off-shore islands between January and March 1902. Canoe traffic on the lake caused the disease to spread rapidly (Willett, 1965). Wijers (1969) gave a very interesting and readable account of the local traditions and beliefs regarding the origins of human sleeping sickness in the Lake Victoria region of Kenya, which put a very human perspective on that epidemic.

The disease spread to all the lakeshore habitat of *G. f. fuscipes* and to its riverine habitat in the Nyando river basin east of Kisumu and the Kuja–Migori river system in S. Nyanza. The disease remained endemic with occasional epidemic outbursts. One of the epidemic periods decimated the population of the shores and islands of Lake Victoria between 1907 and 1908 and a second epidemic arose between 1920 and 1930 near Kisumu. A third epidemic period, between 1940 and 1950, occurred in the Nyando river basin. Tsetse-spraying of the riverine vegetation with DDT emulsion in 1952 (Wilson, 1953) successfully controlled the disease in Nyando and less successfully in Kuja–Migori in 1954–1957 (Fairclough and Thompson, 1958; Morris, 1960).

The first recorded outbreak was assumed to be due to *T. b. gambiense* based on clinical diagnosis (Gray and Tulloch, 1905), but the subsequent discovery of *T. b. rhodesiense* in southeast Uganda (Duke, 1930) raised doubts. The peak of the epidemic was in 1932 (Fairbairn, 1948); subsequently there was low-level endemic *T. b. rhodesiense* in the upper reaches of the Mara river, possibly derived from a Tanzanian outbreak, but this did not spread to Nyanza region.

In early 1942, *T. b. rhodesiense* sleeping sickness spread from southeast Uganda over the Sio river into Kenya, possibly via a *G. pallidipes* infestation at Samia, which was an eastern extension of the south Busoga fly-belt of Uganda. Alternatively, the disease could have been introduced by Samia fishermen who fished in the Ugandan waters of Lake Victoria, rather than simply by an extension of the fly-belts (Robertson, 1963). MacKichan (1944) speculated that the original source of the infection was immigrant labourers from western Tanzania coming to work on the Kakira sugar plantations near Jinja. Labourers on the estate mainly came from Rwanda or Burundi. In Central Nyanza it was only apparent in 1953 that *T. b. rhodesiense* was becoming a problem, as evidenced by failures of treatment with Tryparsamide and by the acute course of the disease. The presence of *T. b. rhodesiense* was first confirmed the following year (Willett, 1955) and in 1956 the parasite was isolated from *G. pallidipes* and from a bushbuck in western Sakwa location (Heisch *et al.*, 1958). The disease was confined to western Central Nyanza from 1954 to 1960, after which it appeared in and around the Lambwe valley. *Trypanosoma rhodesiense* was then also detected in south Nyanza, from where *T. b. gambiense* had been eradicated.

The Lambwe valley is a frequently occurring name in the literature on tsetse and sleeping sickness control, having been the site for much research work and many varied tsetse control operations over the years. The valley is small, with an area of about 100 km², and part of it has been declared a game reserve by the Kenyan government. *Trypanosoma b. rhodesiense* sleeping sickness has been recorded in the area since 1959 (Willett, 1965). The average number of sleeping sickness cases between 1959 and 1984 was about 36 per year. Baldry (1972) gave a thorough history of sleeping sickness in the Lambwe valley and this has been followed by more recent reviews by Wellde *et al.* (1989a–d).

The 1964 epidemic in Alego and Yimbo, Central Nyanza

A sleeping sickness epidemic that occurred in Alego location in 1963–64 was totally unexpected, according to Wijers (1974a–d). The epidemic resulted in 319 cases and was quickly brought under control by use of insecticides. Alego was a densely populated area with the original savanna mostly replaced by cultivated fields. Wild animals were rare but cattle were numerous. Only *G. f. fuscipes* was known to occur in Alego, but the area of infestation was contiguous with an area infested by *G. pallidipes*. One factor contributing to this epidemic was that *G. f. fuscipes* colonized new vegetational areas away from the rivers, particularly *Lantana* thicket and the rings of vegetation around Luo Bomas (homesteads). Heavy rainfall in the years preceding the epidemic may have produced suitable conditions of humidity and temperature for the species to spread from its usual riverine habitat. There was intimate human–fly contact, similar to that described by Nash (1960) in northern Nigeria, in which flies repeatedly fed on the same restricted human population during the drier season. It is almost certain that the cases were only due to *T. b. rhodesiense*, making this the first recorded case of an epidemic of *T. b. rhodesiense* sleeping sickness transmitted by tsetse of the *palpalis* group.

Onyango *et al.* (1966) demonstrated that cattle could act as reservoir hosts of *T. b. rhodesiense* in Alego location, and Baldry (1972) confirmed that the parasite was being transmitted by peridomestic *G. f. fuscipes*. Van Hoeve *et al.* (1967) showed that the strain could be cyclically transmitted to a human, a cow and sheep, substantiating evidence that cattle were reservoirs for the disease. Furthermore, the parasite was resistant to the drug normally used to treat cattle trypanosomosis in the area and no clinical symptoms appeared in infected cattle (Van Hoeve *et al.*, 1967).

A study of the epidemiology of *T. b. rhodesiense* sleeping sickness along the shores of Lake Victoria in Kenya and Uganda suggested that nearly all women, children, most cultivators and some fishermen became infected by *G. pallidipes* near their homes (Wijers, 1974c). Furthermore, transmission at the lakeshore itself was rare and *G. f. fuscipes* only played a minor role. Humans were apparently poor reservoirs for the disease and human–fly–human transmission was thus considered rare. Bushbuck and duiker were believed to be the main source of infection of flies. In the light of current knowledge, some of these ideas would now have to be modified. Whilst *G. pallidipes* may have been responsible for some transmission during the epidemic referred to by Wijers (1974c), *G. f. fuscipes* is believed to have been the most important vector in the area during subsequent epidemics, and cattle are suspected to be important reservoirs of the parasite (Wijers, 1974a; I. Maudlin, Nairobi, 1998, personal communication). Willett *et al.* (1964) isolated *T. b. rhodesiense* from *G. pallidipes* from Sakwa in Central Nyanza and confirmed the presence of sleeping sickness by isolating ten strains from human patients. The parasite had previously been isolated from *G. pallidipes* by Jackson (MacKichan, 1944) and subsequently by Willett (1956). Sleeping sickness caused by *T. b. rhodesiense* occasionally spread from the endemic *G. pallidipes* areas along the lake shore into the *G. f. fuscipes* areas towards the north, which led to minor outbreaks or a few sporadic cases (Wijers, 1974c). The 1963–64 Alego epidemic and the Busesa epidemic in 1971 (only 169 cases) (Kangwagye, 1975) were thought to be exceptional. Wijers reported the invasion of hedges around homesteads in Alego by *G. f. fuscipes* but still thought this might not have been the direct cause of the epidemic. He believed cattle to be the main reservoirs for infection during outbreaks in *G. f. fuscipes* areas.

12.14.3 Tanzania

As sleeping sickness in Tanzania was believed to have been introduced from Uganda or the Democratic Republic of the Congo, the same confusion over the parasite causing the disease arose, and therefore the initial outbreak was believed to have been caused by *T. b. gambiense*. The first outbreak of *T. b. gambiense* was recorded on the western (Congolese) shore of Lake Tanganyika in about 1901 and soon reached the eastern (Tanzanian) shore, where it quickly spread.

Origins of the disease and attempts to control it

TRYPANOSOMA B. GAMBIENSE SLEEPING SICKNESS. Evidence cited for the introduction of *T. b. gambiense* from the Democratic Republic of the Congo or Uganda includes the detection of trypanosomes in ten out of 16 people in a caravan of porters from the Democratic Republic of the Congo. In a review of trypanosomosis control in Tanzania, Laufer (1955) stated that the disease caused by *T. b. gambiense* had spread from Uganda from the western shore of Lake Victoria and from the Congo along the western shore of Lake Tanganyika. In 1907, after the disease appeared in Tanzania, the Germans established sleeping sickness camps at Niansa at the north end of the lake, at Kigarama on the western shore, just south of the Uganda border, and at Shirati on the east side, near the Kenyan border. In 1908 cases of sleeping sickness were discovered in the area from Usumbura to the Malagarasi river. Cases were detected on Ukerewe island at the south of the lake, in inhabitants of the islands of Bumbire and Iroba, near the eastern shore, and at Ihangiro on the mainland opposite those islands. In 1908, 1517 patients were under treatment, including cases around Lake Victoria.

The southward spread was stopped by extensive bushclearing, especially along the River Mori, south of Shirati. The 1902–03 epidemic and its history were described further by Davies (1962a,b). It was while working in the River Mori area that Kleine (1909) discovered the cyclical transmission of *T. b. gambiense* by *G. f. fuscipes*. Later, in 1911, Taute showed that *G. m. centralis* was also capable of cyclical transmission in Niansa. By 1914 the disease had been successfully controlled in the Tanzanian Lake region but the lakeshore was still considered to be a dangerous source of infection. The introduction of fresh cases was prevented by controlling movement of the local people and by bushclearing. No Tanzanians were allowed to act as porters in the Democratic Republic of the Congo.

TRYPANOSOMA B. RHODESIENSE SLEEPING SICKNESS. The epidemiology and history of sleeping sickness in Tanzania were described by Fairbairn (1948). *Trypanosoma b. gambiense* was believed to cause the disease until, in 1911, a new focus of *T. b. rhodesiense* was detected in Songea district, on the headwaters of the Rovuma river, at the border of Tanzania and Mozambique.

Lamborn and Howat (1936) give some details of the patient who, it is speculated, carried the disease to this area. By 1912/13, 3303 cases were under observation on the upper Rovuma, mostly coming from Rwanda (Maclean, 1926). From there it spread northward and eastwards and by 1917 had crossed the river Mbemkuru to Kilwa. It occurred at resting places, near riverbanks and water-holes, used by labourers of the Yao tribe, passing from plantations at Lindi/Kilwa. Sleeping sickness cases were not reported away from the routes of the Yao labourers. There was substantial circumstantial evidence that human–fly–human transmission occurred at these sites and that wild animals played a less important role. Human-to-human transmission was considered particularly important in maintaining the epidemic, and a major contributing factor was thought to be visits to sick people (Swynnerton, 1923b). However, there was some suspicion that the disease was transmitted mechanically from human to human by tsetse, rather than by cyclical transmission (Duke, 1923).

The initial doubt about whether or not this epidemic was due to *T. b. gambiense* or *T. b. rhodesiense* was resolved by Taute (1913), who established that the disease here was caused by *T. b. rhodesiense,* and the parasite was identified in 1922 in Maswa district. By the end of 1946, 23,955 cases had been diagnosed; many more people probably died undiagnosed. The outbreak of *T. b. rhodesiense,* transmitted by *G. swynnertoni,* probably began in Maswa district in 1918 and peaked in 1925. The disease spread from the 1922 Maswa outbreak to Ikoma district in 1925.

Following the outbreak of rinderpest in 1891, the distribution of *G. swynnertoni* in the lake region of Tanzania receded. However, recovery of host animals was rapid, and by 1913 flies started spreading again and invaded previously tsetse-free country that had been abandoned by the human population. The evacuation of people as a means of disease control further accelerated the spread of the fly, which was finally stopped by organized bushclearing between 1923 and 1930 in Sukumaland. In 1932 and 1931, respectively, famine and a rinderpest epizootic caused a decline of the previously increasing cattle population.

In 1922, in the vicinity of Mwanza, there was a severe form of hookworm (ankylostomiasis) known locally as *safula*, which had similar symptoms to trypanosomosis (Duke, 1923). This disease outbreak was initially confused with trypanosomosis and coincided with the end of a serious famine in 1918. The famine may have resulted in ecological conditions facilitating the transmission of sleeping sickness.

Before the First World War, very few cases of sleeping sickness were reported in Tanzania and these were confined to the south (Maclean, 1929). After the war, three large outbreaks occurred in Mwanza, Ufipa and Tabora and in Liwali.

An important epidemiological factor contributing to these outbreaks was the establishment of new villages in tsetse-infested forests following the suppression of hostilities in those areas (Maclean, 1929). Secondly, the loss of manpower from forest villages to supply labour for development elsewhere resulted in increased vegetation cover, which led to an expansion of tsetse distribution and increased sleeping sickness transmission.

In other areas, developing agriculture resulted in a decreased tsetse distribution (Maclean, 1929). Fairbairn (1943) discussed sleeping sickness settlements in Tanzania in relation to agricultural development. Before colonization, people lived in large compact settlements formed as defensive communities with large herds of cattle, and these settlements were associated with bushclearing to provide the necessary agricultural land. Following colonization, intertribal warfare ceased and the settlements tended to break up. The people were then less able to keep cattle and were scattered in the bush. Large-scale epidemics of sleeping sickness then occurred (Maclean, 1929). Maclean estimated that a human population density of 0.2–10 persons km^{-2} was the most favourable for the epidemic spread of the disease; a density of < 1 km^{-2} could not sustain an epidemic, and a density of > 10 km^{-2} caused enough bush destruction to result in a reduction in tsetse populations. A density of 20–30 km^{-2} resulted in practically fly-free country. Fairbairn (1943) agreed with this theory, but only if 'person' was changed to 'taxpayer' and therefore represented a household with a family farm.

The spread of *T. b. rhodesiense* sleeping sickness in Ufipa district was thought to have been due to the movements of infected people from village to village and not due to reservoirs of infected animals (Maclean, 1926). The First World War, and an outbreak of influenza, led to disorganization and movements of the population and may have contributed to the outbreak at Ufipa (Maclean, 1926). Most epidemics in Tanzania could be traced back to humans as the source of infection (Laufer, 1955) and a proposal for disease control in Tanzania was to eliminate this source, through detection and supervision of infected people. There was no evidence of symptomless carriers (Laufer, 1955).

In 1924 an outbreak of *T. b. rhodesiense* sleeping sickness occurred along the fertile Liwali river valley in Tanzania, causing the evacuation of villages, although the inhabitants later attempted to return (Dye, 1927). Dye argued that people from all villages were exposed to the same level of tsetse challenge during activities in the bush away from their villages; however, the trypanosomosis prevalence differed amongst the villages, as did the tsetse challenge. The disease outbreak therefore appeared to result from human-to-human contact, rather than transmission from wild animal reservoirs to humans, and was therefore thought to be transmitted by flies living in the villages rather than by those in the surrounding bush (Dye, 1927).

Sleeping sickness caused by *T. b. rhodesiense* in Tanzania temporarily disappeared by 1954, coincident with the closing of the Kilimafedha and other gold mines, and evacuation of people from those areas. Thorough bushclearing at a width not less than 150 m on either side of roads passing through the area and around villages was recommended in order to protect people who had to pass through infected areas.

In 1964 a new outbreak started in association with the creation of the Serengeti National Park and Ikoma Game Reserve. As this was considered a threat to tourism, it received prompt attention; however, people were incidental hosts and no large-scale epidemic occurred.

Similar genetic strains of *T. b. rhodesiense* have persisted in Tanzania for over 30 years, suggesting that clonal rather than sexual reproduction is the norm (Gashumba *et al.*, 1994), agreeing with a model proposed for trypanosome population structure by Tibayrenc *et al.* (1990).

12.14.4   Ethiopia

Trypanosoma b. rhodesiense sleeping sickness was detected in Ethiopia in 1967 (Balis and Bergeon, 1968, 1970; Baker and McConnell, 1969; McConnell *et al.*, 1970). This is the northernmost limit of the parasite's known range.

The disease was first detected in Kaffa province of southwest Ethiopia, at Maji and in the Gambela area, particularly along the Gilo, Akobo and Baro rivers. This was a small epidemic and was probably detected not long after it started (Baker *et al.*, 1970; McConnell *et al.*, 1970; Hutchinson, 1971). Up to 1970, most cases occurred in the Anuak tribe from near the river Gilo in Illubabor province. All the rivers in the affected area are tributaries of the Blue Nile. In the 1960s there was an influx of Anuak refugees from Sudan, who may have been the source of the epidemic. The people inhabiting the affected area were Nilotics whose areas of occupation extended southwest into Sudan.

There is some doubt as to whether or not the disease was present in the area much earlier. According to Hutchinson (1971) there is no evidence for this, but Langridge (1976) referred to observations by Kunert in 1956 suggesting that the disease occurred, and to suspicions that the disease occurred in the Mursi people, along the lower Omo river, in the early part of the century. The disease may have reached Ethiopia from Sudan, from where cases had occasionally been reported (e.g. Archibald and Riding, 1926), as *G. m. submorsitans* exists in the border area of each country.

Tsetse in southwest Ethiopia may also have extended their range at about the time of the epidemic in the late 1960s and early 1970s (USAID, 1976). Reasons for this expansion are complex and may be related to rainfall distribution, and possibly rinderpest and yellow fever epidemics. Malaria and trypanosomosis in domestic livestock and humans have caused significant periodic movements of human settlement in Ethiopia from the 1960s up to the present day. These movements themselves are also likely to have influenced tsetse distribution. For example, *G. m. submorsitans* extended its range in southwest Ethiopia (Leak and Mulatu, 1993), possibly by moving into areas where suitable habitat has been created by bush encroachment into previously cultivated land. Alternatively, climatic changes, resulting in a warmer climate, may have enabled tsetse flies to occupy land at higher altitudes than was previously possible.

Again, it has been questioned whether sleeping sickness in Ethiopia is caused by *T. b. rhodesiense* or *T. b. gambiense*, as the original diagnosis was based on clinical signs (Gibson *et al.*, 1980). There is no *T. b. rhodesiense* in neighbouring Sudan, which is the closest sleeping sickness focus to the Ethiopian outbreak, although two suspected cases were reported some time previously in the Bahr-el-Ghazal province. The closest *T. b. rhodesiense* focus is in the Busoga area of Uganda. Whilst not questioning that the epidemic was caused by *T. b. rhodesiense*, Baker (1974) noted that the epidemiology of the disease in Ethiopia more resembled that of *T. b. gambiense*, in that peridomestic infection occurred, producing similar rates of infection in men, women and children.

12.15   DISEASE CONTROL – CHEMOTHERAPY

The World Health Organization Division of Control of Tropical Diseases coordinates sleeping sickness control in 22 African countries and has adopted the following strategy (WHO, 1995):

- Mobile medical surveillance of the population at risk by specialized staff using the most effective diagnostic tools (serology and parasitology) available. Patients are sent to specific referral centres for determination of the stage of the disease and treatment, and for post-therapeutic follow-up.
- Fixed post-medical surveillance delivered at dispensaries, health centres or hospitals where blood samples are taken and analysed at reference centres. All patients or suspected cases are sent to special centres for confirmation of diagnosis, determination of the stage of the disease and treatment, and for post-therapeutic follow-up.
- Vector control using screens and traps.

It is widely recognized that *eradication* of tsetse flies is usually impractical, even if technically feasible, and elimination of wild animal reservoirs is unacceptable. Control of human sleeping sickness therefore relies on surveillance and treatment of the disease to reduce the human reservoir of infection, the prevention of epidemics through diagnosis and treatment, and vector control to break disease transmission (Molyneux, 1983). Human trypanosomosis is usually fatal if not treated, although there have been reports of healthy carriers. The drugs that have been used to treat the disease in humans are suramin and pentamidine for the early stage before involvement of the cerebrospinal fluid (CSF), and Melarsoprol for the later stages when trypanosomes are detected in the CSF. Melarsoprol is very expensive, and there has been little progress in developing new drugs or change in treatment strategies until recently. Pentamidine, administered by intramuscular injection every 6 months, was also used as a prophylactic in francophone African countries during epidemics. However, *T. b. rhodesiense* developed resistance to the drug in some areas, and the drug is now only used to treat *T. b. gambiense*.

Diminazene aceturate, a trypanocide for livestock, has been used experimentally to treat *T. b. gambiense* infections in humans in the Democratic Republic of the Congo and northern Nigeria (Hutchinson and Watson, 1962). In the Congo, 17 cases were treated using a dose of 2 mg kg^{-1} body weight. The treatment was successful except for one patient who relapsed and was cured with Antrypol® and tryparsamide.

A recently developed drug for treatment of late-stage *T. b. gambiense* sleeping sickness is difluoromethylornithine (DFMO), which is much safer than current drugs – it was given the popular name of 'resurrection drug', as it was very effective in patients with late-stage infections of *T. b. gambiense* (Pépin *et al.*, 1987). The drug is also effective against arseno-resistant strains of the parasite, which would otherwise invariably be fatal. Drawbacks to DFMO are that it is expensive (US$250 for the drug alone), large doses are needed and it produces some adverse side-effects.

Chapter 13

Epidemiology of Trypanosomosis in Domestic Livestock

The percentage of the world's population living in sub-Saharan Africa is increasing, as is the percentage living in developing countries worldwide. Consequently, world food production must more than double to meet the needs of the extra 2–2.5 billion people in the world's population by 2025 (McCalla, 1994). It is accepted that this doubling must come predominantly from increased productivity of crops and livestock. Trypanosomosis of domestic livestock has been described as the 'scourge of Africa' as it allows infected cattle to consume resources – fodder and water – without being productive, in contrast with some other diseases from which animals die quickly, leaving resources available for other animals (Murray and Dexter, 1988). Emphasis is currently placed more on increased crop production than on livestock, which is sometimes seen as detrimental to the environment and as providing a food which is not a staple for much of the developing world's population. This argument ignores the great importance of livestock to crop production, due to inputs both from manure and nutrient recycling and from animal traction. It has been argued that the carrying capacity of rangelands has been exceeded (Brown and Kane, 1994), and therefore increased production will come from removal of constraints to productivity by diseases such as trypanosomosis.

About 10 million km^2 of sub-Saharan Africa are infested by tsetse, extending through 38 countries, and 30% of the total of approximately 147 million cattle in the continent are said to be at risk from trypanosomosis (Murray and Gray, 1984). Mortelmans (1984) stated that in some tsetse-infested regions of Africa, three to four times more livestock could be carried if trypanosomosis did not exist. However, such an increase in live-stock populations is unlikely, and would incur a high cost. Urquhart (1980) pointed out that the 120 million extra cattle, 150 million extra sheep and 250 million extra goats would face constraints such as the availability of water and the presence of other diseases. Nonetheless, he acknowledged that

279

great improvements in livestock production and productivity could be achieved. Ford (1969) also referred to the potential increase in the cattle population in the absence of tsetse, but thought that this would result in periodic starvation and commented that a doubling of the livestock population of Nigeria might result in a 20% mortality from starvation every 5 or 10 years. He discussed the effects of such an increase on the ecology and of the loss of acquired resistance to trypanosomosis.

In addition to the effect of trypanosomosis on livestock raised for food production, the effect on horses has frequently been referred to as one of the significant factors in isolating Africa from other parts of the world and retarding the early development of the continent. An example of the effect of trypanosomosis on horses was reported regarding an epidemic in Sierra Leone. Horses were commonly used in the first half of the 19th century, but between 1856 and 1858 an outbreak of trypanosomosis 'annihilated' the horse population (Dorward and Payne, 1975). It was speculated that this epidemic occurred as a result of an expansion of the tsetse population and possibly the introduction of *T. b. brucei* in imported cattle.

There are few data on trypanosomosis in small ruminants, which have received much less attention from scientists and veterinary researchers compared with cattle. This may be due to their smaller size and lesser apparent importance, but may also be because they are less susceptible to trypanosomosis. For this reason, the emphasis in this chapter is on trypanosomosis in cattle.

13.1 DISTRIBUTION OF CATTLE AND THEIR ORIGINS IN AFRICA

A short review of the origins of cattle is given in order to provide some background information on the relationships between tsetse and trypanosomosis and the cattle breeds occurring in tsetse-infested areas. The development of varying degrees of trypanotolerance can then be better understood. Table 13.1 lists the origins and migration routes of cattle in Africa.

Theories on the origins of cattle are speculative, based on inadequate or incomplete archaeological evidence; however, molecular genetic techniques are helping to provide additional evidence to clarify some of these ideas. *Bos primigenius opisthonomus* was the original North African wild cattle species and is thought to have become extinct in Egypt by the 14th century BC. *Bos taurus* and *B. indicus* cattle are all descended from these original animals.

The centre from which wild cattle originated was thought to be in Asia, from where they spread into Europe and Africa during the Pleistocene. However, analyses of cattle DNA suggest that wild oxen were domesticated in at least two separate regions, approximately 10,000 years ago: southwest Turkey, and east of the Iranian desert near the Pakistan–India border. Some

Table 13.1. Origin and migration routes of domestic cattle in Africa.

Breed (origin)	Date of entry into Africa	Place of entry	Routes of migration and dates
Bos taurus			
Hamitic longhorn (Turkestan)	5000 BC	Northern Egypt Nile valley Ethiopia	West along the Mediterranean littoral South along the Nile Southwest to Sudan (Displaced by the shorthorns) (Evidence from Tassili and Tibesti rock engravings, and Mount Elgon rock paintings)
Shorthorns (Turkestan)	2750–2500 BC (3000 BC)	North Egypt from Isthmus of Suez Horn of Africa East coast	West and south from 2500 BC onwards
Bos indicus			
Zebu (Western Asia)	2000–1500 BC AD 700	Egypt Horn of Africa	AD 700 North, west and south, widely distributed
Sanga (East Africa)	AD 1000	East Africa Ethiopia	AD 1400 North, west and south Ethiopia and Kenya are the sites of origin of the Sanga type

of the earliest evidence of domestication of cattle was found at a site in southern Turkestan, dating back to 8000 BC (sheep and goats are said to have been domesticated in the Nile valley 9000 years ago). These cattle were probably the ancestors of the Hamitic Longhorn types, but there is evidence that in 6000 BC *B. brachyceros* shorthorn types also originated from this area.

The main African cattle types are shown in Table 13.2. Hamitic Longhorn and shorthorn types are believed to be the ancestors of all *B. taurus* in the world. Zebu cattle (*B. indicus*) are now thought to originate from western Asia, and not, as previously believed, from the Indian subcontinent. Mitochondrial DNA (mtDNA) sequences from European, African and Indian cattle fall into two distinct geographic lineages (Loftus *et al.*, 1994). The humped *B. indicus* and non-humped *B. taurus* can be included in one grouping whilst the humped Indian cattle appear distinct. It was suggested that the two groups might have diverged between 200,000 and 1 million years ago, and the authors interpreted the divergence as evidence of

Table 13.2. Main African cattle types.

Humpless cattle (*Bos taurus*)		Humped cattle (*Bos indicus*)	
Type	Examples	Type	Examples
Longhorns	Kuri, N'Dama	West African zebu	Fulani
Shorthorns	West African Shorthorn	East African zebu	East African Shorthorn
		Sanga	Ankole
			Nilotic Sanga
			Afrikander

two separate domestication events. Cattle may, therefore, have been domesticated independently in Africa (Wuethrich, 1994; Loftus *et al.*, 1994). The reason for the similarities between African *B. indicus* and *B. taurus* may have been due to ancestral cross-breeding in Africa. Loftus *et al.* believed that domestication in southwest Turkey led to the non-humped taurine, while a second domestication, at about the same time and possibly in the Indus valley, led to the humped zebu cattle.

Cattle were brought into Africa by migratory peoples, probably first entering tropical Africa from western Asia, not more than 7000 years ago. Their first owners would have been proto-Hamites (people inhabiting North Africa and highland areas of eastern Africa).

Three major types of cattle were imported or migrated into northeast Africa with nomadic people:

- humpless Hamitic Longhorn – ancestral *B. taurus*
- humpless shorthorn – ancestral *B. taurus*
- humped zebu - *B. indicus.*

From the first domesticated longhorn cattle, the Hamitic pastoralists of North Africa developed a giant horned type. Rock engravings of longhorn cattle exist in Ethiopia and in Nigeria. Cattle probably reached Ethiopia from Arabia via the horn of Africa, as animals with very long lyre-shaped horns are portrayed in rock engravings of central Arabia dated to the 3rd and 2nd millennia BC. Mount Elgon, on the Kenya/Uganda border, seems to be the southernmost point at which their presence is recorded.

Cattle with long, lyre- or crescent-shaped horns extended from the eastern shores of the Mediterranean into Mesopotamia, where they are depicted in numerous works of art. Ancient traces of these longhorn cattle are to be found along the shores of the Mediterranean from Palestine through Egypt to northwest Africa. There is little difference in conformation between the modern longhorn cattle of Portugal and the breed of the ancient inhabitants of the Nile valley. Furthermore, similarities occur between the longhorned N'Dama cattle of West Africa and silver figurines of longhorn cattle from Mycenae. Following the spread of longhorn cattle from

Asia to Africa, *B. p. opisthonomus* cows and calves were doubtless incorporated into herds of domesticated longhorn in various regions and times, thus extending their range.

The tsetse belt in West Africa is presumed to have halted the westward migrations of the Hamitic cattle. There is no evidence that the West African Shorthorn (WAS) originated from Europe, as had once been suggested. A theory that they were domesticated in Africa in the Sahara region also has no support.

13.1.1   Origin of the N'Dama breed

The precise origin of N'Dama is not known. One theory that they are descended from European cattle (Iberian) as crossbreeds of *B. taurus* and *B. aegyptacus*, and were introduced from Portugal into the Mandingo country by early Portuguese seafarers, has no supporting evidence and is considered unlikely. There are early records of their occurrence in the Nile valley.

13.1.2   Hamitic Longhorn

Earliest records of Hamitic Longhorns in Africa are from Egypt, dating to around 5000 BC in the Badarian period. The Badarian people originated from Asia and therefore may have brought the cattle with them. Their continued presence in the area can be dated up until 30 BC. No traces of Hamitic Longhorns remain in Egypt today and one theory for their disappearance is that they interbred with wild cattle along the Nile valley.

From 2750 to 2500 BC, humpless shorthorn cattle (*B. brachyceros*) were first introduced into Egypt by invaders from the east Mediterranean littoral and the indigenous people migrated west and southwards with their Hamitic Longhorn cattle. Three main migrations of the latter are believed to have taken place across, or round, the Sahara. The first of these migrations was westwards along the Mediterranean littoral. When this migration reached Morocco a split probably occurred, with one branch going north to Spain and southern Europe, and the other west and south into West Africa. There may also have been a migration across the Sahara from the Mediterranean littoral to Chad. The N'Dama, in the Gulf of Guinea, are more or less pure descendants of Hamitic Longhorn cattle from the original migration. The Kuri cattle, found around the islands of Lake Chad, are a related descendant breed.

The second route was southwestwards across the Sahara via the Tibesti and Tassili highlands in Algeria where pre-3000 BC rock engravings provide evidence of the migration. It is speculated that the Kuri breed could have arisen from this migration but there is insufficient evidence. The rock paintings at Tassili are one of the largest collections of such paintings that have been found and illustrate cattle, horses and camels. The earliest date for human occupation of this area, coming from radiocarbon dating, is 5450 ± 300 BC although the paintings may not be so old. Radiocarbon dating of a second site south of Tassili, where bones of the extinct buffalo (*Bubalus antiquus*) and domestic cattle were found, gave a date of 3460 ± 300 BC. It is speculated that this period was a transitional phase in which people had started keeping domestic cattle but still hunted wild animals: it could be argued that this 'transitional phase' is still in progress in some areas. A third site, Titerast-n'-Elias, where bones and wall paintings of cattle were found, was dated 2610 ± 250 BC. The cattle paintings showed pastoral scenes, and artefacts such as pottery, stone axes, arrowheads and grindstones were also found.

The third migration was southwards up the River Nile to the Ethiopian and Kenyan highlands. These cattle reached Mt Elgon on the Kenyan–Ugandan border, where there is evidence in the form of 1000 BC rock paintings, and possibly went as far as Zimbabwe, where they still occur as the Binga breed (Payne, 1990). For some time the longhorn cattle predominated over shorthorns but then the shorthorns grew in number. Rock paintings in Ethiopia and Somalia also provide evidence of the third migration.

13.1.3 Shorthorns

Asia Minor was the starting point of the extension of shorthorn cattle not only to Africa but also, it is believed, to the Channel Islands and other parts of Europe. The cervico-thoracic humped cattle were probably introduced into Egypt from the region of the Somali coast, which they had reached by sea from the coastlands of the Persian Gulf. Immigrants brought the cattle from Asia, entering Egypt by way of the Isthmus of Suez, towards the middle of the third millennium BC. Between 1700 and 1580 BC, as remnants of dispersed Asiatic tribes settled in the Nile valley, shorthorns became predominant over the longhorn cattle in lower Egypt. The introductions of shorthorn cattle also resulted from military expeditions of the Pharaohs against rulers of Syria and Palestine. Every successful military operation brought new shorthorn herds into the Nile valley. Shorthorn cattle then moved southwards along the Nile to upper Egypt and Sudan. They also migrated along the Mediterranean littoral and are believed to have split into two branches in Morocco. Earliest records from Morocco date from 2750 to 2625 BC. In North Africa, shorthorn cattle eventually began to replace the original longhorn type at about the same time as in southern and central Europe. One migrating branch went northwards to the Channel Islands, whilst the others moved south and west to the forest region of West Africa until they reached the barrier of the rain forest zone. This branch is believed to have become the West African Shorthorn. Although the West African Dwarf Shorthorn were very susceptible to rinderpest, Lewis (1953) stated that they were little affected by the bites of Tabanidae.

Shorthorn cattle may also have been independently imported by alternative routes from 2200 to 1780 BC into the Horn of Africa, the east coast of Africa and offshore islands (Socotra, Pemba, Mafia, Madagascar), perhaps from western Asia. This alternative importation of shorthorn cattle probably reached Ethiopia via Arabia and the Horn of Africa by 3000–2000 BC. They had already penetrated to the Ethiopian highlands by the middle of the 2nd millennium BC. Evidence for this comes from the earliest paintings of such cattle, which date from that time. In contrast, evidence of thoracic-humped cattle does not appear in Africa prior to the first centuries AD, except for one depicted on an Egyptian bronze weight. Thoracic-humped zebu cattle reached Africa at a relatively late period. In Somalia and Ethiopia they occur, along with the camel and horse, not before the 4th century AD. Broadly, it is accepted that thoracic-humped cattle represent the zebu type, whilst cervico-thoracic-humped cattle represent the sanga type.

From Egypt the shorthorn cattle migrated southwards down the Nile into the Nuba mountain area of Sudan, where they have now been replaced by zebu stock, and into Ethiopia. They reached the Koalib hills in Sudan and as far as Lake Victoria and Karagwe in Uganda.

There is some historical interest in accounts of trypanosomosis and trypanotolerant cattle in the area of the Koalib hills. These hills, covered mainly by deciduous trees, are situated in the most northern tsetse fly belt in Sudan, and are a focus of tsetse infestation. In 1889 tsetse were recorded in the southern Koalib hills (the largest mass of hills) but were not recorded in the northern Koalib, nor in the surrounding bush. Traditional legends, supported by some evidence, say that the Nuba inhabitants introduced tsetse (*G. m. submorsitans*) into the northern Koalib hills as protection against the Arabs. Because of its 'evil reputation', the neighbourhood of the hills was avoided by nomadic Arabs who drove their stock southwards for grazing purposes. In the past, the Nuba occasionally raided the stock-keeping Arabs on the plains. The Nuba cattle were small (dwarf) and black in colour: an early photograph of a small but fully grown bull shows that the animals referred to as West African Shorthorns were humped (Balfour, 1913). They were also trypanotolerant, according to local belief, described in Balfour's report. The cattle were in relatively good condition, suggesting that they were indeed capable of withstanding trypanosomosis challenge (Archibald, 1927; Archibald and Riding, 1926), although there was no scientific evidence for this. In contrast, cattle on the plains were less capable of thriving under trypanosomosis risk. The Koalib tsetse population was described by Ruttledge (1928), who was one of the first people to report the use of tsetse traps and even suggested the use of odour attractants in the form of central bands, or wicks, soaked in a solution of human or animal sweat. He also referred to the possibility of using live animal baits.

In Ethiopia a small number of humpless shorthorn cows are kept by Sheko tribesmen for milk production. The Sheko cattle arose from the early migrations of shorthorn cattle into Ethiopia and occur in a mountainous, rugged and almost completely isolated territory west of Shewa Ghimirra, adjacent to the Sudan border (Alberro and Haile-Mariam, 1982a,b). West of Egypt, shorthorn breeds are found from Libya to Morocco down the west coast to latitude 14° N. The continuity of their range is interrupted by the intrusion of humped cattle, but prior to this they had succeeded in establishing a hold in West Africa, where they extend from Guinea Bissau to Nigeria and northern Cameroon. In that region they have become fully adapted to the unfavourable climatic conditions, but in several areas they have become dwarfed.

13.1.4   Zebu

Zebu cattle were the most recent introductions, brought in small numbers in successive waves between 2000 and 1788 BC, to Egypt and the Horn of Africa, probably from somewhere in Asia. They were only imported into eastern Africa in significant numbers after the Arab invasion of AD 669. They then spread rapidly to inhabit West Africa north of latitude 14° N, northeast Africa, East Africa, and as far south as Zambia and Malawi (Payne, 1970). They are still expanding and replacing sanga and other types in Uganda, the Congo and in Ethiopia.

13.1.5   Sanga

The term 'sanga' has been used to describe a particular type of cattle such as the West African sanga, or, in the broadest sense, to describe any breed that is a result of crosses between zebu, Hamitic Longhorn and/or shorthorn cattle. Typical sanga cattle have long horns and small cervico-thoracic humps. East Africa and Ethiopia were probably the major centres of origin in which the early sangas zebus, imported in the 1st millennium BC from India via southern Arabia, evolved from crosses with original longhorn humpless cattle of the Hamitic pastoralists. The sanga cattle then accompanied the Bantu people in their migrations to southern Africa and probably to the north and west (Epstein, 1971). At one time sanga cattle were ubiquitous from Ethiopia through the lake region of East Africa south to Namibia and South Africa. The comparatively recent southerly migrations are listed below:

- AD 700 Bantu crossed the Zambesi.
- AD 900 Remains of sanga cattle at Zimbabwe.
- AD 1400 Hottentots crossed the Limpopo and Orange rivers.
- AD 1652 Dutch first bought cattle from the Hottentots.

The Bantu migrations took sanga cattle to Zambia, Angola, Botswana and Namibia as well as westwards from Ethiopia to Uganda to West Africa. The present Afrikander breed consists of sanga cattle developed from the Hottentot cattle.

Ford (1960) also reviewed the origins of African cattle, adding more detail (some of which may be rather speculative) to the previous descriptions. He described a route in the tsetse zone south of the Sahara between latitudes 10° and 15° N, followed from west to east first by the longhorn and then by the shorthorned *Brachyceros*, and from east to west by zebus and sanga cattle. A second route, from Ethiopia and the upper Nile through the centre of the continent, was followed first by the Hottentot ancestors, with their zebus, and secondly, by the sanga stocks, which now populate the northern part of southern Africa between latitudes 12° and 25° S. Thirdly were the routes by which the chest-humped zebu spread west throughout Kenya and most of Uganda, and south through Tanzania to Malawi and parts of Zambia. These waves of cattle migrations obviously did not come against an impenetrable wall of tsetse, but either found routes through the infested zone, or possibly evolved what Ford called 'protective adaptation against infection'. Ford's review is an interesting one giving a historical background to cattle distribution in Africa in relation to tsetse and trypanosomosis.

13.2 ASPECTS OF THE EPIDEMIOLOGY OF BOVINE TRYPANOSOMOSIS

It is not my intention to review here in detail aspects of epidemiology of trypanosomosis in domestic livestock from a veterinary viewpoint (for which the reader is referred to Stephen's *Trypanosomiasis – A Veterinary Perspective*, 1986) but to concentrate on aspects associated with transmission by the vector.

Nagana, a word traditionally used for trypanosomosis in cattle, was referred to by Bruce (1895), who noted that *nagana* was caused by the bite of the tsetse fly. *Nagana* was a Zulu word, meaning to be depressed or in low spirits. Other Zulu names for the disease were *injoko* or *munca*, the latter name meaning 'sucked out'. Bruce cited three theories on the cause of the disease:

- That the disease was caused by the bite of the tsetse fly (which he called the European theory).
- That the disease was caused by the presence of wild animals, which in some way contaminated the grass or drinking water.
- That the disease could be some kind of malaria or vegetable poison produced by the physical conditions of a tropical situation.

Bruce suspected that the tsetse fly was responsible but did not know how, although he suggested that it could transmit a living virus or small parasite and he attempted transmission experiments between tsetse flies and domestic animals.

Ford (1964) referred to 'the *Glossina*–trypanosome–wild-game complex', which resulted in two paradoxical consequences. The first was that the one class of country in which we can be certain of never finding domestic bovine trypanosomosis is inside a tsetse fly-belt; there are no domestic bovines there to contract the disease. The second was that where we *do* find domestic bovine trypanosomosis it is commonly difficult, if not impossible, to find *Glossina*. Similarly, Gillman (1936) put forward the idea that it is not that tsetse keep out humans, but that tsetse flourish where humans cannot. The advent of trypanocides has altered this situation to some extent. It is, however, true to say that bovine trypanosomosis is predominantly an ecotonal, or 'edge' problem. The natural 'equilibrium' situation is one of low endemicity interspersed with a series of localized epidemics of varying intensity. This will be disrupted when tsetse belts advance or recede. The use of drugs allows a situation of relatively high endemicity but without disastrous epidemics.

In southern and East Africa, trypanosomosis partly determines cattle distribution in the *Brachystegia–Isoberlinia* zone (as cattle are absent in most areas where tsetse occur) but has little effect on cattle numbers where cattle are raised at the edge of a tsetse infestation, under a low challenge. In contrast, trypanosomosis has a large direct effect on cattle *numbers* in West Africa in locations where cattle have to go into tsetse infested areas for grazing, but transhumance, which takes place on a large scale, minimizes the effect on cattle distribution (Ford, 1969).

13.2.1   Patterns of disease incidence – the *vivax*-ratio

In tsetse, diagnosis of trypanosome infection type by the commonly used technique of dissection and examination of the various tsetse organs may be less reliable than some of the more recent techniques described later in this chapter. *Trypanosoma vivax*-type infections generally predominate in species of all three tsetse subgenera, although *palpalis* group flies generally have the higher proportion of these infections. The reasons why *T. vivax*-type infections predominate in tsetse have been discussed in relation to barriers to infections in Chapter 11.

In cattle, the '*vivax*-ratio' (Ford, 1971) is affected by the species of tsetse to which they are exposed, the influence of chemotherapeutic drugs, potential reservoir hosts for trypanosomes and host immune responses, which can be determined by breed. As a general rule, it has often been stated that *T. vivax* infections predominate in cattle in West Africa and are rapidly fatal, whilst *T. congolense* causes chronic disease. In contrast, *T. vivax* may be commonly encountered in East and Central Africa but causes a mild disease in cattle in comparison with *T. congolense*. There are exceptions to this rule: for example, the haemorrhagic *T. vivax* infections that occasionally break out in Kenya are rapidly fatal. The *vivax*-ratio is usually high where the overall incidence of trypanosomosis in cattle is low. Why this is so is not clear. A possible explanation is that a greater proportion of infections are transmitted mechanically rather than cyclically in such areas and *T. vivax* is more readily transmitted in this manner than other trypanosome species. Such situations may occur at the edges of tsetse distribution, when drug treatment reduces overall incidence of trypanosomosis, when cattle are removed from tsetse areas, or where *palpalis* group flies (especially *G. fuscipes*) are the main, or sole, vectors.

Models of trypanosomosis transmission predict that, in cattle, *T. vivax* infections would be more common as the basic reproduction rate of the parasite is 150 times greater than that of *T. brucei* and it has a lower transmission threshold. *Trypanosoma congolense* occupies an intermediate position (Rogers, 1988a).

13.2.2   The influence of trypanocidal drugs on trypanosome infection type

It has been observed that *T. congolense* infections often predominate in cattle except outside tsetse belts (Godfrey *et al.*, 1964). Godfrey *et al.* suggested that, in Nigeria, this was because drug treatments did not control *T. congolense* to the same degree as *T. vivax*. The predominance of *T. congolense* infections in livestock may also be due to the development of a better immune response to *T. vivax* in infected animals.

Inoculation of cattle blood into rats was used by Godfrey and Killick-Kendrick (1961) for diagnosing trypanosomosis in cattle in the Donga valley, Benue Province, Nigeria. This, together with blood smear examination, was a more sensitive technique than the standard method of blood smear examination alone. In this study, *T. congolense* infections predominated in cattle, in contrast with most previous studies, in which a higher ratio of *T. vivax* infections was found. Godfrey and Killick-Kendrick believed that the predominance of *T. congolense* infections reflected the true trypanosome infection in cattle and was not simply due to the fact that rat inoculation was more sensitive for *T. congolense* than for *T. vivax*, which does not grow well in rats. They speculated that the predominance of *T. congolense* infections in cattle in Benue Province, Nigeria, could have resulted from a lower sensitivity of this species to phenanthridinium drugs than *T. vivax* (Stephen and Gray, 1960). Homidium, for example, may control *T. congolense* less effectively than *T. vivax*. Thus, *T. congolense* infections may dominate when homidium has been in general use for several years. Conversely, *T. vivax* may dominate after widespread and continued use of diminazene aceturate. If blood examinations are made sufficiently often and the appearance of trypanosomes is followed by cure, *T. congolense* and *T. b. brucei* will not have the opportunity to show themselves. In Kenya, diminazene treatment increased the prevalence of *T. vivax* infections in cattle (Wilson *et al.*, 1975, quoting Boyt *et al.*, 1962). It was suggested that this was a result of the longer prepatent period of *T. congolense*. The *vivax*-ratio can therefore be influenced by both the type and the frequency of treatment. It is not clear what effect, if any, such treatment regimes may have on trypanosome infections in tsetse.

In the 1960s, Whiteside (1962a) stated that trypanosomes in cattle become resistant to the action of almost every drug rather easily. However, he believed that resistance to diminazene did not exist at that time and suggested that resistance had not yet developed because this drug was very quickly excreted. Whiteside considered that chemoprophylaxis provoked some degree of immunity in cattle exposed to tsetse flies. The higher the incidence (mean time in weeks between diminazene treatments) the shorter was the period of prophylaxis. However, cows heavily protected by drugs developed less immunity than those lightly protected (Whiteside, 1962a).

13.2.3   Livestock management practices and trypanosomosis

In traditional cattle-keeping areas of sub-Saharan Africa it is common for nomadic livestock owners, such as the Fulani, to reduce contact between tsetse and their livestock by grazing management, avoiding tsetse-infested areas whenever possible. Nomadic cattle owners often have a good knowledge of such areas. The best-known nomadic cattle keepers of Africa are the Maasai in Kenya, the Nuba in Sudan and the Fulani, who inhabit countries of the northern Guinea zone of West Africa, notably those countries bordering the Sahel region, close to the northern limits of distribution of tsetse flies. The nomadic way of life of the Fulani has been adopted to cope with the seasonal shortages of grazing and water. There are fewer tsetse in those northern areas and the Fulani grazing practices avoid contact with them as far as possible. Constraints of grazing and water availability compel them to move south into tsetse-affected areas in the dry season. At this time, risk of trypanosomosis transmission is decreased as tsetse populations are usually lower. These areas are avoided in the rainy seasons, when tsetse challenge is usually higher and grazing is more readily available in non-infested areas. Avoidance is carried out either by grazing cattle at particular hours when tsetse are less active, or by migration to uninfested areas. Peuhl cattle owners interviewed by my colleagues and I in Burkina Faso in 1995 recognized the risk of taking their animals into tsetse-infested areas during the dry season but considered this to be preferable to the risk of their animals dying from starvation. They are also able to use trypanocidal drugs to give some degree of protection to their animals before moving into infested areas.

At the time of day when tsetse are most likely to be actively feeding, Fulani take their cattle from the grazing areas into encampments, where they may be protected from insect attack by smoke from fires. The use of smoke to protect cattle from biting and nuisance flies is also practised by cattle owners in southern Sudan (Lewis, 1953). Fulani cattle owners of West Africa graze their herds during the dry season in pastures as far south as the permanent tsetse belts permit, before retreating northward when approaching seasonal rain encourages the spread of the fly.

In southern Sudan, cattle tend to migrate northwards in the rainy season between May and September, away from the main rivers and tsetse-infested areas (Lewis, 1953). These migrations are also undertaken to avoid Tabanidae, which cause considerable nuisance and disruption to grazing by cattle through their bites.

The practice of grazing cattle at different times of day to avoid fly activity (as described above) may be only partially effective in reducing trypanosomosis risk, as fly activity appears to be stimulated by the presence of cattle (Rawlings *et al.*, 1994).

13.3   CATTLE TRYPANOSOMOSIS IN WEST AFRICA – NIGERIA

Tsetse flies have had a marked influence on cattle distribution and production systems in West Africa and the situation in Nigeria is representative of West Africa. The bulk of Nigeria's human trypanosomosis, and much cattle trypanosomosis, is transmitted by the riverine species of tsetse. *Glossina tachinoides* was especially significant, as a most important vector of pathogenic trypanosomes to cattle in West Africa. As in most West African countries, the cattle-keeping areas are mainly in the northern regions where savanna woodland provides grazing areas. The southern parts of Nigeria and neighbouring countries are more forested and are infested by the *palpalis* and *fusca* group flies. The northern savanna regions are mainly infested by *G. m. submorsitans* of the *morsitans* group and some of the *palpalis* group flies such as *G. palpalis* and *G. tachinoides*. Cattle keepers in these regions avoid tsetse-infested areas as far as possible in ways that have already been described. Trypanosomosis was highly prevalent in cattle in the northern areas of Nigeria until recent times when population pressure, ground spraying of insecticides, and habitat destruction resulted in eradication or decline of *G. m. submorsitans* populations.

In Nigeria, and other parts of Africa, indigenous cattle have traditionally been trekked as rapidly as possible along well-defined routes to marketing places in the south and to seasonal grazing areas. These treks could take up to 28 days and were of epidemiological and economic significance as cattle were exposed to varying degrees of tsetse challenge, and some mortality or loss of condition occurred before the destination was reached. Jordan (1965c) described the tsetse challenge to cattle along one of these routes. *Glossina m. submorsitans* was the main vector, occurring in the savanna woodland, whilst challenge from *palpalis* group tsetse was light. By the 1970s the more frequent use of lorry or rail transport was associated with a reduced prevalence of trypanosomosis in cattle (Kilgour and Godfrey, 1978).

In 1969 roughly 200,000 trade cattle were railed southward annually and approximately the same number trekked the same distance, passing through tsetse-infested country (Na'isa, 1969). Consequently, trypanosome prevalence in trekked cattle was studied from early times (e.g. Macfie, 1913b) until the mid 1970s, when Riordan (1976) estimated the rate of linear advance of *G. m. submorsitans* along one of these routes in southwestern Nigeria. His estimate of about 5.4 km per year was about half the rate proposed by other workers. Riordan attributed this relative slowness to the equability of the climate. Isometamidium prophylaxis is important in maintaining productivity of trade cattle (Na'isa, 1969).

During the rainy season 76% of the cattle population are found in five partly tsetse-free northern provinces of Nigeria, occupying 44% of the region. In the remaining parts, if fly-free plateaux are excluded, 54% of the area holds only 14% of the cattle population. In the dry season, the cattle have to leave tsetse-free zones to find grazing and water, and are subject to considerable losses due to trypanosomosis. Most zebu cattle go to an area largely free of savanna tsetse, in which there is severe seasonal pressure on grazing and water (MacLennan, 1963).

Clearly, the situation in Nigeria has changed considerably over the last few decades as a result of an increasing human population and consequent reduction in the area of suitable habitat for tsetse flies. Increased settlement of land is also altering the traditional nomadic practices of the pastoralist peoples, such as the Fulani (Jordan, 1986).

According to Ikede *et al.* (1987), the role of trypanosomosis in limiting livestock production in the derived savanna zone has been exaggerated as the prevalence of both the disease and its vector are now low. *Glossina palpalis* is normally an unimportant vector of bovine trypanosomosis in southern Nigeria as both its density and rate of infection with trypanosomes are very low; furthermore, it feeds primarily on humans and reptiles, which are not important reservoirs of animal trypanosomes (Jordan, 1965c). Strains of trypanosomes transmitted by riverine flies are frequently only slightly pathogenic, so cattle can live in fairly close association with them (MacLennan, 1970).

13.4   PATHOLOGY OF TRYPANOSOMOSIS IN LIVESTOCK

No attempt is made here to describe pathological effects of trypanosomosis in domestic livestock in any detail. Some of the characteristic symptoms and consequences of infection are referred to as more detailed information can be obtained elsewhere (e.g. Stephen, 1986).

13.4.1   The chancre

From the 1950s it was recognized that a period of development took place between the infective bite of a tsetse fly and the detection of patent parasitaemia. During this period, metacyclic trypanosomes multiply and transform to procyclic bloodstream forms. Within a few days after the susceptible host has been bitten by infected tsetse a nodule, called a chancre, appears; it increases in size and becomes a raised painful swelling, reaching its maximum thickness during the second week, before declining to undetectable levels during the third week. In humans there is, at first, no reaction except perhaps for a transitory urticarial weal. Puncture of this nodule yields trypanosomes of the long bloodstream form some days before they can be found in the blood. (In humans, the metacyclics remain at the site of the bite, where they develop into bloodstream forms, which produce the chancre reaction; Fairbairn and Godfrey, 1957.) The trypanosomes then pass from the chancre into the blood circulation. In Zimbabwe, chancres were present in 70% of Europeans infected with trypanosomosis, but were less frequent in infected Africans (Gelfland, 1966). Gelfland suggested that this might have been because African patients were seen at a later stage of the disease, after the chancre had disappeared.

The chancre, which is the first site at which an immune reaction in the host can be stimulated, is a local skin reaction in a susceptible host at the sites of deposition by tsetse flies of metacyclic proliferating trypanosomes. It is an established feature of the pathology of both human and animal trypanosomosis (several references supporting this are given by Taiwo *et al.*, 1990). The chancre not only forms a site for the establishment of infection but also is a focus for multiplication and persistence of trypanosomes before their dissemination into the bloodstream (Akol and Murray, 1982; Luckins *et al.*, 1994). When an infected tsetse fly feeds, it 'injects' a relatively small number of metacyclics, with a limited number of metacyclic variant antigenic types (M-VATs). In the chancre, the number of trypanosomes and M-VATs increases, giving the parasite a better chance of survival in the vertebrate host. Godfrey and Fairbairn (1958) found a few trypanosomes from a chancre in a large lymphatic vessel embedded in the subcutaneous fat in a chancre and suggested that this was a possible route for the trypanosomes to pass from the chancre into the general circulation.

13.4.2   Anaemia

One of the major effects of infection with pathogenic trypanosomes is anaemia. In acute infections, packed red cell volume (PCV) falls rapidly due to erythrophagocytosis, but an equally rapid recovery takes place following trypanocidal drug treatment. Measurement of anaemia gives a reliable indication of disease status and productivity performance, although its severity is affected by virulence of the infecting trypanosome species and host factors such as age, nutritional status and breed (Murray and Dexter, 1988). In chronic infections, the bone marrow function becomes impaired and haemopoeisis can no longer compensate for erythrophagocytosis. Following treatment, PCV either recovers slowly or remains at a low level. Anaemia caused by trypanosomosis in cattle was reviewed in detail by Murray and Dexter (1988).

Haemorrhagic syndrome

Certain strains of *T. vivax*, mainly in East Africa, cause a haemorrhagic syndrome in cattle associated with a high parasitaemia, in which haemorrhages, particularly of the gastrointestinal tract, occur. This can result in a high level of mortality (Gardiner, 1989).

13.4.3   Effect on fertility

Trypanosomosis affects livestock productivity not only through mortality, abortion, decreased growth rates and stunting, and loss of efficiency in animal traction, but also through effects on fertility.

Infections of cattle with *T. vivax* or *T. congolense* cause lesions in the male reproductive organs of cattle. *Trypanosoma congolense* appears to cause more severe effects than *T. vivax* (Sekoni *et al.*, 1990). Novidium treatment was sometimes ineffective, leading to regeneration of the lesions caused, and infertility in bulls could persist following chronic trypanosomosis (Sekoni, 1990). Male goats may also become infertile in the course of experimental infection of *T. evansi* (Ngeranwa *et al.*, 1991). *Trypanosoma vivax* infection in West African Dwarf ewes has been shown to cause anoestrus for up to 3 months after infection, but with no persistent effects following treatment (Elhassan *et al.*, 1994).

13.5   TRYPANOTOLERANCE AND TRYPANOTOLERANT CATTLE

Innate resistance, or trypanotolerance, has been recognized since 1906 when the ability of indigenous taurine cattle in West Africa to survive and be productive under trypanosomosis risk was observed (Pierre, 1906). Both acquired and innate resistance to African trypanosomosis can occur in cattle. The two most important trypanotolerant breeds are the *B. taurus* subtypes, N'Dama and Baoulé, whilst a degree of trypanotolerance has also been shown to occur in some *B. indicus* zebu breeds – for example, the Orma Boran (Njogu *et al.*, 1985) and the Maasai zebu (Mwangi *et al.*, 1993). The tolerance to trypanosomosis of N'Dama was compared with zebu and with zebu/N'Dama crosses by Chandler (1952, 1958) who believed this to be an inherent quality of the breed, not merely a partial immunity acquired as the result of previous infection. The N'Dama only account for 5% of the total cattle population of sub-Saharan Africa, and there is now considerable interest in their conservation.

13.5.1   Comparisons between trypanotolerant and susceptible cattle

An experiment in which an artificial challenge (using *G. m. submorsitans* collected from various locations) was continuously administered to N'Dama calves led to a degree of stunted growth, clinical illness, retardation of sexual development and hypoplasia of adrenals and pancreas (Stephen, 1966b). Nonetheless, they could be maintained in a much better condition than zebu cattle.

Roberts and Gray (1973a,b) compared N'Dama, Muturu and zebu cattle kept under trypanosomosis risk in Nigeria, and found N'Dama to have more qualities of potential practical value than Muturu. They suggested that the breed could be used more intensively in tsetse-infested areas of West Africa than was the case at that time.

On the Madina-Diassa ranch in Mali, subject to a high tsetse challenge from three species, trypanosomosis was a serious problem in N'Dama calves from birth to 3 months, resulting in a high infection rate and mortality (Diall *et al.*, 1992).

High prevalences of trypanosomosis in trypanotolerant cattle have sometimes been reported. In a herd of N'Dama and N'Dama × Muturu crosses in Nigeria, between 70 and 90% of the cattle examined were infected (Godfrey *et al.*, 1964). No clear explanation was given for the high trypanosome prevalence in these cattle, although their poor condition resulting from other diseases such as intestinal and tissue helminth infestations might have increased their susceptibility.

Young animals are more resistant to trypanosomosis than adults, possibly due to transmission of maternal antibodies (Whiteside, 1962a; Fiennes, 1970, in Wellde *et al.*, 1981). Wellde *et al.* (1981) found that 'an appreciable immunity to *Trypanosoma congolense*' could develop. They also found increasing resistance to the parasite with age; younger animals survived better than those over 2 years of age. N'Dama cattle are less susceptible to *Amblyomma* and *Hyalomma* tick species than Gobra zebu cattle, with significantly fewer ticks attaching, and a significantly lower prevalence of *Anaplasma marginale* infection (Mattioli *et al.*, 1995).

Trypanotolerance has since been demonstrated in a more precisely quantified way. In The Gambia, N'Dama village cattle can survive and produce meat and milk even under a high tsetse challenge (Dwinger *et al.*, 1994). Where both zebu and N'Dama cattle were kept in the same localities and under the same management in three areas of The Gambia, the trypanosome prevalence in zebu, at 6.2%, was significantly higher than the prevalence of 1.9% in N'Dama ($P < 0.001$; Leperre and Claxton, 1994).

In addition to the innate trypanotolerance of such cattle, immunity acquired after repeated infections contributes to their greater resistance to the disease compared with susceptible breeds. Research is being carried out to establish the means by which trypanotolerant animals control parasitaemia, with the aims of producing a vaccine or the production, through genetic engineering, of resistant cattle (Murray and Black, 1985). Dolan (1987) suggested that resistance to trypanosomosis might be under the control of several genes rather than a single one.

Njogu *et al.* (1985) reported that the East African Orma Boran cattle are trypanotolerant compared with improved Kenya Boran. In some areas zebu cattle have been under selection pressure from trypanosomosis in Africa for perhaps several thousand years and some degree of trypanotolerance would be expected; however, throughout most of their range they have been kept outside the distribution of tsetse and have not been subject to any selection pressure.

13.5.2   The mechanisms of trypanotolerance

At least two main characteristics of the host animal are involved in resistance to trypanosomosis:

- The ability to regulate parasite population expansion.
- The capacity to resist anaemia.

Murray and Black (1985) reviewed the means by which those factors operate in the host animal. The trypanotolerance observed in wild animals, particularly the African buffalo, may be different to that observed in trypanotolerant breeds of cattle. In the African buffalo, xanthine oxidase in the serum and plasma kills bloodstream stages of all pathogenic trypanosome species (Muranjan *et al.*, 1997). The serum and plasma of a variety of other animals tested, including N'Dama, waterbuck and yellow-fronted duiker, does not kill the trypanosomes (Reduth *et al.*, 1995). Laboratory experiments indicate that the killing of the trypanosomes results from inhibition of the trypanosome glycolytic pathway and of ATP production (Muranjan *et al.*, 1997); this only happens with the serum and plasma, not with whole blood.

Investigations into the genetic basis for trypanotolerance have shown that the ability to control parasitaemia and the capacity to resist anaemia during infection in trypanotolerant N'Dama cattle are highly heritable criteria and genetically correlated with production (Murray *et al.*, 1990). This indicates a possibility for selection based on PCV when animals can be detected parasitaemic (Trail *et al.*, 1991).

13.5.3   Exploitation of trypanotolerant cattle

A number of studies have shown that trypanotolerant cattle are more productive than previously thought, and that trypanotolerance is an innate characteristic, not lost when the animals are moved to new areas. Efforts have therefore been made to exploit the trait, both in existing livestock production systems, through breeding programmes, and through improved management. As conventional selection programmes are very slow – probably too slow to be of use in a programme to produce trypanotolerant phenotypes – the production of transgenic animals is seen as a possible means of rapid multiplication, using multiple ovulation and embryo transfer techniques (Jordt and Lorenzini, 1990). The goal of the International Livestock Research Institute's research in this area is 'to provide animal breeders with the means of combining these genes in new ways to provide livestock owners with fitter farm animals that remain productive under stressful conditions' (ILRI, no date). This may be difficult if indeed there are multiple genes for the trypanotolerance trait; furthermore, the time period between identification of genes responsible for the trypanotolerance trait and any impact arising from the application of this knowledge to produce productive and disease-resistant livestock may be rather long. Genetic markers to distinguish *B. taurus* from *B. indicus* cattle could benefit efforts to conserve trypanotolerant cattle (Kemp and Teale, 1994; Teale *et al.*, 1995).

Some critics of the use of trypanotolerant cattle, which only account for 5% of Africa's cattle population, ask why, if there is nothing 'wrong' with trypanotolerant cattle, their innate resistance to the disease has not resulted in their becoming widespread throughout the continent in the last 10,000 years? In response, it has been suggested that they may have been more widespread, but are slightly more susceptible to rinderpest than zebu cattle. This resulted in their numbers and distribution being reduced during the recent pandemic rinderpest epidemics (J. Trail, Nairobi, 1989, personal communication). N'Dama cattle are also apparently susceptible to East Coast fever (ECF) although they appear to be less susceptible to tick attachment than zebu cattle (Mattioli *et al.*, 1993). Ford (1971) voiced an opinion widely held before more detailed research on trypanotolerant cattle had been conducted, when he stated that trypanotolerant cattle 'are of little commercial importance and in any case are tolerant of infection by local strains of trypanosomes'. Despite evidence that trypanotolerant cattle can be as productive as susceptible breeds under trypanosomosis risk, they are not popular with the Fulani in West Africa, who regard them as small and inferior cattle. Nonetheless, mixtures of trypanotolerant and trypanosusceptible breeds are often kept by pastoralists in areas of intermediate challenge, although they usually introduce as much zebu blood as circumstances permit. The Fulani prefer to keep zebu cattle if possible, and to manage them, with respect to trypanosomosis risk, as described in section 13.2.3. Where necessary, they will cross their zebu cattle with Baoulé or N'Dama. They often keep animals for milk, or to sell to provide for their other needs, and they find that trypano-tolerant breeds, particularly the Baoulé, give very little milk compared with zebu. Furthermore, they do not fetch good prices in markets where, due to their small size, they are hidden by zebu cattle, which may fetch 2–3 times the price of a Baoulé. When grazing is scarce and cattle have to go on transhumance, trypanotolerant cattle cannot compete with zebu for grazing because of their small size. Sedentary cattle owners have commented that a single zebu/Baoulé cross or a zebu can be used for traction where two pure Baoulé oxen would be needed.

The present population of trypanotolerant cattle in West and Central Africa is estimated at 4,862,700 N'Dama, 1,963,900 shorthorns and about 3 million crosses. Recent introductions of these cattle, from commercial ranches in the Democratic Republic of the Congo and from The Gambia, have produced 600,000 animals originating from about 34,000 introduced stocks (Shaw and Hoste, 1991a). The exported animals were mainly introduced to ranch management systems whilst originating from village production systems. There was a small surplus of heifers from the West African village sector, relative to the requirements for replacing breeding stock, which was large enough to meet the current needs for different African countries (Shaw and Hoste, 1991b).

It has been argued that the supply of trypanotolerant cattle is too small to allow their rapid and successful spread throughout tsetse-affected areas of the continent. Dehoux (1990) reported that the transfer of N'Dama from Senegambia and the Democratic Republic of the Congo to Gabon induced a deficiency in their trypanotolerance which had to be compensated for through a programme of chemoprophylaxis. Results of experiments with N'Dama imported from The Gambia using embryo transfer techniques (Paling *et al.*, 1991a,b) make Dehoux's findings questionable. Dehoux (1990) gave no indication of the level of tsetse challenge under which the imported N'Dama were kept and, as trypanotolerance is a tolerance, not resistance, under high tsetse challenge it would be expected that some chemotherapy might be necessary. Furthermore, imported cattle often need a period of adaptation to their new environment.

A study of the economics of production of N'Dama cattle recently introduced in a village farming system in the Democratic Republic of the Congo was carried out by Itty *et al.* (1995). This showed that even in a system with no tradition of cattle keeping, the introduction of trypano-tolerant cattle in a tsetse-infested area was successful in terms of profitability and integration of cattle into the farming system.

13.5.4   Tsetse challenge and trypanotolerance

Trypanotolerance is a relative rather than an absolute trait. Thus, trypano-tolerant breeds of livestock can be severely affected in high challenge situations or when under stress. In order to compare trypanotolerance and productivity from different localities and between breeds, it is therefore essential to estimate the level of tsetse challenge or trypanosomosis risk under which these livestock are kept, however inadequate the available methods for assessing it may be (Murray *et al.*, 1981a). In studies of trypano-tolerance and livestock productivity at a network of sites across Africa, tsetse challenge (defined as the product of the relative densities of tsetse, their trypanosome infection rates and the proportions of feeds that they have taken from domestic livestock) was used as an index of the numbers of infected tsetse feeding on cattle. These data were correlated with trypanosome prevalence in trypanotolerant and trypanosusceptible cattle (Leak *et al.*, 1990). Figure 13.1 shows the predicted trypanosome prevalence for any given level of tsetse challenge for each breed type. At a given level

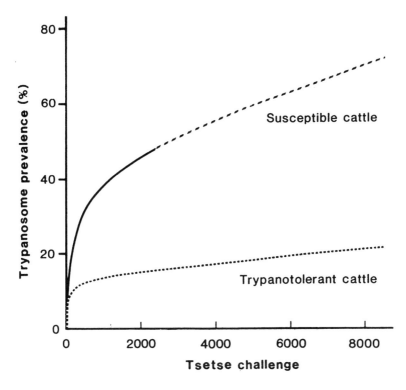

Fig. 13.1. Relationships between tsetse challenge and trypanosome prevalence in trypanotolerant and susceptible cattle. (The dashed section of the curve for susceptible cattle is an extrapolation as there were no sites with susceptible breeds kept at very high levels of tsetse challenge.)

of tsetse challenge a lower trypanosome prevalence is predicted in trypanotolerant, compared with trypanosusceptible, cattle. The rate at which trypanotolerant cattle appear to acquire infections is lower and, in the localities studied, susceptible cattle could not be found in areas with as high a tsetse challenge as trypanotolerant cattle. The curves also demonstrate the high degree of control of tsetse populations needed before significant decreases in trypanosome prevalence can be achieved.

Small ruminants

There have been few experiments to assess the trypanotolerance of small ruminants. Whilst sheep and goats can apparently be kept under low tsetse challenge with little apparent disease, under a high challenge they may suffer a high degree of mortality (MacLennan, 1970). Mutayoba *et al.* (1989) reported comparative trypanotolerance of small East African goats to *T. congolense.* The West African Djallonke sheep and dwarf goats are also reputed to be trypanotolerant. Sheep are less likely to be fed on by tsetse where there is a choice of alternative hosts, but otherwise sheep will be fed on to a greater extent. In West Africa, *T. b. brucei* caused fatal infections of Cameroon dwarf goats, whereas few clinical signs accompanied *T. vivax* or *T. congolense* infections (Bungener and Mehlitz, 1976). Generally, *T. vivax* is a more important cause of trypanosomosis in sheep and goats in West Africa than *T. congolense* (Murray *et al.*, 1981b). Infection rates in tsetse with *T. b. brucei* are lower, so if the former situation is correct, it is likely that fatal or serious cases of trypanosomosis would be less frequently seen than in cattle even if biting rates were the same for sheep as for cattle.

Mawuena (1987) reported that, in contrast with trypanotolerant cattle, in which parasitaemias may be lower than in susceptible animals, Djallonke sheep and goats had high parasitaemias of pathogenic parasites, yet could be kept in areas where trypanotolerant cattle could not, without showing effects of disease. This observation was supported by Osaer *et al.* (1994), who suggested that the mechanism for trypanotolerance in small ruminants differs from that in cattle.

13.5.5   Some factors affecting the expression of trypanotolerance

There remain many questions regarding the causes of the difference observed in trypanosome prevalence, or in other words, the expression of trypanotolerance.

Some of the entomological questions relating to trypanotolerance that have been asked in the past, and which have still not been completely answered are as follows:

- Is apparent trypanotolerance expressed in trypanotolerant breeds a result of these cattle not being fed upon by infected tsetse to the same extent as susceptible breeds?
- Are trypanotolerant cattle fed on to the same extent as susceptible cattle but require a higher infective dose of trypanosomes in order for an infection to become established?
- Are trypanotolerant cattle infected with trypanosomes at a similar level of trypanosome prevalence as susceptible cattle, but are able to control parasitaemia at a level less frequently detectable by standard diagnostic techniques?

Experiments have shown that trypanotolerance is an innate characteristic, not simply a reflection of a lower attractivity to tsetse. There is no evidence that trypanotolerant cattle require a higher dose of trypanosomes in order to become infected (Paling *et al.*, 1991a,b).

Differential susceptibility to tsetse bites

Little research has been conducted on the differential susceptibility of cattle to tsetse bites; however, studies on horn flies (*Haematobia irritans*) and cattle have suggested some possible mechanisms for this. There is individual animal variation in the numbers of horn flies and of *Stomoxys calcitrans* that are attracted to cattle and these differences are possibly under genetic control (Warnes and Finlayson, 1987; Brown *et al.*, 1992). Estimates of heritability suggested that selection procedures could be used to control these horn flies, raising the possibility that this could apply to trypanotolerant livestock and tsetse flies. Pinder *et al.* (1987) examined experimental and cyclical infection or field challenge of *T. congolense* to zebu and Baoulé cattle. Trypanotolerant cattle showed little or no parasitaemia during natural fly challenge but became parasitized when *Glossina* were forced to feed on them, suggesting that part of their tolerance could be due to their being bitten less often. This might be due either to a lower attractivity to tsetse or, alternatively, to more efficient mechanisms to prevent flies from feeding, such as tail flicking or neuromuscular twitching of the skin.

The few studies conducted on behaviour of tsetse flies towards trypanotolerant and susceptible cattle have given conflicting results. In Burkina Faso, the numbers of tsetse that fed on Baoulé and zebu cattle were counted in fly-proof chambers. There was some evidence that tsetse fed more on susceptible cattle but the results were inconclusive (Bauer *et al.*, 1987). As the number of bites determines the number of trypanosomes injected into the skin of an animal, and this number may affect the severity of the disease, lower infective doses could contribute to trypanotolerance under natural exposure (Authié, 1994).

Factors that could influence the feeding habits of tsetse in these circumstances include:

- Physical – skin thickness.
- Visual – coat colour, size.
- Olfactory – breath/urine/skin secretions.
- Behavioural – tail flicking, head swings, skin rippling.

PHYSICAL FACTORS. Skin thickness may determine feeding success, although tsetse are able to feed from very thick-skinned animals by choosing specific sites with thin skin, such as the eyelids of crocodiles. Carr *et al.* (1974) established a link between skin thickness in zebu and the prevalence of *T. congolense* infections, the latter being more common in thinner-skinned animals. Skin thickness might also influence the capacity of metatrypanosomes to become established following the bite of an infected tsetse fly (Murray *et al.*, 1981b).

VISUAL FACTORS. Tsetse have a preference for alighting on dark surfaces (Vale, 1982; Green, 1989) and therefore one might expect dark animals to be fed upon more frequently than light animals. However, unless selected for, there is no significant breed difference in coat colour between trypanotolerant and trypanosusceptible animals. Similarly, in a study to assess the effect of coat colour and pattern on trypanotolerance in Senegal, no significant differences in PCV, parasitaemia or antibody levels were found between white, black or piebald N'Dama cattle (Touré *et al.*, 1981). Certain tsetse species have a preference for large animals (*Glossina longipennis* had a preference for rhinoceros, and *G. brevipalpis* for hippopotamus), and the larger size of zebu compared with N'Dama might result in more feeds being taken from them (higher challenge) than from the latter.

OLFACTORY FACTORS. In preliminary studies, a 'global odour' (all of the body smells coming from either Baoulé or zebu cattle) from zebu cattle proved more attractive to tsetse, although the differences were low and therefore inconclusive (Filledier *et al.*, 1988; Filledier and Mérot, 1989a). In contrast, initial experiments suggested that tsetse were more attracted to Baoulé urine than zebu urine. This was an unexpected finding, as Baoulé are trypano-tolerant. Further experiments with phenolic fractions of urine confirmed that Baoulé urine was more attractive to *G. tachinoides*, mainly due to the phenolic fraction. N'Dama urine did not increase trap catches of *G. longipalpis* in Guinea-Bissau (Jaenson *et al.*, 1991), although catches of the closely related *G. pallidipes* in East Africa are greatly increased when traps are baited with zebu urine. No comparison was made between zebu and N'Dama urine in that study, so it may be that *G. longipalpis* is simply not attracted by cattle urine.

BEHAVIOURAL FACTORS. Warnes and Finlayson (1987) reported on the effects of host behaviour on host preference in *Stomoxys calcitrans*. Few examples of behavioural influences on tsetse feeding behaviour towards trypano-tolerant or trypanosusceptible livestock have been reported, although such effects have been recorded for other hosts such as monkeys, which can kill most tsetse that attempt to feed on them. Vale (1977b) showed that far fewer tsetse were able to feed successfully from a goat than from an ox unless the goat was sedated, in which case the number of fed flies coming from the goat was 15 times higher than from the ox. Skin rippling is an important factor in deterring tsetse from feeding from some antelope species such as impala (Bursell, 1980).

Establishment of trypanosome infections

The need to assess and compare trypanosome transmission from tsetse to trypanotolerant/susceptible livestock and from these livestock breeds to tsetse has been emphasized (Rogers *et al.*, 1986). Figure 13.2 shows a theoretical risk of infection. Is there a different, lower risk of infection in trypanotolerant cattle (Fig. 13.3)? Little work has been carried out to address these questions and, in one study, an unexpectedly higher parasitaemia resulted in needle-challenged Baoulé than in zebu cattle (*T. vivax* and *T. congolense*) (Guidot and Roelants, 1982).

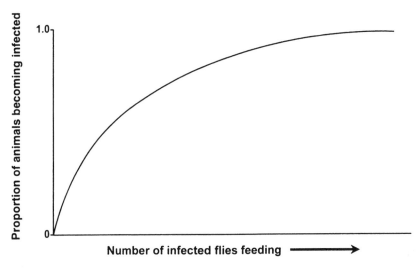

Fig. 13.2. A theoretical relationship between the proportion of feeds from infected tsetse and the proportion of susceptible cattle becoming infected. (From D.J. Rogers, J. Hargrove and A.M. Jordan, unpublished report.)

The bite of a single infected tsetse fly can infect N'Dama cattle with trypanosomes under field conditions (Dwinger *et al.*, 1990). Therefore there is no evidence that a higher challenge or infective dose of trypanosomes is necessary to infect trypanotolerant cattle.

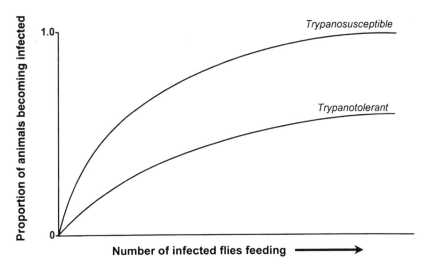

Fig. 13.3. Theoretical relationships between the proportion of feeds from infected tsetse and the proportion of animals becoming infected for trypanosusceptible and trypanotolerant cattle.

N'Dama cattle are capable of acting as reservoirs for trypanosome infections, although they may not be as efficient reservoirs of *T. vivax* as zebu cattle. There is some evidence of a lower transmission rate from tsetse to N'Dama and from N'Dama to tsetse for *T. vivax*, although differences in transmission of *T. congolense* were not significant (Moloo *et al.*, 1992b, 1993).

Whilst it would seem that there are lower transmission potentials between trypanotolerant livestock and tsetse, the main differences between tolerant and susceptible breeds of livestock appear to be determined by the way in which they deal with the trypanosomes that are transferred from the bite of an infected tsetse. Once infected, at least with some trypanosome parasites, trypanotolerant cattle can undoubtedly control the disease better than susceptible breeds.

13.6 EFFECTS OF TRYPANOSOMOSIS ON LIVESTOCK PRODUCTION

In addition to the effect of trypanosome infections on productivity of livestock, the presence of tsetse and consequent risk of trypanosomosis has an effect on the suitability of infested areas for livestock rearing. As 46% of tropical sub-Saharan Africa is infested, this can be a very important effect. It is estimated that between 46 and 62 million cattle are at risk of trypanosomosis in Africa (Reid *et al.*, 1999; P.N. De Leeuw, Nairobi, 1997, personal communication) and Table 13.3 shows the stocking density of cattle in three regions of the continent.

Clearly other factors besides tsetse infestation are responsible for the lower stocking densities but there are strong correlations with tsetse infestation, particularly in East Africa (Pingali *et al.*, 1987).

There have been surprisingly few studies on the direct effects of trypanosomosis on livestock productivity. This is partly because such studies take a long time if calving intervals and conception rates are to be measured. Furthermore, these data have to be collected under trypanosomosis risk and also when that risk has been controlled. Some data collected before and after

Table 13.3. Effect of tsetse infestation on cattle productivity: stocking density (from Jahnke, 1982; De Leeuw and Rey, 1995).

Region	Cattle density (km^{-2})	Area (km^2) $\times 10^6$	Area infested by tsetse (%)	Cattle under trypanosomosis risk
Humid tropics	< 3	4.14	90	7.92
Sub-humid	7	4.86	68	22.28
Semi-arid tropics	11	4.05	50	22.72
Arid	4	8.33	12	3.76
Highlands	29	0.99	0.19	5.51
Sub-Saharan Africa	6.6	0.99	46	62.19

tsetse control interventions in Côte d'Ivoire and Ethiopia are shown in Table 13.4 (G.J. Rowlands, 1998, unpublished data). Clearly, changes in body weight and growth rates are small and not always significant. The most significant increase in productivity is probably from increased numbers of animals, resulting from higher calving rates and lower mortalities.

In the Didessa valley, southwest Ethiopia, Jemal and Hugh-Jones (1995) examined the effects of tsetse control through insecticide-impregnated targets and odour-baited traps on productivity of zebu cattle in four villages: one within the controlled area and three outside. Although the amount of data before and after control were limited, the results obtained showed that mortality in the unprotected villages was one to nine times higher than in

Table 13.4. Effects of tsetse control on productivity.

Parameter	Côte d'Ivoire[1] *Jan 87–Feb 97 [†]Impregnated traps		Ethiopia, Ghibe[1] *Mar 86–Dec 95 [†]Pour-on cypermethrin		Mkwaja, Tanzania[2] *1989–Mar 1991 [†]Deltamethrin dip	
	Before control	After control	Before control	After control	Before control	After control
Trypanosome						
prevalence: Cows	12.6	1.2	39.1	14.4	10.5	3.0
Calves	21.4	1.8	20.9	7.1		
Mean cow PCV	29.8	32.4	24.3	25.9	26	30
Mean calf PCV	30.4	32.7	27.7	28.9		
Diminazene treatments (cows)			39.3	19.8	3.9[§]	0.9[§]
Diminazene treatments (calves)			17.9	8.1		
Calf growth rate (g day[−1], wet season)	243	269	216	243		
Cow body weight (kg)	223	237	198	204		
Cow conception rate	28[‡]	51	41	50		
Mean annual calving rate			62	71	58	77
Calf mortality (%, including stillbirths)			21.0	6.9		
Overall mortality (trypanosomosis)					0.4%	0.1%
Mean no. cows year[−1]			17	30		
Mean no calves year[−1]			16	34		
Mean ratio of calves to adult females			0.49	0.73		

[1] Rowlands, 1998 (unpublished data); [2] Fox *et al.*, 1993.
* Study period; [†] type of tsetse control; [‡] increase in numbers of cows conceiving within 6 months of calving; [§] annual cost per head.

the village where tsetse control had taken place. Recorded calving rates were 81% in the protected village cattle whilst the highest calving rate in the unprotected villages was 64%.

On Mkwaja ranch in Tanzania, following a more than 90% reduction in tsetse apparent density brought about by the use of a deltamethrin pour-on insecticide applied to cattle (see Chapter 17), cattle mortality was reduced by 66% and calving percentages and weaning weights increased (Fox *et al.*, 1993).

Table 13.5 shows some of the productivity traits measured in village herds comprising approximately 4000 N'Dama cattle at six sites in The Gambia subjected to differing trypanosomosis risk. The variation in calf mortality appeared to be correlated with both mean monthly trypanosome prevalence and mean PCV recorded at each site (Agyemang *et al.*, 1998).

There were significant differences among sites for PCV, calf mortality, annual mortality and lactation offtake. The productivity index at Missira and Bansang, where there was the highest trypanosome prevalence in cattle, was below the mean of 94. A major limiting factor to increased production, particularly at the four sites where tsetse challenge was low (Keneba, Pirang, Gunjur and Nioro Jattaba), was poor nutrition.

13.7   MECHANICAL TRANSMISSION

The mechanical transmission of trypanosomes by Tabanids, *Stomoxys* spp. or other biting flies was usually inferred where trypanosomosis of livestock occurred but tsetse were not detected. Although *T. vivax* is the species most likely to be transmitted mechanically, other species pathogenic to cattle may also be transmitted (Persoons, 1967). *Glossina* spp. may also act as mechanical as well as cyclical vectors (Roberts *et al.*, 1989).

Table 13.5. Mean productivity traits of N'Dama cattle in The Gambia.

Trait	Missira	Bansang	Keneba	Pirang	Gunjur	Nioro Jattaba
Trypanosome prevalence (%)	8.4	3.6	1.8	1.2	0.4	0.3
PCV (%)	25.1	26.7	28.0	27.8	28.8	27.2
Calf mortality (%)	26.3	15.7	16.2	8.7	12.8	10.8
Annual mortality (all cattle) (%)	20.0	11.4	2.4	2.4	4.5	3.7
Lactation offtake (kg)	203	278	421	495	405	322
Herd growth per year (%)	3	−7	12	20	9	14
Productivity index*	75	71	100	124	96	99

* Calculated as {[(12 month calf body weight + lactation offtake/9) $\times$ 365/calving interval]/cow body weight $^{0.73}$ $\times$ 100} $\times$ (100 − annual cow mortality)/100 $\times$ (100 − calf mortality to 12 months of age)/100.

There is very little good evidence for the occurrence of mechanical transmission in Africa, except for the proven transmission of *T. evansi* by Tabanidae. What evidence there is falls into two categories:

● Circumstantial field evidence based upon occurrence of trypanosomosis in the apparent absence of tsetse.
● Experimental evidence from laboratory infection of animals by mechanical means.

13.7.1   Field evidence for or against mechanical transmission

Trypanosomosis in cattle caused by *T. congolense* was reported from Zanzibar at a time when no tsetse had been detected on the island even after repeated searches (Mansfield-Aders, 1923). The disease was common only where Tabanidae were dominant and was uncommon where Stomoxidae were dominant. Tabanidae were therefore implicated as mechanical vectors; indeed, it may be that habitats suitable for Tabanidae were also those most suitable for tsetse. At the time, a large number of cattle, sheep and goats were imported into Zanzibar from Somalia, and to a lesser extent Tanzania, many of which were infected with trypanosomes. *Glossina austeni* was later detected on Zanzibar, invalidating Mansfield-Aders' conclusions.

In southern Sudan, where trypanosomosis in cattle occurred outside the known distribution of tsetse, the main species detected was *T. congolense*, followed by *T. vivax* (Lewis, 1953). This is contrary to the accepted belief that *T. vivax* predominates outside tsetse areas, and may simply reflect the inadequacies of sampling methods for low density tsetse populations at the time. In a subsequent survey in southern Darfur District of Sudan, *G. m. submorsitans* were found beyond the previous northern limits of their distribution (Yagi and Razig, 1972).

In Ghana, an outbreak of *T. b. brucei* infection in polo ponies was attributed to biting flies rather than tsetse (Turton, 1953). In support of this, Turton cited observations by Hoare (1936a,b) regarding the transmission of *T. simiae* by *Stomoxys* to pigs after original introduction by tsetse. Again, due to the inefficiency of sampling methods, it cannot be concluded that tsetse were absent.

In field observations over a 2-year period, a herd of ten uninfected cattle were grazed together with two steers infected with *T. vivax* and *T. congolense* in an area of Zimbabwe in which no tsetse could be caught (Boyt *et al.*, 1970). Only one case, of *T. vivax*, was detected in the control (uninfected) herd, during the second year, at a time when tsetse-transmitted trypanosomosis had advanced to neighbouring farms. The authors concluded that 'at most, direct (mechanical) transmission played a very minor part' in trypanosomosis transmission in the area.

Because *T. vivax* is frequently more prevalent in livestock at some distance from known tsetse-infested areas, mechanical transmission was often proposed as the means of transmission (Ford, 1964; Folkers and Jones-Davies, 1966). Alternative explanations for the predominance of *T. vivax* infections at a distance from *G. m. submorsitans* concentrations (Leeflang, 1975) are as follows:

- *T. vivax* has a higher infection rate than *T. congolense* in *Glossina m. submorsitans*.
- *T. vivax* has a relatively short development period in *G. m. submorsitans*.
- A close contact exists between dispersed tsetse and cattle.

13.7.2 Experimental evidence for or against mechanical transmission

The following attempts to demonstrate mechanical transmission have been reported:

- Early experiments on mechanical transmission of *T. congolense* in Uganda showed no evidence that transmission occurred in the absence of tsetse (Bruce *et al.*, 1911c; Lucas, 1955).
- Close to Entebbe, in Uganda, a study by Dixon *et al.* (1971) of *Stomoxys* or *Tabanidae* feeding on cattle was confounded by the fact that tsetse also occurred in the area and no evidence that biting flies actually transmitted the disease was obtained.
- In recent laboratory attempts to transmit pathogenic trypanosomes of domestic livestock by *Stomoxys* spp., *T. brucei*, *T. vivax* and *T. evansi* could be transmitted to mice at success rates of 11.5%, 3.4% and 0.9%, respectively, but *T. congolense* could not be transmitted in 129 attempts (Mihok *et al.*, 1995b).

The latter experiment was carried out using mice with high parasitaemia to infect *Stomoxys* spp.; the feeding of *Stomoxys* spp. on infected mice was interrupted and they were allowed to continue feeding immediately on an uninfected mouse. The data were used only if the fly successfully fed after being transferred to the new host. Few conclusions can therefore be made about mechanical transmission under natural conditions.

In Burkina Faso, savanna-type *T. congolense* was detected in the midguts of tabanidae using the PCR technique (Solano and Amsler-Delafosse, 1995), suggesting that Tabanidae could be potential vectors of *T. congolense*. However, in that study, although it was stated that the flies had not fed for several hours, the trypanosomes may have been digested during feeding. There was no evidence that they were viable, transmissible trypanosomes.

13.7.3 General observations

In many places where tsetse were initially not found, improved sampling methods have revealed their presence. Despite the lack of evidence, mechanical transmission was often assumed. In the early 1950s Lewis (1953) referred to *Tabanus thoracinus* as being capable of transmitting *T. congolense*. Unfortunately he quoted no evidence or reference for this statement, although he noted that cattle trypanosomes appeared less suited than *T. evansi* to mechanical transmission.

Before it became clear that methods of detecting tsetse were inadequate, Hoare (1947) reviewed reports of trypanosomosis occurring in areas outside areas of known tsetse distribution in the Democratic Republic of the Congo, Uganda, Sudan and Zanzibar. Stating that the 'foregoing examples leave no doubt that in certain tsetse-free areas of Africa' mechanical transmission takes place, Hoare conjectured that trypanosomes were able to extend far beyond their natural boundaries by adapting to new insect vectors. A consequence of this was a loss of the ability for cyclical development, which was replaced by mechanical transmission.

Wallace (1962) stated that trypanosomes of the *T. brucei* complex were not transmitted mechanically in nature and 'never occur naturally in the absence of tsetse flies'. He concluded that *T. congolense* and *T. vivax* could both be mechanically transmitted, because their infections occurred in livestock where *Glossina* were absent but Tabanidae were abundant. Wallace noted that such transmission would take place only if the interval between bites did not exceed a few minutes, and pointed out that, for successful experimental transmission, numerous flies were required.

It is difficult to assess the importance of mechanical transmission under natural conditions, but several researchers have observed that where tsetse are eliminated, or suppressed, trypanosomosis ceases to be a problem (MacLennan, 1974; Gruvel, 1980). This observation was supported by an investigation conducted at the Shika research station in Nigeria in an area where tsetse flies had been controlled (Folkers and Mohammed, 1965). Despite creating favourable conditions for mechanical transmission, no evidence of it was found, although this research station had previously suffered serious losses due to trypanosomosis. In such situations it might be expected that *T. vivax* or *T. evansi* would remain a problem if a significant level of mechanical transmission were to occur.

In a review of the subject, Krinsky (1976) repeated that evidence for mechanical transmission is based largely on circumstantial field observations and, to a much lesser extent, on laboratory transmission studies.

Methods of detection of tsetse at low densities have been insufficiently sensitive for establishing the presence of tsetse in some areas in which cases of trypanosomosis in cattle could be found. Ford (1964) postulated that *G. morsitans* could extend 16 km beyond any area where they could be detected by conventional entomological methods then available. The importance of such dispersed tsetse may not have been fully recognized. In Busoga district, Uganda, careful surveys detected *G. f. fuscipes* beyond the northern limits of their previously known range, in an area where bovine trypanosomosis was previously presumed to be mechanically transmitted

(Wain *et al.,* 1970). One site where tsetse were detected was 26 km north of the previously recorded distribution.

Although there is evidence for the mechanical transmission of *T. evansi,* the significance of transmission in Africa, of *T. vivax* or other salivarian trypanosome species, by tabanids or other biting flies, is still unknown. Associations can be made between populations of biting flies and trypanosome prevalence in livestock where tsetse do not occur, but the demonstration of living trypanosomes in the mouthparts of biting flies in field situations has not been adequately proven, and only recently was *natural* mechanical transmission shown to take place (Ferenc *et al.,* 1988).

13.7.4   Trypanosome transmission outside Africa

Despite being among the major pests of humans and animals worldwide, tabanid vectors of disease agents are among the least studied dipterans (Foil, 1989) and there is little information about mechanical transmission by these flies even from countries where it is accepted to occur. Trypanosomosis, both African and South American, is said to be transmitted mechanically by Tabanidae. Trypanosomosis in cattle, almost certainly *T. vivax,* was first reported in South America (French Guyana) in 1919 (Leger and Vienne, 1919). *Trypanosoma vivax* is believed to have been introduced to South America from West Africa, a hypothesis supported by molecular biology studies showing genetic similarities between parasites from the two regions (Dirie *et al.,* 1993). Of the three main salivarian trypanosomes pathogenic to cattle, only *T. vivax* has become permanently established outside Africa where it is transmitted mechanically.

Following extensive studies in different countries, conclusive evidence for mechanical transmission by Tabanidae of *T. evansi,* causing the disease surra, has been obtained (Foil, 1989). In Mauritius, where there are no tsetse flies, *T. evansi* was transmitted principally by *Stomoxys nigra.* Surra was introduced to Mauritius in 1901 from India, by way of imported cattle, and the parasite caused a high mortality in both equines and cattle the following year. *Trypanosoma vivax* is also believed to have been introduced to Mauritius, in cattle imported from Africa (Adams, 1935a,b, 1936). A 2% trypanosome incidence was detected in stained blood films examined from 2170 cattle at that time (1930s). Adams interpreted the wide distribution of the parasite on Mauritius as evidence that the disease was probably present but unnoticed previously; therefore, suggestions that the disease was recently introduced were incorrect.

Transmission of *T. vivax* by *Tabanus nebulosus* has been suggested in South America (Otte and Abuabara, 1991) and tabanids have experimentally transmitted the parasite in Colombia and elsewhere by *Cryptotylus unicolor* (Ferenc *et al.,* 1988). Despite these examples, and strong positive correlations demonstrated between the incidence of *T. vivax* and tabanid populations (e.g. in Colombia), as recently as 1991, there was no conclusive evidence for regular transmission by any particular Tabanid species (Otte and Abuabara, 1991). Otte and Abuabara were still able to note that the 'existence of a biological vector of *T. vivax* in the New World cannot be discounted'. Hoare (1965) suggested that vampire bats (*Desmodus rotundus*) could act as vectors and hosts of trypanosomes infecting horses and cattle in South America.

13.7.5 Mechanical transmission of *Trypanosoma theileri*

Trypanosoma theileri is a normally non-pathogenic trypanosome transmitted to cattle by Tabanidae throughout the world. In Nigeria, *T. theileri* was relatively common (Gray and Nixon, 1967), and it is probably common throughout Africa. However, despite being larger in size than pathogenic trypanosome species infecting cattle, it is little known and infrequently reported, and may therefore be confused with other trypanosome species in the field. It has been speculated that the occurrence of infective forms in the peripheral blood of hosts during periods when vectors are active is an adaptation by *T. theileri* for mechanical transmission. This is a form of circadian or perhaps seasonal periodicity, which has resulted in difficulties in isolating *T. theileri* from the blood of known infected cattle during winter in Europe. In tropical areas the organisms have more frequently been reported in blood smears during the rainy season than during the dry season.

Experimental transmission of *T. theileri* has been achieved in Germany using infected *Haematopota pluvialis, H. italica, Hybomitra micans* and *Tabanus bromius*. Infective stages of *T. theileri* could be found in the guts and faeces of the Tabanidae used in transmission experiments. The minimum pre-patent period was less than 4 days and the parasites were not pathogenic to cattle (Bose *et al.*, 1987).

Experimental transmission of trypanosomes by ticks has been carried out in the laboratory (Shastri and Deshpande, 1981; Morzaria *et al.*, 1986) but there is insufficient evidence of the role of ticks in natural transmission. *Hyalomma anatolicum anatolicum* may transmit *T. theileri* to calves under experimental conditions as the developing stages of the trypanosome have been found in engorged nymphs, freshly moulted adults and mature adults partially engorged on rabbits (Morzaria *et al.*, 1986). Other evidence that *T. theileri* can be transmitted by ticks was reported by Burgdorfer *et al.* (1973).

In summary:

1. Some past reports of mechanical transmission of trypanosomes of the *T. brucei* group and of *T. congolense* may have resulted from inadequate means of detecting tsetse flies.

2. Mechanical transmission of *T. vivax* almost certainly takes place, but in Africa it may only be responsible for spreading the parasite within a herd of cattle. In such circumstances, multiple interrupted bites may occur at short intervals. Introduction of the infection to a herd may initially be via the bites of infected tsetse flies.

3. In Africa (but not in South America) there is insufficient evidence that pathogenic trypanosomes of the *T. brucei* group, *T. congolense* or *T. vivax* are sufficiently adapted to mechanical transmission for infections to be maintained for long periods in areas clearly outside the limits of tsetse distribution.

4. *Trypanosoma theileri, T. evansi* (and possibly *T. simiae* after introduction by *Glossina*) are transmitted mechanically.

5. Despite the potential importance of mechanical transmission of pathogenic trypanosomes to domestic livestock, the extent and frequency with which it takes place are still very poorly understood.

13.8   DIAGNOSIS

Current techniques for diagnosis of trypanosomosis in tsetse, humans and domestic livestock are briefly reviewed here, in view of their importance in studying disease epidemiology. Diagnosis of trypanosomosis in tsetse, humans or domestic livestock is a basic requirement for epidemiological studies as well as for planning and implementing chemotherapy and for monitoring vector control operations. Whilst relatively insensitive but easy-to-use techniques may be adequate for some purposes, more sensitive and specific tests may be required for epidemiological studies and in areas with a low prevalence of infection. Xenodiagnosis, in which the insect vector is fed upon a host animal with a suspected infection and subsequently dissected and examined for infection, is the most sensitive method for detecting trypanosomes but is not a practical method for general use as it requires a colony of uninfected tsetse and is time consuming.

13.8.1   Diagnosis of infections in the vertebrate host

Trypanosomosis in domestic livestock is frequently diagnosed by owners or by veterinary staff simply based on clinical symptoms, which are often well known to farmers living in affected areas. This type of diagnosis is a practical necessity in rural situations where equipment such as microscopes or centrifuges is absent, but for research purposes, or for large-scale control campaigns, more precise diagnostic techniques are required. Until relatively recently, adequate diagnostic techniques were not available for epidemiological studies of trypanosomosis. However, advances have been made with a variety of serological tests and these, together with direct methods of diagnosis, are summarized in Table 13.6.

Sensitivity and specificity are commonly used terms for describing diagnostic tests and are sometimes misunderstood. Sensitivity of a diagnostic test is the proportion of true positives that are detected by the method, and specificity is the proportion of true negatives detected (Thrusfield, 1995).

Table 13.6. Techniques for diagnosis of trypanosomes in mammalian hosts.

Method	Advantages	Disadvantages	References
Microscopical examination			
Wet blood smears	Simple, cheap, rapid	Tedious	
Giemsa-stained smears (thick and thin)		Low sensitivity	
Haematocrit centrifugation (Woo)	Simple, cheap, rapid, PCV can also be measured	Tedious	Woo, 1969, 1971
	Higher sensitivity than examination of smears	Low sensitivity	
DG or phase contrast (BCT)	Simple, cheap, rapid, PCV can also be measured	Tedious	Murray et al., 1983
	Higher sensitivity than examination of smears	Low sensitivity	Paris et al., 1982
Indirect serological tests			
Antibody-based			
IFAT test	Moderate sensitivity and specificity	Do not distinguish current and previous infections	Bailey et al., 1967
CATT tests			Magnus et al., 1978
Antibody ELISA	Automatable	IFAT requires sophisticated fluorescent microscope	Luckins and Mehlitz, 1978
Antigen-based			
Antigen ELISA	Theoretically able to distinguish current and previous infections	Requires a more sophisticated, well equipped laboratory	Rae and Luckins, 1984; Nantulya, 1989, 1990
		Problems with sensitivity and specificity	
Latex agglutination	Simple and rapid to use	Sensitivity and specificity still have to be determined	Nantulya, 1993, 1994
DNA based tests			
PCR	Theoretically PCR has a high sensitivity and specificity	Poor sensitivity and specificity (DNA probes)	Majiwa et al., 1994; Masiga et al., 1992, 1996b
DNA probes	Automatable	Expensive	
		Require a sophisticated 'high tech' laboratory, susceptible to contamination	

Direct (traditional) methods

The microscopical examination of blood samples and various modifications of this technique are described here as traditional diagnostic techniques to distinguish them from more recent immunological and DNA-based methods. They may also be referred to as direct methods as they depend on actual detection of the parasite, whereas immunological methods detect parasites indirectly, from antibodies or antigens.

The three basic direct methods are:

- Examination of wet blood smears and stained thick and thin smears.
- The haematocrit centrifugation technique (HCT or Woo method).
- Dark ground/phase contrast buffy coat method (BCT method).

The latter two methods concentrate parasites by centrifugation of blood, collected with heparin or EDTA, in microcapillary tubes. Trypanosomes are found in the buffy coat (the thin layer of white blood cells at the interface of the red cells and the plasma). They may be examined either within the capillary tube (Woo technique) or by making a smear of the buffy coat after carefully cutting the capillary at the interface (DG technique). For the DG technique, the buffy coat smear is examined using dark ground or phase contrast microscopy under which live motile trypanosomes show up more clearly. Trypanosomes can be identified and the level of parasitaemia estimated using a scoring system (Murray *et al.*, 1983). The packed red cell volume of the animal is usually measured before examination of the blood using the Woo and DG methods.

An additional diagnostic technique, more commonly used for research purposes, is the inoculation of blood into mice or rats and subsequent examination of the blood for parasites (Godfrey and Killick-Kendrick, 1961). This technique is unsuitable for *T. vivax* infections as mice are normally refractory to infection with most strains of this species.

For diagnosis in humans, identification of trypanosomes in the blood or lymph is the most specific procedure but can be difficult as parasitaemias are generally low. Concentration of trypanosomes in blood samples, using the Woo (Woo, 1969, 1971) or mini-anion exchange centrifugation techniques improve the sensitivity.

Indirect serological methods

IMMUNOFLUORESCENT ANTIBODY TEST (IFAT). The IFAT developed for trypanosomosis (Bailey *et al.*, 1967) uses known trypanosome antigens to coat glass slides or microtitre plate wells. Blood samples to be tested are added to the wells and incubated, and then conjugate is added. After washing, the wells are examined with a fluorescence microscope. Apple-green fluorescence occurs with positive blood samples.

Drawbacks to antibody tests, such as IFAT, are that antibodies remain in an animal's circulation for some months, and therefore the test does not distinguish between past and current infections. Consequently they may reveal what appear to be very high rates of infection. As an example, in a serological survey of trypanosome infections in domestic animals in Liberia, Mehlitz (1979) found infection rates of 80.4% in N'Dama cattle, 76% in pigs, 48.5% in dogs, 35.1% in goats and 28.1% in sheep.

CARD AGGLUTINATION TEST FOR TRYPANOSOMOSIS (CATT). The CATT test for human trypanosomosis relies on the agglutination of formalin-fixed, stained and freeze-dried trypanosomes by antibodies in the patient's serum. The major antigen contributing to the test is the variant surface glycoprotein (VSG).

Evaluation of a CATT test was carried out at a sleeping sickness focus at Daloa, in Côte d'Ivoire, comparing tests on dried blood samples on filter papers with tests on diluted blood and on whole blood. The results obtained were similar with all three methods, but collection of blood on filter papers for the first method could be carried out by non-specialized staff and, because of the smaller volumes required, the reagents could be used to test a greater number of people (Miezan *et al.*, 1991).

Trypanosoma b. gambiense is difficult to diagnose by other techniques, such as examination of blood or lymph, because parasitaemias tend to be extremely low. Furthermore, it has been shown that a variable antigen type (VAT), designated LiTat 1.3, which is used as an antigen for detecting *T. b. gambiense* antibodies in the CATT test, is absent from some *T. b. gambiense* stocks in Cameroon (Dukes *et al.*, 1992). Thus, using the standard CATT test, cases occur of patients whose lymph node puncture results are positive yet the CATT test is negative. The problem of interpretation of CATT results was discussed by Penchenier *et al.* (1991) in relation to data obtained from field tests in the Congo. Used in the Democratic Republic of the Congo, the CATT test was reported to have a sensitivity of 98.9–100% but a specificity of only 78–88% (Pépin *et al.*, 1986). The predictive value of the test was considered too low to justify its use except in areas of high disease incidence. In the Democratic Republic of the Congo, up to 75% of people were said to remain positive for as long as 3 years after initial treatment and therefore the test would not be suitable for screening for recurrent infection or reinfection. As Dukes *et al.* (1992) suggested, the drawbacks to the CATT test could have important implications for epidemiological studies and control of sleeping sickness where the test is used. New CATT tests have been developed making use of additional antigens, which improve its sensitivity.

ANTIBODY ELISA. The antibody ELISA (Luckins and Mehlitz, 1978) was one of the first serological tests for trypanosomosis generally applicable to the pathogenic trypanosomes affecting cattle. Trypanosome antigens are adsorbed on to micro-ELISA plate wells; in the presence of an enzyme-conjugated antiglobulin, the test serum reacts after incubation to give a chromogen-mediated colour change normally read with an ELISA reader.

The test does not distinguish between current and previous infections as circulating antibodies remain in the host for some time after an infection is cured.

ANTIGEN ELISA. In an attempt to identify current trypanosome infections more accurately, tests were developed to detect circulating antigens released by parasites in the blood of infected animals rather than antibodies produced against them (Nantulya, 1990). An antigen-trapping test used monoclonal species-specific antibodies to coat ELISA plates or polystyrene tubes to which test serum was added. The coating antibody captured the antigen in the serum. A second enzyme-labelled antibody was then added which binds to free combining sites of the captured antigen. Excess labelled antibody was washed off and the reaction was revealed by addition of a substrate and chromogen. Nantulya *et al.* (1989) described antigen-detection ELISAs for the diagnosis of *T. evansi* infections in camels and for *T. b. rhodesiense* sleeping sickness in humans (Nantulya, 1989). The test for *T. evansi* was a refinement of a similar double antibody sandwich ELISA used to test for *T. evansi* and *T. congolense* infections in rabbits and goats (Rae and Luckins, 1984). However, the test cross-reacted with *T. b. brucei* and *T. vivax*. An experimental antigen-detection ELISA using a monoclonal antibody against a plasma membrane antigen of *T. b. rhodesiense* was used to detect *T. evansi* circulating antigen in camel sera from endemic areas in Mali and Kenya. The antigen ELISA was proposed as a potentially suitable tool for the field diagnosis of *T. evansi* infections, which are difficult to diagnose in the field by other means. No evidence was found of cross-reaction with *T. theileri* or other trypanosome species (Delafosse *et al.*, 1995).

In comparisons of the antigen-detection ELISA and the Woo test for *T. vivax* in French Guyana, the mean sensitivity of the antigen detection test was very low: 2.1%, compared with 54% for the Woo test (Desquesnes and de La Rocque, 1995). Similar results were obtained for *T. vivax* and *T. congolense* in Burkina Faso (Kanwe *et al.*, 1992) and in laboratory studies (Eisler *et al.*, 1998).

A simplified sandwich antigen-detection ELISA was used to detect an invariant trypanosome antigen of *T. b. rhodesiense* in serum or cerebrospinal fluid, in patients from clinics in Tanzania, Uganda and Zambia (Komba *et al.*, 1992). The test had an overall sensitivity of 91.5% and there was no evidence of cross-reactivity with common bacterial, viral or parasitic pathogens.

A 'dip-stick' antigen-detection ELISA was recently reported for diagnosis of trypanosomosis in cattle (Kashiwazaki *et al.*, 1994). This test is a dipstick colloidal dye immunoassay for multiple antigen detection, and could be used in the field with little sophisticated equipment.

Antigen-detection tests are not in routine use in Africa and they are unlikely to be used as a basis for treatment, as false positives are detected in patients who show no clinical symptoms. Drugs that are currently in use for treating sleeping sickness are potentially dangerous and patients should not be treated on the basis of antigen ELISA data alone.

PCR (POLYMERASE CHAIN REACTION). The PCR technique for amplifying DNA samples has been developed as a diagnostic test for a number of parasites and details of the technique have been well described (e.g. Moser *et al.*, 1989; Weiss, 1995). Trypanosome infections can be detected in both tsetse (described below) and cattle using PCR and DNA probes.

In an evaluation of a PCR test for detecting *T. vivax* in cattle in Guyana, sensitivity was not greater than parasitological techniques and gave false negative results when parasitaemia was high (Desquesnes, 1997). This was thought to be due to the presence of a component in the test sera that inhibited the PCR reaction. Dilution of the test serum with distilled water 1:20 gave positive results for those samples.

A PCR test has been developed for quantification of parasitaemia of *Trypanosoma cruzi* using a competitor DNA (Centurion-Lara *et al.*, 1994). The amount of parasites is estimated from the quantity of competitor DNA at the equivalency point when equimolar ratios are used when performing the test. Such a test adapted for African trypanosomes of livestock would be a useful development for research purposes.

PCR diagnostic tests are at present not sufficiently simple or reproducible for widespread use and are likely to be too expensive and technically demanding for general use in developing countries.

DNA PROBES. The principle of DNA hybridization is that a single-stranded DNA fragment containing the specific DNA sequences is identified and preferably purified. It is then labelled, for example with a radioisotope, and used to probe whole parasite DNA or whole organisms. Prior to application of the probe, the test parasite DNA is denatured and split into single strands. When the probe is applied, the sequences in the probe will hybridize with complementary DNA sequences of the parasites. The results (the bound label) are then revealed by autoradiography. The sensitivity can be increased using an amplification step that incorporates the polymerase chain reaction.

Parasite DNA is made of two broad classes of nucleotide sequences: those sequences that exist in a single copy and those that are in multiple copies within the parasite genome. The trypanosome genome is relatively small and a high proportion is composed of repetitive DNA sequences, of which most may be unique to each trypanosome species. Repeated DNA sequences are known as satellite DNAs. Such repetitive sequences are the most convenient ones for diagnostic purposes, giving good sensitivity as hybridization probes for detection of complementary sequences. Trypanosome DNA has inherent stable characteristics that can be used to identify the trypanosomes (Ole-Moi Yoi, 1987). In general, kDNA minicircles of kinetoplastid organisms seem to be good candidates for use in a sensitive

DNA test, but in *T. b. brucei* these minicircles undergo rapid evolutionary diversification and therefore even closely related strains will appear to be different (Ole Moi Yoi, 1987).

DNA probes were used for the identification of trypanosomes in cattle in Uganda; the test revealed a predominance of mixed infections in cattle (Nyeko *et al.*, 1990), whereas by microscopy, mixed infections are less commonly detected. The test was carried out on samples spot-blotted on to nylon filters, which could be kept for 1 month at room temperature.

LATEX AGGLUTINATION ANTIGEN TEST. Nantulya (1993, 1994) described latex agglutination tests for detection of trypanosome antigens. These tests are simple to use under field conditions and are claimed to provide accurate diagnosis. The principle of the test is that latex particles are sensitized with species- or subgenus-specific anti-trypanosome antibodies. If the relevant antigens are present in whole blood, plasma or serum, the antibodies capture the antigens. As there are several combining sites on the antigen molecules, aggregates form, which can be detected as an agglutination reaction. These tests could theoretically provide accurate diagnosis within minutes of a blood sample being taken in the field, but still require rigorous field testing. If successful, the test would have a significant advantage over the antigen-detection ELISA, which requires laboratory facilities and consequently takes longer to produce results.

In conclusion, indirect diagnostic tests still need to be evaluated, particularly to determine what a positive result actually means in terms of epidemiology, disease control and effect of infection on livestock productivity. It may be that many animals would test positive whilst showing no disease symptoms. Such animals would not necessarily require treatment, and their status as reservoirs for transmission of parasites needs to be further elucidated. Furthermore, whilst a crush-side test may be useful for practical decision making by livestock producers – enabling decisions to be made quickly regarding treatments – for epidemiological studies it may be more useful to take serum samples to a laboratory where a whole range of analyses for a broader range of diseases could be carried out. Examples of PCR and DNA probes developed for diagnosis in vectors or hosts are given in Table 13.7.

13.8.2   Identification of trypanosome infections in tsetse

The method most commonly used for detecting trypanosome infections in tsetse is that of dissection. However, this method has a number of disadvantages and potential sources of error, so new techniques are being developed. It has been demonstrated that whole tsetse can be frozen, together with their parasites, in liquid nitrogen, for storage and subsequent examination for trypanosome infections or other studies (Minter and Goedbloed, 1971). Trypanosomes preserved in tsetse in this way survive the

Table 13.7. PCR and DNA probes developed or used for identification of trypanosomes in Africa.

Test	Trypanosome species	Host	Reference
PCR	*T. godfreyi*	Tsetse	McNamara *et al.*, 1994
	Nannomonas	Tsetse, *G. longipalpis*	Solano *et al.*, 1995
	Nannomonas	Tabanidae	Solano and Amsler Delafosse, 1995
	T. congolense, T. vivax, T. brucei	Tsetse	Masiga *et al.*, 1992
	T. vivax	Cattle, tsetse	Masake *et al.*, 1997; Almeida *et al.*, 1997
	T. b. gambiense	Humans	Schares and Mehlitz, 1996
	T. simiae	Tsetse	Majiwa and Webster, 1987
	T. simiae	White rhinoceros	Mihok *et al.*, 1994b
DNA probes	*T. godfreyi*	Tsetse	McNamara *et al.*, 1994
	T. grayi	Tsetse, *G. f. fuscipes*	Gouteux and Gibson, 1996
	T. congolense, T. simiae,	Tsetse	McNamara and Snow, 1991
	Trypanozoon		McNamara *et al.*, 1989
	T. b. gambiense	Humans	Mathieu-Daudé *et al.*, 1994
	T. brucei, T. congolense, T. vivax	Cattle, sheep, goats	Kayang *et al.*, 1997

freezing and retain viability and infectivity, but this potentially useful technique is not commonly used. Table 13.8 summarizes techniques for the diagnosis of trypanosome infections in tsetse flies and lists some of their advantages and disadvantages.

Tsetse dissection (Lloyd and Johnson's technique)

Lloyd and Johnson's technique for determining the trypanosome infection rates in tsetse was first described in 1924, based on work carried out in Nigeria. With a few modifications, it has remained the standard technique until relatively recently. The technique consists of dissecting out the organs of the tsetse fly in which trypanosomes develop, then examining them by microscopy. These organs are the midgut, salivary glands and mouthparts (specifically the labrum and hypopharynx). The type of trypanosome present is usually determined solely according to the locations within the fly in which they are found. Thus, an infection in the labrum or hypopharynx

Table 13.8. Techniques for diagnosis of trypanosomes in tsetse flies.

Method	Advantages	Disadvantages	References
Dissection and microscopy	Relatively cheap, and easy Suitable for field use	Subgenus specific identification only Complications with mixed infections	Lloyd and Johnson, 1924 Simmonds and Leggate, 1962
Examination of saliva deposited on to glass slides	Useful for experimental work	Mainly for laboratory use Not efficient for *T. vivax*	Bruce *et al.*, 1915 Burtt and Fairbairn, 1946 Burtt, 1946c
DNA probes	High specificity	Difficulty in distinguishing mature and immature infections Sub-species specific Complicated procedure High cost	Kukla *et al.*, 1987 Majiwa and Otieno, 1990 McNamara *et al.*, 1989 McNamara and Snow, 1991
PCR	High specificity	Complicated procedure High cost Difficulty in distinguishing mature and immature infections	Masiga *et al.*, 1992 Majiwa *et al.*, 1994
Dot-ELISA	High specificity	High cost Not possible to test proboscides more than once	Bosompen *et al.*, 1995 Bosompen *et al.*, 1996

alone (the proboscis) is considered to be '*vivax*-type', infections in the proboscis and midgut '*congolense*-type', and an infection of the proboscis, midgut and salivary glands a '*T. brucei* complex' infection. It is recognized that mixed infections, which have been shown to occur in tsetse (e.g. Clarke, 1969), will not be detected as such by this method, nor can *T. simiae* infections (which are not pathogenic to cattle) be distinguished from *T. congolense* infections. Dissection and examination are best carried out with freshly killed flies, although infection rates have been determined from flies preserved in Machado's fluid prior to dissection. This is likely to be less reliable than dissection of fresh flies, since, in fresh flies, viable trypanosomes can be seen in motion, which makes their detection easier. Staining of trypanosomes within the relevant organs has also been carried out and attempts have been made to distinguish trypanosome species on morphological grounds, but these modifications of the technique are less commonly used. For determining overall trypanosome infection rates the technique is reasonably reliable, particularly when data obtained are used for comparative purposes. The main drawback of the technique has been the inability to determine trypanosome species accurately. More modern techniques, discussed below, have been developed with this problem in mind.

In a modification of Lloyd and Johnson's method, Simmonds and Leggate (1962) carefully cut the heads off field-caught tsetse and stored them in bottles of methyl alcohol for subsequent dissection in the laboratory. They stained the tsetse proboscis with Giemsa stain. This was thought to have the advantage of avoiding the difficult process of dissecting and examining flies in the field. A disadvantage was that the method provided no information on immature gut infections or infections of the salivary glands.

Probing

Infected flies may be examined for trypanosomes without killing them by dissection, by causing them to salivate whilst probing on to warmed glass slides. This is particularly useful for experimental work with laboratory colonies of flies rather than for determining infection rates of wild flies. The method was used as early as 1914 by Bruce *et al.* (1915) who fixed and stained saliva with Giemsa's stain. Burtt and Fairbairn (1946) also experimented with the technique and Burtt (1946c) used it to investigate the numbers of trypanosomes a fly could eject when it fed. Burtt and Fairbairn improved the method of Bruce *et al.* (1915) by smearing the glass slides with albumen and allowing it to dry. The fly was then contained in a small bottle or tube closed with mosquito netting and induced to probe by placing it close to a guinea-pig.

Probing of tsetse on to warmed slides, examined by dark-ground microscopy without staining, is insufficiently sensitive to be recommended for the determination of *T. vivax* infections (Moloo, 1982a). Youdeowei (1975) described a method using bat wing membranes for probing, and differences between tsetse species and age groups in probing activity were reported by Gidudu *et al.* (1995).

DNA probes

DNA probes for detection of infected vectors could be very useful for epidemiological surveys. Such diagnostic tests potentially fulfil the criteria of being highly specific and sensitive and, it is claimed, simple to perform.

These techniques have been developed for identification of trypanosomes both in tsetse (Kukla *et al.*, 1987; Gibson *et al.*, 1988; Majiwa and Otieno, 1990; Majiwa *et al.*, 1994) and in vertebrate hosts. Trypanosomes can be detected by DNA probes in the organs where parasite development characteristically occurs for each subgenus and can also distinguish different strains of parasite – for example, savanna and forest types of *T. congolense.*

Species- and strain-specific DNA probes have been used in The Gambia to identify midgut infections in *G. m. submorsitans* and *G. p. gambiensis. Trypanosoma simiae* accounted for the majority of identified infections in *G. m. submorsitans,* indicating the importance of distinguishing this species from *T. congolense* (McNamara *et al.*, 1989). Further development of DNA probes was undertaken in The Gambia to improve identification of *Nannomonas* infections in tsetse (McNamara and Snow, 1991). Probes specific for *T. congolense* savanna type, *T. congolense* riverine-forest type, *T. simiae* and *Trypanozoon* parasites showed that *T. simiae* and savanna-type *T. congolense* could only be found in *G. m. submorsitans.* Furthermore, a new *Nannomonas* species was found occurring commonly in tsetse in The Gambia and was named as *Trypanosoma godfreyi* (McNamara *et al.*, 1994). The parasite appears to be restricted to suid hosts and is of intermediate pathogenicity compared with other *Nannomonas* species. A PCR test has also been developed for identifying the parasite and it has since been detected in Zambia, Zimbabwe and Côte d'Ivoire (Masiga *et al.*, 1996a; Woolhouse *et al.*, 1996). A second stercorarian trypanosome, probably *T. grayi,* was also found in *G. p. gambiensis* in The Gambia.

A dominant repetitive DNA sequence identified in the genome of a *T. simiae* clone (Majiwa and Webster, 1987) has been successfully used in hybridization analyses to distinguish *T. simiae* from clones or stocks of *T. congolense* or DNA from any other trypanosome species examined. This is a particularly useful probe as *T. simiae* cannot be distinguished from *T. congolense* morphologically, and in tsetse fly dissections it is always classified as 'congolense-type', which leads to inaccuracies when assessing tsetse challenge to cattle. This probe revealed simultaneous infections of tsetse in the Lambwe valley, Kenya, with *T. congolense* (savanna and Kilifi types), *T. b. brucei* or *T. simiae* (Majiwa and Otieno, 1990). A DNA probe has also been developed to identify *T. grayi* in *G. f. fuscipes*; again, this will allow more accurate estimation of tsetse challenge (Gouteux and Gibson, 1996).

In addition to identification of trypanosomes in tsetse, DNA probes have been used for identification of plague bacillus (*Yersinia pestis*) in fleas (Thomas *et al.*, 1989). *Onchocerca volvulus* in *Simulium* vectors (Unnasch, 1987), and *Brugia* spp. (Piessens *et al.*, 1987). Total parasite DNA probes have been used for the detection of *T. cruzi* in South America (Greig and Ashall, 1987). DNA probes have also been used to examine and detect genomic diversity in *Theileria parva* (Conrad *et al.*, 1987).

DNA probes have also been used to identify insect vectors. This has been particularly useful with *Anopheles gambiae*-complex mosquito vectors of malaria, but is less likely to be applicable to tsetse which can normally be identified fairly easily on morphological criteria (Hill and Crampton, 1994).

Dot-ELISA

A monoclonal antibody-based dot-ELISA has been developed for detecting and identifying trypanosomes in tsetse (Bosompem *et al.*, 1995, 1996). The test, based on nitrocellulose membrane sample dots, had a high sensitivity and specificity in laboratory trials but was less sensitive when used for detecting infections in tsetse mouthparts. It would therefore be less useful for determining mature trypanosome infection rates, the parameter most required for epidemiological studies. A further and major drawback to the test was an inability to test each proboscis more than once.

PCR

Amplification of satellite DNA sequences by PCR can be a simple and rapid technique for detecting trypanosomes from the proboscides and salivary glands of tsetse, and is highly sensitive, although midgut samples require more complicated purification of the DNA (Masiga *et al.*, 1992). In addition to the PCR test for *T. godfreyi* referred to earlier (McNamara *et al.*, 1994), PCR tests have been used to detect trypanosome infections in *G. longipalpis* in Côte d'Ivoire (Solano *et al.*, 1995; McNamara *et al.*, 1995). One of the disadvantages with current PCR tests for use with tsetse flies is that they do not easily distinguish between mature and immature infections of *T. congolense*.

Chapter 14

Estimation of Disease Risk: Models of Disease Transmission

In order to assess the losses associated with trypanosomosis, and to estimate the effectiveness of control measures, methods to estimate trypanosomosis risk and tsetse challenge are required. Assessment of disease risk is also needed for planning control strategies for use against both trypanosomosis in livestock and sleeping sickness in humans. Methods for estimating disease risk and a brief outline of mathematical models of disease transmission are given in this chapter.

14.1   THE BERENIL INDEX

Whilst complex models (e.g. Habtemariam, 1987a,b, discussed below), or even estimates of tsetse challenge based on a few parameters, may be useful in understanding the dynamics of trypanosomosis transmission, they will not be of immediate beneficial use to livestock producers (Wilson *et al.*, 1986), who may benefit more from less precise, rapid appraisal of disease risk.

The Berenil index (Whiteside, 1962a) is a relatively simple way of measuring trypanosomosis risk by measuring the frequency of infections in susceptible zebu cattle when each infection, as soon as it is detected, is treated with the trypanocidal drug, diminazene aceturate (Berenil®). The number of treatments required per head per year by zebu cattle continuously exposed gives an estimate of risk, but is only a suitable method in the proven absence of drug-resistant trypanosomes. Where there is drug resistance, then the Berenil index and trypanosomosis risk would be over-estimated. Whiteside (1962a) had initially proposed an index of trypanosomosis challenge, defined as the product of the number of non-teneral male tsetse per 10,000 yards (9144 m) (apparent density) and their infection rate. The challenge index, or 'transmission rate', did not prove to be very useful (Smith and Rennison, 1958). At about the same time, Fairclough (1962) advocated the use of diminazene-treated animals in a tsetse-infested area as an indirect means of estimating tsetse challenge.

In his work on the relationships between tsetse challenge and drug treatment, Whiteside (1958) observed that the greater the challenge, the shorter was the period of prophylaxis obtained in cattle injected with a prophylactic drug. He concluded that quantification of the Berenil index would allow a regime of chemotherapy for a given area to be planned and would also provide an estimate of trypanosomosis incidence.

The Berenil index has been used in a number of situations and compared with tsetse challenge estimates as a means of determining disease risk (Cawdery and Simmons, 1964; Wilson *et al.*, 1986). In The Gambia, the Berenil index was used as a means of assessing trypanosomosis risk in experimental herds of zebu and N'Dama cattle in three villages (Claxton *et al.*, 1992b). In this work the relationships between Berenil index, trypanosome incidence and tsetse challenge were examined. Tsetse challenge was more strongly correlated with incidence of parasitaemia in zebu than in N'Dama cattle. There was a strong correlation between trypanosome prevalence and incidence of infection in the N'Dama but no correlation between trypanosome prevalence and tsetse challenge unless the data were offset by 3–5 months. It was therefore concluded that an assessment of tsetse challenge might be more useful for assessing risk to susceptible rather than trypanotolerant breeds of cattle.

14.2 TRYPANOSOMOSIS RISK AND FACTORS AFFECTING TSETSE CHALLENGE

Despite the development of sophisticated models, Smith and Rennison's (1958) definition of challenge as 'the number of infective bites from tsetse which a host receives in a unit of time' can still be regarded as a satisfactory and practical summary of the concept. Nonetheless, it remains difficult to obtain an accurate measure of this apparently simple definition of tsetse challenge.

Whiteside (1958) listed 18 determinants of tsetse challenge under the following four headings:

- Intensity of infection.
- Trypanosome characteristics.
- Susceptibility of cattle.
- Factors modifying susceptibility.

The first two of these headings were considered most important whilst the others were 'accessory factors'. Aspects most directly concerning tsetse come under the first heading and included:

- Species of fly.
- Numbers of fly.
- Disposition to feed on cattle.
- Infection rate of fly.
- Mechanical transmission.

They remain the principal parameters determined in the estimation of tsetse challenge today.

14.2.1   Components of tsetse challenge

Tsetse challenge, estimated as the number of infective bites received per animal in a given time period, is difficult to measure directly, although Leggate and Pilson (1961) attempted to do this by catching tsetse from a bait ox. They determined the trypanosome infection rates in the flies and estimated the number of infective bites per hour as an index of challenge. This might have provided accurate estimates of challenge provided the repellent effect of humans did not affect numbers of flies approaching the ox, but was impractical for large areas.

Initially, indirect methods of estimating risk were attempted. As the number of infective bites will depend upon the number of flies in a given area, the density of tsetse and their trypanosome infection rates are clearly key factors.

Tsetse density

Tsetse density is usually estimated as apparent density, often expressed as the number of tsetse caught per trap per day, although in some studies, estimates of actual population size have been made using mark–release–recapture techniques (Chapter 6, section 6.5.1). Tsetse population densities are less variable than those of many other insects (for example, mosquito vectors of malaria); thus, field estimates of apparent density can be used for estimating disease risk. However, as described in Chapter 6, methods for estimating apparent density are biased, and early methods using fly-rounds were particularly inaccurate (Smith and Rennison, 1958; Leggate and Pilson, 1961). Estimates based on fly-rounds did not take into account the female population and under-represented the feeding proportion of the populations, as most flies caught were not hungry. Finally, no account was taken of the diurnal changes in activity of tsetse, as it was a measure taken only once.

Biconical trap catches seem to be more closely related to accurate estimates of tsetse population densities than those obtained by earlier sampling devices, although they are also subject to some unexplained variability (Chapter 6, section 6.4). Indeed, it was known in the 1940s that some species of tsetse such as *G. pallidipes* were reluctant to come to humans (Vanderplank, 1944) and observations that some species of tsetse were repelled by humans cast doubt on the validity of data obtained by man fly-round catches.

In order to overcome some of the disadvantages associated with apparent density measurements, Leggate and Pilson (1961) stated that estimates of challenge should be based on:

- a feeding portion of the population;
- activity in relation to a given host;
- the total period of tsetse feeding activity;
- the trypanosome infection rate.

The development and use of the odour-baited biconical trap went some way towards satisfying those conditions.

Trypanosome infection rates

Trypanosome infection rates in tsetse are easier to determine than density, and are subject to less intrinsic variability. However, their estimation is subject to some degree of error, depending upon which technique for identification is used. The sources of error in identifying trypanosome infections are discussed in Chapter 13.

Biting rates and infective biting rates

Rather than directly observing numbers of flies feeding on an ox (the method of Leggate and Pilson, 1961), biting rates, or proportions of feeds taken from livestock, may also be estimated indirectly by analysis of bloodmeals taken from recently fed flies. This requires that fly samples are representative of the population and are taken from the area in which challenge is to be assessed.

Each successive probe by an infected tsetse during an attempt to feed can potentially transmit a trypanosome infection (Corson, 1932b). Wild tsetse may therefore cause more infections than would be represented by the number of times that they feed if feeding is interrupted and recommenced on a different host. Calculations of challenge based on trypanosome infection rates, apparent density and average feeding interval would therefore result in underestimates of the risk of infection to humans or domestic livestock.

Feeding behaviour of partially fed tsetse is important in the estimation of challenge and has been investigated by Pilson and Leggate (1962a) and Bursell (1958b) in Zimbabwe. As cattle react to attempts by tsetse to feed on them, feeding is likely to be interrupted frequently. Thus, trypanosome challenge might exceed values based solely on tsetse density and hunger cycle.

Bursell (1958b) suggested that the proportion of small bloodmeals in recently fed flies caught in the wild should be investigated. He found that if a fly took more than a third of its normal meal there would be no tendency for it to attempt to feed again until the bloodmeal had been digested. If less than a third of its normal meal had been taken, probing would be repeated in a proportion of individuals, particularly if an opportunity arose before the first meal had passed into the midgut.

In their study, Pilson and Leggate (1962a) found that the diurnal pattern of feeding activity of tsetse in Zimbabwe was similar in all seasons, with a well-marked peak of activity at, or shortly before, sunset. Trypanosome infection rates in flies were determined and combined with the numbers of flies attacking a bait ox to obtain an estimate of the seasonal trypanosome risk. This was greatest in November (the late dry season) and smallest in February (during the rainy season).

Experimental observations show that the mean number of bites by infected female tsetse received by one ox in 1 day varies from month to month, with more frequent bites shortly after periods of high rainfall (Harley, 1966b). These changes are mainly due to changes in fly density rather than to changes in infection rates. Harley estimated the number of bites from an infected fly from the product of the mature-infection rate in the flies and the number caught per day. This is obviously an over-simplification. It would be better referred to as an estimate of challenge rather than as the number of bites per day, although the flies were caught from an ox and so it was presumed that they had been going to feed on that ox.

In Kenya, Snow and Tarimo (1983) quantified the number of trypanosome infected bites received by cattle in an area by estimating the following:

- The total tsetse population size – from mark–release–recapture techniques.
- The proportion of feeds from cattle – determined by bloodmeal analysis.
- Trypanosome infection rates in tsetse – determined by dissection and examination of samples of the tsetse population.
- Tsetse feeding frequency.
- The number of cattle in the area.

Feeding frequency was measured indirectly from the interval between peaks in the frequency of recaptures of marked flies and was estimated to be about 4 days during a first experiment and 3 days during a second. It was assumed that the interval was the same for males as for females. A transmission potential (defined as the proportion of bites from infected tsetse that can give rise to trypanosome infection in mammalian hosts) of 0.2 was used for both *T. vivax* and for *T. congolense*.

The final estimation of challenge came from a simple multiplicative model (Snow and Tarimo, 1983), which they compared with that of Rogers (1979). As an example, he estimated that 3400 male flies were feeding every 4 days, of which 60% fed on cattle; the infection rate in the flies was 2.22% with *T. congolense* and only 20% of the infective bites were assumed to give rise to infections. Forty-six cattle were present in the area at that time.

These data were used to calculate the number of infective bites per day as follows:

$$[(3400/4) \times 0.6 \times 0.0222 \times 0.2]/46 = 0.05 \text{ infective bites per day} \quad (14.1)$$

Obviously there are sources of error at every stage of this calculation, except perhaps the number of cattle in the area. The biggest sources of error are probably the total tsetse population size, the proportion of feeds from cattle and the proportion of infective bites giving rise to infections. There is some evidence that estimates of *T. congolense* infection rates in tsetse could also be inaccurate, as a significant number of infections with *T. simiae*, which is non-pathogenic to cattle, may occur in tsetse on the south coast of Kenya.

In the Diani area of the Kenya coast, Snow and Tarimo (1983) estimated that each cow received from *G. pallidipes* one infective inoculum (one infective tsetse bite) of *T. congolense* every 5.8 days during one experiment, and every 5.0 days in a second. For *T. vivax*, the estimates were 3.2 and 79.1 days respectively.

Estimates of the probability of tsetse acquiring trypanosome infection from a single bloodmeal at various locations in Kenya are summarized in Table 14.1. This table includes estimates of the probability of tsetse acquiring an infection from a single bloodmeal in a sleeping sickness focus in the Lambwe valley (Tarimo *et al.*, 1991). The latter estimates were from data obtained from biconical trap catches, dissections to determine trypanosome infection rates and types, ovarian ageing and examination of cattle blood. The blood incubation infectivity test (BIIT) was used to identify human-infective strains of *T. brucei* whilst a previously published method was used for estimating the probability of a tsetse picking up an infection in one bloodmeal (Rogers and Boreham, 1973; Tarimo *et al.*, 1985). Additional data on the proportions of tsetse bloodmeals from different hosts came from England and Baldry (1972).

In their calculations, Tarimo *et al.* (1985) assumed that the probability of acquisition of trypanosomes by tsetse of different ages was constant, although they recognized that this might not be so.

Table 14.1. Probability estimates of tsetse (*G. pallidipes*) acquiring a trypanosome infection from a single bloodmeal in Kenya (from Tarimo *et al.*, 1985, 1991).

Trypanosome species	Rural situation with domestic animals	Natural situation with wild animals	Domestic animals with trypanocidal drugs	Human sleeping sickness area (Lambwe valley)
T. congolense	0.0077	0.0019	0.0013	0.0028
T. vivax	0.0010	0.0024	0.0021	0.0092
T. b. brucei	–	–	–	0.00097
T. b. rhodesiense	–	–	–	0.00024

One source of error in estimates of tsetse challenge (flies/trap/day ×
trypanosome infection rate) arises from the assumption that trap catches
reflect fly densities equally well at all times of the year, in all situations, and
for different species (W. Snow, unpublished, British Society for Parasitology
Seminar, 1989). Snow preferred a cruder definition of challenge based on
the number of infective bites received by livestock per unit time, or the
reciprocal of this value, as the interval between infective bites. Some adjust-
ment was considered necessary as not all infected bites give rise to an
infection in cattle. The value for infective bites each day on cattle divided
by the number of cattle regularly grazing the infested areas gives an estimate
of the number of infective bites per bovine per day. The reciprocal of this
figure is the interval between infective bites, and provides an alternative
way of visualizing this representation of challenge. Such estimates can form
a basis for assigning values to rankings of challenge.

In The Gambia, seasonal patterns of tsetse abundance were consistent
and rates of trypanosome prevalence in cattle were relatively unchanging
(Rawlings *et al.*, 1991). Overall, annual trypanosome prevalence rates and
tsetse challenge indices showed a significant, linear relationship.

Cattle–tsetse contact

There have been relatively few studies on the movements of cattle in
relation to tsetse distribution and activity, although such studies enable tsetse
challenge to be more accurately quantified. The most detailed studies of this
aspect have been carried out in The Gambia. Movements of cattle from
selected herds were monitored throughout the grazing periods in different
seasons, and were mapped in relation to time spent in different areas. These
movements were then related to tsetse distribution and abundance and
seasonal changes of those tsetse populations. Much improved correlations
between tsetse challenge and trypanosome prevalence were obtained
compared with estimates based simply on apparent tsetse density, their
trypanosome infection rates and proportion of feeds from cattle (Wacher *et
al.*, 1994). Cattle owners in The Gambia followed a migratory strategy that
allowed them to maintain a high quality of grazing throughout the year; this
sometimes required that their cattle were taken into areas of high tsetse
density. The risk entailed from this strategy was minimized because the high
numbers of cattle taken into tsetse areas at a particular time of the year had
a 'dilution effect' in terms of the tsetse challenge per head. Estimates of tsetse
challenge were therefore corrected to allow for stocking density (Wacher *et
al.*, 1993).

In simultaneous studies in The Gambia, daily flight activity of *G. m.
submorsitans* was monitored in relation to seasonal contact with cattle
(Rawlings *et al.*, 1994). There were discrepancies between data on tsetse
activity obtained from traps, from hand-net collections and from observations of contact with cattle herds in the dry season. Thus, trap data
suggested that there was negligible activity from midday to late afternoon in

the dry season, whilst other observations indicated no decrease in attack rates. These apparent differences in activity may have been due to different sections of the tsetse population being monitored by the different methods, trap catches of male flies being particularly hungry flies with low fat reserves (Rawlings *et al.*, 1994). It appeared that traps were sampling spontaneous flight activity at that time of day, whilst hunger-related activity took place towards dusk. Fly movement close to cattle herds was greater, as the presence of cattle activated flies.

From these studies, Rawlings *et al.* suggested that management of grazing times would be unlikely to eliminate the risk of trypanosome transmission regardless of season. Although activity patterns derived from trap catches would not give accurate estimates of attack rates on cattle, Rawlings *et al.* (1994) pointed out that mean tsetse abundance from trap catches and trypanosome infection rates still provide a useful challenge index closely correlated with trypanosome prevalence in cattle.

14.2.2 Measurement of risk of human sleeping sickness

Laveissière *et al.* (1994) described a method for estimating risk of human sleeping sickness in endemic areas in Côte d'Ivoire. This index of risk was based on apparent density of teneral flies, longevity of the tsetse population and human–fly contact. An estimate of the density of the teneral, rather than the adult, population was taken because they argued that a fly must be teneral to become infected with *T. b. gambiense*. They used a formula for the relationship between apparent density and actual density, derived by Gouteux and Buckland (1984):

$$N = k(ADT)^a \tag{14.2}$$

where the constants $k = 632$ and $a = 1.23$. N is the actual density and ADT is the apparent density estimated by trapping. For teneral flies:

$$T = k(t + 1)^a/(pj) \tag{14.3}$$

where t is the apparent density of teneral flies caught by p traps during j days. T is the actual density of teneral flies. The term $t + 1$ is used to prevent the estimate dropping to zero where no teneral fly is caught. Longevity was included in the equation because the teneral fly that feeds on an infected host has to survive long enough to acquire a mature infection, and the longer infected flies live, the greater is the chance of them transmitting the infection to another human. Longevity was based on the daily survival rate (Dsr) and is proportional to the fraction that survives after 20 days (taken as the time for an infection to mature), denoted as Dsr_{20}, and the average remaining lifespan of the tsetse fly, $-1/\log (Dsr)$. Human–fly contact (P) has to be sufficiently personal and close, as discussed in Chapter 12, and is determined from estimates of the proportion of feeds taken by the flies from humans:

$$P = N^n/C \qquad (14.4)$$

where N is the size of the tsetse population and n is the number of human bloodmeals found in the catch C. From this equation the formula:

$$P = knC^{0.23}/(pj)^{1.23} \qquad (14.5)$$

can be derived for human–fly contact.

The index of risk, r, is calculated from a combination of these formulae to give the equation:

$$r = (t + 1)^{1.23} \times n^2/(pj)^{3.69} \times C^{0.46} \times -(Dsr_{20})/\log (Dsr) \qquad (14.6)$$

Estimates of risk thus obtained (after multiplying r by 10^4 to obtain a workable figure) correlated well with epidemiological field observations. Laveissière *et al.* (1994) suggested that this type of entomological estimation of risk was necessary to determine the extent of transmission rather than simply using disease prevalence.

14.3   MODELLING TRYPANOSOMOSIS TRANSMISSION – THEORETICAL CONSIDERATIONS

New approaches to modelling have become possible through advances in computer technology, which enable simulation models to be developed. The use of models to describe the transmission of parasites by bloodsucking insects was reviewed by Dye (1992), who encouraged 'a comparative approach to quantitative medical entomology'. This review mainly referred to malaria modelling, which is at a more advanced stage than modelling of tsetse and trypanosomosis. However, many aspects of modelling these two diseases are similar.

Theoretically, mathematical models could allow the prediction of long-term parasitological consequences of a rapid change in vectorial capacity, induced, for example, by application of insecticide, although this may be difficult in practice. Neither the total size of a vector population nor the ratio of vectors to people need be known in epidemiological studies; the number of bites per person per day (m) is the most useful parameter. Accurate estimates of vectorial capacity of anopheline mosquito vectors of malaria can be difficult to obtain because of random variations and systematic errors in sampling methods, even though its quantification is relatively simple compared with vectorial capacity of trypanosomosis and other parasitic diseases. Dye (1992) suggested that the following 'more explicitly comparative questions' could be more easily answered and would be sufficient for most practical epidemiological studies:

- What is the most important vector in an area?
- During what season is transmission rate at a maximum?
- What explains geographic variation in the prevalence of infection?

- Why did vector control have no impact on incidence of infection?
- Which of alternative control methods will have the greatest impact on disease prevalence?

Advantages of this approach were that: (i) not all components of transmission need to be estimated and (ii) the bias associated with various parameter estimates can be eliminated or reduced by comparing measurements biased in the same or a similar way.

As the model of Rogers (1988b) demonstrates (section 14.3.1), entomological variables are sometimes more sensitive indicators of epidemiological change than parasitological variables, and it is, therefore, often useful to monitor vector populations and determine the key variables rather than just to determine the incidence of disease or infection in hosts (Dye, 1992).

Biases in measurement of vectorial capacity are multiplicative rather than additive. In practical epidemiological studies it may be more useful to concentrate on changes in parameters – for example, percentage change in vectorial capacity – rather than attempt to obtain precise estimates for the values of those parameters, as the errors for estimates of change are likely to be much smaller (Dye, 1992). Although problems of precision remain, these can be handled either by calculating standard errors or by treating entomological measurements as explanatory and predictive of epidemiological change. This is possible especially when vector studies are carried out in conjunction with parasitological ones.

In principle, the relation between transmission index and trypanosome prevalence should be curvilinear and pass through the origin. In some examples of this procedure, points are too scattered to indicate significant curvature.

Key concepts for the development of vector-transmitted disease models are the basic rate of reproduction and disease transmission thresholds. Models for malaria transmission have been developed based on those concepts and have formed the basis for models of trypanosomosis. Causal models concentrate on behaviour, to the exclusion of dynamics, whilst analytical models concentrate on dynamics to the exclusion of behaviour. Analytical models, expressed in mathematical terms, allow predictions to be made concerning disease dynamics. The main question addressed by models can be put simply as: 'What is the rate of increase of the number of hosts affected?' (Dietz, 1988).

Reviewing mathematical models for disease transmission, Gettinby (1989) noted that the term 'model' could be used to describe any equation or representation attempting to relate the constituent parts of a biological system. Animal and vector populations are divided into classes and subclasses and differential equations are used to express the rate of change over time of the numbers within each class or subclass. Field workers establish the rate parameters used, which are important for validating the models.

Although differential equations are commonly used to model animal disease systems, statistical (often regression) models have been used to establish relationships between disease variables such as disease prevalence and vector intensity; the latter approach was taken by Leak *et al.* (1990, 1991). Estimates of disease risk over a number of sites were made using only a few critical parameters (Leak *et al.*, 1990). This was the most feasible approach to obtain comparative data with the limited resources available. From these data a relationship was derived between entomological and parasitological variables.

The statistical regression models are mainly directed towards investigation of field-collected data whilst stochastic simulation models have been used to mimic the interaction between hosts, vectors and pathogens (Gettinby, 1989). Fitted regression lines can also be used to predict fall in trypanosome prevalence due to vector control.

According to Gettinby (1989), stochastic simulation allows a realistic framework for modelling diseases, not constrained by the limitations of calculus matrix or network techniques. Causal models for trypanosomosis developed by the Antwerp Trypanosomiasis Causal Modelling Group are described in some detail, together with a review of models, in the proceedings of a workshop held at Antwerp (Rogers and de Muynck, 1988). Details of trypanosomosis models are not given here although some are mentioned and described in general terms, with references provided for further details. Two classes of analytical techniques are regression analysis and dynamic modelling. Gettinby listed examples of what he considered to be the three types of model investigation:

- The systems analysis/simulation model of Habtemariam (1987a,b).
- The vector model of Rogers (1988a,b) and Randolph and Rogers (1984a).
- Class models (Milligan and Baker, 1988; Rogers, 1988b; Milligan 1990).

14.3.1   Rogers' model (Rogers, 1988b)

Rogers (1988b) described a general model for trypanosomosis involving two vertebrate host species and one insect vector species. The model was developed from the single-host model for malaria (Aron and May, 1982) and allowed for incubation and immune periods in the two host species and for variable efficiency of transmission of different trypanosome species from the vertebrates to the vectors and vice versa. An early model for malaria was based on the idea that in a normal situation the level of infection in mosquitoes, or 'malaria risk', is at an equilibrium with the level of disease prevalence in the human populations (MacDonald, 1952). In Rogers' model, differential equations were derived for equilibrium disease prevalence in each of the species involved. Using data for a 'typical West African village situation', Rogers' model predicted equilibrium prevalences for *T. vivax*, *T. congolense* and *T. b. brucei* of 47.0%, 45.8% and 28.7%, respectively, in the animal hosts; 24.2%, 3.4% and 0.15% in the tsetse vectors; and a 7.0% infection in humans with *T. b. gambiense.* Variables and parameters used in Rogers' model are shown in Table 14.2, and Table 14.3 shows predictions arising from the model. One of the conclusions was that sleeping sickness could not be maintained in humans alone, and some animal reservoir was crucial for maintenance of the disease. Another significant conclusion from Rogers' model was that sudden removal of domestic animals from a village could result in a 'mini-epidemic' of sleeping sickness as infected tsetse turned to feed more frequently on humans. This epidemic would be short-lived as trypanosome infections in tsetse would tend to fall to a lower equilibrium level after a period of feeding on humans.

The model just described was particularly sensitive to changes in vector-related parameters (feeding behaviour, survival rate, and the transmission

Table 14.2. Variables and parameters of a two-species trypanosomosis model (from Rogers, 1988b).

	Parameter	Value
(a) General symbol		
N_1	Number of animals, species 1 (e.g. humans)	300
N_2	Number of animals, species 2 (e.g. livestock)	50
V	Number of tsetse	5000
p_1	Proportion of tsetse blood meals from species 1	0.3
p_2	Proportion of tsetse blood meals from species 2	0.7
U	Daily mortality rate of flies	0.030
D	Duration of feeding cycle in flies	4 days
$a_1 = p_1/d$; $a_2 = p_2/d$; $m_1 = V/N_1$, $m_2 = V/N_2$		

	(b) Disease specific	*T. vivax*	*T. congolense*	*T. brucei*
$1/i_1$	Incubation period in species 1	–	–	12 days
$1/i_2$	Incubation period in species 2	12	15	12 days
$1/r_1$	Duration of infection in species 1	–	–	70 days
$1/r_2$	Duration of infection in species 2	100	100	50 days
$1/v_1$	Duration of immunity in species 1	–	–	50 days
$1/v_2$	Duration of immunity in species 2	100	100	50 days
T	Incubation period in tsetse	10	20	25 days
b_1	Probability of infected fly bite producing infection in species 1	–	–	0.62
b_2	Probability of infected fly bite producing infection in species 2	0.29	0.46	0.62
c	Probability of any infected bloodmeal eventually giving a mature infection in a fly	0.177	0.025	–
c'	Probability of first feed only (taken during first day) eventually giving a mature infection in a fly	–	–	0.065

Table 14.3. Predictions from the trypanosomosis model (Rogers, 1988b) (using parameters and variables shown in Table 14.2).

Parameter	T. vivax	T. congolense	T. brucei Species 1 + 2	T. brucei Species 1 or 2
Equilibrium trypanosomosis prevalence in species 1 (human)	–	–	7.0%	–
Equilibrium trypanosomosis prevalence in species 2 (animals)	47.0%	45.8%	28.7%	–
Equilibrium trypanosomosis prevalence in tsetse	24.2%	3.4%	0.61%	–
Basic reproductive rate	388.2	64.4	2.65	Species 1 = 0.11 Species 2 = 2.54
Threshold for transmission (flies/host), species 1	–	–	6.29	153.0
Threshold for transmission (flies/host), species 2	0.26	1.55	37.7	39.3

potential of the vector) and Rogers (1988b) therefore suggested that research should be concentrated in those areas, particularly to determine their variability.

14.3.2 Milligan's transmission model

Milligan (1990) outlined a compartmental model for tsetse-transmitted trypanosomosis of cattle, modified from Anderson (1981). Among the parameters included in this model (Fig. 14.1) were trypanosome prevalence in the vector, its expected life span, influence of infection on the vector, disease incubation periods and their relationship with the expected life span of the host. An earlier model (Milligan and Baker, 1988) was tested using data on productivity and chemoprophylaxis at Mkwaja ranch in Tanzania (Trail *et al.*, 1985). Differential equations were used to relate different classes of host and vector populations in a similar approach to that of Habtemariam (1987a,b).

14.3.3 Other models for trypanosome transmission

Following the discovery that RLOs affected susceptibility of tsetse to trypanosome infections, Baker *et al.* (1990) developed a model of human and animal trypanosomosis based on Ross's equation for disease transmission, supplemented by a differential equation for inheritance of susceptibility of the vector to trypanosome infection. The model predicted cycles of RLO prevalence in tsetse, which could affect their susceptibility to trypanosome infection, thus resulting in the occurrence of periodic epidemics of human sleeping sickness at roughly the observed intervals.

Differential equation models have been developed for analysing economic impact of trypanosomosis as well as for trypanosome transmission between cattle and tsetse populations (Habtemariam *et al.*, 1982a,b, 1986). These are based on sets of interrelating classes of the host and tsetse populations. Habtemariam's 'linear programming model for trypanosomosis' was later modified to give a less complex set of equations (Koen, 1990). The outcome was a very theoretical mathematical model that has not been field-tested. In another model, Habtemariam *et al.* (1986) estimated the probability of trypanosome infection in cattle from the probability of effective transmission of trypanosomes from *Glossina* to susceptible cattle. This probability was a function of the contact rate between cattle and tsetse and was calculated using a Poisson distribution. The product of seven parameters provided an estimate of the probability of effective transmission. His calculations gave figures of seven to nine infections per 10,000 susceptible cattle per year!

Dransfield and Brightwell (1989) described attempts to test models in the field, firstly by seeing how well they mimic the performance of a natural population, and secondly by manipulating the population to see how well

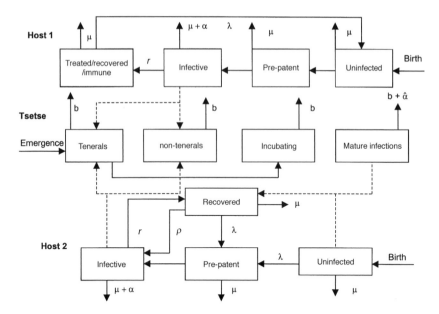

Fig. 14.1. Compartmental model of animal trypanosomosis (from Milligan, 1989). λ, force of infection = $Ys\beta f/K$; $\hat{\alpha}$, parasite-induced mortality of tsetse; μ, host mortality rate; N_i, size of the ith host population; K, normalization constant, $\Sigma N_i s_i$; f, rate of relapse to parasitaemic state; f_i, probability of transmission by a bite from a fly with a mature infection; r, recovery rate; s_i, feeding preference for host i; Y, number of flies with mature infections; b, tsetse mortality rate; ρ, rate of relapse to parasitaemic state; α, parasitic-induced mortality of vertebrate host; β, biting rate.

the model could predict the behaviour of the system. These field tests were performed at Nguruman, Kenya. A simulation model appeared to describe seasonal fluctuations of *G. pallidipes*, although when the tsetse population was suppressed by trapping, it was revealed that these fluctuations were affected by tsetse dispersal rather than resulting solely from changes in mortality rates.

Rawlings *et al.* (1991) used the same approach as that used by Leak *et al.* (1990) to examine relationships between trypanosome prevalence in trypanotolerant cattle and tsetse challenge/trypanosomosis risk in The Gambia. A comparison of results from these studies showed small differences in the regression equations between their data and those of Leak *et al.* (1990). Rawlings *et al.* (1991) suggested that these were due to the fact that the latter work was based on results from a number of sites with different breeds of cattle and species of tsetse whilst their own was derived for one breed of cattle and one tsetse species.

A stochastic model developed to study spatial dispersal of tsetse has been integrated with an epidemiological model for cattle trypanosomosis and was used to determine the theoretical size of a trap or target barrier to prevent dispersal or spatial progression of the disease (Yu *et al.*, 1995). The authors concluded that to protect an area in which tsetse control was being carried out, a barrier 1000 m wide would be effective in stopping disease spread. This estimate might seem rather low compared with other estimates recommended for the width of target barriers described in Chapter 19 (Hargrove, 1993).

14.4 MODELS FOR OTHER VECTOR-BORNE PARASITIC DISEASES

Models for transmission of trypanosomosis by tsetse flies have been derived to some extent from models for malaria transmission by mosquitoes. Many of the principles and applications are relevant to descriptions of transmission of other vector-borne parasitic diseases. Some examples of such models are therefore discussed here.

14.4.1 Malaria

Mathematical models have been used for a considerable time in malariology to describe dynamic interactions between vector populations and prevalence of infection.

In contrast to trypanosomosis, malaria is a disease of high endemicity and vector populations can vary greatly in space and time; nonetheless, there are similarities in the development of mathematical models for the two diseases that are of interest.

Prior to vaccine trials in Kenya, malaria risk was estimated from transmission indices calculated as follows. Abundance of mosquitoes was expressed as the number of female *Anopheles* per human per night for all night biting collections (Beier *et al.*, 1990). Monthly estimates of human-biting rates were calculated as arithmetic means of weekly collections, and sporozoite rates were calculated as the proportion of infected mosquitoes per month. For analysis of variance (ANOVA) and correlation analysis, human-biting rates (MBR) were transformed to $\log_{1n}(n + 1)$ to normalize the distribution and to control the variance. Overall monthly mean MBRs were presented as back-transformed values (termed Williams' mean, M_w; Williams, 1937). Sporozoite rates (SR) were transformed to arcsines before analysis (Sokal and Rohlf, 1981) and entomological inoculation rates (EIR) were calculated as the product of the monthly MBR and the SR (MacDonald, 1957); EIR data were analysed using $\log_{1n}(n + 1)$ transformed values. In all analyses the degrees of freedom represented the number of sampling months. The theoretical time needed to experience one infective bite was calculated as the inverse function of the EIR, (1/EIR), with an upper limit of 30 days to allow for independent monthly estimates. Intensity of transmission is determined most effectively by calculating the EIR, which is the only comprehensive indicator of transmission (Onori and Grab, 1980). For both malaria and trypanosomosis, some of the parameters required are difficult to measure accurately in the field and, as Dye (1992) remarked, 'exceptional quantities of data and unusually careful analysis' would be required if they are to be exact estimates.

Smith *et al.* (1995) used a conditional autoregressive (CAR) log linear model for interpreting mosquito population densities within a village. This was based on the assumption that mosquito densities are related to an underlying variable, the value of which for any house, although it cannot be measured directly, depends on that in neighbouring houses and hence shows a degree of spatial smoothness.

Recently, studies have been carried out to examine the value of estimates for malaria transmission based on estimates of key components (Burkot and Graves, 1995).

14.4.2 Onchocercosis

Modelling was an important component of the Onchocercosis Control Programme (OCP) in West Africa, which started in 1974. This was claimed to be the first parasitic disease control programme to actively support epidemiological modelling (Habbema *et al.,* 1992). A force-of-infection model was used for epidemiological evaluation of the OCP, and showed key factors determining epidemiological trends (Remme *et al.,* 1986). These were the initial age and the initial endemicity level of the human population. As a result of this model an index, called the community microfilarial load, was developed for routine epidemiological evaluation of the control programme. Habbema *et al.* (1992) further described the development of this epidemiological model and its role as an aid for evaluating vector and disease control. The force-of-infection model was developed into a simulation program used as a tool for analysis of epidemiological data. A full transmission model, called ONCHOSIM, was also developed for understanding the epidemiology of savanna onchocercosis. Stochastic microsimulation, in which individual human life histories were simulated, proved the most appropriate modelling technique. Their model was intended to be of direct practical use in the planning and evaluation of disease control programmes in addition to the study of structural and quantitative aspects of epidemiology and transmission dynamics. The researchers who developed this model suggested that their approach could also be used as a tool for decision making for other diseases such as trypanosomosis.

Following successful control in savanna regions, where the ONCHOSIM model was used, the OCP is moving into the forest areas. As onchocercosis epidemiology in forest areas is very different, a further simulation model for forest onchocercosis was developed, called SIMON (Davies, 1993). This model was developed based on epidemiological data collected in Sierra Leone, and has different parameter values, but is otherwise developed along the same lines as ONCHOSIM.

14.5 REMOTE SENSING FOR QUANTIFYING OR PREDICTING TRYPANOSOMOSIS RISK

Aerial photographs have been used for studying tsetse habitats and distribution and for the planning of tsetse control activities for many years. For example, in the 1960s they were used in Zambia for vegetation surveys related to tsetse control activities (Verboom, 1965). With improvements in technology allowing detailed satellite images, there are new possibilities for using remotely sensed data as tools for epidemiological studies. There has recently been considerable interest in the possibility that remote sensing will identify microclimates/habitats of parasites and their vectors to enable mapping, in a probabilistic way, of their distribution. Changes in parasite and vector populations could then be predicted using data obtained from satellite images on quantitative and qualitative changes in habitats, and potential distributions could be indicated. For small areas, light aircraft and infra-red film may be the best means for remote sensing surveys, but for large areas satellite imagery is more appropriate.

In 1976 remote sensing was suggested as an aid to tsetse fly control in Mali, to map seasonal habitats related to burning of vegetation and to identify areas for insecticide spraying. Subsequently, it was shown that the location prediction was correct but not the intensity of fly infestation (Hugh-Jones, 1989). A comparison was made by Okiwelu *et al.* (1981) between tsetse distribution in Mali determined largely by biconical trap catches, and a distribution map based solely on interpretations of earth satellite photography of the area. The earth satellite distribution map was based on predictions of tsetse distribution in certain zones by MacLennan (1977), use of satellite imagery and brief field observations. *Glossina m. submorsitans* were captured outside the areas predicted by MacLennan (1977), notably in an area predicted to be tsetse free. Predictions of the distribution of *G. p. gambiensis* were more accurate. Okiwelu *et al.* cited Ford's view that too much emphasis has been wrongly placed on vegetation in explaining aspects of tsetse ecology (Ford, 1971), stating that their findings of discrepancies with MacLennan's predictions support that view.

There have been several studies using remote sensing techniques to predict disease occurrence through vector distribution and abundance. The relationship between tsetse populations and climatic and vegetation measurements is reflected in a close correlation between the normalized differential vegetation index (NDVI) and tsetse abundance. This is based on the close relationships between limits to tsetse distribution and saturation deficit, and between saturation deficit and vegetation. There are also significant correlations between mortality rates (estimated using Moran curve techniques), mean monthly NDVI values, saturation deficit and physical size of the vectors (Rogers, 1991; Rogers and Randolph, 1991). It may, therefore, be possible to use satellite data to predict both mortality rates and abundance of tsetse over large areas of Africa and to produce maps of high-risk areas of disease transmission. This has already been done for Togo (Hendrickx *et al.*, 1997). Biological characteristics of trypanosomosis may be monitored by meteorological satellites to provide an understanding of the spatial distribution of tsetse and disease prevalence, and these data can be statistically analysed using discriminant analysis to predict changing distributions, as described by Rogers and Williams (1993).

Correlations between size of adult tsetse and estimates of mortality derived from satellite data were detected in a study carried out over a 700 km transect of a range of ecoclimatic conditions in West Africa. The same study provided possible explanations for the localization of human sleeping sickness (Rogers, 1991). One explanation was that flies in the south of the transect, although not having a high mortality rate, were not sufficiently stressed to bite humans, whilst in the north the mortality was too high for the few flies present to be a risk. In the intermediate zone flies were sufficiently stressed to be a risk and mortality rates were low enough to allow sufficient survival of the tsetse population. Rogers also noted that seasonal changes in numbers of human sleeping sickness cases in Uganda and Kenya were correlated with mean monthly NDVI from those areas.

Although remote sensing may be usefully applied to the study of trypanosomosis epidemiology in Africa, there will still be a need for ground-based studies to verify satellite data and to provide a full understanding of disease transmission.

Geographical information systems (GIS), in which data layers are built up on a computer database incorporating geographical information, have been used to study the epidemiology of *Theileria parva* causing disease in cattle (Lessard *et al.*, 1990). The same approach is currently in use for the study of epidemiology of trypanosomosis (Lessard *et al.*, 1990). In the Lambwe valley, Kenya, density of *G. pallidipes* was highly correlated with finer resolution (Landsat thematic mapper) satellite data for moisture content of the soil and vegetation. Satellite-sensed data accounted for as much as 87% of the variance in tsetse catches (Kitron *et al.*, 1996). Related applications of remote sensing have been to identify villages at high risk of malaria transmission – for example, in Mexico (Beck *et al.*, 1994).

In contrast to the potential benefits from the GIS tool which are frequently referred to, a warning has been given that, unless properly used, this approach could result in 'graveyards of databases' (Mott *et al.*, 1995).

Part III

Vector Control

Chapter 15

Insecticidal Spraying

Jordan (1986) stated in the preface of his book that 'it is only by removal of the vector ... that a disease-free environment can be created'. Complete eradication of the vector is rarely achieved, yet in the absence of a vaccine for trypanosomosis, and as drug resistance to available trypanocidal drugs increases, control of the tsetse fly remains the most theoretically desirable means of controlling the disease. Despite the known limitations, the use of trypanocidal drugs, generally administered by the livestock owners themselves, is the major method by which animal trypanosomosis is today controlled in most African countries. It has frequently been stated that we now have the technical means of either controlling, or even eradicating, tsetse flies. The major difficulty lies in successfully implementing control in a sustainable way. Ikede (1986) therefore suggested that emphasis should be on the application of existing knowledge to the problems of livestock production rather than research. The current status of tsetse control methods is presented in this and following chapters.

Based on the population dynamics of tsetse, the most successful attempts at controlling tsetse flies are likely to be at the extreme limits for survival of the fly (Rogers, 1979). A summary of control methods is presented in Table 15.1.

15.1 INSECTICIDES

Table 15.2 lists the commoner insecticides that have been used for tsetse control. Until the mid-1970s the insecticides used were almost exclusively organochlorine compounds, including DDT and dieldrin. Most organo-phosphorous compounds and carbamates are more toxic than organochlorine compounds or endosulfan, which was later used for aerial spraying. In the 1970s environmental considerations led to a search for alternative insecticides, especially synthetic pyrethroids, to which tsetse flies are highly susceptible.

Table 15.1. Summary of methods for the control of tsetse flies.

Control method	Insecticides/specific method	Examples	References
Groundspraying			
Knapsacks	DDT, dieldrin (synthetic pyrethroids)	Nigeria, Zambia, Zimbabwe, Uganda, Côte d'Ivoire	MacLennan, 1967; Hocking, 1961 (Sékétéli and Kuzoe, 1986a; Woof, 1964)
Mistblowers	DDT, dieldrin	Kenya, Nigeria	Thomson et al., 1960; Koeman et al., 1971
Unimogs	DDT, dieldrin	Zambia, Botswana	Graham, 1964
Thermal fogging	DDT, dieldrin		
Aerial spraying			
Fixed wing residual	BHC, DDT, dieldrin, deltamethrin	Zambia, Zimbabwe, Botswana, Somalia	Hocking et al., 1954a,b, 1966
Fixed-wing non-residual	Endosulfan	Nigeria, Niger	Le Roux and Platt, 1968; Spielberger, 1971; Bauer, 1971
Helicopters	Deltamethrin, endosulfan		Cockbill, 1971; Chorley, 1947; Robertson and Bernacca, 1958; Potts and Jackson, 1952
Wild animal elimination	Selective hunting, e.g. warthogs, antelopes	Zimbabwe, Uganda, Tanzania (Shinyanga)	
Bushclearing	Manual felling, tractor-assisted clearing, chemical (e.g. Tordon)	Zambia, Tanzania, Nigeria	
Bait technology	Using odour attractants (Chapter 10). Usually impregnated with synthetic pyrethroid insecticides		Green 1994; Laveissière et al., 1990; Wall and Langley, 1991
Traps	See Table 6.1	Kenya, S. Africa (Zululand), Côte d'Ivoire	Harris, 1938; Dransfield et al., 1990
Targets	Combinations of blue/black cloth and flanking netting	Zimbabwe	
Pour-ons		Ethiopia, Burkina Faso	Leak et al., 1996; Bauer et al., 1995; Thomson and Wilson, 1992
Repellents	DEET, permethrin	Laboratory test	Wirtz et al., 1985
Sterile-insect technique (SIT)	Irradiation	Zanzibar, Burkina Faso, Tanzania	Vreysen et al., 1995; Politzar and Cuisance, 1982; Oladunmade et al., 1990
Chemosterilants	Tepa, Metepa	Experimental trials	Dame and Schmidt, 1970
Biological control		Only experimental trials	
Insect growth regulators	Diflubenzuron, pyriproxyfen	Zimbabwe	Clarke, 1982; Hargrove and Langley, 1990, 1993
Ivermectin	Drug given to host animals	Experimental trials	See Table 17.2

Table 15.2. Compounds used for the control of tsetse flies (*Glossina* spp.).

Organochlorines	Pyrethroids	Avermectins
DDT	Natural pyrethrum	Ivermectin
Dieldrin	Permethrin	
BHC (Gammexane)	Deltamethrin	
Propoxur (a methylcarbamate)	Alphamethrin	
Dimethyl phthalate (indalone)	Flumethrin	
Diethyl toluamide (DEET repellent)	Lambda-cyhalothrin	
Endosulfan	Cyfluthrin – 'SOLFAC/Bayofly'	
Ethylhexane-diol (Formula 22)	Alpha-cypermethrin (Cypermethrin)	
	• 'Flectron' eartags	
	• 'ECTOPOR'	

Susceptibility to all insecticides can vary from one species to another, and between the different classes (age, sex, physiological state) of a species. Teneral male and female flies and older, fed males are generally similar in susceptibility to organochlorine insecticides, but fed pregnant females are much less susceptible by a factor of four to nine (Burnett, 1962; Irving, 1968). Toxicity of insecticides can also be temperature dependent. Endosulfan has a positive temperature coefficient of toxicity whilst that of deltamethrin is negative (Hadaway, 1978; Smith *et al.*, 1994). Deltamethrin is more toxic than endosulfan at any temperature, and can be up to 300 times more toxic at 10°C but the degree of greater toxicity varies widely with temperature.

DDT, dieldrin and BHC are potent and persistent lipophilic insecticides, which can penetrate the cuticles of insects. Pyrethroids have low mammalian toxicity, whereas carbamates and organophosphates (developed after organochlorine use was restricted) are highly toxic to mammals and other vertebrates.

Pyrethroids exhibit two distinct effects; one is lethal and the other is 'knock-down'. In insects, pyrethrins combine with lipids in the nerve cell membranes and disrupt cationic conductance. This leads to excitation of the nervous system, which induces knock-down and then paralysis. Insects can recover from the knock-down and paralysis (provided they are not eaten, which at least 90% of them are; Laveissière *et al.*, 1984b). The two effects are not necessarily linked and little further is known about pyrethroid action. Deltamethrin kills insects after very brief contact – as brief as 5 milliseconds – yet it is environmentally safe as it is rapidly degraded in soil and, in contrast with DDT, does not accumulate in food chains. In laboratory experiments deltamethrin had a longer knock-down effect than some other pyrethroids, permethrin, bioresmethrin and tetramethrin, and the duration of knock-down was longer in male *G. m. morsitans* than in females (Quinlan and Gatehouse, 1981). These experiments also showed that sublethal doses of insecticides can reduce lipid accumulation from a bloodmeal, which could render flies more susceptible to death from starvation.

15.2 GROUND SPRAYING

The first attempts at control of tsetse using insecticides were by ground spraying, in which the resting sites of tsetse are sprayed with a residual insecticide. This technique became widespread after the Second World War, following the discovery of cheap, persistent insecticides such as DDT and dieldrin.

15.2.1 Knapsacks

Pressurized knapsacks were one of the commonest methods for ground spraying insecticides to tsetse resting sites, particularly in anglophone Africa (Fig. 15.1). Much research on the day and night resting sites of tsetse was undertaken (Chapter 8), so that insecticides could be directed efficiently at those sites using knapsack sprayers. Discriminative spraying of just the resting sites of tsetse would reduce costs, cause less environmental pollution and would be easier to carry out as only a small percentage of the total tsetse habitat would be sprayed. The technique was highly labour intensive, demanded a high level of supervision and only allowed limited areas to be covered during the spraying season. Spraying was usually carried out during the dry season after first burning the grass and other dry vegetation to facilitate access, drive away snakes and, to some extent, reduce the area of possible tsetse resting sites. The technique was particularly widely used in Nigeria, Uganda (Hocking, 1961), Zambia and Zimbabwe.

Fig. 15.1. Knapsack spraying, Zambia.

Over 200,000 km² of tsetse-infested land was cleared by ground spraying in West Africa, mainly in Nigeria, and proved particularly successful and cost-effective (Barrett, 1997). One of the early control schemes using knapsacks in Nigeria was in 1956, when a 5% DDT wettable powder suspension was applied to dry season resting sites of *G. m. submorsitans* in the northern Sudan vegetation zone. The area freed of tsetse infestation was significantly greater than the area actually sprayed and the cost of control was estimated to be only $38 km^{-2} of tsetse habitat. In a human sleeping sickness focus in Nigeria, knapsacks were used to apply 5% DDT to tsetse resting sites in riverine vegetation in combination with an application of 20% dieldrin to create barriers against reinvasion. Over 10,000 litres of DDT and 870 litres of dieldrin were applied to 528 and 88 km of river, respectively, apparently successfully (although it is not known for how long). In total, over 745 metric tonnes of DDT and dieldrin were applied, mainly by ground spraying in Nigeria between 1955 and 1978.

Synthetic pyrethroids have also been tested as residual insecticides applied by ground spraying; for example, alphamethrin 10% emulsifiable concentrate was used to control *G. p. palpalis* in a human sleeping sickness focus at Bouaflé, Côte d'Ivoire (Sékétéli and Kuzoe, 1986a). Applications of four different concentrations were made over 40 linear kilometres and assessed for efficacy. All applications gave a short-term reduction in tsetse density between 96 and 100% and even the lowest (12 g active ingredient ha^{-1}) gave more than 98% reduction 3–5 months after spraying. The higher application rates (up to 48 g a.i. ha^{-1}) did not seem to increase the insecticidal persistence or the mortality rate. A maximum dose of 12 g a.i. ha^{-1} at a concentration not exceeding 0.016% a.i. was recommended, as higher application rates caused effects such as irritation to the skin of the face and mucosa and temporary loss of appetite in people exposed to the insecticide. Trials with three other synthetic pyrethroids were conducted in the same area (Sékétéli and Kuzoe, 1986b). Deltamethrin (decamethrin) has been applied by ground spraying to control *G. tachinoides* in Chad but despite a good immediate toxicity it had little residual effect (Gruvel and Tazé, 1978).

15.2.2 Mistblowers

In order to treat larger areas by ground spraying, portable motorized mist sprayers were occasionally used in the past. One example of their successful use was in 1959, on the shores of Lake Victoria in Kenya. In this case, a 1.8% emulsion of dieldrin in water was used to suppress a dense population of *G. pallidipes* from what was believed to be an isolated infestation on a peninsula near Homa bay (Thomson *et al.*, 1960). A side-effect of the spraying was the death of a number of bushbucks. It was thought that they had probably been eating vegetation sprayed with dieldrin and that they may have been particularly susceptible. The area cleared of tsetse was only

500 acres (200 ha) and the cost was KSh 26.59 per acre (KSh 66.5 ha^{-1}) which probably represents about US$12–16 ha^{-1} today. Mistblowers have also been used to control *G. p. palpalis* and *G. tachinoides* in northern Nigeria.

15.2.3   Unimogs

A major limitation of knapsacks, or even motorized mistblowers, was the limited area that could be covered by people carrying this equipment plus insecticide, walking through the bush. In an attempt to clear larger areas, vehicle-mounted spray pumps have been used, notably in Zambia, although this technique never became widespread. The vehicles used were large four-wheel-drive Mercedes Unimogs, and DDT or dieldrin remained the insecticides used (Fig. 15.2). Large numbers of people were required for preparing access tracks through the bush and for servicing the Unimogs. Mechanical difficulties, and the high cost, restricted the use of this means of spraying to Zambia, Zimbabwe and Botswana. Although, in theory, tsetse resting sites were sprayed selectively, the powerful sprayers mounted on the vehicles resulted in considerable environmental contamination and the technique was eventually abandoned. Direct environmental consequences were reported from Botswana, where dieldrin was sprayed using a Unimog, together with knapsack sprayers in areas inaccessible to the Unimog, to control *G. m. centralis* close to the town of Maun (Graham, 1964). Teams who were ring-barking *Acacia* trees, as an additional component of the tsetse eradication attempt, collected dead birds and other animals. Despite the difficulties in adequately monitoring the effects, 109 dead birds of 21 species were detected over a period of 10 days after spraying ceased.

Fig. 15.2. Unimog spraying, Zambia.

15.3   AERIAL SPRAYING

The need to control tsetse over larger areas resulted in the development of aerial spraying techniques. Attempts were made to use aircraft both for applying residual insecticides (the same principle as knapsack and Unimog ground spraying), and to apply non-residual, contact insecticides. The technique evolved from the former to the latter, particularly in response to environmental considerations. Both fixed-wing aircraft and helicopters have been employed for aerial spraying, although fixed-wing aircraft were most commonly used. Apart from the use of aircraft to control tsetse successfully in South Africa in the 1940s (Du Toit, 1954), the technique was developed largely at the then Colonial Pesticides Research Institute in Tanzania in the early 1970s. Arising from this work, large-scale control was carried out in Zambia, Zimbabwe and Botswana in southern Africa, in Nigeria, and more recently in Somalia. In Zambia, the minimum area usually sprayed by air was 2000 km^2. Aerial spraying was used in attempts to control sleeping sickness in the Lambwe valley of Kenya and in Uganda in the late 1980s and early 1990s. Female flies, particularly pregnant ones, are less susceptible to endosulfan than males and this has to be considered when determining application dose rates (Fenn, 1992).

15.3.1   Aerial spraying of residual insecticides with fixed-wing aircraft

One of the first examples of aerial spraying for the control of tsetse and animal trypanosomosis was in the Umfolosi area of northern Zululand in 1947 (Fiedler *et al.*, 1954). This was also the first use of DDT and BHC on an extensive scale, applied as a smoke or aerosol. Initially, spraying was carried out with fixed-wing aircraft, supported by ground 'spraying' using smoke generators and dusting in mountainous areas. Later, in December 1948, helicopters were used to spray in areas of rougher terrain unsafe for fixed-wing aircraft.

Perhaps not surprisingly, aerial spraying resulted in the disappearance of the parasites of tsetse before the tsetse flies themselves (Fiedler *et al.*, 1954). This is critical where eradication is not achieved, under which circumstances the absence of parasites and predators can allow the tsetse populations to return rapidly to levels above those before control was applied.

Much of the developmental research into aerial spraying at the Colonial Insecticide Research Unit (now the Tropical Pests Research Institute, TPRI) in Tanzania is now only of historical interest as the insecticides (DDT, Dieldrin and BHC) used at that time are no longer acceptable. Aerial spraying of gamma BHC in Zimbabwe and of DDT in Tanzania was described by Hocking *et al.* (1954a,b). The trials with BHC proved to be rather unsatisfactory, whilst aerial spraying of a coarse aerosol of DDT was quite effective in eradicating *G. pallidipes* and substantially reducing populations of *G. morsitans* and *G. swynnertoni* in Tanzania.

One of the first trials using the more acceptable insecticide, endosulfan, took place in Tanzania, where a 20% w/v solution was injected into the twin exhausts of a Cessna aircraft (Hocking *et al.*, 1966). An emission rate of 2.27 l min^{-1}, swathe width of 64 m and an aircraft speed of 185 km h^{-1} resulted in an average application rate of 0.124 l ha^{-1}. The operation, in which there were four applications, had not aimed at complete eradication although this was almost achieved, even though the second application was ineffective. The costs were estimated at just under US$55 km^{-2} including 'transit costs' and salaries. More refined techniques were subsequently used to apply ultra-low-volume applications of endosulfan using Micronair atomizers.

The TPRI also experimented with aerial spraying of pyrethrum to control *G. pallidipes* in Tanzania. Three applications of 0.4% w/v active pyrethrins were used in the first two sprays, whilst a 5% DDT solution was added in the third application (Tarimo *et al.*, 1970). Two per cent piperonyl butoxide was used as a synergist in all three applications. The effectiveness of this experiment was assessed using either Langridge traps or man fly-rounds with a black cloth screen, and showed that the spray was ineffective after the third application. The two sampling techniques gave quite different results, as the fly-round technique suggested a reduction in apparent density of about 64% after the first application, whilst the Langridge traps showed little or no reduction in density from all three applications. Langridge traps were probably better than fly-rounds for assessing the trial and, therefore, previous reports in which tsetse control trials were assessed by fly-rounds may have been inaccurate (Tarimo *et al.*, 1970).

When the Kariba dam was constructed across the Zambezi river, between Zimbabwe and Zambia, and the valley was flooded, the inhabitants were resettled elsewhere. One resettlement area, the Lubu valley, had been a sleeping sickness focus and was first sprayed to reduce the tsetse population to a level that would allow the Batonga people to be resettled by late 1957. Gamma-BHC was sprayed as a coarse aerosol over 250 km^2, and the tsetse population was reduced to between 81 and 100% of previous levels, but not eradicated (Cockbill *et al.*, 1963).

15.3.2   Aerial spraying of non-residual contact ultra-low-volume insecticides

Aerial spraying with non-residual insecticides is carried out in a number of cycles, and is sometimes termed sequential aerial spraying or ultra-low-volume (ULV) spraying, as the insecticide is broken into small droplets of the correct diameter by an atomizer and is sprayed in very small quantities. The cycles are timed to coincide with the emergence of adult flies from underground puparia, which are protected from insecticide contamination during the first spray cycle. The timing is such that flies emerging the day after the first cycle will be killed by a second cycle before being able to deposit a third-stage larva ready for pupariation. The final cycle will theoretically kill any adults emerging from puparia arising from larvae deposited immediately before the first spray cycle. The final cycle will therefore be roughly 30 days after the first, depending upon the temperature conditions. Timing of the cycles is critical and estimations of puparial duration from temperature data (Chapter 3, section 3.4.2), or by field experiments, are used to determine the cycle interval. Up to nine cycles of spraying were commonly used, although theoretically this is more than required.

The insecticide kills tsetse on contact and is only effective whilst it remains in the air and impacts on tsetse in flight or whilst resting. The optimum droplet size has to be small enough to allow them to remain suspended in the air as long as possible rather than sinking quickly to the ground. They must, however, be sufficiently large so as not float upwards or be rapidly blown away from the spray area. Droplet sizes of 30–35 μm volume median diameter (vmd) are optimal for application of endosulfan at a rate of 6–12 g ha^{-1}. Atmospheric conditions are also important in determining the efficiency of the method. Aerial spraying is generally carried out at night during temperature 'inversion' conditions, when a layer of denser cool air close to the ground provides a cushion in which insecticide drops can remain suspended. Such conditions occur from the early evening to dawn, and this is therefore the most suitable time to carry out spraying. Aircraft use navigational aids to fly in parallel paths 200–300 m apart at 280 km h^{-1} to produce swathes of 100–300 m. As the aircraft have to fly very low, just a few metres above the height of the canopy, flying can be dangerous and very powerful headlights are fitted. A detailed handbook for aerial spraying has been produced (Cooper and Dobson, 1994) which provides information on droplet sizes, spray volume and quality and the monitoring of aerial spraying operations for tsetse control.

A computer model has been used to predict that a 99.9% mortality could be achieved using an application of 30 g endosulfan ha^{-1} (Johnstone and Cooper, 1986), which is slightly higher than the 25 g ha^{-1} that has been used in Zimbabwe.

Between 1974 and 1976 in Gokwe district, Zimbabwe, an area of over 210 km^2 was sprayed with a solution of 20% a.i. delivered at 5.62 l min^{-1} to a Micronair atomizer from a height of 25–30 m above ground level at 200 m swathe intervals (Chapman, 1976). Six applications at a rate of 14 g of a.i. ha^{-1} were made initially, at roughly 18-day intervals. However, the intervals between applications were too long, and a second series of applications was required in 1975. It was claimed that eradication was achieved at the end of the operation, but as the area was not isolated from adjacent infestations, it was probably subsequently reinvaded.

15.3.3 Helicopter spraying

Helicopters have been used to apply insecticides for tsetse control in Zululand (South Africa), in trials in the Lambwe valley (Kenya) (Le Roux and Platt, 1968) and in West Africa, where human trypanosomosis vectors inhabit linear, winding, riverine vegetation less suitable for fixed-wing aerial spraying (Molyneux *et al.*, 1978a,b). Advantages of helicopter spraying over fixed-wing aircraft are the manoeuvrability and the capability of landing almost anywhere. In addition, the rotor downwash from helicopters produces a greater penetration of the vegetation than aerial spraying from fixed-wing aircraft, which is an advantage when spraying densely canopied gallery forest.

The first tsetse control trials using helicopters took place in 1969 along the Niger river, in Niger (Spielberger, 1971), and in Nigeria (Bauer, 1971), using a mixture of DDT, lindane and dieldrin. The cost per hectare actually sprayed in Nigeria was approximately $12, equivalent to approximately $200 km^{-2} reclaimed (Spielberger *et al.*, 1977). Reinvasion took place between 1971 and 1974 at a rate of approximately 25% per year and these areas were re-sprayed.

In Burkina Faso, aerosol applications of endosulfan and synthetic pyrethroids have been made using similar methods to those used for fixed-wing aircraft (Molyneux *et al.*, 1978a,b). A helicopter fitted with one of two systems was used. The first was a low-speed system with eight electrically driven rotary atomizers capable of rotating at up to 9000 rpm and producing droplets > 100 μm vmd. The second system consisted of six high-speed rotary atomizers reaching speeds of 16,000 rpm and producing a droplet size of 50 μm vmd. The highest mortality achieved was 91.3% for *G. tachinoides*, using endosulfan, at a dose of 9.0 g ha^{-1}. Helicopters were extensively used in the 1970s to apply residual insecticides to the dry season habitats of tsetse in Nigeria (Spielberger and Na'isa, 1977). In those habitats tsetse rested at night in the tree canopy, which was more easily treated from the air (Scholz *et al.*, 1976). Baldry *et al.* (1978a,b, 1981) described helicopter spraying of riverine populations of *G. tachinoides* in Burkina Faso with endosulfan and deltamethrin and gave information on costs. Aerially applied barriers of dieldrin were used, but were ineffective. There was no evidence of long-term environmental effects of the spraying, only a short-term effect on freshwater invertebrates (Takken *et al.*, 1978). Similarly, side-effects of endosulfan applied from helicopters in Côte d'Ivoire in 1978/79 were not apparent 2 years later (Everts *et al.*, 1983).

Chapter 16

Traps and Targets

16.1   GENERAL PRINCIPLES

Tsetse flies are *k*-strategists, having a low rate of reproduction: they produce a single egg at a time and one egg every 9 or 10 days. The eggs and larvae produced are retained in the female uterus until just before the 3rd-stage larva is ready to pupate, and this high level of parental 'care' ensures a high probability of each individual surviving. This differs from most insects, which are *r*-strategists, producing large numbers of eggs each having a low chance of survival to adulthood. Because of the stability of tsetse populations and the low reproductive rate, little sustained mortality pressure (additional to natural mortality) needs to be exerted on a population to bring about extinction (Weidhaas and Haile, 1978). From estimates of the degree of trapping necessary to control a variety of insects, Weidhaas and Haile concluded that tsetse would require the lowest daily trapping rates, because of their low reproductive capacity and long-lived adults. An additional mortality of 4% per day imposed on female flies was estimated to be sufficient to result in extinction in the absence of any immigration (Hargrove, 1988). Where reinvasion or immigration takes place, a tsetse population will decline to a new lower equilibrium level (asymptote). This theory is behind the use of traps and targets used for the control of tsetse. The basic idea is simple: tsetse are visually attracted to a trap or target; this attraction may be augmented by the use of olfactory attractants. When flies land on a trap or target they either receive a lethal dose of insecticide through their tarsi, or are caught in the trap and subsequently die.

A comprehensive review of traps and targets for the control of tsetse populations was published by Green (1994). Although attempts were made to control tsetse using targets impregnated with insecticide many years ago, successful application of this technique followed the production of the second-generation synthetic pyrethroid insecticides and the development of potent odour attractants in the last 10–20 years.

16.2   EARLY EXAMPLES OF THE USE OF TRAPS AND TARGETS TO CONTROL TSETSE POPULATIONS

The concept of using traps for controlling tsetse flies is by no means new. Balfour (1913) envisaged using traps baited with odour attractants for tsetse control, although his trap (obtained from manufacturers of a mosquito trap) did not turn out quite as he had wished. He planned to bait the trap with a central wick soaked in a solution of human or animal sweat. He even thought of putting citrated blood in the trap, which, he said, could be poisoned if desired.

Vanderplank (1947b) mentioned the possibility of using cheap tsetse traps sprayed with DDT and placed about the bush for controlling or eradicating tsetse. He also suggested the possibility of dipping or spraying cattle with DDT (the 'pour-on' technique, discussed in Chapter 17). A field trial in which DDT was sprayed on cattle or attracting screens killed a high percentage of flies alighting on them. *Glossina pallidipes* and *G. morsitans* both landed more readily on black than on white cloth targets, and in the laboratory the landing response was most marked when the flies were most active (Barrass, 1970). It was observed that tsetse resting on black cloth usually rested longer than those resting on a white surface.

Principe

The first attempt to control tsetse by catching out a population was on the island of Principe in 1910, using sticky traps on the backs of plantation workers. Tsetse flies may not have been introduced to Principe until 1825, when they came with cattle imported from West Africa. There was no apparent problem of trypanosomosis either in cattle or in humans until 1877, when human sleeping sickness was probably introduced with imported labourers. This resulted in a serious epidemic around the turn of the century. The attempt to control this, using sticky targets to control the tsetse population, has often been referred to in reviews of tsetse control; the most detailed recent account was by Busvine (1993). Principe was a suitable location to attempt such a technique, being an island of only about 50 km^2. In order to improve the efficiency of the sticky traps to catch tsetse, the numbers of alternative potential hosts were first reduced by killing as many wild pigs and dogs as possible.

Zululand

One of the first relatively successful attempts to control tsetse by trapping on a larger scale was in Zululand, using the Harris trap (Harris, 1938). The Harris trap was large and unwieldy, measuring 2 m long by 1 m wide and 1 m deep, covered by hessian. In the Umfolosi Game Reserve a mean of 81.1 *G. pallidipes* per trap per day were caught in 1931. As the population was suppressed, the number of flies caught decreased to 0.2 flies trap^{-1} day^{-1} in 1934 and to 0.0003 in March 1938. From January 1931 to April 1938 it was estimated that over 10.5 million tsetse were caught. The effect of this control of tsetse flies on trypanosomosis prevalence in cattle was an unspecified but rapid reduction. The trap also was said to work successfully against *G. palpalis* in the Democratic Republic of the Congo but not against *fusca* group flies. Dr Marchi, working in the Congo, reported in a personal communication to Harris that just 12 traps used at Kanda-Kanda village reduced the number of new cases of sleeping sickness from 20 to zero in a period of several months. It is unclear why the trap was not more widely used since it produced such an effect.

Hand-catching

In the 1940s, an outbreak of human sleeping sickness in the south Kavirondo district of Kenya was controlled by first creating a barrier of bushclearing to isolate the area, and then reducing the population of *G. f. fuscipes* by hand-catching the flies with nets (Glasgow and Duffy, 1947, 1951). Although eradication of the flies was claimed, it is likely that more efficient sampling methods would have shown that a low-density population remained.

16.3   RECENT DEVELOPMENTS

Laveissière *et al.* (1990) and Wall and Langley (1991) reviewed the use and development of traps and targets for tsetse control. As more powerful tsetse attractants and more effective formulations of insecticides, particularly the synthetic pyrethroids, are being produced, this method of tsetse control is developing rapidly. There have been important advances in the production of synthetic attractants, and various phenol mixtures are in use either in trials or for practical control operations against several tsetse species (Hall *et al.*, 1984; Vale *et al.*, 1988a,b). Traps and targets are a more acceptable means of controlling tsetse than either ground or aerial spraying of insecticides in terms of the direct ecological and environmental impact they might have. However, they require a considerable amount of costly maintenance, and their sustained use for tsetse control has, in many cases, been severely compromised by theft of trap or target materials. Consequently, attempts are being made to develop easily maintained or disposable targets, which could be partially managed by beneficiary communities.

Most development of this method of tsetse control has concentrated on improved and cheaper designs of the target and odour attractants in order to attract as many tsetse as possible and to increase the number of tsetse actually landing on a target. More efficient odour attractants would allow fewer targets to be deployed per km^2 and would reduce costs. At present attractants are most effective for *morsitans* group tsetse whilst further improvements are needed for attractants of *palpalis* group flies. Almost no work has been done with attractants for *fusca* group tsetse.

Because of the differences in the habits and habitats of the three groups or subgenera of tsetse, whilst control using impregnated targets and traps is successful against species of the *palpalis* group (Küpper *et al*, 1984; Laveissière *et al.*, 1981; Laveissière and Couret, 1981) and the *morsitans* group (Vale *et al.*, 1988b; Willemse, 1991), control of *fusca* group tsetse, which are less attracted by synthetic phenols or other attractants, is more difficult. They generally occur in areas without cattle and are therefore economically less important for Africa as a whole, although locally they can be important vectors of trypanosomosis to cattle. Most interest in tsetse control using bait technology in West Africa has been directed at the control of human trypanosomosis as the control of animal trypanosomosis generally necessitates the treatment of much larger areas.

16.4   INSECTICIDES

Synthetic pyrethroids are now widely used for tsetse control on impregnated traps, screens and targets. The most commonly used pyrethroid is deltamethrin, and the literature reporting control operations is growing rapidly (Hargrove and Vale, 1979; Laveissière *et al.*, 1981, 1985; Gouteux *et al.*, 1982b, 1986; Schönefeld, 1983; Küpper *et al.*, 1984, 1985; Torr, 1985; Vale *et al.*, 1985, 1986; Douati *et al.*, 1986; Gouteux and Lancien, 1986; Gouteux and Noireau, 1986; Takken *et al.*, 1986; Vale, 1986; Willemse, 1991).

There are three possible effects of insecticides used to impregnate traps and targets:

- A lethal effect – direct mortality.
- A 'knock-down' effect – leads to 90% or more indirect mortality (Laveissière *et al.*, 1984b).
- A repellent effect.

The knock-down effect (Chapter 15, section 15.1), results from paralysis caused by binding of pyrethrins to lipids in the nervous tissues and is highly significant as at least 85% of knocked-down flies will die, mainly through predation. There have been suggestions recently that the insecticides used for impregnating traps and targets, or for use as pour-on insecticides on cattle (Chapter 17), may not have a large direct impact on mortality. In contrast to the knock-down effect, any possible repellency of pyrethroids has not been satisfactorily demonstrated. If there is a significant repellent effect of deltamethrin at the concentrations used, this may diminish the efficacy of targets by reducing the number of flies landing. In contrast, flies entering a trap are forced to come into contact with the insecticide for a longer period.

16.4.1   Insecticide impregnation and re-impregnation of targets

Initially, trials with different insecticides and dosage rates were aimed at reducing quantities of insecticide used, both for economic and environmental reasons. When synthetic pyrethroids were first developed they were particularly expensive, which made it important to attempt to use as low dosage rates as possible. Subsequently, following experiences of the difficulties in maintaining targets on a large scale in the field, and as insecticide costs came down, a different strategy was adopted and recommended by some of the insecticide companies. Economic analyses showed that savings in costs of insecticides had little impact on cost–benefit analyses as they were small compared with the cost of maintenance. The new strategy thus consists of using higher insecticide impregnation rates so that targets can be deployed without the requirement for frequent maintenance and re-impregnation. It has been claimed that deltamethrin formulations applied to cloth screens are sufficiently persistent to maintain efficacy for up to one year.

Typical of the impregnation strategy initially recommended was a trial conducted by Takken *et al.* (1986) using blue cotton screens 70 cm × 110 cm. These were soaked in 0.05% a.i. deltamethrin leaving a deposit of 150 mg m^{-2} and were re-impregnated 4 months later. These proved effective for reducing, but not eradicating, populations of *G. p. palpalis* and *G. tachinoides* over a short period of time.

In Zimbabwe, most *G. pallidipes* landed on the bottom half and centre of targets, and so financial savings could be made by impregnating only this portion of the target with insecticide for control purposes (Vale, 1993a). A similar observation was made in Zanzibar, where *G. austeni* land mainly on the bottom half of sticky targets.

Following a review of insecticides for trap or target impregnation Torr (1985) concluded that they should: (i) prove lethal to a fly briefly in contact; (ii) be resistant to weathering; and (iii) not adversely affect the visual and olfactory characteristics of the target.

16.4.2 Types of fabric – target–insecticide interactions

The persistence of insecticides on cotton or synthetic materials for making screens or traps has been determined using chromatographic analysis of insecticide residues and bioassays (Laveissière *et al.*, 1984b). Ninety-day-old gravid female tsetse are most resistant to insecticides and Laveissière *et al.* used this group of *G. p. gambiensis* for their tests. Residues on cotton/polyester cloth were more persistent than on pure cotton. Emulsifiable concentrates adhered best, whilst no insecticide tested remained effective beyond 45 days in the rainy season or beyond 75 days in the dry season. More than 50% of the active ingredient was removed shortly after even a small amount of rainfall; a complication in evaluating the effects lay in the high percentage of 'knock-down' rather than mortality.

The characteristics of fabrics used subsequently proved to be as important as their dyes (Laveissière *et al.*, 1987). Again, polyester/cotton fabric was highly efficient for screens, depending upon the weave of the cloth. Fly mortality could be maintained at a high level for more than 4 months. A closely woven fabric with thin thread allowed a good fixation of the insecticide but prevented tsetse from receiving a lethal dose. Blue fabrics can lose their colour and become inefficient after a short time, depending on the dye used and method of fixation.

16.4.3 Persistence of insecticides on traps and targets

Torr (1985) examined the susceptibility of wild *G. pallidipes* to DDT, dieldrin and deltamethrin on cotton canvas and Terylene netting. He used 3% dieldrin wettable powder (wp) and emulsifiable concentrate (ec), 5% DDT wp, and 0.0625% deltamethrin flowable concentrate (fc) sprayed at $0.36 \, l \, m^{-2}$. Over 95% of flies resting for 45 seconds died within 72 h. Deposits exposed to the sun for 140 days of the dry season produced mortalities of 100, 0 and 10% compared with 100, 75 and 100% for shaded deposits. Dieldrin produced photodieldrin on exposure to the sun.

Küpper *et al.* (1985) used biconical traps, rather than targets, impregnated with 400 mg a.i. of deltamethrin per trap, positioned at 300 m intervals in the southern Guinean zone of Côte d'Ivoire. The insecticidal effect of the impregnated cloth persisted for at least 4 months, with a final fly mortality still over 80%. Several campaigns to control riverine tsetse with traps in West Africa were carried out without odour attractants, as these tsetse species were only slightly attracted to the odours available (Küpper *et al.*, 1984; Mérot *et al.*, 1984; Laveissière *et al.*, 1986a). In a previous trial comparing screens impregnated with either dieldrin (4 g a.i. per screen) or deltamethrin to control *G. tachinoides* and *G. p. gambiensis* in Côte d'Ivoire, dieldrin was ineffective due to a slower effect and 'bad impregnation' (Laveissière and Couret, 1983).

Early studies showed that deltamethrin but not dieldrin deposits appeared to be resistant to weathering by rain (Torr, 1985). However, those laboratory studies did not adequately simulate field conditions and deltamethrin on targets is probably less persistent during the wet season than previously reported (Torr *et al.*, 1992). In the wet season, targets impregnated with 0.1% deltamethrin fc consistently produced mortalities of only about 10%. The shortcomings of targets during the wet season could be overcome to some extent by using black, all-cloth targets and increasing the concentration of insecticide to 0.6–0.8% suspension concentrate (Torr *et al.*, 1992). A 0.6% suspension produced mortalities of > 90% for 300 days even in the wet season; therefore targets thus treated could be deployed effectively for up to 1 year.

Oil formulations of pyrethroids are much more persistent than wettable powder formulations. An oil formulation of lambda-cyhalothrin gave 100% knock-down for up to 10 months and 100% mortality after 24 h for 8 months, compared with periods of 4 months and 3 months, respectively for the wettable powder formulation (Langley *et al.*, 1992). Similar results were obtained in tests with deltamethrin in coconut oil (Hussain *et al.*, 1994).

The relative topical toxicity of a range of insecticides is shown in Table 16.1.

16.5 TRAP AND TARGET DESIGN

The use of traps and screens for controlling, or sampling, tsetse populations, including designs and background details, was reviewed by Cuisance (1989). Some of the widely used traps for control purposes are shown in Fig. 6.2 (Chapter 6, section 6.1.2).

Some of the monoconical 'trap' designs widely used for control in West Africa could be regarded as targets as they do not always have devices for retaining the flies attracted to them. Dagnogo and co-workers (1985b,c), comparing monoconical traps, biconical traps and blue/black targets impregnated with 300 mg a.i. deltamethrin m^{-2} for controlling *palpalis* group tsetse in the Congo and Côte d'Ivoire, found that targets were the least effective. Best results were obtained with a monoconical trap with crossed blue/black panels. A similar trap with a blue/black combination of intersecting panels and a plastic roof was even more effective (Gouteux and Noireau, 1986) and, when fitted with a permanent collecting device, the pyramidal trap was more effective than traps impregnated with deltamethrin every 2 months (Gouteux *et al.*, 1986). The standard trap was effective in controlling *G. palpalis* peridomestic populations in riverine habitat within savanna areas of the Congo (Gouteux and Sinda, 1990).

Laveissière and Grébaut (1990) designed the Vavoua trap (named after the town in Côte d'Ivoire where it was developed) for control of human sleeping sickness. This was a modified form of the biconical trap, with the same upper net cone, but with three vertical panels of black and blue cloth hanging down at 120° angles to each other, in place of a cone, on the bottom half. This trap, designed for control with community participation rather than for sampling purposes, is quite similar to the monoconical trap of Lancien (1981). Although about half the cost of a biconical trap, the

Table 16.1. Relative topical toxicity of insecticides to *Glossina austeni* (from *FAO Tsetse Training Manual*, Vol. 5).

Insecticide	Relative toxicity
Malathion	< 2
DDT	3
Dieldrin	26
Natural pyrethrins	20
Permethrin	87
Cypermethrin	350
Alpha-cypermethrin	1100
Deltamethrin	3300

efficiency of the Vavoua trap was about the same against *G. p. palpalis* (Laveissière and Grébaut, 1990).

A mono-screen trap was made from indigenous plant materials in Uganda for tsetse control (Okoth, 1985). This was constructed from bark cloth (traditionally prepared for making clothing) manufactured from fig trees (*Ficus natalensis*), with bamboo (*Oxytenanthera abyssinica*) for the poles, rattan cane (*Calamus deeratus*) for the apical cone and swamp palm (*Phoenix reclinata*) for the upper cone and collecting cage. The trap was used on a small scale by rural communities for controlling *G. f. fuscipes* in a sleeping sickness epidemic area (Okoth, 1991). Compared with biconical, pyramidal and Vavoua traps, it was claimed that the mono-screen trap performed best and was cheap, costing approximately US$4.7 (Okoth, 1991).

16.5.1   Trap/target colour

Bax (1937) observed that *G. swynnertoni* was attracted (activity was caused) to a small black target at a distance of about 45 m. At about the same time it was observed that more tsetse were attracted to a black ox and a white ox than to a yellow/red ox (Moggridge, 1936b). These observations indicated that colour was an important factor in attracting tsetse and more sophisticated experiments and analyses of colour have since been made. Responses to colour can be analysed according to attractivity (number of flies approaching) and the landing response.

Light wavelength

The responses of tsetse to ultraviolet (UV) light and the reflectivity of cloth targets to UV light are complex, as attraction and alighting responses differ. Whilst UV light attracts tsetse, UV-reflecting surfaces repel them. Under certain circumstances a highly UV-reflecting surface, such as a trap or target, can increase the landing response of tsetse flies (Green, 1993).

Tsetse are attracted to near-UV light, followed by blue light, and appear to have a spectral range and sensitivity similar to that of other Diptera (Green and Cosens, 1983). Because of the sensitivity of *Glossina* to UV light as well as visible wavelengths, Green and Cosens concluded that it is inadequate to relate attractiveness of coloured surfaces to their reflectivities in only the visible range. Visual responses of *G. m. morsitans* to stationary targets are greatest in the UV and visible light wavelengths (Jordan and Green, 1984). When odour baits are not used together with traps or targets, the proportion of attractive blue (400–500 nm) to unattractive green–yellow (500–600 nm) in trap material is the most important determinant of reflectance. Reflectivity of red (above 600 nm) is also a positive factor in trap attractiveness. Effectiveness of the F3 trap for catching *G. pallidipes* and *G. m. morsitans* depends mainly upon reflectivity in four different wavelength bands, two of which (blue–green and red) are positively correlated with trap score; and two of which (UV and green–yellow–orange) are negatively correlated with trap score (Green and Flint, 1986). The most important effects came from the blue–green and green–yellow–orange bands. The best trap material was royal-blue cotton, which reflected blue–green strongly but very little UV or green–yellow–orange. Green and Flint (1986) showed that tsetse employ colour information and not intensity-contrast information alone in trap-orientated behaviour. There appeared to be no differences in preference between sexes or species.

Green (1986) caught tsetse flying around different coloured traps and targets to examine the effect of colour and odour on attraction of *G. pallidipes* and *G. morsitans*. The results for relative attractivity, efficiency and landing responses of tsetse in these experiments are shown in Table 16.2. Yellow and green were unattractive and inefficient, black and red attractive but inefficient, white moderately attractive and very efficient and blue traps attractive and efficient; the order was the same for targets. Landing responses were strongest on black surfaces and weakest on white, whilst results with blue were variable. The relative performances of traps coloured differently on inside surfaces were not affected significantly by odours. Black or red target areas inside the trap were required for optimum performance. Later work on the effect of colour on target and trap orientated responses showed that pthalogen blue targets were best for *G. p. palpalis* in Côte d'Ivoire, compared with black for *G. pallidipes* in southern Africa (Green, 1988).

Further field and laboratory studies produced fairly consistent results. Landing responses on blue and black targets increased in the presence of CO_2, whilst there was no effect with white or yellow targets. Green (1986) showed that external colour affected both attractiveness and efficiency of targets and the order of attractiveness was similar to that of traps. Green (1989) also examined the use of two-coloured targets for the control of *G. p. palpalis* with and without flanking mosquito netting (section 16.5.3). No target with flanking netting caught more flies than an all-blue single-coloured one. Blue and black was the best combination for male flies. Blue and white was a good combination, with (in this case) landing response being elicited by the white. In similar work on the effect of colour on the

Table 16.2. Responses of *G. pallidipes* and *G. morsitans* to differently coloured traps and targets (from Green, 1986).

Colour	Attractivity	Efficiency	Landing response
Green	–	–	
Yellow	–	–	
Black	+++	–	+++
Red	+++	–	
White	++	+++	– – –
Blue	+++	+++	+ –

attractiveness of targets for *G. tachinoides*, blue was the most attractive and yellow the least, with black and white intermediate (Green, 1990). The strongest landing response by females was with UV-reflecting white cloth. The response of males was high with all colours except yellow. In the absence of netting panels, blue-and-white targets were twice as good as all-blue targets.

In Somalia, blue targets caught 1.7 times as many tsetse as all-black ones (Torr *et al.*, 1989).

16.5.2 Response of tsetse to pattern and orientation of targets

Zebras may be protected from being fed on by biting flies, including tsetse, by their striped pattern (Waage, 1981). Studies using stationary targets in Zimbabwe supported this, and indicated that the orientation of the stripes is an important factor in determining attractiveness to tsetse (Gibson, 1992). Grey and vertically striped targets caught significantly fewer flies than black or white targets, whilst horizontally striped targets caught less than 10% of the number of flies caught by any other target tested. Stripes appear only to affect attraction of flies from a distance and apparently do not interfere with the landing response. Interpretation of observations on landing responses to vertical and horizontal objects by tsetse is rather confusing. In the field *G. pallidipes* and *G. m. morsitans* show a better landing response to horizontally orientated objects in preference to vertical ones, whilst laboratory studies with the latter species show the opposite. Furthermore, this behaviour is altered in the presence of odours.

16.5.3 Flanking netting

Panels of flanking mosquito netting were formerly added to cloth targets in Zimbabwe, after observations that many tsetse circled but did not come into contact with plain cloth targets. When insecticide-impregnated flanking netting is added, many flies hit this and are killed, significantly improving the efficiency of targets. This led to the development, in Zimbabwe, of the 'type S' target of black cloth and netting. Subsequently, because of the added expense of making targets with flanking netting and its poor durability, a slightly larger cloth target was suggested as a more economical and sufficiently effective alternative.

Other studies confirmed the improved efficiency of targets with flanking netting; for example, Mérot and Filledier (1985) found that to control *G. m. submorsitans* in Burkina Faso, the addition of black terylene netting panels to a blue and black target increased catches. This target was at least twice as effective as the plain blue target without netting panels previously used in West Africa.

In Zimbabwe, black cloth targets appear superior to blue ones for control purposes: although blue cloth is more attractive, the black elicits a better landing response (Vale, 1982). Earlier, it was suggested that blue or blue/black could be a little superior to black for *G. pallidipes* when the targets are large (Vale, 1993b).

Differences observed between targets in West Africa (*G. p. palpalis*) and Zimbabwe (*G. pallidipes*) could be explained by differing behaviour of the two tsetse species. The preferred alighting position of *G. p. palpalis* is on the sides of the target rather than on the centre, which is preferred by *G. m. morsitans* and *G. pallidipes*. Thus, the West African target has blue at the centre, not at the sides (Fig. 16.1).

16.6 EXPERIMENTAL TRIALS

16.6.1 Traps and targets with insecticides

Vale *et al.* (1986) reported on the experimental control of *G. m. morsitans* and *G. pallidipes* on an island in Lake Kariba initially using six traps with carbon dioxide and acetone as attractants. Subsequently, 20 targets coated with dieldrin (later replaced by deltamethrin), with acetone and octenol as attractants, were used. It was estimated that 2% and 5% of the *G. m. morsitans* and *G. pallidipes* populations were killed per day, respectively. Both populations declined rapidly, the former disappearing in 9 months and the latter in only 11 weeks.

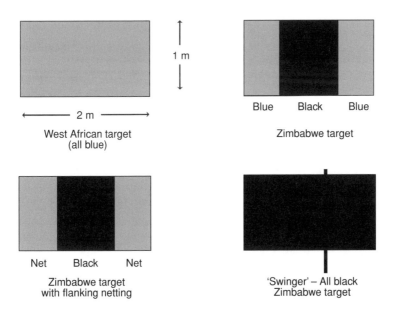

Fig. 16.1. Diagram of target designs.

This was followed by a larger scale tsetse control operation in the Zambezi valley, Zimbabwe, using targets consisting of black cloth and netting, baited with octenol released at 0.5 mg h^{-1}, and either acetone or butanone released at 100 mg h^{-1} or 15 mg h^{-1}, respectively (Vale *et al.*, 1988b). The targets were coated with deltamethrin and deployed at a density of 3–5 km^{-2}. It was estimated that about 3,000,000 adult *G. pallidipes* and 200,000 *G. m. morsitans* were killed by targets or removed by sampling between March and December 1984, causing a population decline of at least 99%. Eighteen months later, the targets were badly faded and as they became less efficient they were sprayed with a black dye and a UV light absorber that protected the dye and the insecticide. Decreased maintenance of targets resulted in a lower efficiency and subsequent reinvasion of the area. Later developments avoided having to re-dye the targets and thus reduced maintenance requirements.

In northern Côte d'Ivoire, simple 1.17 m × 0.75 m blue cloth targets impregnated with deltamethrin e.c. 5% at a rate of 100 mg a.i. of deltamethrin per target were used to control *G. palpalis* and *G. tachinoides* (Douati *et al.*, 1986). A total of 1630 targets were deployed at 100 m intervals over 163 km of gallery forest, covering a total area of 1200 km^2. Simultaneously, traps impregnated with insecticide were used as barriers to prevent reinvasion. The apparent densities of both species were reduced by about 90%. Changes of trypanosome prevalence in cattle were approximately correlated with changes in the apparent tsetse density. Comparing results from an area where, in 1983, both traps and targets were used with results from an area where only traps were used, traps alone appeared more efficient, possibly because in addition to flies entering the trap being caught, those which only alighted on the trap but did not enter were impregnated with insecticide. Coulibaly *et al.* (1995) described the impact of continued tsetse control using impregnated traps and targets in northern Côte d'Ivoire on livestock productivity.

Traps and targets are much less effective against *fusca* group tsetse. In the southern Guinean zone of Côte d'Ivoire, trapping reduced populations of *G. palpalis* and *G. fusca* by approximately 97% and 67%, respectively, and *G. medicorum* and *G. longipalpis* populations by 35% along an 82 km stretch of gallery forest (Küpper *et al.*, 1985). *Glossina palpalis* was most affected and breeding was interrupted for about 2 months. The poorer control of *fusca* group flies may have been due to their different activity pattern (Küpper *et al.*, 1985); furthermore, odour attractants are less effective against these species and the nature of their habitat might decrease the range of visual attraction. Deployment and maintenance of traps could only be carried out effectively during the dry season; therefore, in the wet season, reinvasion and increased disease transmission could take place.

There are now many examples of the use of targets for tsetse control in human sleeping sickness foci. At Vavoua, in Côte d'Ivoire, impregnated traps and targets were used to control human sleeping sickness over a long period and Laveissière published many reports on the development of the technique and its effects. Rural communities were successfully mobilized and local people participated in setting up targets at previously identified sites. A reduction in the population of *G. palpalis* of over 90% was achieved after a week and was maintained at a high level for 8 months.

In trials using the pyramidal trap (Gouteux and Lancien, 1986) to control *G. p. palpalis* in 35 villages in a savanna region of the Congo (Brazzaville), traps with a permanent collecting device were more effective than traps impregnated with deltamethrin every 2 months (Gouteux *et al.*, 1986). Insecticide impregnation before deploying the trap appeared to be unnecessary and fly populations were reduced substantially.

16.6.2   Traps without insecticides

Traps have been developed to control tsetse without using insecticides, as these are costly, often difficult to obtain in African countries, and there is concern over their possible adverse environmental effects. The NGU trap (Brightwell *et al.*, 1987) and its subsequent modifications, including the NG2B trap, were successfully used to suppress, but not eradicate, populations of *G. pallidipes* and *G. longipennis* at Nguruman, Kenya (Dransfield *et al.*, 1990). The trap, used with acetone and cow urine and a plastic-bag catching device, requires less material than biconical or F3 traps and is more efficient in trapping *G. pallidipes* at Nguruman. The Maasai community was involved in the tsetse control and made some traps in their homesteads. In one trial, in which 190 traps were deployed over an area of 100 km², the efficacy of the operation was monitored using mark–release–recapture estimates of the tsetse population size. After 10 months a reduction of 98–99% was achieved relative to estimates of the numbers of flies 3 km outside the suppression zone. Seasonal population increases resulting from tsetse that reinvaded during the short rains in November were soon brought down by trapping. Ovarian age data were used for estimating mortality using a maximum likelihood method (Dransfield *et al.*, 1986a). Average mortality rates due to trapping were estimated at 4–5% per day. Together with natural mortality, it was estimated that the adult population of *G. pallidipes* was reduced at a rate of about 2.6% per day during the harsh dry season. During the wet season, when climatic conditions were more favourable to *G. pallidipes*, the population at Nguruman was able to recover to some extent. Even the population of *G. longipennis*, which is less easily caught in traps, was reduced by up to 90% in the dry season. The effects of the tsetse population suppression using the 'low technology' NG2B traps were modelled, providing a very detailed assessment of the use of traps for tsetse control (Dransfield *et al.*, 1986b).

The efficiency of the NG2B trap was subsequently improved by raising it 15–20 cm above the ground, using a new cage design and adding a 1 m wing of blue cloth to one side of the trap, which increased catches of females of both species by about 60%. Decreasing the amount of acetone used reduced costs. The cost of the trap/odour bait system was estimated at US$8.5 per unit per annum. The white netting was replaced by a blue material, which lasts longer and resulted in 40–50% more flies being caught (R. Dransfield, Nairobi, 1993, personal communication).

16.7   RATES OF DISPERSAL (REINVASION) IN RELATION TO RATES OF POPULATION DECLINE

One of the factors responsible for the lack of success in sustaining control of tsetse flies is their high mobility, resulting in continual invasion pressure against cleared areas. It has been calculated that *G. morsitans* and *G. pallidipes* can move with an average displacement of up to 1 km per day in random steps. Overall displacement in 100 days is closer to the square root of 100 km, which is about 10 km per day (Laveissière *et al.*, 1990).

In a study (referred to in Chapter 9) from which the following is drawn, Williams *et al.* (1992) modelled tsetse dispersion in order to examine the effect of parameters related to tsetse dispersal on the rate of decline of tsetse populations and distribution of traps or targets used for control. The important parameters related to dispersal and rate of decline of a population are: r, the innate rate of increase of the population; λ, the root-mean-square displacement per day; a, the range of attraction of a trap or target; and δ_a, the additional mortality rate imposed. Allowing for puparial and pre-adult periods, they estimated the maximum rate of increase (r) at about 2% per day, which would necessitate an additional mortality imposed by traps or targets of greater than that figure per day.

The distance a fly can move will depend on the speed and duration of flight; estimates from the field and the laboratory show that tsetse fly at an average speed of 5 m s^{-1} and for a relatively short duration of 1–2 min per flight. Total flying time is estimated at about 15–30 min per day; thus, each flight will have a range of 300–600 m, and the total distance per day would be between 4.5 and 9 km. The number of flights and duration will vary according to the hunger state of the fly. Given the random walk model with a step length of 50 m (Williams *et al.*, 1992), a root-mean-square displacement per day of 167 m would be derived for a flight distance of 4.5 km, and a root-mean-square displacement of 1.3 km in a day for a step length of 200 m and total flight distance of 9 km.

In Kenya, ranges of attraction for unbaited biconical traps were 15–20 m for *G. pallidipes* and 10–15 m for *G. brevipalpis* (Dransfield, 1984). Williams *et al.* (1992) assumed a radius of attraction of between 50 and 100 m for baited traps (based on various field studies), and mortality rates, within that range, of 0.3–3.0% per day. It was then deduced mathematically that if the effective trapping mortality rate exceeds the population growth rate, the population will converge exponentially to zero at a rate s equal to $\delta - r$. To achieve a 99% reduction in a year, s must be 1.9% per day; and if r is 1.7% per day, δ_t must be 3.6% per day and δ_a, the trapping mortality applied to the adult flies only, must be 6% per day. For a situation of infinite diffusion, where flies are infinitely mobile, the probability that a fly is within the radius of attraction at any time is $\pi a^2 n$, where n is the density of traps. The effective mortality is then $0.6\delta_a \, \pi a^2 n$ and the density of traps that is needed to reduce a population at a rate s is:

$$n = (r + s)/(0.6\delta_a \, \pi a^2 n).$$

If the mortality rate within the radius of attraction is 1.0 per day, population growth is 1.7% per day and the desired reduction rate is 99.9% per year (so that s is 1.9% per day), the equation can be applied to show that trap density should be greater than $0.019/a^2$ or the average distance between traps should be about $7.2a$. Thus, if the radius of attraction for baited traps is 100 m, 1.9 traps km^{-2} or a mean trap spacing of about 720 m would be required. If the radius of attraction for unbaited traps comes down to 10 m, 190 traps km^{-2} would be required, which would not be economically viable. Williams *et al.* (1992) also modelled other possible scenarios, including one for relatively immobile flies such as *G. f. fuscipes*.

Glasgow and Wilson (1953) reviewed methods of controlling tsetse populations by simulating artificial predation by humans, such as trapping or using cattle coated with insecticide as impregnated screens. They concluded that, theoretically, reduction of an animal population by artificial predation would follow a geometric progression. However, for tsetse populations that breed throughout the year, rather than having a breeding season, and with puparia remaining in the soil for 4–5 weeks, the geometric fall in population would only occur for the first 3 weeks and would then stabilize during the following weeks. Later there would be a rapid fall followed by a repetition of the above pattern. The duration of the falling and stable periods would depend upon the exact duration of the puparial period.

16.8   PROSPECTS FOR THE FUTURE

Tsetse control using impregnated targets or traps, together with odour attractants where appropriate, appears to be a very promising method: it is technically fairly simple yet efficient and relatively cost-effective. The techniques are also attractive because of the minimal environmental pollution that they are likely to cause. However, unless good supervision is maintained, traps or targets may become ineffective after a few months; reimpregnation may not be carried out adequately, or at necessary intervals, and target or trap cloth material can quickly rot, particularly in the hot and humid regions of West Africa. Successful control has frequently been demonstrated in trials throughout Africa, yet examples of long-term control by national institutions or government departments, unaided by outside organizations, are rare at best. Management, or supervision, of any tsetse control scheme thus remains one of the main constraints to their widespread successful use.

A second aspect which has proved to be a widespread and serious obstacle to successful, sustainable use of traps and targets for tsetse control is that of theft. In contrast to experimental trials, in many larger-scale target control schemes (where, of necessity, target supervision is reduced) theft of traps or targets has, after an initial successful period, become so prevalent as to render the tsetse control scheme ineffective. An example of this was in Somalia in a tsetse control scheme carried out in the Webbe Shebelle area in the 1980s. Theft of targets also rendered tsetse control ineffective in the Ghibe valley, southwest Ethiopia, after an 18-month period of successful control (Leak *et al.*, 1996).

Community participation schemes (discussed further in Chapter 19), in which affected communities are educated about the target or trap schemes and their potential beneficial effects, have often been introduced to attempt to overcome this problem, sometimes with a degree of success, but theft remains a major drawback to this technique. An alternative approach, which is discussed in the next section, is to apply the insecticides to cattle using the 'pour-on' technique to create living, moving targets with built-in odour attractants.

Among other possibilities for control is the feasibility of using odour-baited trees as targets, and the responses of tsetse to baited trees have been studied (Vale, 1991). Neither *G. m. morsitans* nor *G. pallidipes* were caught landing on the bases of trunks but they were caught flying within 1 m of the trunk. Catches increased further when the trunks were blackened and baited with odour attractants but were still only about 30% of the catches from odour-baited black cloth targets. Vale *et al.* (1994) subsequently showed that the presence of tall tree trunks had an inhibitory effect on numbers of flies contacting short insecticide-impregnated trunks. Only 20% of attracted flies landed on the base of trunks, and Vale concluded that the use of odour-baited trees for tsetse control would be uneconomical.

With the same objective, Vale experimented with insecticidal paints to produce cheaper, longer lasting targets than the cloth ones currently used. It was suggested that painted targets could last for at least 18 months compared with 3–6 months for cloth targets before they need either reimpregnating or replacing entirely.

Plastic materials have been used to construct more durable bipyramidal traps for Mbororo livestock owners to control tsetse in the Central African Republic. The Mbororo are an important group of nomadic cattle keepers who recently started keeping cattle in tsetse-infested areas of the country (Gouteux and Le Gall, 1992). Since 1963, the number of cattle, mainly Mbororo zebu, has increased from 400,000 to approximately 2 million, associated with a decreased distribution of *G. m. submorsitans* in the Central African Republic (Gouteux *et al.*, 1994).

Chapter 17

Application of Insecticides to Livestock

17.1 POUR-ON INSECTICIDES

As with so many ideas for tsetse control that are currently being developed, that of control using cattle covered in insecticide is not new. Recent developments in the use of pour-on insecticides came too late to receive attention in reviews of tsetse control methods by Allsopp (1984) and Jordan (1986) but are referred to in volumes 4 and 5 of the FAO tsetse training manuals (FAO, 1992a,b).

17.1.1 General principles of the technique

The principle is simply that tsetse coming to feed on cattle or other treated domestic livestock will be killed by picking up a lethal deposit of insecticide on the ventral tarsal spines and on pre-tarsi whilst feeding. Alternatively, they will be repelled by the insecticide and will therefore not attempt to feed. Repellents used as such against tsetse are discussed in Chapter 16. Whilst a repellent effect may protect treated livestock, control of the tsetse population depends upon: (i) a relatively large proportion of feeds being taken from domestic rather than wild animals; (ii) a sufficient proportion of the livestock population being treated; and (iii) a sufficiently low level of reinvasion. The treated livestock will then be equivalent to moving insecticide-impregnated targets, complete with built-in odour attractants. As cattle tend to aggregate in herds, a higher number of treated livestock may be required than the number of cloth targets that would be required in an area. None of these parameters has yet been adequately defined and there are, therefore, no clear guidelines to determine how many cattle need to be treated in a given area. There are rough estimations of the number of cattle requiring treatment in relation to the total number of potential hosts in an area, and of the theoretical time that would be taken for the tsetse population to be

371

eradicated. In practice, control rather than eradication is likely in most situations, due to problems with reinvasion from untreated areas.

17.1.2   Early attempts to use the pour-on technique

Whiteside (1949) pioneered the pour-on technique, by treating cattle with a 9% w/v solution of pure DDT mixed with a resin based on groundnut oil. The solution was applied at a rate of 110 cc per ox (450 mg 0.09 m^{-2} body surface). This is equivalent to 9.5–10 g of pure DDT per ox. Three hundred and forty oxen were treated, herded at a density of 176 km^{-2}, or 1 ox per 4 ha. The rate of reduction of the tsetse population achieved was slow, though the apparent density of female *G. pallidipes* was reduced by 70% in 3 months. The tsetse population returned to its normal level 3 months after stopping the experiment. The mixture quickly lost its insecticidal activity, but the cattle showed visibly improved condition and none was lost from trypanosomosis. No ill effects of the DDT were noticed after weekly treatment for a 2-month period. Whiteside considered that the rapid recovery was partly due to reduced density-dependent factors affecting the tsetse population. The results fell below expectations because of the low persistence of the insecticide and because of the small proportion of feeds believed to have been taken by tsetse from the oxen. Whiteside concluded that under those circumstances control would prove too lengthy and expensive, but thought it could be more successful against *G. morsitans* and *G. swynnertoni.*

Burnett (1954) repeated Whiteside's experiment against *G. swynnertoni* in Tanzania using a 10% solution of DDT. His experiment was carried out on a small scale, however, in which a maximum of only 26 cattle were treated. He failed to achieve effective control despite a promising beginning. The efficiency of cattle as poisoned bait decreased, possibly due to loss of insecticide from the skin, and also because the tall grass in the area might have reduced the visibility of cattle to tsetse. Finally, it was believed that tsetse flies immigrated into the area.

In America a similar method had been used earlier, against the horn fly, *Siphona irritans*, and against ticks, apparently successfully (Matthysse, 1946).

In addition to external application of DDT to cattle as a means of control, experiments in which DDT and Gammexane (Gamma BHC) were fed to cattle have been carried out, to protect against both trypanosomosis and East Coast fever (Wilson, 1948). Gammexane reduced both tick and tsetse numbers for up to 25 days, but DDT was less effective. These insecticides and their method of application could no longer be considered due to their persistence and potential environmental consequences, although studies based on the same principle have continued, with the systemic use of ivermectin described in section 17.4.

17.1.3   Later developments resulting in 'revival' of the technique

The initial pour-on trials came shortly after the discovery and mass production of DDT, one of the first cheap, persistent insecticides. However, the residual effect, which was desirable from the control entomologist's point of view, subsequently resulted in its withdrawal from widespread use for environmental reasons. The development of synthetic pyrethroids revived the pour-on technique. The so-called first-generation pyrethroids were rather unstable, being broken down readily by UV light, whilst the newer third-generation pyrethroids (Table 17.1) are more stable and can have useful spreading properties. Tsetse flies are very susceptible to pyrethrins (100–500 times more so than the housefly, *Musca domestica*) with LD_{50}s at doses of 0.002 and 0.0012 µg per female of unsynergized and synergized pyrethrum, respectively (Burnett, 1961a). Used as a residual application against tsetse, deltamethrin is hundreds of times more toxic than dieldrin (Zerba, 1988). Van den Bossche (1988) fed *G. tachinoides* on pigs that had been dipped in deltamethrin and obtained promising results with a long-lasting knock-down effect. Mammalian toxicity of the insecticides used is low and data from the manufacturers indicate that deltamethrin residues in meat of treated animals are not detectable at the level of 0.001 ppm, nor are these insecticides accumulated in the food chain. Despite that encouraging evidence, concern is still raised in some quarters about the application of insecticides directly to domestic livestock. Applications of pour-ons to wild hosts as well as to domestic livestock would obviously help to control tsetse enormously if that were possible, and it is therefore interesting to note that a self-medicating applicator has been described for applying permethrin acaricide for tick control on deer in North America (Sonenshine *et al.*, 1996).

The new formulations of pyrethroids for pour-on application move quickly through the skin, so that an application to the backline will soon spread to other areas such as the lower parts of the body, where tsetse flies are more likely to feed. Studies using radio-labelled cypermethrin to investigate movement in sheep showed that cypermethrin moves radially within the stratum corneum of the epidermis at more than 11 cm h^{-1} (McEwan Jenkinson *et al.*, 1986).

Table 17.1. Pour-on insecticides.

Insecticide	Trade name	Manufacturer	Persistence*
Deltamethrin	Spot On®	Cooper/Pitman-Moore	60–90 days
Flumethrin	Bayticol®	Bayer	< 20 days
Cypermethrin	Ectopor®	Ciba-Geigy	14 days
Cypermethrin	Barricade®	Shell	
Cyfluthrin	Bayofly®	Roussel-Uclaf	< 40 days
Alphacypermethrin	Renegade	Shell	60 days

* Data from Bauer (1993).

Cyfluthrin, marketed as Solfac® or Bayofly®, is produced to control nuisance flies and there is little published information on its efficacy against tsetse flies.

To be economically viable, residual deposits of synthetic pyrethroids from dipping or spraying would have to control tsetse (or at least trypanosomosis in cattle) effectively, with little modification or additional expense to existing dipping or tick control strategies. Although pyrethroid formulations for treating cattle by dipping or spraying are much cheaper than pour-on formulations, a major difficulty (other than on commercial ranching systems) is that in many African countries dips and dipping services are non-existent or do not function well. The example given below of a trial in Zimbabwe may be exceptional, as dips and veterinary services have been well maintained and organized, but is an important consideration in many traditional cattle-keeping areas elsewhere.

In many situations, pour-on insecticides may effectively complement, or replace, control using odour-baited impregnated traps and targets. Such situations are likely to be on large commercial, well-managed ranches or dairy farms. In peri-urban dairy systems, pour-on insecticides are already being used by small-scale farmers as acaricides also providing some degree of biting and nuisance fly control. However, there are difficulties of organization and administration of the treatments for sustained use in rural village production systems, despite the success of experimental trials.

17.1.4   Recent examples of pour-on trials

Control of Glossina pallidipes *in Zimbabwe using deltamethrin as a dip or spray*

Field trials using deltamethrin were carried out in Zimbabwe after it was noticed that dip washes containing deltamethrin for tick control resulted in prolonged knock-down or mortality of tsetse flies even at very low concentrations. The formulation used for tick control by dipping or spraying, with an advertised effect against biting flies, was originally marketed by Coopers as Decatix® and has been available for some years. A small-scale trial was conducted in an area of increasing trypanosome prevalence with 350 cattle dipped weekly in deltamethrin (Thomson *et al.*, 1991). There was a marked decline in trypanosome prevalence in the herd compared with a threefold increase in herds from adjoining dips.

The results of an experiment using an untreated control ox, an ox sprayed with deltamethrin wash and an ox treated with a 1% deltamethrin pour-on, marketed as Spot-on®, were as follows:

- The mortalities of fed *G. pallidipes* from the deltamethrin-sprayed ox and the pour-on ox were 97% and 80% from days 1–4 post treatment, respectively.

- Knock-down of fed *G. pallidipes* from the deltamethrin-sprayed ox and the pour-on ox at 12 h exceeded 90% for 15 days and 40% for 45 days.
- There was evidence of knock-down in non-fed flies from the sprayed ox (7%) and the pour-on ox (20%) at 70 days post treatment.

The effects of deltamethrin Decatix® spray and the pour-on were similar but the latter has a longer residual effect (Bauer *et al.*, 1992, 1993). Spot-on® is a more expensive formulation, but because of its greater residual effect its use at longer intervals, up to 90 days, might be a more economic treatment strategy. There are indications from unpublished work that its persistence has been overestimated.

In a large-scale trial in a 2500 km² area of Zimbabwe, where trypanosomosis was endemic, 13 dip tanks were filled with deltamethrin at a concentration of 0.00375%, replacing dioxathion, which was formerly used for tick control (Thomson and Wilson, 1992a,b). Approximately 80% of the 22,000 cattle from herds in the area were dipped every 14 days. Following this treatment, and the cessation of prophylaxis with isometamidium, the number of cases of trypanosomosis fell from 257 in May/June 1986 to about 35 in August 1987. Subsequently, monthly treatments with the pour-on formulation were applied rather than 2-weekly sprayings (Thompson *et al.*, 1991). Again, the knock-down effect contributed to the treatment's success (Thomson and Wilson, 1992a). The deltamethrin pour-on (applied as a dip) finally eliminated trypanosomosis in cattle over a wide area (Thomson and Wilson, 1992b). Deltamethrin was more effective than either alpha-cypermethrin or flumethrin. Thomson and Wilson (1992b) believed that if all cattle were adequately treated with deltamethrin, eradication could be achieved in 3 months. More conservatively, Hargrove and Vale (1979) calculated that if 2.5% of female tsetse in a population were killed per day, 95% of the population would be eradicated in 12 months. In areas where cattle distribution is irregular, it was suggested that the method be used together with impregnated targets in an integrated control approach.

Trials with pour-on insecticides in Burkina Faso and Kenya

Flumethrin (Bayer) was laboratory tested for pour-on application against *G. p. gambiensis* in Burkina Faso (Bauer *et al.*, 1988). After feeding tsetse 3 times on a zebu cow treated with flumethrin at a dose of 1 mg a.i. kg⁻¹ body weight, mortality and abortion rate increased significantly. An initial 90% knock-down effect decreased to 40% during the first 15 days. Flumethrin was then tested in a field trial in which approximately 2000 cattle were treated at monthly intervals. The cattle and tsetse populations were monitored to determine the impact on trypanosome prevalence and on tsetse density and tick populations (Bauer *et al.*, 1992). Three applications of flumethrin were sufficient to reduce trypanosome prevalence in cattle below 5% and no more tsetse could be caught after six treatments. Tick infestations were 3–10 times lower than in a control site at which another acaricide was repeatedly used during the trial. The control campaign was continued through participation of the local community when the field trial stopped.

A flumethrin pour-on (Bayticol®) has also been tested in an area of high challenge from *G. pallidipes, G. brevipalpis* and *G. austeni* at the coast of Kenya, where 40% of bloodmeals were from cattle. A flumethrin dose of 1 mg kg^{-1} was applied fortnightly to a total of 2000 cattle and after 1 year the apparent tsetse density was reduced to 87.5% of that in the 2 weeks before pour-on application. Trypanosome prevalence in cattle was reduced by about 70% (Löhr *et al.*, 1991).

In laboratory tests of deltamethrin Spot-On® in Burkina Faso, more than 90% mortality was achieved for up to 20 days post treatment and over 50% for up to 59 days post treatment (Bauer *et al.*, 1992). A significantly high knock-down effect was observed for up to 75 days. Sunlight did not significantly affect performance of the insecticide and flies repeatedly landed on treated cattle (Bauer *et al.*, 1992).

In the field trials that followed, Spot-On® was applied to 1500–2000 cattle in an area of high densities of *G. m. submorsitans* and some *G. p. gambiensis,* first at monthly intervals and later at 2-monthly intervals. This resulted in a decrease in apparent tsetse density from 54.2 to 0.06–2.0 flies trap^{-1} day^{-1} after 11 months. Most of the tsetse caught after 11 months were *G. p. gambiensis*, which survived in some areas where it was feeding mainly on monitor lizards (Bauer *et al.*, 1995).

Control of Glossina austeni *on Zanzibar using deltamethrin pour-on insecticide*

Deltamethrin pour-on was used to treat livestock on Zanzibar island (Schönefeld, 1988) for the suppression of *G. austeni*. A 1% w/v formulation was applied to all available cattle older than 6 months and to goats and donkeys in five consecutive cycles at intervals of 15–18 days. The intended dose rate was 10 ml 100 kg^{-1} estimated body weight. A dose of 5 ml was applied to each animal and small additional quantities were applied to certain areas of the body. There was an immediate and marked effect on the *G. austeni* population and 3 months after the start of the project no tsetse were caught at the monitoring sites. The transmission cycle appeared to be broken and a reduction in trypanosome prevalence in the cattle was achieved. However, trypanosomes were later detected in cattle even though no tsetse flies were initially caught. Eventually, *G. austeni*, which is difficult to catch, was detected. An area of inadequate contact between cattle and tsetse in Jozani forest was believed to be the focus of a residual fly population.

Control of G. pallidipes, G. f. fuscipes, G. m. submorsitans *and other biting flies with cypermethrin pour-on insecticide in Ethiopia*

Like deltamethrin, cypermethrin (a lyophilic synthetic pyrethroid, insoluble in water and non-volatile) is strongly bound to the sebum in an animal's coat. Movement of the ears, rubbing of animals against each other and grooming will disperse the active ingredient over the body within 24–48 h. Cypermethrin is claimed to have a touch-dependent repellency (Küpper and Harbers, 1987); thus, flies can land on an animal but they will normally fly away before they feed and will die later. Trials have been carried out in Ethiopia with a cypermethrin-based pour-on (Ectopor®, Ciba-Geigy Ltd, Switzerland), at a site with a high prevalence of multi-drug resistant trypanosomes. Because of the failure of trypanocidal drug treatments, trypanosome prevalence in cattle was steadily increasing before the trial (Leak *et al.*, 1995). The effect of a monthly application of cypermethrin was a 98% decrease in apparent density of the main vector, *G. pallidipes*, and a 70% reduction in trypanosome prevalence in cattle. In addition to the reduction in trypanosome prevalence, farmers reported a significant improvement in the condition of their cattle and that their cattle were able to graze much more satisfactorily, as other nuisance flies bothered them less. At that site, a significant reduction in numbers of Tabanidae and *Stomoxys* spp. was observed (Leak *et al.*, 1995). The improvement in body condition and productivity of cattle was monitored and is referred to in Chapter 12.

17.1.5   Control of biting and nuisance flies

'Rub-on' application of permethrin for controlling nuisance flies in the Philippines

Permethrin has been used as a 'rub-on' for control of biting flies on horses in the Philippines (Lang *et al.*, 1981). Treatments of 0.5, 1.0 and 1.5% permethrin emulsions gave 7–10-day control of *Stomoxys calcitrans* on about 100 horses at an American air base in the Philippines. *Tabanus megalops* and other horseflies were controlled for 3–6 days following rub-on or spray-on treatments of 1% permethrin. These results would not seem encouraging for tsetse control because of the relatively low period of protection, which would make its use both impractical and uneconomic. They observed 'no obvious permethrin repellency to flies'.

As already referred to, the pour-on insecticides are effective in controlling biting and nuisance flies which may transmit trypanosomes mechanically, but also contribute significantly to production losses in domestic livestock throughout the world due to disturbance of grazing cattle. Effects of these flies on domestic livestock production were reviewed by Steelman (1976) who cited losses to cattle production caused by the horn fly *Haematobia irritans* alone, estimated at US$179,000,000 in 1965.

17.1.6   Advantages and disadvantages compared with other methods

Pour-on treatments provide an extra benefit, compared with targets, as nuisance flies and ticks may also be controlled. In some cases there appears to be an improvement in the body condition of cattle treated with pour-on insecticides even when the prevalence of trypanosome infections has not been significantly reduced (Baylis and Stevenson, 1997; Leak *et al.*, unpublished data). This could be due to a decrease in numbers of other biting and nuisance flies which are considered to be major pests causing production losses (Drummond, 1987). The pour-on technique is relatively recent and its success depends upon the specific ecological and epidemiological conditions of the area where it is to be used. Consequently, a considerable amount of further research is required to determine the numbers of cattle that would need to be treated in a given situation, and how the technique may be integrated with other techniques for controlling or preventing reinvasion from tsetse-infested areas where cattle/fly contact is low.

The technique may be less susceptible to disruption than target or trap methods, as there are no targets or traps to be stolen. Additionally, the livestock owners are likely to have a greater interest in ensuring that their animals are treated. On the other hand, the loss of immunity to tick-borne diseases has also to be considered, as serious consequences could result from the breakdown of a pour-on treatment regime, particularly in areas where East Coast fever is endemic. Such a loss of endemic stability could result from any acaricide used, not specifically to pyrethroids used for tsetse control. Furthermore, as some deltamethrin pour-on tsetse control formulations may be persistent enough to give an adequate knock-down effect for up to 90 days, their less frequent use at a lower dose than that recommended for tick control may not reduce tick numbers to levels which would threaten endemic stability. This is particularly so as calves under one year of age would not normally need to be treated with the pour-on, and these are the animals most at risk from tick-borne diseases.

The possibility of ticks developing resistance to acaricides used for tsetse control is another concern. Resistance to a number of acaricides has already developed (Lourens, 1980). All but one occurrence of resistance to acaricides in ticks has been due to single genes, and both very low and exceptionally high selection pressure may tend to delay the appearance of resistance to DDT (Stone, 1995).

If domestic livestock were infrequently fed upon by tsetse, pour-on application could not be expected to have a long-term effect. However, there is growing evidence that where many cattle are kept in tsetse-infested areas they are fed upon to a considerable extent (e.g. Van den Bossche and Staak, 1997). Synthetic pyrethroids, particularly pour-on formulations, which are manufactured using costly vegetable oils such as coconut oil, are generally expensive. The cost may be lowered if the market expands, and investigations are under way to look for cheaper formulations.

Additionally, lack of availability of pour-ons, due to import controls and lack of foreign exchange or due to government policies on insecticides or acaricides, has often prevented their more widespread use. The requirement for communal use of pour-on insecticides on cattle in order to have an impact on tsetse populations means that there may not be a significant rise in the quantity of these insecticides used in Africa through private consumption from local retail outlets. An exception could perhaps occur in some areas in which grade livestock kept under intensive zero-grazing conditions might benefit from such treatments, or where they are primarily used for tick control. Any significant increase in their usage is likely to come from large-scale, donor-funded tsetse control projects such as the EU-funded Regional Tsetse and Trypanosomiasis Control Project (RTTCP) in southern Africa. However, pour-on insecticides still seem to be looked upon unfavourably, along with other insecticides, by the donor community.

17.2 REPELLENTS

17.2.1 Repellency of pour-on insecticides

A repellent effect of pour-ons may still lead to fairly high mortality in a tsetse population if most livestock in the area are treated and there are few or no alternative hosts, as flies would either die of starvation or be forced to land on a treated surface. The possible repellent effect of pour-ons, sometimes termed the 'hot-foot effect', could be quite significant. Any significant repellent effect which prevented flies from feeding on a treated animal would confer protection to such animals whilst not necessarily reducing the overall tsetse population in the area (provided sufficient alternative hosts were available to prevent death by starvation). Alternatively, if the repellent effect allowed tsetse to attempt to feed but not successfully complete feeding, trypanosome transmission could be increased as flies making many attempts to feed before being successful could transmit trypanosomes in their saliva each time they probed. Recent work on Galana ranch in Kenya showed no significant hot-foot effect, as no effect of pour-on application, either deltamethrin or cypermethrin, was observed on the proportion of *G. pallidipes* coming to treated oxen that successfully fed. There was a small difference between the proportions of flies that successfully fed, however, in comparison with the proportion of flies successfully feeding on untreated control oxen (Baylis *et al.*, 1994). As only eight oxen were treated, it is possible that the difference would have been significant had a larger number of animals been used. No significant difference in the numbers of flies coming to feed at treated or untreated oxen was observed in that study, nor was there any difference in the rate of anti-fly movement by the treated oxen. Baylis *et al.* (1994) did show that the feeding success of *G. pallidipes* was density dependent, because as the numbers of tsetse approaching the oxen rose, their rate of anti-fly movement increased and the proportion of flies that successfully fed decreased.

A repellent effect of some insecticides, including some of the synthetic pyrethroids and natural pyrethrum, may be useful to prevent disturbance from biting flies and to give temporary protection against tsetse; but alone, it is unlikely to provide adequate long-term control. There is, at present, little evidence for the repellency of some of the pour-on formulations of synthetic pyrethroids, although there does seem to be some repellency from high doses of deltamethrin (at least for a short period). Some formulations of cypermethrin are said to have a contact-dependent repellency. Where there is a repellent effect cattle would be protected as long as the active compound is in use. There may be no long-term effect on tsetse populations with repellents unless the treated animals were the predominant hosts and tsetse mortality were to rise due to starvation.

17.2.2 Early experiments using repellents

The repellent activity of pyrethrum against tsetse was recognized and tested along with over 140 other compounds in the 1940s (Hornby and French, 1943; Holden and Findlay, 1944). Apart from the necessity for a repellent substance to be persistent and biologically active for a sufficient length of time, the main requirement noted by Hornby and French (1943) was that the substance must be cheap, non-toxic to mammals and easily applied as a dip or a spray. Of the substances tested, only pyrethrum used as a 2% emulsion was an effective repellent.

Early experiments gave disappointing results. Those described below, with now obsolete insecticides, all suffered from the lack of persistency of the repellent or insecticidal activity observed. Thus, a 2% pyrethrum emulsion sprayed on donkeys to protect against *G. morsitans* gave promising but inconclusive results. Formulated in a cream, pyrethrum gave protection to humans for up to 6 h against bites of *G. palpalis* (Holden and Findlay, 1944).

A mixture of indalone, dimethyl phthalate (DMP) and ethylhexane-diol-1,3 (known as 'Formula 622') significantly reduced the number of tsetse that settled on and bit treated animals. However, even 1 h after application several flies did bite (Findlay *et al.*, 1946). None of the repellents offered complete protection against tsetse bites, and only for a few hours against other haematophagous insects.

Exposure to sunlight and to sweat could reduce their efficiency. To test this, Findlay *et al.* made their 'flyboys' dance strenuously in the shade for 30 min, but observed no loss of efficiency.

Traditionally, oil from the seeds of the African mahogany (*Khaya senegalensis*) has been used by cattle owners in southern Sudan as a repellent to protect against Tabanidae (Lewis, 1953). Lewis also referred to an experiment using dimethyl phthalate 'liberally smeared' on a human arm, upon which a container of *Tabanus taeniola* was placed. He stated that the tabanids immediately bit through the repellent!

17.2.3 Recent experiments and experiences with insecticidal repellents

Experimental trials with repellents carried out by Wirtz *et al.* (1985) showed that permethrin had both an insecticidal and repellent activity and resulted in a 96.7% mortality in *G. m. morsitans* after 24 h, compared with around 10% for the other compounds tested. Among the compounds tested, none was significantly more effective than DEET (*N,N*-diethyl-*m*-toluamide). Diethyltoluamide is one of the most widely used and effective insect repellents for other insects, especially mosquitoes. In Ethiopia, Scholdt *et al.* (1975) tested five repellents, including DEET, on wide-meshed jackets as 'space repellents' rather than contact repellents. They obtained better results with a di-isopentyl malate compound than with DEET, but the former compound gave only an average 83% protection against *G. m. submorsitans* for 8 days.

Trials using permethrin-impregnated clothing (0.125 mg a.i. cm^{-2}) in combination with formulations of diethyltoluamide repellent were also carried out in Zambia for protection against *G. m. centralis*. People wearing treated clothing were driven slowly through the bush in a vehicle and a 91% protection against bites was reported. However, the results were not significantly different from using the repellent alone (Scholdt *et al.*, 1989).

In an investigation of insecticides used for impregnating traps, no repellency was reported at the dosages normally used. At higher concentrations, two synthetic pyrethroids and two organochlorine compounds exhibited repellence (Dagnogo and Gouteux, 1983). It was initially thought that tsetse (*G. morsitans*) did not have contact chemo-receptors on the tarsi, which might be important for repellency (Dethier, 1954), but these were later reported from the tarsi and other sites on the legs (D'Amico *et al.*, 1992).

The economical application of expensive pyrethroids depends on the frequency of application and the protection offered to the animal from other important pests. Pyrethroids are used for tick control and some are repellent to some tick species. Pyrethroids decompose very fast due to UV radiation from sunlight and they also oxidize very readily. To overcome this, Galun *et al.* (1980) suggested microencapsulation of pyrethrum, an idea which has been tried with deltamethrin for mosquito control (Taylor *et al.*, 1986). Microencapsulation may cut down absorption by the skin as well as photo-decomposition and oxidation.

Torr *et al.* (1996) tested synthetic olfactory repellents, amongst which pentanoic acid or hexanoic acids, acetophenone or 2-methoxyphenol reduced catches at an unbaited trap by 40%, 75% and 60%, respectively. They concluded that it was unlikely that any repellents would provide a useful degree of protection against trypanosomosis where tsetse were abundant. However, repellents, including the currently unidentified ones from human odour, could be useful in areas where tsetse density or trypanosome infection rates were low when used together with trypanocidal drugs.

17.3   IMPREGNATED EAR-TAGS AND TAPES

Ear-tags impregnated with insecticides have mainly been used for preventing disturbance by nuisance and biting flies to particular groups of animals (for example, horses or commercial dairy cattle) rather than for large-scale control of vectors transmitting disease. However, attempts have been made to assess the feasibility of impregnated ear-tags for control of ticks and tsetse flies, so far with little or only partial success. Some examples of these attempts are described here.

17.3.1   Experimental use of ear-tags for control of tsetse flies

There have been few trials with insecticide-impregnated ear-tags, perhaps because of the initial lack of success. This is not unexpected as tsetse feed on the lower half of cattle and are therefore not likely to come into close contact with ear-tags.

In Zimbabwe, ear-tags impregnated with 4% deltamethrin proved ineffective in preventing *Glossina pallidipes* or *G. m. morsitans* from landing and feeding on oxen (Thomson, 1987). Knock-down of *G. pallidipes* did not exceed 41% throughout the study, whilst mortality was below 16%, and it was concluded that the failure of ear-tags to control tsetse was due to poor translocation of the insecticide to the preferred landing sites of the flies.

In a similar trial on the Marahoue ranch, in Côte d'Ivoire, a single ear-tag impregnated with 1.2 g of photostable permethrin a.i. per ear-tag 'did not improve the general state or the rate of *Trypanosoma* induced parasitaemia in N'Dama cattle' (Mayer and Denoulet, 1984). The Marahoue ranch is infested by *G. p. gambiensis*, *G. tachinoides*, *G. longipalpis*, *G. m. submorsitans* and species of the *fusca* group. The action on tsetse seemed to be negligible.

Cypermethrin-impregnated 'Flectron' ear-tags, tested in West Africa, appeared to bring about a reduced trypanosome prevalence and an improvement in condition in N'Dama cattle, compared with untreated control animals (Küpper and Harbers, 1987).

Dolan *et al.* (1988) used fenfluthrin ear-tags on Boran steers on Galana ranch in Kenya and obtained similar results to those of Küpper and Harbers (1987): a 39% reduction in trypanosome prevalence was observed at a period of high challenge. Two ear-tags gave better results than one. Treated cattle had fewer infections than the group with no tags.

As tsetse usually land and feed on the lower parts of a cow – the belly and legs – it is unlikely that impregnated ear-tags would be effective unless the insecticide also had a very strong repellent effect or could efficiently disperse from the ear-tag through the skin of the animal. Insecticide may spread by diffusion via sebaceous secretions as well as by self-grooming and contact with other treated cattle (Taylor *et al.*, 1985). Despite the reported success in some instances, it is unclear how this approach could be successful. A problem with the use of ear-tags was that the available methods of application resulted in repeated perforation of the animals' ears.

17.3.2 Control of biting and nuisance flies

Ear-tags have been used successfully against biting flies (Elger and Liebisch, 1982; Liddel and Clayton, 1982) and for control of haematophagous flies on horses in India (Parashar *et al.*, 1989). In the latter trial, ear-tags impregnated with 10% (w/w) permethrin were effective for 1–2 months against *Stomoxys irritans* and *Haematopota dissimilis*, but completely ineffective against *Hippobosca maculata*. This was claimed to be the first reported use of ear-tags for control of haematophagous flies on horses. In contrast to tsetse, nuisance flies frequently congregate around the heads of livestock, close to the ears.

Permethrin tapes applied to the tails of beef and dairy cattle have also been tested for controlling stable and horn flies (Hogsette and Ruff, 1987). These tapes contained ampoules of viscous preparations of permethrin. *Haematobia irritans* were controlled for 9 weeks on beef cattle and 4 weeks on dairy cattle. When applied via the tail, the same permethrin preparation was more efficacious than when applied via the ear.

17.4 IVERMECTIN

The use of ivermectin for tsetse and trypanosomosis control falls into the same category as other insecticides described in this chapter, as it is applied to livestock and has a potentially lethal effect on tsetse. Ivermectin is one of a family of recently discovered drugs, the avermectins – fermentation products isolated from microbial cultures of *Streptomyces avermertilis*. These drugs are active against a broad spectrum of nematode and arthropod parasites. Ivermectin is also effective against *Onchocerca volvulus* (the parasite causing river-blindness in humans) in a slow-release form. The effectiveness against arthropods has not been as fully investigated as against nematodes although activity against a wide range has been demonstrated (Strong and Brown, 1987). For tick control, treatment of a host animal with ivermectin does not cause prompt detachment but disrupts their physiology.

A summary of experimental trials with ivermectin is given in Table 17.2. A critical aspect of the use of ivermectin for control of tsetse flies is that of dosage, as concentrations sufficient to cause mortality of tsetse flies are close to the limits that can be used before toxic effects are observed in the host. Furthermore, treatment at frequent intervals, at doses higher than those normally recommended, would be costly.

Preliminary experiments showed high mortality in *G. p. palpalis* and *G. m. morsitans* after feeding on treated hosts, as well as a dose-dependent

Table 17.2. Experimental data from trials with ivermectin for tsetse control.

Tsetse species	Stage	Host	Dosage	Mortality/fecundity	Remarks	Reference
G. p. palpalis	Teneral males	Goat	10 mg kg⁻¹ s.c.	100%	Died within 5 days	Van den Abbeele et al., 1986
G. morsitans	Males / Females	Defibrinated pig blood	0.1–1.6 mg ml⁻¹	100%	Dose-dependent decline in female fertility	Langley and Roe, 1984
	Teneral males		0.04 mg ml⁻¹		Repeated feeding	Distelmans et al., 1983
Tsetse		Goats and guinea pigs	2 mg kg⁻¹ / 2 mg kg⁻¹	Some mortality		Distelmans et al., 1983
Tsetse		Goats and guinea pigs	10 mg kg⁻¹	100%	Died within 5 days (recommended therapeutic dose)	Distelmans et al., 1983
G. p. palpalis	Females		0.2 mg kg⁻¹	No significant effect on fecundity		Van den Abbeele et al., 1986
G. p. palpalis	Females		1–2 mg kg⁻¹	Reduced fecundity		Van den Abbeele et al., 1986
G. p. palpalis	Teneral males	Pigs	3 mg kg⁻¹	100%	Up to day 8 p.t.	Van den Bossche and Geerts, 1988
G. tachinoides	Females	Pigs	3 mg kg⁻¹ up to 22 d.p.t.	Much lower mortality	Fecundity affected	Van den Bossche and Geerts, 1988

d.p.t., days post treatment; s.c., subcutaneous.

decline in fertility (Langley and Roe, 1984; Van den Abbeele *et al.*, 1986). However, subcutaneous injection of concentrations greatly in excess of the recommended clinical dose appear to be required to achieve levels lethal to feeding flies following a single bloodmeal (Distelmans *et al.*, 1983). Drug concentrations in the host blood of about 0.005 μg ml^{-1} (the maximum achieved in a cow after treatment with injectable ivermectin) reduced female fecundity by 50–100% but had no effect on tsetse longevity (Distelmans *et al.*, 1983; Van den Abbeele *et al.*, 1986; Van den Bossche and Geerts, 1988). Ivermectin formulated as Ivomec® (Merck Sharp and Dohme, USA), containing 1% ivermectin, is designed to be given to cattle subcutaneously at doses as high as 6.0 mg kg^{-1} with no ill effects; however, an acute toxicity syndrome in cattle has been elicited with ivermectin administered orally as a drench at a dose of 4.0 mg kg^{-1} (Campbell and Benz, 1984).

Where domestic animals constitute the major hosts of tsetse, ivermectin treatment might achieve a measure of fly population reduction through an effect on fecundity but would be unsuitable for use as a direct insecticide. High doses in cattle could be more effective but financial constraints and possible toxic effects would limit its use for killing adult tsetse directly.

Chapter 18

Non-insecticidal Methods of Tsetse Control

For environmental reasons there is greater interest at present in searching for methods of tsetse control that either use lower quantities of insecticide or are based on techniques which do not require them at all. The use of traps without insecticides was described in Chapter 16 (section 16.6); some other approaches are described in this chapter. Whilst some techniques, such as the sterile male release method, have been attempted in field trials, others are still very much in the research stage and may prove to pose insurmountable problems.

Biological control of tsetse, as for other insects, would be an attractive option as an alternative to the use of insecticides. Successful biological control of a number of insect pests has been achieved; for example, the cottony-cushion scale (*Icerya purbasi*) pest of citrus fruit was controlled by a coccinellid beetle predator, *Rodolia cardinalis* (Doutt, 1958; Van den Bosch and Messenger, 1973). Approaches that can be classed as biological control include genetic control, the use of natural enemies (parasites, predators, bacteria and fungi), bushclearing and elimination of wild hosts. Sustained attempts to control tsetse have only been made using the latter two methods, which are now out of favour for environmental reasons.

18.1 BUSHCLEARING

Theoretically, habitat alteration as a means of tsetse control could be more lasting than other methods, particularly if the altered habitat is maintained in a state unsuitable for tsetse by appropriate land use, such as cultivation. The disappearance of *G. m. submorsitans* from much of northern Nigeria, where it appears unable to exist above certain densities of human population, indicates that it is particularly sensitive to habitat changes (Jordan, 1986). *Palpalis* group tsetse, on the other hand, seem able to coexist with high human populations and to inhabit small areas of woodland and gallery forest. Bushclearing has been widely employed in the past, and more recently it has been a component of tsetse control schemes based on other techniques, in order to make an unsuitable tsetse habitat so that reinvasion will be prevented.

West Africa

In West Africa, partial bushclearing was used for tsetse control close to the northern climatic limits of tsetse distribution at Gadau, Northern Nigeria, where high mean temperatures and low relative humidity confine tsetse to small areas of forest within the savanna (forest islands). The limited extent of the forest islands and the unfavourable climate resulted in the species inhabiting those areas (*G. m. submorsitans* and *G. tachinoides*) being susceptible to control by bushclearing. Nash (1940) carried out an experiment in which thickets were cleared around a forest island and thickets acting as windbreaks were removed. No trees were cut down in the 'true' forest. This partial clearing altered the forest microclimate, increasing both the evaporation rate and temperatures. *Glossina tachinoides* were more susceptible to these climatic changes than *G. m. submorsitans* and when partial clearing was carried out on a large scale the former species was almost exterminated. At around the same time, partial clearing was used to control *G. tachinoides* and *G. palpalis* in the 'Anchau corridor' in Nigeria, the location of a settlement scheme 112 km long and an average of 16 km wide. All thickets and low-branching shade trees were removed, to allow the hot winds to penetrate streambeds and make the microclimate unsuitable for those species of tsetse. At the ends of stream systems within the area it was 'essential to put in a mile-long, ruthless, barrier clearing ... not a shrub is left standing' (Nash, 1948). Nash (1948) believed that whilst partial bushclearing was satisfactory in controlling *palpalis* group tsetse in the savanna zone of Nigeria it would not be suitable in the forest zone.

Discriminative bushclearing was used in Ghana to control human sleeping sickness around villages where close human–fly contact was a key factor in transmission (Morris, 1949). This bushclearing was thought to have been sufficient to cause the disappearance of *G. tachinoides* and *G. palpalis* from infested river systems.

East Africa

Bushclearing projects in East Africa were reviewed by Lewis (1953). Discriminative clearing was used experimentally for the control of *G. m. centralis* in Uganda, exploiting the knowledge that tsetse flies concentrated in certain areas of woodland during the dry season (Harley and Pilson, 1961), and based on the hypothesis that removing taller trees would render the habitat unsuitable for tsetse. The selective clearance of *Acacia* trees over 6 m tall from an area in Ankole district, Uganda, initially resulted in a substantial reduction of the tsetse population, but this was not maintained and in the southern part of the experimental area the reduction was less marked. Subsequently all the upper-storey trees were removed. Even this was unsuccessful, possibly because *Acacia hockii,* which only grew to 4–5.5 m high, was not removed and seemed to provide sufficient cover for a suitable tsetse habitat (Harley and Pilson, 1961).

In Tanzania, between 1923 and 1930, bushclearing was used to stop the spread of a sleeping sickness epidemic in Maswa district, the vector of which was *G. swynnertoni* (Chapter 12, section 12.11).

Bushclearing was also used in Tanzania to control tsetse on the Malialuguru (Malya) ranch after it became infested with *G. swynnertoni* and *G. pallidipes* that advanced into areas where there was extensive encroachment of thickets. Deaths from trypanosomosis in cattle rose from 0.4% (5/1187 head) in August 1931, to 8.5% (50/591 head) in November. As a result of this, most of the stock were moved from the ranch. Between 1932 and 1934 approximately 83 km² of bushclearing was carried out, allowing temporary reclamation of the ranch.

Despite the apparent technical success of some trials with bushclearing, Lloyd *et al.* (1933) concluded that the method was unsuitable as a control measure due to the expense and speed of reinvasion. Arguments against bushclearing are that it can encourage soil erosion, decrease soil fertility and adversely affect water supplies.

18.2   ELIMINATION OF WILD ANIMAL HOSTS

In the late 19th century, it is likely that in Zimbabwe and South Africa hunting resulted in an unintended contraction of tsetse distribution before the dramatic effects of the rinderpest pandemic (Phelps and Lovemore, 1994). Wild animal elimination was introduced thereafter as a deliberate measure to control tsetse in Zimbabwe from 1919. Initially, attempts were made to eliminate all potential hosts in an area. However, this strategy was refined after tsetse bloodmeal analysis techniques (Chapter 8, section 8.5) were used to determine the predominant host species. Instead of elimination of all large wild animals in an area, there was more selective hunting of only those animals constituting a significant part of the tsetse diet.

Attempts at tsetse control by elimination of their hosts started after Jack (1914) observed that tsetse density was closely related to that of wild animals. As vertebrate blood is the sole source of tsetse nutrition, the destruction of all vertebrates would inevitably lead to the disappearance of the fly through starvation (Bursell, 1961b). Bursell (1961b) regarded the population dynamics of tsetse to be a balance between two opposed rate processes:

● The rate of utilization of food reserves.
● The rate of replenishment of food reserves, which may be described as the frequency or the probability of host encounter.

The theoretical basis of control through elimination of potential hosts relies on reducing the probability of tsetse flies encountering and successfully feeding from a host animal. For a population to maintain itself at a steady level, females must live long enough to produce at least one female puparium; as the puparial sex ratio is approximately 1:1, this means that at least two puparia must be produced. The 'replacement threshold' theoretically requires a probability of encountering a host of 0.4–0.6. If an encounter probability in the region of 0.5 is applicable to the natural situation, on average a period of 2 days would have to be spent before a suitable host is encountered. Bursell (1970) made the distinction between the hunger cycle (the interval between successive bloodmeals) and the feeding span (the interval between starting to search for a host and death of the fly from starvation). Bursell derived a relationship between the daily probability of host encounter (P) and the average expectation of life. This simplified relationship did not take into account other causes of mortality. Taking into account the estimates of the rate of dispersal of tsetse populations, Bursell speculated that the probability of host encounter would be low, even in areas of high host density. Wild animal elimination could, therefore, be an extremely effective method of tsetse control (Bursell, 1970) and a number of extensive control operations using this technique were carried out, particularly in Zimbabwe, until this approach became unacceptable for environmental reasons.

Despite its theoretical feasibility, elimination of wild animals in areas where cattle are kept is unlikely to be useful. In Zimbabwe, cattle and donkeys were highly attractive to *G. m. centralis*; thus, this species could not be eradicated by the use of selective wild animal elimination alone in areas where those hosts were common (Robertson, 1983). Furthermore, there is unlikely to be a large population of wild animals in areas where domestic livestock are kept, and those that are present are likely to be hidden and difficult to eliminate. Removal of wild animals in these areas might increase the probability of domestic livestock being fed upon. In such a situation, it would be necessary to remove all domestic livestock before attempting to control the tsetse population through elimination of wild hosts.

18.2.1 Field operations

The Shinyanga game experiment in Tanzania (Harrison, 1936; Potts and Jackson, 1952), went on for many years and was a comprehensive study of relationships between wild animals, tsetse and their habitats. Herds of wild animals were followed and their movements mapped, in a similar way to that used by Wacher *et al.* (1994), described in Chapter 13. Vegetation characteristics were recorded and feeding habits of tsetse determined. The objective of this was to determine whether or not tsetse could be controlled by elimination of the major hosts of tsetse without killing all wild animal species.

Wild animals were shot in an isolated area of about 1550 km² over a period of about 5 years. The original density of wild animals was about 4 km⁻², which is low compared with densities in national parks. Populations of *G. m. centralis* and *G. swynnertoni* were eradicated from the area and the population of *G. pallidipes* was greatly reduced. Experimental cattle in the area remained free of trypanosomosis for 11 months, after which they were removed. The cost was US$32 km⁻². Despite the favourable results obtained, Potts and Jackson (1952) did not recommend the method.

The presence rather than absence of wild animals, particularly elephants, has been cited as a means of controlling tsetse populations (Ford, 1966) through their effect on the vegetation. In Murchison National Park in Uganda and in Tsavo Park in Kenya, bush destruction by elephants resulted in reduction of tsetse populations and the creation of grazing pastures. Obviously this is not an observation with very great practical application. Numbers of elephants are declining and they generally occur in national parks, where agricultural development is not going to take place.

Elimination of wild animals was widely practised in Zimbabwe from 1919 until the 1960s, when successful tsetse control by ground spraying with dieldrin replaced the technique. It is claimed that, by this means, tsetse were eradicated from about 26,000 km² of southern Zimbabwe between 1925 and 1933 (Cockbill, 1971), 11,600 km² in Uganda between 1946 and 1951 (Chorley, 1947; Robertson and Bernacca, 1958) and 1550 km² in Shinyanga (Potts and Jackson, 1952). Lewis (1953) described its use in these countries.

Selective shooting of the main hosts of tsetse, warthog, bushpig, bushbuck and kudu was undertaken at Nagupande, Zimbabwe (Cockbill, 1967), in an attempt to control trypanosomosis in cattle. Initially the numbers of tsetse caught rose, possibly because they were hungrier and therefore more likely to be attracted to the traps used for monitoring the tsetse population. The apparent density then fell by over 95%, remaining low for some years and associated with a decrease in trypanosomosis prevalence in cattle. The success of this experiment resulted in the temporary reintroduction of selective shooting in 1964.

In a similar experiment in Zimbabwe, to determine the effect of selective wild animal elimination, an area was enclosed by a warthog-proof fence and all warthog within the fence were removed over a short period (Vale and Cumming, 1976). The tsetse diet switched from 80% warthog to 40–80% Bovidae (mainly kudu), 20–50% elephant and 0–20% warthog. The following year, after elephants were driven out of the block, Bovidae formed 90% of the tsetse diet. The changes had little effect on the nutritional state of the tsetse compared with a control population.

Some unexpected effects of hunting, for tsetse control purposes, were observed with a duiker population in Zimbabwe. The level of hunting carried out actually stimulated growth of the duiker population (Child and Wilson, 1964a). In a similar wild animal control operation over a 2-year period in eastern Zambia, in which duiker (*Silvicapra grimmia*) were selected as the target host, hunting failed to remove more than the annual increment to the duiker population. It was therefore ineffective as a means of tsetse control (Wilson and Roth, 1967). Duikers are small antelopes that would be extremely difficult to eradicate.

Extensive shooting of wild animals in Uganda drew considerable opposition from conservationists, but Wooff (1967, 1968) argued that conservationists paid insufficient attention to the economic losses due to tsetse-transmitted trypanosomosis affecting cattle in the country. In Ankole, in a 7-year period starting in 1958, 44,973 animals of 12 species were killed, representing a density of 10 km^{-2}. Of these, 79% were small species, duiker, reedbuck and bushbuck. Wooff (1967, 1968) claimed to have eradicated *G. m. centralis* and *G. pallidipes* from between 21,000 to 27,000 km^2 of Ankole in southwestern Uganda through hunting, between 1946 and 1966. After 1963 more use was made of insecticidal techniques.

18.3   BIOLOGICAL CONTROL – NATURAL ENEMIES

Parasites and predators of tsetse flies were long ago observed in the field, leading to various experiments to investigate their usefulness for control purposes. Nolan (1977) tabulated pathogens of tsetse other than arthropods, with references and abstracts. Examples of arthropod parasites and predators (Table 18.1) are described below.

To be successful, biological control organisms generally have to originate from a different geographical or ecological area from the potential pest to be controlled. Otherwise it would be expected that the control and target organisms would be sufficiently well adapted to each other that no significant degree of control would easily be maintained over a long period. Most trials of parasites of tsetse flies have used insects that occur naturally in tsetse habitats, which may partly explain the lack of success.

18.3.1   Hymenopteran and dipteran parasites

Two species of a parasitic wasp of the genus *Nesolynx* (formerly classified as *Syntomosphyrum*; Boucek, 1976), *N. glossinae* and *N. albiclavus,* have been recorded as parasites of tsetse puparia. The former has an east–west distribution across Africa from Senegal to the Indian Ocean, whilst the latter has a north–south distribution along the eastern side of the continent. Both species occur in Kenya, Tanzania and Malawi. *Nesolynx glossinae* was first found by Fiske in Uganda in 1913 parasitizing puparia of *G. f. fuscipes.* The second species (named by Kerrich, 1961) in puparia of *G. m. morsitans* was found in Malawi (Waterston, 1916). In West Africa, *N. glossinae* was first reported by Nash (1947).

The life cycles of the two species and the biology of *N. albiclavus* were described by Saunders (1961a,b,c). The parasite was also described in some

Table 18.1. Arthropod parasites of tsetse.

Parasite	Country	Host	Reference
Chalcidoidea			
Eupelminus tarsatus	Malawi	*G. morsitans*	Lamborn, 1916
Anastus viridiceps	Zambia	*G. morsitans*	Waterston, 1915; Lloyd, 1916
Anastatus spp.	Nigeria	*G. palpalis*	Baldry, 1969b
Dirhinus inflexus	Ghana		Waterston, 1917
Chalcis amenocles	Ghana		Waterston, 1917
Mutillidae			
Mutilla glossinae	Zambia	*G. morsitans*	Turner, 1915; Eminson, 1915
M. glossinae	Malawi	*G. morsitans*	Lamborn, 1916
M. benefactrix	Zululand	*G. pallidipes*	Fiedler *et al.*, 1954
M. auxiliaris	Zululand	*G. brevipalpis*	Fiedler *et al.*, 1954
M. auxiliaris	Zululand	*G. austeni*	Fiedler *et al.*, 1954
Braconidae			
Coelalysia glossinophaga	Ghana		Turner, 1917
Nesolynx (*Syntomosphyrum*)			
glossinae	Malawi	*G. morsitans*	Lamborn, 1925
N. (*S.*) *glossinae*	Uganda	*G. fuscipes*	Fiske, 1913
N. (*S.*) *glossinae*	W. Africa		Nash, 1947
N. (*S.*) *glossinae*	Zululand	*G. pallidipes*	Fiedler *et al.*, 1954
N. (*S.*) *glossinae*	Zululand	*G. brevipalpis*	Fiedler *et al.*, 1954
N. (*S.*) *albiclavus*	Malawi	*G. morsitans*	Waterston, 1916; Kerrich, 1961
Mormoneilla vitripennis			
Bombyliidae			
Villa lloydi	Zambia	*G. morsitans*	Austen, 1923; Lloyd, 1913
Exhyalanthrax abruptus	Zambia	*G. morsitans*	Lloyd, 1916
E. abruptus	Tanzania	*G. morsitans*	Nash, 1930
E. abruptus	Nigeria	*Calliphoridae*	McDonald, 1957
E. abruptus	Zululand	*G. pallidipes*	Fiedler *et al.*, 1954
E. abruptus	Zululand	*G. brevipalpis*	Fiedler *et al.*, 1954
E. argentifrons	Nigeria	*Glossina* spp.	McDonald, 1957; Taylor, 1926
E. brevifacies	Zululand	*G. pallidipes*	Fiedler *et al.*, 1954
E. brevifacies	Zululand	*G. brevipalpis*	Fiedler *et al.*, 1954
E. brevifacies	Zululand	*G. austeni*	Fiedler *et al.*, 1954
Proctotrupidae			
Trichopria capensis	Kenya	*G. fuscipleuris*	Lewis, 1939
T. capensis		*G. brevipalpis*	
T. capensis	Zululand	*G. pallidipes*	Fiedler *et al.*, 1954
T. capensis	Zululand	*G. brevipalpis*	Fiedler *et al.*, 1954
Acari			
Erythraeoidea			
Leptus spp.	Kenya	*G. fuscipes*	Krampitz and Persoons, 1967

detail by Nash (1933b), including its laboratory rearing for biological control experiments. Immature stages of *Nesolynx* live as ectoparasites in the space between the pupa and puparium of the host. The life cycle lasts from 20 to 40 days and about 30 adults emerge from each parasitized puparium.

Nesolynx spp. were released in field trials to control tsetse in Malawi, Nigeria (Lloyd *et al.*, 1927) and Tanzania before the Second World War. Although in one of these trials an astonishing 13,750,000 parasites were released in 14 months, little was achieved in terms of control, probably because the parasite is unable to burrow sufficiently deeply in soil to parasitize tsetse puparia (Onyiah and Riordan, 1978) and, consequently, natural rates of parasitism do not exceed 3%. *Nesolynx* species can be highly effective at destroying tsetse in laboratory colonies where puparia are easily accessible.

After Lamborn (1925) reported *N. glossinae* and *Eupelmella tarsata* parasitizing tsetse, Ferriere (1935) suggested their possible role in biological control, but acknowledged that they would play only a minimal role in the natural control of tsetse populations. He also commented on another common parasite of Muscidae, *Mormoniella vitripennis,* which could not be used against tsetse, as the adults are unable to emerge from the hard tsetse puparia. Ferriere (1935) tabulated details of hymenopteran parasites of tsetse puparia known at that time; these are updated in Table 18.1.

A hymenopteran parasite, *Anastus* sp., of the family Eupelmidae, was found parasitizing the puparium of *G. palpalis* in West Africa (Baldry, 1969b). However, tsetse were not believed to be its usual hosts and the parasite would therefore be unsuitable for biological control.

Several species of Mutillidae have been described, but the taxonomy of this family has since been revised (Brothers, 1971) to leave just three species, from the genera *Chrestomutilla* and *Smicromyrme,* that have been detected parasitizing *Glossina.* The species formerly identified included *Mutilla glossinae,* parasitic on *G. morsitans* in Zambia (Turner, 1915) and *M. benefactrix* (Lamborn, 1916). *Mutilla glossinae* was found in flies collected by Eminson (1915) when the wingless wasp emerged from one tsetse puparium out of a batch of 258 collected. The wasp broke open the puparium in the same way that the tsetse fly would itself. Other specimens were subsequently collected in Zambia. The life cycle, described by Lloyd (1916), lasts 2–4 months, and although natural rates of parasitism are higher than *Nesolynx,* its fecundity is lower, as it produces only one parasite per puparium. Lamborn (1916) obtained many *M. glossinae* from puparia of *G. morsitans* in Malawi and described how, when the tsetse puparia were parasitized whilst on the soil surface, the female mutillid would cover the puparium with earth subsequent to oviposition. Lamborn, describing the parasitism of tsetse puparia by *M. glossinae,* suggested that *Nesolynx glossinae* was hyperparasitic on *M. glossinae.* Lamborn found one specimen of a much less common species, *Eupelminus tarsatus,* from 1210 living tsetse puparia examined, and also found five or six large chalcid wasps from the same collection.

An attempt was made to control wild *G. m. morsitans* around Lake Malawi in the 1920s by releasing *N. glossinae* mass-reared in blowflies (*Sarcophaga* spp.) (Lamborn, 1925). Only 0.4% of tsetse puparia were parasitized in year 1 and 0.6% in year 2 but by year 3 the proportion of parasitized puparia had risen to 8.7%.

Austen (1914, 1929) described dipteran parasites of *G. morsitans,* discovered in Zambia by Lloyd in 1913, and species of *Bombyliidae* from Zimbabwe, Kenya, Uganda and Malawi which were commonly found parasitizing tsetse puparia. The bombyliid parasite from Zambia, named *Villa lloydi,* was the first recorded dipteran parasite of tsetse. Another species, *Exhyalanthrax abruptus,* first found in Zambia (Lloyd, 1916), occurred at a rate of 9.7% in *G. morsitans* puparia collected in Tanzania (Nash, 1930), and was reported parasitizing a calliphorid fly in northern Nigeria in 1957 (McDonald, 1957). *Exhyalanthrax argentifrons* was found parasitizing *Glossina* in Nigeria, but at very low infection rates (Taylor, 1926, in McDonald, 1957).

Parasites of tsetse in Zululand, South Africa, were documented by Fiedler (1954), who described correlations between size of the tsetse puparium and length of parasites that emerged from them. The size of adult *Mutilla glossinae* and *Exhyalanthrax abruptus* reared from tsetse puparia varies with the size of the host: larger tsetse species produce larger parasites (Heaversedge, 1968).

In Kenya, Lewis (1939) found a proctotrupid, *Trichopria capensis,* naturally parasitizing *G. fuscipleuris.* It could also parasitize *G. brevipalpis* and in laboratory studies it showed a predilection for larger tsetse species.

18.3.2   Parasitic mites

Ectoparasitic mites of the family Erythraeoidea (probably *Leptus* species) have been reported on *G. fuscipes* in Kenya (Krampitz and Persoons, 1967). There is no indication that mite infestations are harmful to tsetse or that they have a potential as control agents.

18.3.3   Helminth parasites

Nematode parasites of tsetse from the family Mermithidae have been recorded since the beginning of the century. The first detailed description was from *Mermis* spp. parasitizing *G. p. gambiensis* in Liberia (Foster, 1963a) but the first record was from a wild *G. f. fuscipes* from Entebbe, Uganda, in 1910. Subsequently, nematodes were found parasitizing *G. m. centralis* in the Democratic Republic of the Congo (Rodhain *et al.*, 1913), Tanzania (Thomson, 1947) and Zambia (Lloyd, 1913). Carpenter (1912) described finding a minute larval nematode in a laboratory-bred *G. f. fuscipes.*

The likely method of infestation is thought to be by 2nd-stage larvae, which migrate to the surface of the soil and climb up grass and vegetation in wet weather. They then penetrate the body of the host through the inter-segmental integument, using the buccal stylet and a chitin solvent from the oesophageal glands (Foster, 1963b, based on descriptions from Christie, 1936).

A mermithid nematode, *Hexamermis glossinae,* parasitizing *G. palpalis* s.l., *G. p. pallicera* and *G. n. nigrofusca* in West Africa (Poinar, 1981), was subsequently detected in *G. pallidipes* and *G. brevipalpis,* but not in *G. austeni,* from an area of the Kenyan coast (Odindo and Hominick, 1985).

18.3.4   Fungi

The effects of entomopathogenic fungi in *G. m. morsitans* have been studied by Kaaya (1989b). The fungi *Beauveria bassiana, Metarhizium anisopliae, Paecilomyces fumosoroseus* and *P. farinosus* were all pathogenic, particularly *B. bassiana* and *M. anisopliae,* which killed up to 100% of infected adult flies. Male tsetse were more susceptible to infection than females and puparia did not appear to become infected. Among the problems of using fungi for tsetse control are those of a suitable formulation and method of dispersal (Kaaya, 1989b). The fungi *Absidia repens* and *Penicillium lilacinum* have been isolated from puparia of *G. f. congolensis* in the Central African Republic and proved pathogenic in experiments. They may therefore contribute to natural mortality, though it appears that entry points such as wounds are necessary for the puparia to become infected (Vey, 1971).

Mass-rearing tsetse for release after contamination with *B. bassiana* has recently been proposed, and termed the 'lethal insect technique' (LIT) (Mahamat *et al.*, 1997). Such tsetse would probably die shortly after release but would spread the contaminating bacteria.

18.3.5   Bacteria

The bacterium *Serratia marcescens* applied to rabbits' ears killed tsetse that fed on them (Poinar *et al.*, 1979) and has been isolated from wild tsetse (Onoviran *et al.*, 1985) in which it invades the haemocoel. The bacteria multiply in the infected host's blood and body cavity and produce a general septicaemia that kills within a few days. It is an unlikely candidate for biological control, however, as it may cause mastitis in cows and pneumonia and septicaemia in humans.

Tsetse flies experimentally infected with live *Escherichia coli, Enterobacter cloacae* and *Acinetobacter calcoaceticus* show a remarkable increase in two haemolymph proteins, while those infected with live *Bacillus subtilis* and *Micrococcus luteus* or with *T. b. brucei* do not (Kaaya *et al.*, 1986). Haemocoelic injections of *E. coli* decrease fecundity of *G. m. morsitans* but

have no effect on longevity (Kaaya *et al.*, 1987). In Chad, experimental
infection of adults or puparia of *G. tachinoides* with *Bacillus thuringiensis*, or
with two other bacteria isolated from bees, was not pathogenic (Maillard,
1974; Maillard and Provost, 1975). *Bacillus thuringiensis* has been used for
biological control of other insect pests.

18.4   STERILE INSECT TECHNIQUE (SIT)

The sterile insect technique (SIT) has been used successfully against certain
insect pests for many years; for example, the screw-worm fly (*Cochliomyia
hominivorax*) was eradicated from the southeastern United States by the
1960s. More recently, the same species was successfully eradicated from
North Africa, where it had been introduced and posed a threat to livestock
in the region (FAO, 1992c). A disadvantage of the technique is its expense;
it was not considered economically viable for control of the codling moth
(*Cydia pomonella*) for this reason (Proverbs, 1982).

The SIT requires large numbers of tsetse of the target species to be
reared in laboratory colonies. The puparia produced by these flies are
sterilized with a source of radiation and the sterile males emerging from
these puparia are released in large numbers, over a long period, into the
area from which tsetse are to be eradicated. Commonly, the sterile males are
given an initial bloodmeal before release so as to reduce the risk of them
subsequently transmitting trypanosomes.

Success of the technique depends upon a high probability of wild
females mating with a sterile male rather than a wild, fertile male. This
requires a much greater number of sterile males to be released than the
number of wild males in the natural population. Successful colonization of
important tsetse species is, therefore, a prerequisite and was initially difficult, partly due to the low reproductive rate of the tsetse fly. As a female
tsetse fly stores sperm in its spermathecae in sufficient quantity to last for
the whole of its reproductive life, females inseminated by a sterile male will
produce no offspring. Consequently, the population will eventually die out,
provided it is not sustained by immigration from neighbouring or
contiguous areas.

The sterile male technique was first considered as a means to control
tsetse by Simpson (1958); it was concluded, from the mathematical model
of tsetse control by SIT referred to in Chapter 9, that the method would be
impractical for control of high-density tsetse populations above about 1000
males per square mile (385 km^{-2}). This was due to the large number of
sterilized males that would be required. In making this comment, Simpson
was considering that the puparia that would provide the sterile males would
have to be collected from the field rather than produced in a colony. USAID
attempted this approach in Zimbabwe, but concluded that colony production of tsetse would be more practical. In practice, numbers of wild tsetse

have to be reduced by other means, such as insecticide-impregnated traps or targets, before the release of sterile males so that the numbers of sterile males required are within acceptable limits.

For a sterile male release campaign to be effective, it has been estimated that the number of sterile male flies released should be sufficient to ensure that they inseminate at least 10% of the females. Rogers and Randolph (1985) showed that to achieve 10% female sterility the number of males released must constitute 80% of the male population. Knipling (1959, 1964) had estimated that an initial overflooding ratio of sterile to wild males of 3:1 with successively smaller releases could eradicate a low-density population of flies in 12 months. Experience has shown that a higher ratio is necessary of between 10 and 50:1. A ratio of up to 100:1 was reached in a control operation in some areas of Zanzibar (section 18.4.3).

SIT has been developed in three main centres: Tanga, in Tanzania, where American aid has been used to develop the technique and plan trials for Zanzibar; CIRDES (CRTA), in Burkina Faso, where, with German and French support, field trials were carried out in the Sidéradougou area; and the Bicot programme in Nigeria. The latter project has received support from various agencies – for example, the International Atomic Energy Agency (IAEA) in Vienna, which has had involvement with most, if not all, SIT tsetse control trials. Some of these trials and technical aspects of this technique will be discussed in more detail.

18.4.1 Technical and biological aspects of SIT

Multiple mating, dispersal and survival
Both laboratory-reared and wild tsetse flies can mate several times. As flies first mated by a sterile male can later re-mate with a fertile male and produce offspring (Dame and Ford, 1968), this could theoretically affect the success of SIT projects. The frequency of multiple mating in nature is unknown but is considered to occur at a sufficiently low level so as not to affect the success of SIT, particularly as the effectiveness of fertile mating (when subsequent to the infertile one) is not high (Dame and Ford, 1968).

It is also important to the success of SIT for tsetse control that the behaviour and survival of released flies is similar to that of wild flies. Studies of wild and colony-produced *G. m. morsitans* released in Zimbabwe showed that the dispersal and survival of colony flies was comparable to that of wild flies even when they had been sterilized with tepa (Dame *et al.*, 1975).

Irradiated tsetse are fully capable of developing and transmitting mature trypanosomes of the three main species pathogenic to cattle (Dame and MacKenzie, 1968; Moloo, 1982b; Moloo and Kutuza, 1984). To avoid an upsurge in trypanosome prevalence in livestock following large-scale releases, tsetse may be fed on either an uninfected bloodmeal or a blood-meal medicated with a trypanocidal drug before release. Tsetse are less likely to become infected (at least with *Nannomonas* and *Trypanozoon* parasites) after they have taken an uninfected bloodmeal, and trypanocidal drugs in medicated meals will reduce establishment of infections in subsequent meals.

18.4.2 Methods of sterilization

Sterilization of tsetse may be carried out by a number of methods:

- Irradiation (e.g. gamma rays).
- Chemosterilization (e.g. bisazir).
- Physiological sterilization (e.g. juvenile hormones).

Compounds used for the irradiation or chemosterilization of tsetse are shown in Table 18.2.

Irradiation is the usual method employed in the SIT. (Chemosterilization and physiological sterilization are described in section 18.5.) Radiosterilization was first attempted by Potts (1958), exposing puparia of *G. morsitans* to 6000 or 12,000 röntgens (R) of gamma irradiation from a ^{60}Co source. He showed that, although length of life of the males was halved, neither puparial mortality nor mating ability was significantly affected; there was, however, about 20% residual fertility in males at an irradiation dose of 6000 R. Higher doses of radiation may be required to sterilize male puparia older than about 14 days, but female puparia were completely sterilized by 1000 R regardless of age. Some mortality occurs with higher doses.

Gamma irradiation has been used to sterilize *G. pallidipes* and *G. m. morsitans*, using puparia collected from the field in the Zambezi valley, with doses of between 4000 and 16,900 rad (Dean and Clements, 1969; Dean and Wortham, 1969). Complete male sterility was rarely obtained with any of the treatments. Beta irradiation has been used to sterilize *G. p. palpalis* (Hamann and Iwannek, 1981): the optimal sterilizing dose was about 7.5 krad; females survived irradiation better than males. Beta irradiation was used as it was

Table 18.2. Compounds for irradiation or chemosterilization.

Irradiation	Chemosterilants
^{137}Caesium	Bisazir
^{60}Cobalt	Metepa
	Tepa
	Apholates
	Phytosterols
	Diflubenzuron (DFB, Dimilin)
	Pyriproxyfen (juvenile-hormone mimic)
	Sulphaquinoxaline
	Chlordimeform

thought more suitable for field use compared with other types of irradiation, which required shielding. Irradiation in a nitrogen atmosphere results in a lower incidence of early puparial death and somatic damage, and an increased average longevity; however, it also reduces the level of sterility achieved (Langley *et al.*, 1974; Vreysen and Van der Vloedt, 1995a).

A further consideration in the use of SIT is that the period between irradiation and eclosion is short, and this may be a constraint if puparia are to be transported long distances before release. The possibility of chilling puparia to delay eclosion has been investigated but does not seem to be a viable solution, as the duration of chilling required is rather long and adversely affects survival (Vreysen and Van der Vloedt, 1995b).

18.4.3 Field trials

Burkina Faso

Temporary eradication of *G. p. gambiensis* was achieved at Sidéradougou in Burkina Faso after a 5-year research programme (Politzar and Cuisance, 1982; Cuisance *et al.*, 1984). Sterile males were released over a 32 km river system and led to the disappearance of tsetse from 100 km^2. Barriers to prevent reinvasion of the control area were created by total bushclearing and regular spraying of persistent insecticide (thiodan, 3.5%) at the outer limits in addition to permanent trapping with biconical traps. Males were sterilized with ^{137}Cs at a dose of 11 krad, producing a sterility of 95% without significant reduction in insemination capacity. Batches of sterile males were released at 200 m intervals. The estimated cost per sterile male released was considered to be too high; therefore, a method of releasing sterile males from ultra-light aircraft was tested and the cost was reduced to $0.19 km^{-1} of flying, compared with $0.5 km^{-1} using a four-wheel-drive vehicle (Politzar *et al.*, 1984).

SIT was used in conjunction with helicopter spraying with deltamethrin to control tsetse in Burkina Faso (Van der Vloedt *et al.*, 1980). An annual release of 150,000 gamma-irradiated male *G. p. gambiensis* was carried out to control human sleeping sickness. Although no costs were published and the area covered is not clear, this must have been a very expensive exercise.

Tanzania – Tanga

In the 1970s a colony of *G. m. morsitans* was established at Tanga, in Tanzania, to produce sterile males for a field trial of SIT in the coastal region of the country. The flies were maintained mainly on goats, and by making use of the earlier emergence of females from puparia (Birkenmeyer and Dame, 1975) the sexes could be separated. After 52% emergence, the remaining puparia (predominantly males) were chilled to prevent further eclosion prior to sterilization by gamma irradiation and release in the trial area. Emerged females were used to replenish the breeding colony of flies.

A total of about 1.3 million puparia were produced in a 15-month period, of which 0.6 million were released, at a cost of $220 per 1000 (Williamson *et al.*, 1983a). The trial took place in a 195 km² area of Mkwaja ranch, close to Tanga, which was encompassed by a barrier of bushclearing 1 km wide to reduce reinvasion of tsetse from surrounding areas (Gates *et al.*, 1983). Prior to the release of sterile males, two aerial applications of endosulfan were applied to the release area to suppress the tsetse population, which resulted in a 100% reduction of *G. m. morsitans* adults (Williamson *et al.*, 1983b). After the aerial spraying, irradiated puparia were released twice weekly at 120 release points throughout the trial area. Over the 15-month experimental period an average of 81% control of *G. m. morsitans* was achieved, whilst *G. pallidipes* populations in the area, which had also been suppressed by aerial spraying, recovered to pre-spray levels within 5 months (Williamson *et al.*, 1983c). Although this trial could be regarded as a research project, from which valuable lessons were learnt, eradication was not achieved, and unless release of sterile puparia was to be continued, *G. m. morsitans* populations could be expected to return to pre-control levels in a short time. Furthermore, apart from *G. pallidipes*, possibly two other tsetse species (*G. brevipalpis* and possibly *G. austeni*) were present on the ranch (Gates *et al.*, 1983).

Zanzibar

In 1994, after many years of research into SIT and the rearing of *G. austeni*, an eradication programme for Zanzibar was started by IAEA and the Zanzibar authorities. The tsetse population was first suppressed using pour-on insecticides and insecticide-impregnated targets, then male *G. austeni* sterilized by gamma irradiation were dispersed by air over the whole island. Over 7.8 million sterile males were released, to give an estimated ratio of 50:1 sterile to wild males. No wild tsetse have been caught since September 1996 and eradication of *G. austeni* from Zanzibar was declared at the end of 1997 (IAEA, 1997). Monitoring of the operation was carried out using approximately 500 sticky panels and by sequential screening of disease incidence in sentinel cattle herds. Buffy coat examination of these cattle showed that by January 1997 disease incidence was less than 1% and was due to *T. vivax* (Saleh *et al.*, 1997).

Nigeria

In central Nigeria, *G. p. palpalis* was temporarily eradicated from 1500 km² of agropastoral land using 1.5 million laboratory-bred sterile males for field release (Oladunmade *et al.*, 1990). The wild tsetse population was first reduced to less than 10% by continuous trapping and placement of insecticide-impregnated targets over a 6–12-week period. Sterile males were then released weekly to maintain a minimal ratio of ten sterile males to one fertile wild male fly for at least three generations. The area was protected against reinvasion by barriers of impregnated targets. They used 1080 targets at 100 m intervals for the barrier but this was affected by theft, up to 80% of the targets being stolen. *Glossina tachinoides* also occurred in some parts of the control area and so some trypanosomosis still occurred in the livestock.

Ethiopia

A SIT project in southern Ethiopia has been initiated following an agreement between the Ethiopian government and the IAEA in which *G. pallidipes* is the target species.

18.5   CHEMOSTERILIZATION

Insect chemosterilants investigated in the 1970s were often carcinogenic or highly toxic to humans. Difficulties in applying them to tsetse in a practical way limited their use mainly to laboratory experiments and curtailed development of their use in the field. Chemosterilization of tsetse using Apholate and Metepa was attempted by Chadwick (1964b). Flies emerging from puparia collected in the field in Tanzania were dosed topically with those substances. Initial experiments showed little promise as the chemosterilants used caused excessive mortality of *G. m. morsitans* at substerilizing doses. Dame and Ford (1966) reported experimental work with tepa on flies caught in Zimbabwe.

As colony-bred flies survive less well in the field and have less well-developed thoracic musculature than wild flies (Dame *et al.*, 1968), alternative methods were investigated for sterilizing wild flies caught in the field using attractants.

One of the first trials was on an island in Lake Kariba in 1967. The survival rate of sterilized males was only 17% that of wild males. The maximum ratio of sterile to fertile males achieved in the field was 12:100 and, because of the poor survival of the sterile males, the programme ended unsuccessfully after 6 months. A second, more successful trial was carried out on another island, which was overflooded with sterile males (Dame and Schmidt, 1970). Chemosterilized males remain effective vectors of *T. congolense* although the rate of transmission is reduced, especially if flies are sterilized after trypanosome infection (Dame and Mackenzie, 1968). Although flies sterilized with either gamma irradiation or chemosterilants have high levels of sterility, trials have suggested that released sterile males might not compete for normal females in the wild as well as untreated males (Dean *et al.*, 1969b).

18.5.1   Autosterilization

Sterilization and subsequent release of laboratory-bred tsetse is a costly and complicated procedure. To reduce costs, experiments have been conducted to develop devices that catch and sterilize wild flies *in situ* before allowing their release into the wild population. House (1982) showed that flies could be automatically sterilized in a trap with Metepa but this was still a complex and expensive process. In laboratory experiments, bisazir can be absorbed on to cellulose fibre discs to sterilize tsetse. This resulted in permanent sterility without affecting survival but it is a volatile and somewhat unstable compound (Langley and Carlson, 1986). No adverse effects on the mating behaviour of *G. m. morsitans* were observed at the doses of bisazir required to cause sterility (Coates and Langley, 1982b).

One problem was the length of time for which tsetse had to copulate in order for a sufficient degree of sterility to be achieved. In order to stimulate a longer period of copulation, tests were conducted using a synthetic contact sex pheromone of *G. m. morsitans* in conjunction with an autosterilizing device. The pheromone induced tsetse to maintain a copulatory response for a sufficient length of time. Further developments of an autosterilizing device in Zimbabwe used bisazir in the vapour phase, to which flies were exposed in a small chamber attached to a standard F2 trap (Hall and Langley, 1987). Despite the favourable results, no significant development of the technique has taken place since then. Bisazir is a rather dangerous genetic alkylating agent.

Langley *et al.* (1982b) tested other, plant-derived benzodioxole compounds as contact sterilants by feeding them to tsetse. As this was not a practical approach, they were also tested as contact sterilants on both male and female *G. m. morsitans*. One of the compounds tested induced sterility in both sexes at low doses, comparable to those required for an insect growth regulator, diflubenzuron, which is referred to in section 18.6.

One sex or both sexes?

Theoretically, sterilizing both female and male tsetse at a certain rate is a more efficient long-term control method than killing both sexes (using insecticides) at the same rate (Langley and Weidhaas, 1986). It is also more efficient than sterilizing just males through autosterilizing devices based on a combination of contact sex pheromone and sterilant on decoys (Langley *et al.*, 1981a; Hall, 1987).

18.6 INSECT GROWTH REGULATORS

The complexity of the reproductive system and the low fecundity of tsetse suggest that they would be vulnerable to control using insect growth regulators (Clarke, 1980). These substances interfere with chitin synthesis and prevent successful reproduction, effectively sterilizing female tsetse. Growth regulators are much safer than the chemosterilizing agents discussed earlier; however, one of the difficulties is the development of mechanisms for getting the substances into the fly, since, to be effective, they have to be taken up through the cuticle.

In experiments with the growth regulator diflubenzuron (DFB; Dimilin), 0.5 μg applied topically to female *G. m. morsitans* prevented the production of viable offspring for more than 70 days, and smaller doses resulted in production of abnormal offspring. Tarsal contact with diflubenzuron was ineffective, and therefore it has to be applied in solubilized form using a solvent such as acetone (Jordan and Trewern, 1978; Jordan *et al.*, 1979). The diflubenzuron was lost by excretion and at each larviposition by a female fly (Clarke, 1982). The rate of loss depended not only upon the amount of the chemical applied to the fly but also upon the site and method of application. This presents difficulties in using the compound for tsetse control, as a technique for applying a sufficient quantity to sterilize the fly throughout its reproductive life is necessary.

Some phytosterols are effective as growth regulators and abortifacients when mixed in pigs' blood and fed to tsetse through membranes (Whitehead, 1981). This effect appeared to result from: (i) retarded synthesis and/or release of the milk secretion from the milk gland; (ii) slow digestion of the bloodmeal; and (iii) a direct effect on the development of the larva.

18.7 JUVENILE-HORMONE MIMIC (PYRIPROXYFEN)

The use of juvenile-hormone mimics for tsetse control exploits the same apparent weakness – the low reproductive potential – as insect growth regulators. Juvenile hormones disrupt the reproductive cycle, causing abortion. A juvenile-hormone mimic tested on laboratory-bred *G. m. morsitans* resulted in abortions for up to 40 days; however, the colony used had a poor survival rate (Meidell, 1982). An alternative substance, pyriproxyfen, was later investigated, which resulted in adult female tsetse producing non-viable offspring throughout their reproductive life (Langley *et al.*, 1990a). A vegetable oil formulation of pyriproxyfen was used to treat black cloth targets from which female tsetse picked up the active ingredient through tarsal contact. Males also picked up sufficient amounts for effective transfer to females whilst mating.

An advantage of pyriproxyfen is that it has a low toxicity compared with most chemosterilants that are highly toxic. The juvenile-hormone mimic described by Langley *et al.* (1990a) was field tested in Zimbabwe (Hargrove and Langley, 1990). This trial was carried out using 41 odour-baited traps treated with 4 g of pyriproxyfen in 12.3 km^2 of woodland habitat of *G. m. morsitans* and *G. pallidipes*. The trap material was dosed with 2 mg pyriproxyfen cm^{-2} and flies were allowed to escape after entering the trap. Emergence rates from puparia of the two species collected in the block fell to 30% and 2.7%, respectively, after 3 months. A major problem was the lack of persistence: after 4 months, 68–85% of the pyriproxyfen on targets had disappeared. Increased persistence is necessary for the technique to be a viable alternative to normal insecticide impregnated targets (Hargrove and Langley, 1993).

Pyriproxyfen was also tested against *G. p. palpalis* in the forest zone of Côte d'Ivoire, using Vavoua traps (Laveissière and Grebaut, 1990) treated with a single dose of 2 mg cm^{-2}; an 87% reduction of apparent tsetse density achieved after 2 months was maintained for 5 months (Laveissière and Sané, 1994).

The technique was thought to be suitable for the control of animal trypanosomosis, but not for human sleeping sickness control, which requires an immediate and more substantial reduction of the vector population. Studies are still under way on the development of the use of pyriproxyfen and of an alternative chitin synthesis inhibitor, triflumeron, for tsetse control (Langley, 1997).

18.8   GENETIC CONTROL

It has been suggested that control of trypanosomosis might be possible by genetically altering the symbionts in tsetse so that they produce anti-trypanosomal products or products which can affect the establishment of midgut trypanosomal infections or trypanosome differentiation (Aksoy *et al.*, 1995; Beard *et al.*, 1993a). Beard *et al.* (1993b) were able to transform cultured midgut symbionts of *G. m. morsitans* genetically, using a heterologous plasmid, and to show the possibility of introducing and expressing desirable genes into those symbionts. After introducing the transformed symbionts into tsetse, assuming that they are able to compete successfully with natural symbionts, the main difficulty lies in being able to promote natural spread of the engineered refractory fly phenotypes (Beard *et al.*, 1993a; O'Neill *et al.*, 1993). Transposable elements, termed *mariner* sequences, could possibly be used to manipulate the tsetse genome as a means of controlling disease transmission. These *mariner* sequences have been found in tsetse of each subgenus (Blanchetot and Gooding, 1995).

The possibility of inserting genes conferring refractoriness to trypanosome infection remains a complicated task, as multiple genes are almost certainly involved and they may have an effect on the viability of the vector. Genetic control of mosquito vectors of malaria has been proposed, but it is a controversial topic and Spielman (1994) produced four arguments against this line of research:

- The vector competence of mosquitoes was not a significant factor in determining force of transmission.
- Where density-dependent regulation of a mosquito population did not occur, the release of transgenic mosquitoes would simply add additional

mosquitoes and would be unsuccessful unless a 'drive mechanism' favoured the released population. Spielman thought the released mosquitoes would be irrelevant to transmission.

- There were ethical issues in releasing large numbers of what would still constitute a pest, and there would be a need to control them continually.
- The 'non-renewable' nature of the proposed drive mechanisms invalidated their usefulness.

There is, nonetheless, increasing interest in the possibilities of expressing foreign genes in tsetse.

18.8.1 Hybrid sterility

The hybrid offspring from fertilization of female tsetse by males of a closely related but different species are usually sterile. Theoretically, by introducing large numbers of one tsetse species into the habitat of a different species, a degree of control could be achieved. This potential of hybrid sterility for control was recognized by Potts (1944), who introduced *G. morsitans* puparia into an isolated area inhabited by a low density of *G. swynnertoni*. One year later the *G. swynnertoni* population had been reduced according to theoretical expectations (Vanderplank, 1947c, 1948). The existence of behavioural barriers to cross-mating could cause this method of control to fail; behavioural barriers have been demonstrated in wild populations of mosquitoes but such barriers appear to be unimportant for tsetse flies.

Another possible mechanism for genetic control is the use of chromosome translocations (Curtis, 1968a,b), which would result in abnormal pairing of chromosomes and could be induced, for example, by low doses of irradiation. Some of the chromosome combinations produced would lack part of the normal chromosome set, and so gametes from these flies, when fertilizing normal gametes, would produce infertile embryos. If successful, this technique would probably require fewer flies to be produced and released than for the normal sterile male technique (Curtis, 1968b). However, this method of control has not been tested in the field or taken up by other researchers.

Genetic susceptibility to infection is characteristic of vector–parasite relationships of some parasitic diseases and consequently genetic approaches to disease control are being explored. Novel methods of vector control emerging from genetic studies would have the dual advantages of being species-specific and of causing little environmental damage. Semi-sterile mutants have been produced for control purposes in the absence of identified genetic markers (Curtis, 1971; Curtis 1972). Recently, microsatellite markers suitable for genetic population studies have been found in *G. p. gambiensis* (Solano *et al.*, 1997).

Chapter 19

General Issues Relating to the Successful Use of Tsetse Control Techniques

A number of problems and issues relating to the use of the tsetse control methods have been referred to briefly in earlier chapters. In this chapter, problems of reinvasion by tsetse flies and resistance to insecticides are discussed. The environmental impact of tsetse control, its effects on land use, economics of control measures and, finally, the development of community participation in tsetse control are also mentioned. The importance of some of these issues could warrant more detailed discussion than presented here, and references to recent review articles are given.

19.1 PROBLEM OF REINVASION OF TSETSE FLIES INTO CLEARED AREAS

As tsetse flies are relatively mobile insects, there is a constant reinvasion pressure against areas from which the fly has been removed or controlled unless these measures are taken up to natural boundaries, or an effective barrier is maintained. The control of reinvasion into cleared areas has been a persistent problem since the first experiments in tsetse control. One solution may be to eradicate tsetse over large areas up to the natural limits of their distribution. However, this necessitates high levels of funding, which can rarely be met by national governments and thus it requires international funding such as that of the EU-funded regional programme for Zambia, Zimbabwe, Malawi and Mozambique. Indeed, this programme has recently changed its goals, to *control* tsetse in areas where livestock are kept, rather than the more difficult and expensive task of eradication.

Several methods of preventing reinvasion have been attempted over the years, often with little long-term success. In a review of those methods,

Wooff (1967) referred to the use of a sheer clearing 3.2 km wide that Jackson (1954b) stated was necessary to stop the passage of flies. Wooff did not believe that this was wide enough, based on his experience in Uganda using barriers 5–6 km wide made with caterpillar tractors. At Shinyanga, *G. swynnertoni* was able to cross an 800 m clearing, although the flies crossing were mainly hungry and presumably were in search of a blood-meal (Lloyd, 1935).

In Nigeria, where *G. m. submorsitans* was removed from the vicinity of a traditional cattle trade route, the rate of reinvasion was about 5.13 km year^{-1} or 100 m week^{-1}, proceeding faster during dry than during wet seasons. This rate of advance was about half the distance proposed by other workers. It was suggested that this was due to the suitability of the climate, which made rapid dispersal unnecessary (Riordan, 1976).

In Burkina Faso, a trap-barrier designed to block reinvasion of the Koba river system by *palpalis* group tsetse consisted of an arrangement of 100 biconical traps placed at 100 m intervals along the river, and protected an area of 1500 km^2 upstream of the barrier (Politzar and Cuisance, 1983). Mark–release–recapture experiments confirmed that tsetse were unable to pass the barrier in either the dry or the wet season, and were used to examine fly movements and invasion capacities. The barrier did not work for *G. m. submorsitans*, which occupies a wider habitat. It was considered essential to place the barrier in a site with few or no affluents and little human activity.

Hargrove (1993) studied barriers in relation to the problem faced by Zimbabwe in preventing reinvasion from Mozambique into areas from which tsetse had been eliminated. He concluded that all currently used barriers of traps or targets 'leak', and that the rate of passage of tsetse flies depends on the barrier's width and the trap or target density. It also depends on factors related to the fly, such as the rate of movement, natural mortality and the level of mortality inflicted by the targets. Hargrove described how the probability of penetration of a barrier varied according to its target density and width. A 1 km wide barrier with a target density of 32 traps km^{-2} would still have a probability of penetration of about 0.1 (10 flies out of every 100). A barrier 8 km wide would have a decreased probability of penetration, and increasing the number of traps km^{-2} would give a significant decrease in penetration.

Hargrove attempted to quantify the rate of movement of tsetse across a barrier based on the assumption that it follows a normal distribution from the starting point. Mathematical models allow a cost-efficient combination of barrier width and target density to be estimated (Hargrove, 1993). Treatment of cattle with Decatix for a distance of between 10 and 30 km of the border is currently used to prevent reinvasion in Zimbabwe, but this is expensive.

At Nguruman in Kenya, female *G. pallidipes* appeared to disperse faster than males (Brightwell *et al.*, 1997). The probability of females moving was

0.055, compared with 0.029 for males, and the authors pointed out that, where there is a suppressed population close to an unsuppressed population, this will inevitably lead to a large net movement from the larger subpopulation to the smaller one.

19.2   RESISTANCE TO INSECTICIDES

From the mid-1940s to the 1970s, when DDT and dieldrin were widely used for tsetse control, no reports of resistance to those insecticides developing in tsetse were made, whereas mosquitoes rapidly developed resistance at a time when there was hope of eradicating malaria. No resistance by tsetse flies to any insecticides has been reported, despite their use for tsetse control over many years. This is probably mainly due to the low reproductive rate of tsetse flies compared with other insects, which results in a low degree of selection pressure for resistant individuals. Additionally, tsetse are highly susceptible to insecticides and often occur at low population densities. Tsetse flies do have the genetic capacity to develop insecticide resistance, as it has been shown that they can metabolize small amounts of DDT and a dieldrin analogue (Brooks *et al.*, 1981). Evidence of decreasing susceptibility to endosulfan and dieldrin in the Lambwe valley, Kenya, was found by comparing the susceptibility of tsetse flies in that area with that of an unsprayed tsetse population at Nguruman, also in Kenya (Turner and Golder, 1986). Wild-caught non-teneral male and female *G. pallidipes* at Lambwe were significantly more tolerant of both insecticides than their counterparts from Nguruman, except for males treated with dieldrin, but the differences were not great.

Pregnant female tsetse are more resistant to chlorinated hydrocarbon insecticides than could be predicted from their gross weights (Burnett, 1961a, 1962). Such tolerance was subsequently shown to occur with some other types of insecticide. It has been suggested that inert storage of the toxic compounds in the larva contributes to this tolerance, rather than simply being an effect of the greater body weight of the pregnant female. One hypothesis is that the developing larva could be a potential site of detoxification (Irving, 1968; Kwan *et al.*, 1982); pregnant females are up to 9 times more tolerant to dieldrin than young flies (Hadaway, 1972). Tolerance to DDT and gamma BHC increased with age for females but not for males (Burnett, 1961b). Male flies are more susceptible to endosulfan than female flies, and females with a second- or third-stage larva in the uterus were significantly more tolerant (Fenn, 1992).

Although no resistance of tsetse to the synthetic pyrethroids has yet been reported, there are now numerous reports of resistance developing in ticks and other insects to these insecticides.

Maudlin *et al.* (1981) developed a model which predicted that development of resistance was a possibility in certain circumstances. Later, Green and Maudlin (1982) modelled the development of resistance to insecticides in tsetse. The results of this suggested that resistance could eventually occur if the insecticide-treated tsetse population was not rapidly eliminated, and provided that the gene for resistance is dominant or co-dominant to its wild-type allele.

19.3   ENVIRONMENTAL IMPACT OF TSETSE CONTROL USING INSECTICIDES

19.3.1   Direct effects

The publication of *Silent Spring* (Carson, 1962) created an increased aware-ness of the dangers associated with the use of insecticides. Since then, there have been a number of studies on the environmental impact of insecticidal methods of tsetse control. In particular, the direct effects to non-target organisms have been investigated; there have been fewer studies of indirect effects. Table 19.1 lists some of the environmental studies that have been conducted. The residual insecticides used in the past for tsetse control have, as might be expected, caused the most problems.

MacLennan (1973) reviewed the direct environmental consequences of tsetse control with insecticides in Nigeria and the effects of increased land use after control. Residual insecticides did not appear to cause changes in the fauna of 'sufficient magnitude to render such usage undesirable', but MacLennan did suggest that regulation of human populations should be introduced to ensure that land use could be suitably managed. Langridge and Mugutu (1968), reporting on the effects of tsetse control using organochlorines on wildlife in Kenya, found a wide range of animals were affected but did not draw any conclusions about the long-term effects.

In an extensive study of the environmental impact of insecticidal tsetse control in Nigeria, Koeman *et al.* (1978) found that certain species of 'fringe forest birds', such as flycatcher species, were very vulnerable and became rare in the treated area. Ground spraying, being more discriminatory than helicopter spraying, was thought less likely to cause irreversible damage; comparable doses of endosulfan, dieldrin and DDT were used for each technique. The effects of endosulfan and deltamethrin applied by helicopter in Nigeria were described by Baldry *et al.* (1981).

In Zambia, dieldrin was used for tsetse control, particularly at the end of the dry season, as it has a long-lasting effect (more than 4 months) when applied to tree trunks (Wilson, 1972). Dieldrin, when absorbed through the skin, is 40 times more toxic than DDT. A 3.7% dieldrin emulsion, applied by motorized knapsack sprayers to potential resting sites in Eastern Province, Zambia, caused numerous deaths of wildlife (Wilson, 1972).

In Somalia, Douthwaite (1986) monitored the feeding behaviour and breeding success of the little bee-eater (*Merops pusillus*) during aerial spraying

Table 19.1. Evaluation of environmental effects of tsetse control.

Location	Type of control	Insecticides	Target species/remarks	Reference
Direct effects				
Nigeria	Aerial spraying (helicopters)	Endosulfan, deltamethrin		Baldry *et al.*, 1981a,b
	Ground spraying	Endosulfan, dieldrin, DDT	Forest-fringe birds (flycatchers)	Koeman *et al.*, 1978
Zambia	Ground spraying (knapsacks)	Dieldrin	Wildlife	MacLennan, 1973
		DDT		Wilson, 1972
Burkina Faso	Aerial spraying (helicopters)	Endosulfan, deltamethrin	Freshwater invertebrates	Takken *et al.*, 1978
Côte d'Ivoire	Aerial spraying (helicopters)	Endosulfan	No long-term effects	Everts *et al.*, 1983
Zimbabwe	Ground spraying	DDT	White-browed sparrow weaver	Douthwaite, 1992b
			White-headed black chats and red-billed wood-hoopoes	Douthwaite, 1992b; Douthwaite and Tingle, 1992; Douthwaite, 1993
			African goshawk and fish eagle	Douthwaite, 1992c
			Long-tailed cormorant and reed cormorant	Douthwaite, 1992a, 1993; Douthwaite *et al.*, 1992
			Birds – general	Douthwaite, 1995
			Lizards	Lambert, 1993
		DDT, deltamethrin	Foraging ants	Tingle, 1993
Kenya			Wildlife	Langridge and Mugutu, 1968
Somalia	Aerial spraying	Endosulfan	Effects on little bee-eaters	Douthwaite, 1986 Douthwaite and Fry, 1982
Botswana	Aerial spraying	Endosulfan, deltamethrin	General	Douthwaite *et al.*, 1981
			Cyrtobagous salviniae (control agent of the water hyacinth, *Salvinia molesta*)	Semple and Forno, 1990
Indirect effects				
Sub-Saharan Africa	General		Desertification, overgrazing	Ormerod, 1976
			Land use control	MacLennan, 1973
			Effect of human populations on land use	
Nigeria	Ground spraying, aerial spraying (helicopters), bushclearing	DDT, dieldrin, endosulfan	Effects on land use	Reid *et al.*, 1997a
			Effects on land use, animal productivity, cultivation	Putt *et al.*, 1980
Ethiopia			Effects on land use, cultivation, biodiversity	Reid *et al.*, 1997c

of tsetse with endosulfan. After the heaviest spraying, day-flying insects virtually disappeared for 24 h, the feeding rate fell and breeding failure at three nests followed. After the spraying, however, breeding success was generally the same as outside the sprayed area.

Douthwaite and his colleagues carried out some of the most comprehensive studies on the effects of insecticides used for tsetse control on non-target organisms in Zimbabwe (e.g. Douthwaite *et al.*, 1981). These studies focused on the effects of DDT used for ground spraying on birds, lizards and some invertebrates. Douthwaite found little effect on the white-browed sparrow weaver (*Plocepasser mahali*), which feeds on seeds and arthropods (Douthwaite, 1992a). In contrast, populations of white-headed black chats (*Thamnolaea arnoti*), which feed only on arthropods, and red-billed wood-hoopoes (*Phoeniculus purpureus*), also feeding entirely on arthropods from sprayed tree trunks, collapsed over a wide area of north-western Zimbabwe after spraying (Douthwaite, 1992b, 1993; Douthwaite and Tingle, 1992). The collapse of the white-headed black chat populations probably resulted from lethal accumulation of DDT residues acquired from their prey. The most important prey were ants of the genus *Camponotus*. There was no apparent effect of starvation due to lack of prey or of reduced fledgling success. In the study areas, sprayed with DDT between 1987 and 1989, the white-headed black chat populations fell by between 74% and 88% and did not recover during the duration of the study published in 1992. The African goshawk (*Accipiter tachiro*), feeding mainly on birds, was almost eliminated from sprayed woodland, whilst the fish eagle (*Haliaeetus vocifer*) was little affected despite some hatching failure and thinning of eggshells (Douthwaite, 1992c). This also probably occurred in long-tailed cormorants (*Phalacrocorax africanus*) and reed cormorants (*P. africanus*) living on Lake Kariba (Douthwaite, 1992b, 1993; Douthwaite *et al.*, 1992). Douthwaite reported that the use of DDT for tsetse control had declined and would cease completely in Zimbabwe in 1995 and he expected bird populations to recover. DDT residues in woodland birds in Zimbabwe reflected feeding site more than diet: highest concentrations were found in the wood-hoopoe (*Phoeniculus purpureus*) and white-headed black chat (*Thamnolaea arnoti*), both of which sometimes foraged on sprayed tree-trunks (Douthwaite, 1995). Populations of those birds declined over 2–3 years in sprayed areas by about 90%.

As part of the same environmental monitoring programme in Zimbabwe, the effect on lizard populations was investigated because of their importance in the food chain with raptorial birds (Lambert, 1993). Previous studies had shown no significant effect of endosulfan spraying or of sequential spraying of deltamethrin on lizard populations. In Lambert's study, dominant species were *Mabuya striata wahlbergii* on trees in mopani (*Colophospermum mopane*) woodland and *M. quinquetaeniata margaritifer* in outcrops of rocks. Mean frequency of *M. s. wahlbergii* fell from 76% at untreated sites to 48% after DDT treatment. No effect was observed on

M. q. margaritifer. Following spraying with DDT or deltamethrin, in the same area of mopani woodland, there were seasonal effects on foraging ants, of *Pheidole* spp., *Platythyrea cribinodis* and *Monomorium opacum*, but no major adverse effects apart from a short-lived reduced foraging success of *Platythyrea cribinodis* immediately after spraying with deltamethrin (Tingle, 1993).

The effect of insecticides for tsetse control on biological control agents

There was an investigation into the interesting question of the effects of tsetse-control spraying on the biological control agent *Cyrtobagous salviniae*, which is a weevil used for controlling the water pest *Salvinia molesta*. This was carried out in relation to tsetse control in the Okavango region of Botswana, where *Salvinia molesta* is a serious pest. The weevil was 36 times more susceptible to deltamethrin than endosulfan and high mortality occurred. There is thus a danger of disruption to the biological control programme against *S. molesta* in areas where tsetse control using sprays of deltamethrin or endosulfan is carried out (Semple and Forno, 1990).

In Uganda, water hyacinth is becoming a serious problem on Lake Victoria, blocking water supplies, electricity generating turbines and access to the lake for fishing boats as well as depleting oxygen in the lake, which will result in fish mortality. Biological control of this pest with an introduced weevil is being considered, and the possible effects of insecticides used for tsetse control around the lakeshore may well have to be investigated.

19.3.2   Indirect effects

The direct long-term adverse effects of insecticides used for tsetse control may be slight, as outlined in the previous section, but the indirect effects, especially in terms of uncontrolled development of cleared areas, are potentially much more serious. This subject was reviewed by Jordan (1986). There have been very few studies of the indirect effects of tsetse control on the environment, particularly land use and biodiversity. One of the most comprehensive studies was that of Putt *et al.* (1980) in which the effects of tsetse control in Nigeria on livestock productivity, crop production, settlement and communications were reviewed. Among ongoing studies are those of Reid *et al.* (1997c, 1998), which have shown that in the Ghibe valley, Ethiopia, there appear to be few negative changes in the vegetation during the process of conversion of wooded grasslands into smallholder fields.

When long-term tsetse and trypanosomosis control schemes were proposed in 1974, studies were initiated into the potential impacts. The main conclusions of one of these studies (Matzke, 1983) were as follows:

1. Extreme caution is needed in relating the absence of development to the presence of the tsetse fly.

2. Promises of benefits forthcoming with the removal of the tsetse fly must be viewed with scepticism.

3. In some cases a case can be made for retaining the tsetse.

4. Fly zone problems can be addressed in many ways other than tsetse removal.

5. The regions most likely to incorporate successful tsetse control efforts will be on the margins of the tsetse range.

6. It is important to define the targeted human population for the designed development benefits. Different strategies should be employed in virgin lands, places with people lacking livestock experience and places intended for expanding pastoralist rangelands.

7. The push for tsetse eradication arises from its role as a disease vector ... and the strong Western bias toward the role of cattle in the modern agricultural mix.

The study by Putt *et al.* (1980) provides a strong argument against points 1–3, and, with the exception of conclusions 5 and 6, the remaining statements could also be considered rather controversial. Certainly the western view of cattle in agricultural development has changed in recent years, resulting in a significant fall-off in funding, although strong arguments are being made in their favour for their role in overall food production, including their use for traction, in African agriculture. Indications of recent economic analyses of tsetse control are that economic benefits from tsetse control using pour-on insecticides may be very significant (Mulatu *et al.*, 1997). There is, however, a move to target tsetse and trypanosomosis control more closely to farming areas or areas of human sleeping sickness.

19.4   TSETSE CONTROL AND LAND USE

This subject has been reviewed by Jordan (1986) and will therefore only be mentioned briefly here.

Ford (1969) considered that successful tsetse control had sometimes been followed by reinvasion causing 'massive epizootics' or, alternatively, by cattle population growth periodically interrupted by 'catastrophic famines'. He therefore cautioned against large-scale tsetse eradication, in favour of carefully planned tsetse control taking the effects on land use into consideration.

Ormerod (1976) is one of the chief proponents of the dangers of tsetse control to African ecology, and suggested the possibility of a connection between the West African cattle industry and the Sahelian drought, but this is still a matter of contention. It has been argued that grazing lands are highly resilient; thus, whilst overgrazing may occur as a result of increases in cattle numbers, the cattle population will be held in check by periodic droughts which will cause cattle mortality. Elsewhere, it has been argued that no long-term effects would occur to pasture, due to their resilience; periodic movements southwards and northwards of the Sahel border are known to occur, resulting from droughts, and these have occurred continually with no apparent long-term trends (Tucker *et al.*, 1991; Winrock International, 1992).

Linear (1981, 1982, 1985a,b) also made frequent attacks on the policy of tsetse control. He argued that it is unnecessary, as meat production is predominantly for the benefit of western countries. Furthermore, western chemical companies were advocating tsetse control in order to make large profits from sales of insecticides. He considered that little benefit has come from the millions of dollars already spent on tsetse control.

Since those arguments were made, the use of insecticides for tsetse control has changed significantly. DDT and dieldrin are no longer in use and the quantities of other insecticides applied for control, already much less than have been used for agricultural crop spraying, have also decreased, as only small quantities of insecticides are applied to impregnated targets. Jordan (1986) countered Linear's attacks on tsetse control by referring to the scientific evidence demonstrating that effects of insecticide deposits from tsetse-control spraying had only transitory effects.

Rural development will take place independently of whether or not tsetse control is carried out, and will have an impact on the environment, and the challenge lies in planning such development in a manner likely to have the least possible detrimental impact.

19.5   ECONOMICS OF TSETSE CONTROL

Although some data on costs of specific tsetse control schemes have been given (e.g. Chadwick, 1964b: ground-spraying costs; Négrin and MacLennan, 1977: economic assessment of aerial spraying; Laveissière and Couret, 1986: costs of targets), there have been relatively few comprehensive reviews of the economics of tsetse or trypanosomosis control. Among such reviews are those of Jahnke (1976, 1982), who pioneered studies in this field, followed by Brandl (1988), working on tsetse control schemes mainly in West Africa, Barrett (1997) in Zimbabwe, and Itty (1992) in the Democratic Republic of the Congo and The Gambia. As new tsetse-control techniques develop, such as improvements in trap and target technology and the development of pour-on insecticides, economic evaluation of the methods available obviously requires updating. For this reason, and because some of the earlier techniques of tsetse control are now obsolete, early studies on economics of tsetse control are outdated and may only provide an indication of *relative* costs of different control methods. Furthermore, costs of specific control methods can vary greatly according to the topography and scale of an operation; for example, costs per km^2 for targets used to control savanna tsetse (Table 19.2) vary from US$150 to US$524. The costs of odour attractants for use with traps or targets

Table 19.2. Comparison of costs of tsetse or trypanosomosis control.

Method	Cost km^{-2} (US$)			Cost km^{-2} year^{-1}	Cost per animal year^{-1}
	(a)	(b)	(c)	(d)	(d)
Aerial spraying	180–515	290–380			
Ground spraying	170–700	167			
Helicopter spraying	380–790		200[†]		
Traps for riverine tsetse	39–54		3.90*		
Traps for savanna tsetse (4 km^{-2})	187–353	150–165	2.70*	524	77
Targets for riverine tsetse					
SIT	626				
Deltamethrin dipping (infrastructure available)	28	16			
Deltamethrin pour-on	108	70			
Chemoprophylaxy				228	34
5 head km^{-2} for 5 years	62–85				
10 head km^{-2} for 10 years	210–284				
20 head km^{-2} for 20 years	580–774				
Trypanotolerant cattle	670–1400				

Sources:
(a) Tacher *et al.* (1988); Shaw and Hoste (1991) (summarized in FAO, 1993).
(b) Thomson and Wilson (1992b).
(c) Spielberger *et al.* (1977). * Vavoua trap, Côte d'Ivoire in 1987 (half the cost of a biconical trap and less than a pyramidal trap). Targets used in Côte d'Ivoire in 1987 (Appropriate Technology in Vector Control, 1992). [†] Nigeria.
(d) Putt *et al.* (1988).

(resulting in a 300–400% improvement in catches of *G. pallidipes* and 50% improvement for *G. morsitans*) has been estimated by Vale *et al.* (1988b) to be about £2 (just over US$3) per trap per year for 3-methylphenol and 4-methylphenol.

Costs of various control measures are relatively easy to obtain, but they are not always collected in a standardized way and therefore some care has to be taken in their interpretation. It is more difficult to determine accurately the benefits of tsetse or trypanosomosis control, as many of the benefits are indirect and will include social benefits, which are particularly hard to quantify.

Brandl's (1988) study compared sterile male release, aerial spraying, trapping with impregnated traps and chemotherapy. The use of insecticide-impregnated traps was the least costly, provided it did not have to be continued for periods longer than 5–10 years. Eradication would be a more economical approach if practically feasible. Sterile male release was competitive only if the sterile male production facilities could be used in big programmes covering large areas. Brandl stated that he knew of no example of West African stock-owners contributing in any way to costs of a tsetse control scheme.

19.6   LOSSES DUE TO TRYPANOSOMOSIS

Jahnke (1976) wrote that 'very little is known ... about the costs of the different control methods, and quantitative estimates of the resulting benefits are virtually non-existent'. Twenty years later it is perhaps true to say that more is known about costs of control but estimates of losses due to trypanosomosis are still poorly known, and the rather imprecise figure of US$5 billion per year of potential capital lost in Africa (FAO/WHO/OIE, 1963) is still frequently quoted. The reason for this absence of quantification is to do with the difficulties of making accurate estimates rather than a lack of interest in obtaining the estimates. Although the direct costs in loss of meat and milk may be relatively easy to estimate, the indirect benefits, which would result from increased cultivation and crop production and manure, are particularly difficult to assess and are sometimes ignored. A further benefit from tsetse and animal trypanosomosis control relates to human trypanosomosis, which is even more difficult to evaluate. Based on a comparison of the carrying capacity of tsetse-infested and non-infested areas of the humid and subhumid regions of Africa, Jahnke *et al.* (1988) estimated that the disappearance of trypanosomosis in Africa could lead to an increase of 1.032 million tons of meat-equivalent per year, the value of which would be approximately US$1.9 billion, which is likely to be a more realistic estimate than that of FAO/WHO/OIE (1963). It must also be noted that there are other disease constraints to livestock production in tsetse-infested areas and the alleviation of trypanosomosis would not necessarily increase productivity to levels achieved in more disease-free environments.

19.7   COMMUNITY PARTICIPATION IN TSETSE CONTROL

While the technical means of controlling tsetse flies are available, the main difficulty lies in sustaining control over a long period. This is especially the case when using traps and targets for control as opposed to aerial or ground-spraying operations. The reasons for this are that aerial or ground spraying of insecticides are usually big, intensive campaigns carried out over a short period which require central direction and a fairly large amount of capital. On the other hand, targets and trapping techniques are long-term operations, requiring some initial organization and direction followed by a long maintenance phase. It is widely felt that successful control using traps and targets should involve the local community. It is believed that if the communities which are the intended beneficiaries of the control are involved in the exercise, the operation will be much more economical and sustainable. In theory, the operation should also stand more chance of success if the beneficiaries, having a vested interest in the success of the operation, are involved in carrying it out. The desirability of involving beneficiary communities in tsetse control is by no means a new idea; as long ago as the 1920s the cooperation of people who were the intended beneficiaries of tsetse control was believed to be at least desirable, if not essential, for successful large-scale tsetse control (Swynnerton, 1925).

The average farm size of two-thirds of Africa's agricultural land is less than 2 ha (Harrison, 1987). In order to cultivate greater areas, which is essential if the increasing population is to be fed, an increase in the extent of draught power will be needed. If trypanosomosis control is going to depend upon support and participation of rural communities, the demand for these techniques has to be known. Furthermore, these control measures would benefit from being integrated with control of other diseases and overall improvement in management of livestock for increased productivity.

Opinions of what constitutes community participation vary, ranging from simply an awareness by the community of tsetse-control activities being carried out in their area, and passive participation in which the community is persuaded to leave targets or traps alone, to contributions of money and labour towards the implementation of control activities, or 'community ownership' of the project, which some people see as the only way of ensuring sustainability.

Several attempts have been made to carry out tsetse control with community participation, some of which have been mentioned earlier. Laveissière *et al.* (1985b) have had success with community participation during control operations against human sleeping sickness in West Africa, but there have been problems with sustainability. Dransfield *et al.* (1991) had partial success in controlling trypanosomosis of domestic livestock by the use of traps at Nguruman, in Kenya, on a Maasai group-ranch community scheme but this eventually failed, for complex reasons (Williams *et al.*, 1995). In Uganda, Lancien (1991) involved the community in control of human sleeping sickness, though it was managed with external direction.

The pyramid trap (Gouteux and Sinda, 1990) has been used by communities in the Congo for control of human sleeping sickness transmitted by *G. palpalis*. The trap was used without insecticide impregnation and was easily operated by villagers, who could see the effectiveness of the trap as tsetse were preserved in a similar way to the NGU2B traps used at Nguruman. However, long-term community participation was limited and would either have to be improved or associated with the work of a specialized team for successful control to be achieved. Education is an important component of a successful community participation programme; Okoth *et al.* (1991) described the development of awareness of trapping technology amongst a rural community in Uganda, and their subsequent involvement in making and setting traps.

As community participation of some kind is clearly desirable for sustainable tsetse control, further research may be required into situations where success is most likely to be achieved and into the constraints that make community participation ineffective. Many attempts at community participation in vector control against other diseases such as malaria and Chagas' disease have failed, mainly because of the difficulty in sustaining motivation but also because of traditional customs. An example is the unpopularity of bed-nets for malaria control in Burkina Faso because of their association with shrouds for dead people (Service, 1993). Elsewhere, the use of bed-nets has not been adopted by communities because although they protect against malaria-transmitting mosquitoes active at night, other species of 'nuisance' mosquitoes active at other times are not affected, leaving people with the impression that the bed-nets do not work as they are still being bitten.

Studies on communities' willingness to contribute either labour or money and their perceptions of trypanosomosis and tsetse control have been conducted in Kenya (Echessah *et al.*, 1997) and in Ethiopia (Swallow and Mulatu, 1994; Swallow *et al.*, 1995). Community participatory tsetse control among nomadic cattle owners will clearly present special difficulties owing to the distances and varying areas to which cattle are taken for grazing.

19.8   PUBLIC AND PRIVATE GOODS AND PRIVATIZATION OF ANIMAL HEALTH CARE

As African economies are generally unable to provide efficient animal health services in the public sector, there is an increasing wish to transfer some of these services to the private sector. This raises questions of how tsetse control, which generally results in a public rather than a private good, would fit into private animal health services. The question of private or public goods is of relevance to community participation, as beneficiaries of tsetse control may be more willing to contribute, either financially or with labour, if they perceive a private good. On the other hand, as successful tsetse control would generally confer a public good, there is a problem of 'free-riders' – people who are unwilling to contribute, in the knowledge that they will benefit from the results of the contributions of others (Umali *et al.*, 1994). In addition to public or private goods resulting from tsetse control, there may also be what economists call 'negative externalities', i.e. spill-over effects. Negative externalities associated with tsetse control by the pour-on technique, for example, may be loss of endemic stability to tick-borne haemoprotozoan disease, or development of resistance by ticks to the chemicals that are used as both insecticides and acaricides. An unanswered question is how pour-on tsetse control will be administered by private sector animal health services, and how the insecticide will be distributed when benefits depend upon widespread community use. It may be that public-good vector control may have to remain the responsibility, and in the control of, public animal health provision.

19.9 CONCLUSIONS

Efforts to develop non-insecticidal methods of vector control reflect a philosophy regarding insecticidal use as unsustainable and environmentally unacceptable. At present, however, insecticides remain the main means of tsetse control. The method of application is changing from ground or aerial spraying techniques to the more environmentally acceptable impregnation of traps or targets and the use of pour-on insecticides.

In a recent review of responses of vectors to control operations, Schofield (1991) referred to sequential aerial spraying of tsetse in the Lambwe valley, Kenya, in 1981, which reduced the population of *G. pallidipes* by over 99.9% in its main habitats. This population recovered to original levels in only one year and Schofield used this, and other examples, to illustrate the difficulty in maintaining control of vector populations and reduction of disease transmission to acceptable levels. He concluded that the biological robustness and complexity of vector populations and disease transmission cycles are often ignored, and that economic constraints are major hurdles to the sustainability of control methods.

Part IV

Control of Trypanosomosis

Chapter 20

Control of Trypanosomosis in Domestic Livestock

The control of trypanosomosis in domestic livestock depends mainly upon the use of curative and prophylactic drugs. Because of the cost of developing new trypanocides, and the relatively small commercial market, there have been no new drugs for about 30 years and there is little prospect of new ones being developed in the near future. As resistance to the available drugs is on the increase and their continued use is expensive for livestock owners, other control strategies have to be sought. In addition to control of the vector (Chapters 15–18), attempts to develop vaccines against either the disease or the vector have been made. A fourth approach to overcome the disease is to exploit the trypanotolerance trait, either by expanding and improving the rearing of trypanotolerant breeds or by manipulating the trypanotolerance trait in cattle – for example, by introducing the genes for trypanotolerance into susceptible but more productive cattle. Trypanotolerance has been discussed in Chapter 13. Other aspects of trypanosomosis control in domestic livestock are reviewed in this chapter.

A traditional method of protecting cattle from trypanosomosis in Zambia, believed to be both prophylactic and curative, consisted of making cattle drink a mixture made from the bark of a specific tree and containing one dead tsetse fly (Montgomery and Kinghorn, 1908). The cattle were then kept overnight in a hut in which they were smoked with a fire made from a particular bush. The effects of this were said to last for 3–4 days. It is possible that the smoke from the bush could have acted as a repellent, and thus could have protected cattle from the bites of flies. The neem tree is known to have a repellent effect against insects. Montgomery and Kinghorn gave local, but not scientific names, for the tree and bush used.

20.1   CHEMOTHERAPY

Chemotherapy is presently the major method for control of trypanosomosis in livestock and some background on this subject is therefore given here. A comprehensive review of chemotherapy and chemoprophylaxis of animal trypanosomosis is given by Leach and Roberts (1981).

Drugs currently recommended for chemotherapy of animal trypanosomosis come from only three closely related groups. These are the phenanthridines, isometamidium and homidium, and the aromatic diamidine, diminazene. Only isometamidium and homidium are recommended for prophylaxis. The incidence of resistance to these drugs is apparently increasing (Peregrine, 1994) and the main means of controlling the disease is therefore under threat. The six compounds currently used for chemotherapy of trypanosomosis in domestic livestock are listed in Table 20.1. The first three are primarily used for the treatment or prophylaxis of cattle, sheep and goats; the last three are used for treatment of *T. evansi* infections in camels, equids and water buffalo.

Diminazene aceturate (Berenil®) was thought to have some prophylactic activity, providing protection against natural infection for up to 3 weeks when used at a dosage of 7 mg kg^{-1} body weight (Van Hoeve and Cunningham, 1964). However, the drug is quickly excreted (Bauer, 1962) and is not used as a prophylactic.

20.2   CHEMOPROPHYLAXIS

Chemoprophylaxis against bovine trypanosomosis has been in widespread use in tropical Africa for many years. Isometamidium chloride (Samorin®) has been marketed since 1961 as a prophylactic and therapeutic drug. Prophylaxis can be useful under high challenge situations and enables cattle to remain productive, as demonstrated on a commercial ranch at Mkwaja, in Tanzania. On this ranch, cattle maintained under isometamidium prophylaxis were 80% as productive as high quality Boran cattle on

Table 20.1. Anti-trypanosomal drugs in current use.

Chemical name	Trade names
Isometamidium chloride	Samorin®, Trypamidium®, Veridium®
Homidium chloride	Novidium®
Homidium bromide	Ethidium®
Diminazene aceturate	Berenil®, Veriben®
Quinapyramine sulphate	Trypacide sulphate®, Antrycide®
Quinapyramine sulphate + chloride	Trypacide Prosalt®, Antrycide Prosalt®
Suramin sodium	Naganol®
Melarsomine	Cymelarsan®

trypanosomosis-free ranches in Kenya (Trail *et al.*, 1985). Protection can be conferred for a period ranging from 2 to 22 weeks at the recommended dose of 1 mg kg^{-1} body weight administered intramuscularly.

A farmer in Kenya reported that intravenous administration of isometamidium chloride was more effective in protecting cattle from trypanosomosis than the recommended intramuscular route (Dowler *et al.*, 1989). This observation was evaluated in a trial with a control group treated intramuscularly with diminazene aceturate (Munsterman *et al.*, 1992). When administered to Boran cattle under a tsetse challenge from *G. pallidipes, G. brevipalpis, G. austeni* and *G. longipennis* at the coast of Kenya, intravenous isometamidium not only had a very good therapeutic effect but also a prophylactic effect of not less than 4 weeks. However, intravenous administration of isometamidium to Boran cattle experimentally infected with a drug-resistant clone of *T. congolense* did not enhance the therapeutic efficacy of the drug, although intravenous treatment did cure infections with a drug-sensitive clone (Sutherland *et al.*, 1991).

The purchase of anti-trypanosomal drugs by African countries requires foreign currency. At present, foreign donors are often only willing to purchase anti-trypanosomal drugs when a system of revolving funds is set up to pay for them (Connor, 1989). Many governments are reluctant to do this. Due partly to difficulties in setting up revolving funds in countries in a donor-funded control programme in southern Africa, sales of trypanocidal drugs dropped as no foreign money was provided (Connor, 1989).

It has been estimated that at least $30 million is spent annually to treat or protect animals exposed to trypanosomosis in Africa (Borne, 1996).

20.3 DRUG RESISTANCE

No new trypanocidal drugs were produced for treating domestic animals for over 30 years, until melarsomine was marketed in the 1980s, primarily for treating *T. evansi* in camels (Raynaud *et al.*, 1989). The use of the same drugs over such a long period has resulted in the widespread development of drug-resistant strains of trypanosomes (Fig. 20.1; Peregrine, 1994). Drug-resistant trypanosomes develop through: (i) under-dosing, which may occur for a variety of reasons (such as underestimation of animal body weight, over-diluted solutions of trypanocides, deliberate under-dosing or incorrectly calculated dose volume); (ii) incorrect injection; or (iii) an incorrect strategy of drug use.

The sensitivity of trypanosomes to a drug is important in determining the period of prophylactic activity that it provides (Peregrine *et al.*, 1991a,b). One test for assessing resistance consists of incubation of trypanosomes in a liquid medium for 24 h at 37°C in the presence of concentrations of diminazene or isometamidium. The trypanosomes are then assayed for infectivity in mice. Drug-resistant strains can be distinguished using this test.

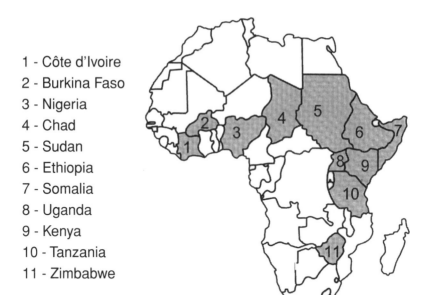

1 - Côte d'Ivoire
2 - Burkina Faso
3 - Nigeria
4 - Chad
5 - Sudan
6 - Ethiopia
7 - Somalia
8 - Uganda
9 - Kenya
10 - Tanzania
11 - Zimbabwe

Fig. 20.1. African countries from which resistance to trypanocidal drugs has been reported.

In vitro techniques for assessing drug susceptibility of trypanosomes have also been examined (Kaminsky, 1990).

Gray and Roberts (1968) reported that, in natural situations, drug-resistant strains undergoing cyclical transmission through tsetse disappear after 6–9 months, provided that infected cattle are withdrawn from the area and the use of the drug concerned is suspended. Smith and Scott (1961) showed that a short period after the administration of prophylactic trypanocidal drugs, apparent relapse infections occurred in animals considered to be protected from reinfection. However, parasitaemias remained at a low level and it was suggested that the drug modified the parasite in some way. One of the earliest reports of drug resistance was cited by Lewis (1953), referring to resistance to Antrypol® by *T. evansi* in camels in Sudan in 1945.

Cross-resistance to isometamidium in *T. congolense* resistant to homidium chloride has been reported in northern Nigeria (Jones-Davies and Folkers, 1966a). This highlights a further complication in which cross-resistance to a drug may occur even in areas where that drug has never been used. Thus, in general:

● Induction of resistance to quinapyramine leads to cross-resistance to homidium, isometamidium and diminazene (Ndoutamia *et al.*, 1993; Whiteside, 1960).

- Induction of resistance to homidium leads to cross-resistance to quinapyramine and isometamidium, but not to diminazene (Whiteside, 1960).
- Induction of resistance to isometamidium leads to cross-resistance to quinapyramine and homidium, but not to diminazene (Whiteside, 1960; Peregrine *et al.*, 1997).
- Induction of resistance to diminazene does not result in cross-resistance to quinapyramine, homidium or isometamidium (Whiteside, 1960, 1963).

'Sanative pairs' of drugs, by means of which induction of resistance to one drug can be eliminated by use of the other (Whiteside, 1960), that can effectively be used are: (i) isometamidium and diminazene; and (ii) homidium and diminazene.

Other measures that may delay the development of drug resistance are to reduce the selection pressure on trypanosome populations by avoiding exclusive reliance on drugs for trypanosomosis control and avoiding mass treatments of livestock at short intervals. As no new drugs are likely to be produced in the near future, better formulations of existing drugs are being sought, as well as improved strategies for their use, in order to avoid sub-therapeutic concentrations, which would select for resistant trypanosomes.

In view of the increasing prevalence of drug resistance, it is important to know how effective tsetse control would be in alleviating trypanosomosis in livestock in situations where trypanocidal drugs alone have become ineffective. Studies carried out at Ghibe, in southwest Ethiopia, where a high prevalence of multiple drug resistance was detected (Codjia *et al.*, 1993), showed that tsetse control, in combination with trypanocidal drug use, can effectively reduce the apparent trypanosome prevalence in cattle (Leak *et al.*, 1996).

20.3.1 Reports of drug resistance

Countries in which trypanosomes resistant to anti-trypanosomal drugs have been reported are listed in Table 20.2. However, there have been few studies to determine its occurrence. Thus, this table under-represents the true prevalence of resistance. Standardized tests for detection of drug resistance are urgently required, especially tests which will distinguish between apparent drug resistance and true drug resistance. Apparent drug resistance can result from the following:

- Fraud, in which the genuine drug has been replaced by an alternative, inactive substance.
- Trypanosomes in sites within the mammalian host not reached by the drug.
- Increased trypanosomosis challenge.

Table 20.2. Reports of resistance to standard doses of anti-trypanosomal drugs.

Country	Drug	Reference
Nigeria	Diminazene	Jones-Davies, 1967, 1968; Joshua, 1988
	Isometamidium	Jones-Davies and Folkers, 1966a
	Homidium chloride	Jones-Davies and Folkers, 1966b
	Multiple	Na'isa, 1967
	Diminazene	MacLennan and Jones-Davies, 1967
	Multiple	Gray and Roberts, 1971; Illemobade, 1979
Chad	Diminazene	Graber, 1968
Uganda	Diminazene	Mwambu and Mayende, 1971
Burkina Faso	Diminazene	Authié *et al.*, 1984; Clausen *et al.*, 1992
	Isometamidium	Authié *et al.*, 1984
	Isometamidium	Pinder and Authié, 1984
Ethiopia	Homidium bromide	Scott and Pegram, 1974
	Multiple	Codjia *et al.*, 1993
Tanzania	Diminazene	Njau *et al.*, 1981; Mbwambo *et al.*, 1988
Somalia	Multiple	Schönefeld *et al.*, 1987
	Multiple	Ainanshe *et al.*, 1992
Sudan	Multiple	Lewis, 1953
	Multiple	Mohamed-Ahmed *et al.*, 1992
	Homidium	Abdel Gadir *et al.*, 1981
	Isometamidium	Abdel Gadir *et al.*, 1972
Kenya	Multiple	Gitatha, 1979; Schönefeld *et al.*, 1987
	Multiple	Röttcher and Schillinger, 1985
Côte d'Ivoire	Homidium/ isometamidium	Küpper and Wolters, 1983
Zimbabwe	Isometamidium	Lewis and Thomson, 1974

20.3.2   Mechanisms of drug resistance

Little is known about the mechanisms of action of anti-trypanosomal drugs or resistance to these compounds. However, it has been shown that isometamidium is transported into *T. congolense* across the plasma membrane by a protein carrier (Zilberstein *et al.*, 1993). Resistance to diminazene and organic arsenicals is associated with a down-regulation of an adenosine transport mechanism, and resistance to isometamidium is associated with the mitochondrial electric potential (Wilkes *et al.*, 1997).

Trypanocides have a high affinity for kinetoplast DNA (Wilkes *et al.*, 1995). The mechanism of isometamidium activity appears to be through disruption of the trypanosome's kinetoplast structure.

20.4   EFFECTS OF ANTI-TRYPANOSOMAL DRUGS FOR CATTLE ON TSETSE FLIES

One of the first experiments to look at the effect of drugs fed to tsetse in their bloodmeal on trypanosome infections involved feeding *G. palpalis* with tryparsamide (Van Hoof *et al.*, 1937c). Unfortunately, as few flies were used, some of the results were contradictory and inconclusive. Nevertheless, it was concluded that the drug did not completely clear infections but that there was some degree of protection to animals bitten by tsetse that had taken a medicated meal. The results suggested that tryparsamide had some effect upon either the maturation or infectivity of trypanosomes.

Quinapyramine, isometamidium and diminazene, at concentrations of 0.1 mg kg^{-1}, eliminate the three trypanosome species pathogenic to cattle (*T. vivax*, *T. congolense* and *T. b. brucei*) within *G. pallidipes* when flies are membrane fed on blood containing the drugs (Hawking, 1963).

The effect of isometamidium chloride on developing and mature *T. vivax* infections in *G. p. palpalis* has been examined by Agu (1984b, 1985). A group of infected flies was fed on animals treated 1 day previously with the drug at a dose of 1 mg kg^{-1} body weight. A second group was fed on untreated animals. Infections in all of the flies that fed on the treated animals were eliminated, whilst 55 out of 127 that fed on untreated animals remained infected. It was concluded that prophylactic treatment of animals with isometamidium could eliminate immature or mature *T. vivax* infections from the vectors. It should be noted that the drug had an effect for up to 20 days when tsetse were fed on treated sheep 1 day after treatment. A shorter period of prophylactic activity against trypanosome infections in tsetse would be expected as the time after drug treatment increased. Consequently, anti-trypanosomal drugs administered to cattle under field conditions are likely to have less effect on trypanosome infection rates in tsetse than under experimental conditions.

Similar experiments have been conducted to determine the effect of isometamidium on infections with the three main trypanosome species in irradiated sterile flies (Moloo and Kamunya, 1987). The infection rate of *T. congolense* in *G. m. morsitans* was reduced when the flies were fed on an infected goat 2 days after they were fed as tenerals on an *in vitro* blood-meal containing 8 µg isometamidium ml^{-1} blood. Infections of *T. vivax* and *T. b. brucei* were completely suppressed at the same dose. When fed on a bloodmeal containing 12 µg isometamidium ml^{-1} blood, and given an infected bloodmeal 10 days later, all mature infections in the flies were completely suppressed. Other studies on the effect of trypanocidal drugs on transmission of *T. b. brucei* by *G. m. centralis* demonstrated that drugs taken early enough during parasite development reduced the prevalence of already established infections and prevented establishment of new infections (Jefferies and Jenni, 1987). In this study, pentamidine and suramin (drugs used for treating human sleeping sickness) had no significant effects, but both diminazene and isometamidium reduced the number of salivary gland infections in comparison with controls. Drug levels in the blood of vertebrate hosts were about the same as used for membrane feeding to flies. Whilst not killing trypanosomes in the midgut, the drugs appeared to prevent their transformation to epimastigote forms which could invade the salivary glands and complete the developmental cycle.

The effect of isometamidium chloride on insect forms of drug-sensitive and drug-resistant stocks of *T. vivax* has been assessed *in vitro* and in tsetse (Kaminsky *et al.*, 1991). The drug-sensitive stock showed reduced growth and died after 6–10 days, depending on the concentration of drug with which it was incubated, whilst the drug-resistant stock continued to grow for 17 days. Trypanosome infection rates in tsetse were greatly reduced when flies infected with the drug-sensitive stock were fed on a Boran steer previously treated with 1 mg isometamidium kg^{-1}; the infection rates of tsetse with the drug-resistant stock were not affected when fed on the same animal. At the doses used for prophylaxis in cattle, isometamidium has no effect on the reproductive performance of tsetse, and therefore no potential for tsetse control (Moloo and Kutuza, 1986). In contrast, in an earlier study the overall reproductive performance of *G. m. morsitans* fed on rabbits injected with the trypanocide was severely affected (Oladunmade and Balogun, 1979). It was suggested that this resulted from an effect on midgut symbionts.

Treatment of cattle infected with diminazene-resistant *T. congolense* at a dose of 3.5 mg kg^{-1} body weight reduced the infectivity of the parasite for tsetse and also impaired the parasite's ability to mature for up to 7 days after treatment (Diack *et al.*, 1997). Thus, in situations with a high prevalence of drug-resistant trypanosome infections, mass treatment of cattle with diminazene could be useful at the onset of a tsetse-control programme. If tsetse predominantly become infected with *T. congolense* on their first feed only (Chapter 11, section 11.3.3) and are thereafter refractory to infection, the usefulness of such a mass treatment would be even greater.

20.5   NATURAL IMMUNE STATUS

Young animals are naturally protected to some extent from trypanosomosis by maternal antibodies in colostrum from the mother (Fimmen *et al.*, 1982a). Acquisition of protective serum-antibody levels from colostrum in young goats is correlated with resistance to infection, and therefore antibodies of maternal origin could influence the response of newborn animals to experimental infection (Whitelaw and Jordt, 1985). Maternal colostrum of N'Dama cattle also contains anti-trypanosomal antibodies (Fimmen *et al.*, 1982a,b). However, the protection given by these antibodies is unclear: an experiment in 1979 showed no evidence of protection, whilst a second experiment in 1981 suggested partial protection against *T. congolense*. There was evidence that protection was specific to trypanosome strains to which the dam had been exposed previously.

Some breeds of livestock have an innate genetic trypanotolerance (Chapter 12), whilst more susceptible breeds can acquire some level of resistance to infection through repeated exposure, particularly if there is a limited number of trypanosome serodemes to which the cattle are exposed.

20.6   VACCINES AND IMMUNOLOGICAL STRATEGIES FOR TRYPANOSOMOSIS CONTROL

Attempts to vaccinate cattle against trypanosomosis started at an early stage in trypanosomosis research; for example, Bevan (1928) tried protecting animals by deliberately infecting them with known strains of trypanosomes and then treating them at various intervals following infection with potassium antimony tartrate. These animals were said to become tolerant, or to some extent 'premunized', but the resistance broke down relatively easily. This 'infection and treatment' technique is currently used for 'vaccination' against *Theileria parva*, the parasite causing East Coast fever (Radley *et al.*, 1975).

Attempts were also made to vaccinate cattle with killed trypanosomes, by part-curative doses of trypanocidal drugs after infection, or by injection of small numbers of living trypanosomes (Schilling, 1935). These attempts failed and it is now recognized that the major obstacle to immunization lies in the phenomenon of antigenic variation in trypanosomes. Trypanosomes are covered by a dense coat of variant surface glycoproteins (VSGs) that stimulate antibody production in the host. The surface coat changes successively during the course of an infection, thus avoiding the immune response of the host. If a drug could be devised to interfere with antigen switching, the prospects of a vaccine for trypanosomosis would be much improved. At present, however, the prospects of such a vaccine are very much in the future.

For the past 25 years or so, much research effort has investigated new approaches of producing a vaccine for trypanosomosis, which would overcome the difficulties presented by antigenic variation. The phenomenon of antigenic variation and its genetic control have been described and it has been concluded that vaccines based on immune responses to the variant surface coat are unlikely to have any impact on trypanosomosis control. This has led to an approach based on molecular biology of the cell, looking at components and products responsible for the development of disease. This approach could lead to the development of vaccines based on immune responses to invariant molecules such as cysteine proteases or cyclophilins produced by trypanosomes (Teale, 1993).

When tsetse were maintained on goats immunized with uncoated forms of trypanosomes grown *in vitro*, parasitaemias of the three trypanosome species pathogenic to cattle were suppressed (Murray *et al.*, 1985). This immunization was species-specific but independent of the stock used.

20.6.1 Immunological control of vectors

For a detailed background of the immunological control of vectors, refer to reviews of this research area (Opdebeeck, 1994; Jacobs-Lorena and Lemos, 1995). Antibodies have been produced against a variety of insect or tick vectors that have an effect on mortality or fertility, but with little practical application so far, other than for *Boophilus microplus*.

Experimentally, *G. m. morsitans* which were fed on rabbits immunized with crude tsetse midgut proteases could not properly digest their blood-meals and larviposited prematurely (Otieno *et al.*, 1984) and immunization using whole *G. m. morsitans* resulted in increased mortality and a slightly decreased fertility of the flies (Nogge, 1978). In a similar experiment, tsetse flies fed on rabbits immunized with thoracic muscles of *Stomoxys calcitrans* were killed or their development was affected (Schlein and Lewis, 1976). Albumin is essential for osmoregulation in *G. morsitans*, and antibodies absorbed through the gut by feeding tsetse on rabbits immunized with human albumin resulted in suppression of crop emptying and primary excretion, and death (Nogge and Giannetti, 1980).

20.6.2 Immunization of hosts to prevent development of trypanosome infections in tsetse

As early as 1912, experiments were carried out to see whether tsetse developed any immunity to trypanosomes (Kleine and Eckard, 1913, in Duke, 1933a). So-called anti-vector vaccines that inhibit transmission of disease organisms have been investigated mainly for malaria transmission by mosquitoes (Billingsley, 1994) but the same approach has been examined for blocking transmission of trypanosomes by tsetse (Maudlin *et al.*, 1984; Murray *et al.*, 1985; Nantulya and Moloo, 1988). These studies showed that ingestion by tsetse of monoclonal antibody to a defined procyclic surface antigen significantly reduced cyclical development of *T. b. brucei*. This approach was suggested as a possible method for the control of sleeping sickness by immunization of livestock hosts of peridomestic disease vectors (Nantulya and Moloo, 1988).

References

Abaru, D.E. (1985) Sleeping sickness in Busoga Uganda, 1976–1983. *Tropical Medicine and Parasitology*, 36, 72–76.

Abdel Gadir, F., Osman, O.M., Abdalla, H.S. and Abdel Razig, M.T. (1981) Ethidium bromide-resistant trypanosomes in southern Darfur. *Sudan Journal of Veterinary Research*, 3, 63–65.

Abdel Karim, E.I. and Brady, J. (1984) Changing visual responsiveness in pregnant and larvipositing tsetse flies, *Glossina morsitans*. *Physiological Entomology*, 9, 125–131.

Abdurrahim, U. (1971) A study of the diurnal resting behaviour of *Glossina palpalis* in southern Zaria northern Nigeria. *ISCTRC 13th meeting, Lagos*, 213–227.

Abedi, Z.H. and Miller, M.J. (1963) Tsetse fly puparia: a new collecting technique. *Science*, 141, 264.

Abubakar, L., Osir, E.O. and Imbuga, M.O. (1995) Properties of a blood-meal induced midgut lectin from the tsetse fly *Glossina morsitans*. *Parasitology Research*, 81, 271–275.

Adams, A.R.D. (1935a) Trypanosomiasis of stock in Mauritius. I. *Trypanosoma vivax*, a parasite of local stock. *Annals of Tropical Medicine and Parasitology*, 29, 1–18.

Adams, A.R.D. (1935b) Trypanosomiasis of stock in Mauritius. II. Observations on the incidence and distribution of trypanosomiasis in cattle. *Annals of Tropical Medicine and Parasitology*, 29, 475–481.

Adams, A.R.D. (1936) Trypanosomiasis of stock in Mauritius. III. The diagnosis and course of untreated *Trypanosoma vivax* infections in domestic animals. *Annals of Tropical Medicine and Parasitology*, 30, 521–531.

Agu, W.E. (1984a) Comparative study of the susceptibility to infection with *Trypanosoma simiae* of *Glossina morsitans* and *G. tachinoides*. *Acta Tropica*, 41, 131–134.

Agu, W.E. (1984b) The effect of Isometamidium chloride on *Trypanosoma vivax* occurring within the insect vector (*Glossina*). *Zeitschrift für Parasitenkunde*, 70, 431–435.

Agu, W.E. (1985) Action of Isometamidium chloride on the insect vector form of *Trypanosoma vivax*. *Research in Veterinary Science*, 39, 289–291.

Agyemang, K., Dwinger, R.H., Little, D.A. and Rowlands, G.J. (1998) *Village N'Dama production in West Africa: six years of research in The Gambia*. International Livestock Research Institute, Nairobi, Kenya and International Trypanotolerance Centre, Banjul, The Gambia, 131 pp.

433

Ainanshe, O.A., Jennings, F.W. and Holmes, P.H. (1992) Isolation of drug-resistant strains of *Trypanosoma congolense* from the lower Shabelle region of southern Somalia. *Tropical Animal Health and Production*, 24, 65–73.

Akol, G.W.O. and Murray, M. (1982) Early events following challenge of cattle with tsetse infected with *Trypanosoma congolense*: development of the local skin-reaction. *Veterinary Record*, 110, 295–302.

Akov, A. (1972) Protein digestion in haematophagous insects. In: Rodriguez, J.G. (ed.) *Insect and Mite Nutrition*. North-Holland, Amsterdam, pp. 531–540.

Aksoy, S. (1995) Molecular analysis of the endosymbionts of tsetse flies: 16S rDNA locus and over-expression of a chaperonin. *Insect Molecular Biology*, 4, 23–29.

Aksoy, S., Pourhosseini, A.A. and Chow, A. (1995) Mycetome endosymbionts of tsetse flies constitute a distinct lineage related to Enterobacteriaceae. *Insect Molecular Biology*, 4, 15–22.

Alberro, M. and Haile-Mariam, S. (1982a) The indigenous cattle of Ethiopia, Part I. *World Animal Review*, 41, 2–10.

Alberro, M. and Haile-Mariam, S. (1982b) The indigenous cattle of Ethiopia, Part II. *World Animal Review*, 42, 27–34.

Allsopp, R. (1972) The role of game animals in the maintenance of endemic and enzootic trypanosomiasis in the Lambwe valley, south Nyanza District, Kenya. *Bulletin of the World Health Organisation*, 47, 735–746.

Allsopp, R. (1984) Control of tsetse using insecticides: a review and future prospects. *Bulletin of Entomological Research*, 74, 1–23.

Allsopp, R. (1985a) Variation in the rates of increase of *Glossina morsitans centralis* and their relevance to control. *Journal of Applied Ecology*, 22, 91–104.

Allsopp, R. (1985b) Wing fray in *Glossina morsitans centralis*. *Bulletin of Entomological Research*, 75, 1–11.

Allsopp, R., Baldry, D.A.T. and Rodrigues, C. (1972) The influence of game animals on the distribution and feeding habits of *Glossina pallidipes* in the Lambwe valley. *Bulletin of the World Health Organisation*, 47, 795–809.

Almeida, P.J.L.P. de, Ndao, M., Van Meirvenne, N. and Geerts, S. (1997) Diagnostic evaluation of PCR in goats experimentally infected with *Trypanosoma vivax*. *Acta Tropica*, 66, 45–50.

Alton, G.G. and Jones, L.M. (1967) *Laboratory Techniques in Brucellosis*. WHO Monograph Series, No. 55, World Health Organisation, Geneva.

Amsler, S., Filledier, J. and Millogo, R. (1994a) Attractivité pour les Tabanidae de différents pièges à glossines avec ou sans attractifs olfactifs. Résultats préliminaires obtenus au Burkina Faso. *Revue d'Élevage et de Médecine Vétérinaire des Pays Tropicaux*, 47, 63–68.

Amsler, S., Filledier, J. and Millogo, R. (1994b) Efficacité comparée de différents pièges pour la capture de *Glossina tachinoides* (Diptera: Glossinidae) au Burkina Faso. *Revue d'Élevage et de Médecine Vétérinaire des Pays Tropicaux*, 47, 207–214.

Anderson, M. and Finlayson, L.H. (1973) Ultrastructural changes during growth of the flight muscles in the adult tsetse fly, *Glossina austeni*. *Journal of Insect Physiology*, 19, 1989–1997.

Anderson, R.M. (1981) Population dynamics of indirectly transmitted disease agents: the vector component. In: McKelvey, J., Eldridge, B. and Maramorosch, K. (eds) *Vectors of Disease Agents*. Praeger, New York, pp. 3–43.

Anderson, R.M. and May, R.M. (1991) *Infectious Diseases of Humans: Dynamics and Control*. Oxford University Press, London.

Añez, N. and East, J.S. (1984) Studies on *Trypanosoma rangeli* Tejera, 1920. II. Its effect on feeding behaviour of triatomine bugs. *Acta Tropica*, 41, 93–95.

Apted, F.I.C., Ormerod, W.E., Smyly, D.P., Stronach, B.W. and Szlamp, E.L. (1963) A comparative study of the epidemiology of endemic Rhodesian Sleeping sickness in different parts of Africa. *Journal of Tropical Medicine and Hygiene*, 66, 1–16.

Archibald, R.G. (1927) The tsetse fly-belt area in the Nuba mountains province of the Sudan. *Annals of Tropical Medicine and Parasitology*, 21, 39–45.

Archibald, R.G. and Riding, D. (1926) A second case of sleeping sickness in the Sudan caused by *Trypanosoma rhodesiense*. *Annals of Tropical Medicine and Parasitology*, 20, 161–166.

Aron, J.L. and May, R.M. (1982) The population dynamics of malaria. In: Anderson, R.M. (ed.) *Population Dynamics of Infectious Diseases*. Chapman and Hall, London, pp. 139–179.

Ashcroft, M.T. (1958) An attempt to isolate *Trypanosoma rhodesiense* from wild animals. *Transactions of the Royal Society of Tropical Medicine and Hygiene*, 52, 276–287.

Ashcroft, M.T. (1959a) The sex ratio of infected flies found in transmission experiments with *Glossina morsitans* and *Trypanosoma rhodesiense* and *T. brucei*. *Transactions of the Royal Society of Tropical Medicine and Hygiene*, 53, 394–399.

Ashcroft, M.T. (1959b) The importance of African wild animals as reservoirs of trypanosomiasis. *East African Medical Journal*, 36, 289–297.

Ashcroft, M.T. (1959c) A critical review of the epidemiology of human trypanosomiasis in Africa. *Tropical Diseases Bulletin*, 56, 1073–1093.

Ashcroft, M.T. (1963) Some biological aspects of the epidemiology of Sleeping sickness. *Journal of Tropical Medicine and Hygiene*, 66, 133–136.

Ashcroft, M.T., Burtt, E. and Fairbairn, H. (1959) The experimental infection of some African wild animals with *Trypanosoma rhodesiense, T. brucei* and *T. congolense*. *American Journal of Tropical Medicine and Parasitology*, 53, 147–161.

Atkins, J. (1978) Sleeping sickness. In: Kean, B.H., Mott, K.E. and Russell, A.J. (eds) *Tropical Medicine and Parasitology, Classic Investigations*, 1. Cornell University Press, Ithaca, New York, pp. 181–182.

Atkinson, P.R. (1971a) A study of the breeding distribution of *Glossina morsitans* Westw. in northern Botswana. *Bulletin of Entomological Research*, 60, 415–426.

Atkinson, P.R. (1971b) Relative humidity in the breeding sites of *Glossina morsitans* Westw. in northern Botswana. *Bulletin of Entomological Research*, 61, 241–246.

Austen, E.E. (1903) *A Monograph of the Tsetse-flies (genus* Glossina *Westwood) based on the collection in the British Museum*. British Museum (Natural History), London, 319 pp. [Johnson Reprint Corporation, New York, 1966].

Austen, E.E. (1914) A dipterous parasite of *Glossina morsitans*. *Bulletin of Entomological Research*, 5, 91–93.

Austen, E.E. (1923) A new East African tsetse-fly (Genus *Glossina*, Wied.) which apparently disseminates sleeping sickness. *Bulletin of Entomological Research*, 13, 311–315.

Austen, E.E. (1929) The tsetse-fly parasites belonging to the genus *Thyridanthrax* (Diptera. Family Bombyliidae), with descriptions of new species. *Bulletin of Entomological Research*, 20, 151–164.

Authié, E. (1994) Trypanosomiasis and trypanotolerance in cattle: a role for congopain? *Parasitology Today*, 10, 360–364.

Authié, E., Somda, B., Dossama, M. and Toutou, P. (1984) Mise en évidence d'une résistance aux trypanocides parmi des souches de *Trypanosoma congolense* récemment isolées au Burkina. *Revue d'Élevage et de Médecine Vétérinaire des Pays Tropicaux*, 37, No. Spécial, 219–235.

Awan, M.A.Q. (1971) The use of human plasma in the blood incubation infectivity test to differentiate *Trypanosoma brucei* and *Trypanosoma rhodesiense*. *Tropical Animal Health and Production*, 3, 183–186.

Awan, M.A.Q. (1979) Identification by the blood incubation infectivity test of *Trypanosoma brucei* subspecies isolated from game animals in Zambia. *Acta Tropica*, 36, 343–347.

Awiti, L.R.S., Odhiambo, T.R. and Ogoma, N.T. (1985) An electron microscopic study of the aortic wall and corpus cardiacum in relation to neurohaemal function in the tsetse *Glossina morsitans centralis*. *Insect Science and its Application*, 6, 489–501.

Bagshawe, A.G. (1908) *Glossina palpalis* Rob, Desv. *Report of the Sleeping Sickness Commission of the Royal Society*, 1, 89–106.

Bailey, N.M. and Boreham, P.F.L. (1969) The number of *Trypanosoma rhodesiense* required to establish an infection in man. *Annals of Tropical Medicine and Parasitology*, 63, 201–205.

Bailey, N.M., Cunningham, M.P. and Kimber, C.D. (1967) The indirect fluorescent antibody technique applied to dried blood, for use as a screening test in the diagnosis of human trypanosomiasis in Africa. *Transactions of the Royal Society of Tropical Medicine and Hygiene*, 61, 696–700.

Bailey, N.T.J. (1951) On estimating the size of mobile populations from recapture data. *Biometrika*, 38, 293–306.

Bailey, N.T.J. (1952) Improvements in the interpretation of recapture data. *Journal of Animal Ecology*, 21, 120–127.

Bajyana Songa, E., Hamers, R., Rickman, R., Nantulya, V.M., Mulla, A.F. and Magnus, E. (1991) Evidence for widespread asymptomatic *Trypanosoma rhodesiense* infection in the Luangwa valley (Zambia). *Tropical Medicine and Parasitology*, 42, 389–393.

Baker, J.R. (1962) Infection of the chimpanzee (*Pan troglodytes verus*) with *Trypanosoma rhodesiense* and *T. brucei*. *Annals of Tropical Medicine and Parasitology*, 56, 216–217.

Baker, J.R. (1963) Speculations on the evolution of the family Trypanosomatidae Doflein 1901. *Experimental Parasitology*, 13, 219–233.

Baker, J.R. (1968) Experimental infections with *Trypanosoma brucei* and *T. rhodesiense* in Chimpanzees. *Transactions of the Royal Society of Tropical Medicine and Hygiene*, 68, Trypanosomiasis Seminar, 138.

Baker, J.R. (1974) Epidemiology of African sleeping sickness. In: *Trypanosomiasis and Leishmaniasis with Special Reference to Chagas Disease*. CIBA Foundation Symposium, pp. 29–50.

Baker, J.R. and McConnell, E. (1969) Human trypanosomiasis in Ethiopia. *Transactions of the Royal Society of Tropical Medicine and Hygiene*, 63, 114.

Baker, J.R. and Robertson, D.H.H. (1957) Experiment on the infectivity to *Glossina morsitans* of a strain of *Trypanosoma rhodesiense* and of a strain of *T. brucei*, with some observations on the longevity of infected flies. *Annals of Tropical Medicine and Parasitology*, 51, 121–135.

Baker, J.R., Sachs, R. and Laufer, I. (1968) Trypanosomes of wild mammals in an area northwest of the Serengeti National Park, Tanzania. *Zeitschrift für Parasitenkunde*, 18, 280–284.

Baker, J.R., McConnell, E., Kent, D.C. and Hady, J. (1970) Human trypanosomiasis in Ethiopia. Ecology of Illubabor province and epidemiology in the Baro river area. *Transactions of the Royal Society of Tropical Medicine and Hygiene*, 64, 523–530.

Baker, R. (1991) Modelling the probability of a single trypanosome infecting a tsetse fly. *Annals of Tropical Medicine and Parasitology*, 85, 413–415.

Baker, R.D., Maudlin, I., Milligan, P.J.M., Molyneux, D.H. and Welburn, S.C. (1990) The possible role of *Rickettsia*-like organisms in trypanosomiasis epidemiology. *Parasitology*, 100, 209–217.

Baldet, T., Geoffroy, B., D'Amico, F., Cuisance, D. and Bossy, J.-P. (1992) Structures sensorielles de l'aile de glossine (*Diptera: Glossinidae*). *Revue d'Élevage et de Médecine Vétérinaire des Pays Tropicaux*, 45, 295–302.

Baldry, D.A.T. (1964) Observations on a close association between *Glossina tachinoides* and domestic pigs near Nsukka, eastern Nigeria. II. Ecology and trypanosome infection rates in *G. tachinoides*. *Annals of Tropical Medicine and Parasitology*, 58, 32–44.

Baldry, D.A.T. (1966a) On the distribution of *Glossina tachinoides* in West Africa. II. An assessment of the probable present distribution of *G. tachinoides* in West Africa and of possible future extensions, based on existing records and recent observations in Southern Nigeria. In: *OAU/ISCTRC 11th meeting, Nairobi*, Publication No. 100, pp. 103–109.

Baldry, D.A.T. (1966b) *Lantana camara* L. as a breeding site for *Glossina tachinoides* Westwood in South-eastern Nigeria. *OAU/ISCTRC 11th meeting, Nairobi*, Publication No. 100, 103–109.

Baldry, D.A.T. (1966c) Variations in the ecology of *Glossina* spp. *Bulletin of the World Health Organisation*, 40, 859–869.

Baldry, D.A.T. (1969a) The epidemiological significance of recent observations in Nigeria on the ecology of *Glossina tachinoides* Westwood (Diptera: Muscidae). *Bulletin of the Entomological Society of Nigeria*, 2, 34–38.

Baldry, D.A.T. (1969b) Variations in the ecology of *Glossina* spp. with special reference to Nigerian populations of *Glossina tachinoides*. *Bulletin of the World Health Organisation*, 40, 859–869.

Baldry, D.A.T. (1969c) Distribution and trypanosome infection rates of *Glossina morsitans submorsitans* Newst. along a trade cattle route in south west Nigeria. *Bulletin of Entomological Research*, 58, 537–548.

Baldry, D.A.T. (1970) Observations on the peri-domestic breeding behaviour and resting sites of *Glossina tachinoides* Westw. near Nsukka, east central Nigeria. *Bulletin of Entomological Research*, 59, 585–593.

Baldry, D.A.T. (1972) A history of Rhodesian sleeping sickness in the Lambwe valley. *Bulletin of the World Health Organisation*, 47, 699–718.

Baldry, D.A.T. (1980) Local distribution and ecology of *Glossina palpalis* and *G. tachinoides* in forest foci of west African human trypanosomiasis, with special reference to associations between peri-domestic tsetse and their hosts. *Insect Science and its Application*, 1, 85–93.

Baldry, D.A.T. and Molyneux, D.H. (1980) Observations on the ecology and trypanosomiasis infection of a relict population of *Glossina medicorum* in the Komoe valley of Upper Volta. *Annals of Tropical Medicine and Parasitology*, 74, 79–91.

Baldry, D.A.T. and Riordan, K. (1967) A review of 50 years entomology of insect-borne diseases of veterinary importance in Nigeria, with special reference to tsetse. *Proceedings of the Entomological Society of Nigeria*, 43–55.

Baldry, D.A.T., Kulzer, H., Bauer, S., Lee, C.W. and Parker, J.D. (1978a) The experimental application of insecticides from a helicopter for the control of riverine populations of *Glossina tachinoides* in West Africa. III. Operational aspects and application techniques. *PANS*, 24, 423–434.

Baldry, D.A.T., Molyneux, D.H. and Van Wettere, P. (1978b) The experimental application of insecticides from a helicopter for the control of riverine populations of *Glossina tachinoides* in West Africa. V. Evaluation of Decamethrin applied as a spray. *PANS*, 24, 447–454.

Baldry, D.A.T., Everts, J., Roman, B., Boon Von Ochsee, G.A. and Laveissière, C. (1981) The experimental application of insecticides from helicopters for the control of riverine populations of *Glossina tachinoides* in West Africa. Part VIII: The effects of two spray applications of OMS-570 (endosulfan) and of OMS-1998 (decamethrin) on *G. tachinoides* and non-target organisms in Upper Volta. *Tropical Pest Management*, 27, 83–110.

Balfour, A. (1913) Animal trypanosomiasis in the Lado (western Mongalla) and notes on tsetse fly traps and on alleged immune breed of cattle in Kordofan. *Annals of Tropical Medicine and Parasitology*, 7, 113–120.

Balis, J. and Bergeon, P. (1968) Etude de la répartition des Glossines en Ethiopie. *Bulletin of the World Health Organisation*, 38, 809–813.

Balis, J. and Bergeon, P. (1970) Etude sommaire de la répartition des Glossines dans l'empire d'Ethiopie. *Revue d'élevage et de Médecine vétérinaire des Pays tropicaux*, 23, 181–187.

Barrass, R. (1960) The settling of tsetse flies *Glossina morsitans* Westwood (Diptera; Muscidae) on cloth screens. *Entomologia Experimentalis et Applicata*, 3, 59–67.

Barrass, R. (1970) The flight activity and settling behaviour of *Glossina morsitans* Westw. (Dipt., Muscidae) in laboratory experiments. *Bulletin of Entomological Research*, 59, 627–635.

Barrett, J.C. (1989) Tsetse control, land use and livestock in the development of the Zambesi valley, Zimbabwe: some policy considerations. *ALPAN Network paper No 19, May 89*, ILCA, Addis Ababa, 22 pp.

Barrett, J.C. (1997) Control strategies for African trypanosomiases: their sustainability and effectiveness. In: Hide, G., Motram, J.C., Coombs, G.H. and Holmes, P.H. (eds) *Trypanosomiasis and Leishmaniasis*. CAB International, Wallingford, pp. 347–362.

Bauer, B. (1971) Helicopter operations for controlling the tsetse fly in Africa. *Agricultural Aviation*, 12, 51–56.

Bauer, B., Petrich-Bauer, J., Pohlit, H. and Kaboré, I. (1988) Effects of Flumethrin Pour-on against *Glossina palpalis gambiensis* (Diptera, Glossinidae). *Tropical Medicine and Parasitology*, 39, 151–152.

Bauer, B., Kaboré, I. and Petrich-Bauer, J. (1992) The residual effect of deltamethrin Spoton when tested against *Glossina palpalis gambiensis* under fly chamber conditions. *Tropical Medicine and Parasitology*, 43, 38–40.

Bauer, B., Amsler, S., Kaboré, I. and Petrich-Bauer, J. (1993) Applications of synthetic pyrethroids to cattle. Laboratory trials and tsetse control operations with specific consideration of extension to rural communities. In: *Organisation of African Unity/ISTRC Nineteenth meeting, Kampala, Uganda, 1993*, pp. 276–279.

Bauer, B., Amsler-Delafosse, S., Clausen, P.-H., Kaboré, I. and Petrich-Bauer, J. (1995) Successful application of deltamethrin pour-on to cattle in a campaign against tsetse flies (*Glossina* spp.) in the pastoral zone of Samorogouan, Burkina Faso. *Tropical Medicine and Parasitology*, 46, 183–189.

Bauer, F. (1962) The development of drug-resistance to Berenil in *Trypanosoma congolense*. *Veterinary Record*, 74, 265–266.

Bauer, J., Pohlit, H., Kaboré, I. and Bauer, B. (1987) Are trypanotolerant cattle less frequently bitten by tsetse flies? *OAU/ISTRC 19th meeting, Lomé, Togo*, 465.

Bax, S. Napier (1937) The senses of smell and sight in *Glossina swynnertoni*. *Bulletin of Entomological Research*, 28, 539–582.

Baylis, M. (1996) Effect of defensive behaviour by cattle on the feeding success and nutritional state of the tsetse fly, *Glossina pallidipes* (Diptera: Glossinidae). *Bulletin of Entomological Research*, 86, 329–336.

Baylis, M. (1997) The daily feeding rate of tsetse (Diptera: Glossinidae) on cattle at Galana Ranch, Kenya and comparison with trypanosomiasis incidence. *Acta Tropica*, 65, 81–96.

Baylis, M. and Mbwabi, A.L. (1995a) Feeding behaviour of tsetse flies (*Glossina pallidipes* Austen) on *Trypanosoma*-infected oxen in Kenya. *Parasitology*, 110, 297–305.

Baylis, M. and Mbwabi, A.L. (1995b) Effect of host packed cell volume on the blood-meal size of male tsetse flies, *Glossina pallidipes*. *Medical and Veterinary Entomology*, 9, 399–402.

Baylis, M. and Nambiro, C.O. (1993a) The nutritional state of male tsetse flies, *Glossina pallidipes*, at the time of feeding. *Medical and Veterinary Entomology*, 7, 316–322.

Baylis, M. and Nambiro, C.O. (1993b) The responses of *Glossina pallidipes* and *G. longipennis* (Diptera: Glossinidae) to odour-baited traps and targets at Galana Ranch, south-eastern Kenya. *Bulletin of Entomological Research*, 83, 145–151.

Baylis, M. and Nambiro, C.O. (1993c) The effect of cattle infection by *Trypanosoma congolense* on the attraction, and feeding success, of the tsetse fly *Glossina pallidipes*. *Parasitology*, 106, 357–361.

Baylis, M. and Stevenson, P. (1997) Trypanosomiasis and tsetse control: Fact and fiction? *24th meeting of the International Scientific Council for Trypanosomosis Research and Control, Maputo, Mozambique* (in press).

Baylis, M., Mbwabi, A.L. and Stevenson, P. (1994) The feeding success of tsetse flies, *Glossina pallidipes* (Diptera: Glossinidae), on oxen treated with pyrethroid pour-ons at Galana Ranch, Kenya. *Bulletin of Entomological Research*, 84, 447–452.

Beard, C.B., O'Neill, S.L., Tesh, R.B., Richards, F.F. and Askoy, S. (1993a) Modification of arthropod vector competence via symbiotic bacteria. *Parasitology Today*, 9, 179–183.

Beard, C.B., O'Neill, S.L., Mason, P., Mandelco, L., Woese, C.R., Tesh, R.B., Richards, F.F. and Askoy, S. (1993b) Genetic transformation and phylogeny of bacterial symbionts from tsetse. *Insect Molecular Biology*, 1, 123–131.

Beck, L., Rodriguez, M.H., Dister, S.W., Rodriguez, A.D., Rejmankova, E., Ulloa, A., Meza, R.A., Roberts, D.R., Paris, J.F., Spanner, M.A., Washin, R.K., Hacker, C. and Legters, L. (1994) Remote sensing as a landscape epidemiologic tool to identify villages at high risk for malaria transmission. *American Journal of Tropical Medicine and Hygiene*, 51, 271–280.

Beier, J.C., Perkins, P.V., Wirtz, R.A., Koros, J., Diggs, D., Gargan, T.P. and Koech, D.K. (1988) Bloodmeal identification by direct enzyme-linked immunosorbent assay (ELISA), tested on *Anopheles* (Diptera: Culicidae) in Kenya. *Journal of Medical Entomology*, 25, 9–16.

Beier, J.C., Perkins, P.V., Onyango, F.K., Gargan, T.P., Oster, C.N., Whitmire, R.E., Koech, D.K. and Roberts, C.R. (1990) Characterisation of malaria transmission by *Anopheles* (Diptera: Culicidae) in western Kenya in preparation for a malaria vaccine trial. *Journal of Medical Entomology*, 27, 570–577.

Bell, H.H. (1909) Report on the measures adopted for the suppression of sleeping sickness in Uganda. *Colonial Reports Miscellaneous* No. 65, 1–28.

Bertram, B.C.R. (1973) Sleeping sickness survey in the Serengeti area (Tanzania) 1971. III. Discussion of the relevance of the trypanosome survey to the biology of large mammals in the Serengeti. *Acta Tropica*, 30, 36–47.

Bevan, Ll.E.W. (1928) A method of inoculating cattle against trypanosomiasis. *Transactions of the Royal Society of Tropical Medicine and Hygiene*, 22, 147–156.

Billingsley, P.F. (1994) Vector–parasite interactions for vaccine development. *International Journal for Parasitology*, 24, 53–58.

Birkenmeyer, D.R. and Dame, D.A. (1975) Storage and sexual separation of *Glossina morsitans morsitans* Westwood puparia. *Annals of Tropical Medicine and Parasitology*, 69, 399–405.

Birtwisle, D. (1974) Sugar phosphate and amino acid concentrations in the thorax of the tsetse fly *Glossina morsitans. Insect Biochemistry*, 4, 63–66.

Blackwell, A., Brown, M. and Mordue, W. (1995) The use of an enhanced ELISA method for the identification of *Culicoides* bloodmeals in host-preference studies. *Medical and Veterinary Entomology*, 9, 214–218.

Blair, D.M. (1939) Human trypanosomiasis in southern Rhodesia, 1911–1938. *Transactions of the Royal Society of Tropical Medicine and Hygiene*, 32, 729–742.

Blanchetot, A. and Gooding, R.H. (1995) Identification of a mariner element from the tsetse fly, *Glossina palpalis palpalis. Insect Molecular Biology*, 4, 89–96.

Bois, J.-F., Challier, A., Laveissière, C. and Ouédraogo, V. (1977) Recherche des lieux de répos diurnes des Glossines (*Glossina palpalis gambiensis* Vanderplank, 1949: Diptera, Glossinidae) par détection de spécimens marqués au 59 Fe. *Cahiers ORSTOM Série Entomologie Médecine et Parasitologie*, 15, 3–13.

Bolland, H.R., Van Buren, A., Van der Geest, L.P.S. and Helle, W. (1974) Marker mutation in the tsetse fly *Glossina morsitans. Entomologia Experimentalis et Applicata*, 17, 522–528.

Boreham, P.F.L. (1972) Serological identification of arthropod bloodmeals and its application. *PANS*, 18, 205–209.

Boreham, P.F.L. (1976) Sterilisation of arthropod bloodmeals prior to blood meal identification. *Mosquito News*, 36, 454–457.

Boreham, P.F.L. (1979) Tsetse feeding patterns and the transmission of trypanosomes. Trypanosomiasis Seminar. *Transactions of the Royal Society of Tropical Medicine and Hygiene*, 73, 130–131.

Boreham, P.F.L. and Gill, G.S. (1972) A method for determining the reptilian hosts of tsetse flies. *Transactions of the Royal Society of Tropical Medicine and Hygiene*, 66, 324–325.

Boreham, P.F.L. and Gill, G.S. (1973) Serological identification of reptile feeds of *Glossina. Acta Tropica*, 30, 356–365.

Borne, P. (1996) *The Hidden Cost of Trypanosomosis*. SANOFI Symposium on the control of trypanosomosis. Bordeaux, France, July 1996.

Bose, R. and Heister, N.C. (1993) Development of *Trypanosoma (M.) theileri* in Tabanids. *Journal of Eukaryotic Microbiology*, 40, 788–792.

Bose, R., Friedhoff, K.T., Olbrich, S., Buscher, G. and Domeyer, I. (1987) Transmission of *Trypanosoma theileri* to cattle by Tabanidae. *Parasitology Research*, 73, 421–424.

Bosompem, K.M., Moloo, S.K., Assoku, R.K.G. and Nantulya, V.M. (1995) Detection and differentiation between trypanosome species in experimentally infected tsetse flies (*Glossina* spp.) using dot-ELISA. *Acta Tropica*, 60, 81–96.

Bosompem, K.M., Masake, R.A., Assoku, R.K.G., Opiyo, E.A. and Nantulya, V.M. (1996) Field evaluation of a dot-ELISA for the detection and differentiation of trypanosome species in infected tsetse flies (*Glossina* spp.). *Parasitology*, 112, 205–211.

Boucek, Z. (1976) Taxonomic studies on some Eulophidae (Hym.) of economic interest, mainly from Africa. *Entomophaga*, 21, 401–414.

Bouvier, G. (1936) Quelques hyménoptères ennemis des Glossines. *Annales de Parasitologie*, 14, 330–331.

Boyt, W.P., Lovemore, D.F., Pilson, R.D. and Smith, I.M. (1962) A preliminary report on the maintenance of cattle by various drugs in a mixed *Glossina morsitans* and *G. pallidipes* fly-belt. *ISCTRC 9th meeting, Conakry*, Publication No. 88, 71–79.

Boyt, W.P., Mackenzie, P.K.I. and Ross, C. (1970) An attempt to demonstrate the natural transmission of bovine trypanosomiasis by agents other than *Glossina* in the Sabi valley of Rhodesia. *Rhodesia Veterinary Journal*, 1, 7–16.

Boyt, W.P., Mackenzie, P.K.I., Pilson, R.D. and Leavis, H. (1972) The importance of the donkey (*Equus asinus*) as a source of food and a reservoir of trypanosomes for *Glossina morsitans* Westw. *Rhodesia Science News*, 6, 18–20.

Boyt, W.P., Mackenzie, P.K.I. and Pilson, R.D. (1978) The relative attractiveness of donkeys, cattle, sheep and goats to *Glossina morsitans morsitans* Westwood and *G. pallidipes* Austen (Diptera: Glossinidae) in a middle-veld area of Rhodesia. *Bulletin of Entomological Research*, 68, 497–500.

Brady, J. (1970) Characteristics of spontaneous activity in tsetse flies. *Nature*, 228, 286–287.

Brady, J. (1972a) The visual responsiveness of the tsetse fly *Glossina morsitans* Westw. (Glossinidae) to moving objects: the effects of hunger, sex, host odour and stimulus characteristics. *Bulletin of Entomological Research*, 62, 257–279.

Brady, J. (1972b) Visual stimulus velocity and 'following' response of tsetse flies. *Transactions of the Royal Society of Tropical Medicine and Hygiene*, 66, 313.

Brady, J. (1972c) Circadian rhythm of spontaneous activity in tsetse flies. *Transactions of the Royal Society of Tropical Medicine and Hygiene*, 66, 312.

Brady, J. (1972d) Spontaneous circadian components of tsetse fly activity. *Journal of Insect Physiology*, 18, 471–484.

Brady, J. (1974) The pattern of spontaneous activity in the tsetse fly *Glossina morsitans* Westw. (Diptera, Glossinidae) at low temperatures. *Bulletin of Entomological Research*, 63, 441–444.

Brady, J. (1975a) 'Hunger' in the tsetse fly: the nutritional correlates of behaviour. *Journal of Insect Physiology*, 21, 807–829.

Brady, J. (1975b) Circadian changes in central excitability – the origin of behavioural rhythms in tsetse flies and other animals? *Journal of Entomology*, 50, 79–95.

Brady, J. (1979) The construction of laboratory studies on tsetse fly behaviour. *Transactions of the Royal Society of Tropical Medicine and Hygiene*, 74, 275–276.

Brady, J. (1987) The sunset activity of tsetse flies: a light threshold study on *Glossina morsitans*. *Physiological Entomology*, 12, 363–372.

Brady, J. (1988) The circadian organization of behavior: timekeeping in the tsetse fly, a model system. *Advances in the Study of Behavior*, 18, 153–191.

Brady, J. (1991) Flying mate detection and chasing by tsetse flies (*Glossina*). *Physiological Entomology*, 16, 153–161.

Brady, J. and Crump, A.J. (1978) The control of circadian activity rhythms in tsetse flies: environment or physiological clock? *Physiological Entomology*, 3, 177–190.

Brady, J. and Gibson, G. (1983) Activity patterns in pregnant tsetse flies, *Glossina morsitans*. *Physiological Entomology*, 8, 359–369.

Brady, J. and Griffiths, N. (1993) Upwind flight responses of tsetse flies (*Glossina* spp.) (Diptera: Glossinidae) to acetone, octenol and phenols in nature: a video study. *Bulletin of Entomological Research*, 83, 329–333.

Brady, J. and Shereni, W. (1988) Landing responses of the tsetse fly *Glossina morsitans morsitans* Westwood and the stable fly *Stomoxys calcitrans* (L.) (Diptera: Glossinidae and Muscidae) to black and white patterns: a laboratory study. *Bulletin of Entomological Research*, 78, 301–311.

Brady, J., Gibson, G. and Packer, M.J. (1989) Odour movement, wind direction, and the problem of host finding by tsetse flies. *Physiological Entomology*, 14, 369–380.

Brady, J., Packer, M.J. and Gibson, G. (1990) Odour plume shape and host finding by tsetse. *Insect Science and its Application*, 11, 377–384.

Brady, J., Griffiths, N. and Paynter, Q. (1995) Wind speed effects on odour source location by tsetse flies (*Glossina*). *Physiological Entomology*, 20, 293–302.

Brandl, F.E. (1988) Costs of different methods to control riverine tsetse in west Africa. *Tropical Animal Health and Production*, 20, 67–77.

Brightwell, R., Dransfield, R.D., Kyorku, C.A., Golder, T.K., Tarimo, S.A. and Mungai, D. (1987) A new trap for *Glossina pallidipes*. *Tropical Pest Management*, 33, 151–159.

Brightwell, R., Dransfield, R.D. and Kyorku, C.A. (1991) Development of a low-cost tsetse trap and odour baits for *Glossina pallidipes* and *G. longipennis* in Kenya. *Medical and Veterinary Entomology*, 5, 153–164.

Brightwell, R., Dransfield, R.D. and Williams, B.G. (1992) Factors affecting seasonal dispersal of the tsetse flies *Glossina pallidipes* and *G. longipennis* (Diptera: Glossinidae) at Nguruman, south-west Kenya. *Bulletin of Entomological Research*, 82, 167–182.

Brightwell, R., Dransfield, R.D., Stevenson, P. and Williams, B. (1997) Changes over twelve years in populations of *Glossina pallidipes* and *Glossina longipennis* (Diptera: Glossinidae) subject to varying trapping pressure at Nguruman, south-west Kenya. *Bulletin of Entomological Research*, 87, 349–370.

Broden, A. (1904) Les infections à trypanosomes dans l'état du Congo. *Bulletin de la Société des Etudes Colon (Brussels)* Feb. 116–139.

Brooks, G.T., Barlow, F., Hadaway, A.B. and Harris, E.G. (1981) The toxicities of some analogues of dieldrin, endosulfan and isobenzan to bloodsucking Diptera, especially tsetse flies. *Pesticide Science*, 12, 475–484.

Brothers, D.J. (1971) The genera of Mutillidae (Hymenoptera) parasitic on tsetse flies (*Glossina*; Diptera). *Journal of the Entomological Society of South Africa*, 34, 101–102.

Brown, A.H., Steelman, C.D., Johnson, Z.B., Rosenkrans, C.F. and Brasuell, T.M. (1992) Estimates of Horn fly resistance in beef cattle. *Journal of Animal Science*, 70, 1375–1381.

Brown, I.R.F., D'Costa, M.A. and Rutesasira, A. (1973) Uric acid metabolism in the tsetse fly, *Glossina morsitans*. *Comparative Biochemistry and Physiology*, 46B, 741–747.

Brown, L.R. and Kane, H. (1994) *Full House: Reassessing the Earth's Population Carrying Capacity*. The Worldwatch Environmental Alert Series, W.W. Norton, New York.

Bruce, D. (1895) *Preliminary Report on the Tsetse Fly Disease or Nagana in Zululand*. Bennett and Davis, Durban, 28 pp.

Bruce, D. (1908) Sleeping sickness in Africa. *Journal of the African Society*, 7, 249.

Bruce, D. (1911) The morphology of *Trypanosoma evansi* (Steel). *Proceedings of the Royal Society B*, 84, 181. *Reports of the Sleeping Sickness Commission of the Royal Society, 1912*, 12, 13.

Bruce, D. (1915a) The Croonian lectures on trypanosomes causing disease in man and domestic animals in Central Africa. Lecture I. *The Lancet*, 4791, 1323–1330.

Bruce, D. (1915b) The Croonian lectures on trypanosomes causing disease in man and domestic animals in Central Africa. Lecture III. *The Lancet*, 4793, 55–63.

Bruce, D. and Nabarro, D. (1903) Progress report on sleeping sickness in Uganda. *Proceedings of the Royal Society London*, 1, 11–88.

Bruce, D., Hamerton, A.E., Bateman, H.R., Mackie, F.P. and Bruce, M. (1911a) Further researches on the development of *Trypanosoma gambiense* in *Glossina palpalis*. *Proceedings of the Royal Society (B)*, 83, 513–527.

Bruce, D., Hamerton, A.E., Bateman, H.R., Mackie, F.P. and Bruce, M. (1911b) Sleeping sickness and other disease of man and animals in Uganda during the years 1908–9–10. *Reports of the Sleeping Sickness Commission of the Royal Society*, 11, 1–244.

Bruce, D., Hamerton, A.E., Bateman, H.R., Mackie, F.P. and Bruce, M. (1911c) Further researches on the development of *Trypanosoma vivax* in laboratory bred *Glossina palpalis*. *Reports of the Sleeping Sickness Commission of the Royal Society*, 11, 50–54.

Bruce, D., Harvey, D., Hamerton, A.E. and Bruce, M. (1913) Trypanosome diseases of domestic animals in Nyasaland. Part III. Development in *Glossina morsitans*. *Proceedings of the Royal Society (B)*, 87, 58–66.

Brues, C.T., Melander, A.L. and Carpenter, F.M. (1954) *Classification of Insects*. Harvard University Press, Cambridge, Massachusetts, 917 pp.

Brunhes, J., Cuisance, D., Bernard, G., Hervy, J.-P. and Lebbe, J. (1994) *Les Glossines ou Mouches Tsetse. Logiciel d'Identification Glossine Expert*. CIRAD-EMVT, Université Paris-VI Editions de l'Orstom, Paris, 160 pp.

Bungener, W. and Mehlitz, D. (1976) Experimental *Trypanosoma* infections in Cameroon dwarf goats: histopathological observations. *Tropenmedizin und Parasitologie*, 27, 405–410.

Burchard, R.P. and Baldry, D.A.T. (1970) Polytene chromosomes of *Glossina palpalis* Robineau-Desvoidy (Diptera: Muscidae). I. The preliminary demonstration. *Proceedings of the Royal Entomological Society of London (A)*, 45, 182–183.

Burgdorfer, W., Schmidt, M.L. and Hoogstraal, H. (1973) Detection of *Trypanosoma theileri* in Ethiopian ticks. *Acta Tropica*, 30, 340–346.

Burke, J. (1971) Historique de la lutte contre la maladie du sommeil au Congo. *Annales de la Société Belge de Médecine Tropicale*, 51, 465–477.

Burkot, T.R. and Graves, P.M. (1995) The value of vector-based estimates of malaria transmission. *Annals of Tropical Medicine and Parasitology*, 89, 125–134.

Burkot, T.R., Goodman, W.G. and Defoliart, G.R. (1981) Identification of mosquito bloodmeals by enzyme-linked immunosorbent assay. *American Journal of Tropical Medicine and Hygiene*, 30, 1336–1341.

Burnett, G.F. (1954) The effect of poison bait cattle on populations of *Glossina morsitans* Westw. and *G. swynnertoni* Aust. *Bulletin of Entomological Research*, 45, 411–421.

Burnett, G.F. (1961a) The susceptibility of tsetse flies to topical applications of insecticides. II. Young adults of *Glossina morsitans* Westw. and organophosphorus compounds, pyrethrins and sevin. *Bulletin of Entomological Research*, 52, 763–768.

Burnett, G.F. (1961b) Effect of age and pregnancy on the tolerance of tsetse flies to insecticide. *Nature*, 192, 188.

Burnett, G.F. (1962) The susceptibility of tsetse flies to topical applications of insecticides. III. The effects of age and pregnancy on susceptibility. *Bulletin of Entomological Research*, 53, 337–345.

Burridge, M.J.M., Reid, H.W., Pullan, N.B., Sutherst, R.W. and Wain, E.B. (1970) Survey for trypanosome infections in domestic cattle and wild animals in areas of East Africa. II. Salivarian trypanosome infections in wild animals in Busoga District, Uganda. *British Veterinary Journal*, 126, 627–633.

Bursell, E. (1955) The polypneustic lobes of the tsetse larva (*Glossina*, Diptera). *Proceedings of the Royal Society (B)*, 144, 275–286.

Bursell, E. (1957a) Spiracular control of water loss in the tsetse fly. *Proceedings of the Royal Entomological Society of London (A)*, 32, Pts 1–3, 21–29.

Bursell, E. (1957b) The effect of humidity on the activity of tsetse flies. *Journal of Experimental Biology*, 34, 42–51.

Bursell, E. (1958a) The water balance of tsetse pupae. *Philosophical Transactions of the Royal Society London (B)*, 241, 179–210.

Bursell, E. (1958b) The effect of partial feeding on the probing reaction of tsetse flies, in relation to trypanosomiasis challenge. *ISCTR/CCTA 7th meeting, Bruxelles*, 227–229.

Bursell, E. (1959a) The water balance of tsetse flies. *Transactions of the Royal Entomological Society of London (A)*, 111, Pts 9–10, 205–235.

Bursell, E. (1959b) Determination of the age of tsetse puparia by dissection. *Proceedings of the Royal Entomological Society of London (A)*, 34, Pts 1–3, 23–24.

Bursell, E. (1960a) Loss of water by excretion and defaecation in the tsetse fly. *Journal of Experimental Biology*, 37, 689–697.

Bursell, E. (1960b) Free amino-acids of the tsetse fly (*Glossina*). *Nature*, 187, 778.

Bursell, E. (1960c) The effect of temperature on the consumption of fat during pupal development in *Glossina*. *Bulletin of Entomological Research*, 51, 583–598.

Bursell, E. (1961a) The behaviour of tsetse flies (*Glossina swynnertoni* Austen) in relation to problems of sampling. *Proceedings of the Royal Entomological Society of London (A)*, 36, Pts 1–3, 9–20.

Bursell, E. (1961b) Post-teneral development of the thoracic musculature in tsetse flies. *Proceedings of the Royal Entomological Society of London (A)*, 36, Pts 4–6, 69–74.

Bursell, E. (1961c) Starvation and desiccation in tsetse flies (*Glossina*). *Entomologia Experimentalis et Applicata*, 4, 301–310.

Bursell, E. (1963a) Aspects of the metabolism of amino-acids in the tsetse fly, *Glossina* (Diptera). *Journal of Insect Physiology*, 9, 439–452.

Bursell, E. (1963b) Tsetse-fly physiology: a review of recent advances and current aims. *Bulletin of the World Health Organisation*, 28, 703–709.

Bursell, E. (1965) Nitrogenous waste products of the tsetse fly, *Glossina morsitans*. *Journal of Insect Physiology*, 11, 993–1001.

Bursell, E. (1966a) Aspects of the flight metabolism of tsetse flies (*Glossina*). *Comparative Biochemistry and Physiology*, 19, 809–818.

Bursell, E. (1966b) The nutritional state of tsetse flies from different vegetation types in Rhodesia. *Bulletin of Entomological Research*, 57, 171–180.

Bursell, E. (1967) The conversion of glutamate to alanine in the tsetse fly (*Glossina morsitans*). *Comparative Biochemistry and Physiology*, 23, 825–829.

Bursell, E. (1970) Theoretical aspects of the control of *Glossina morsitans* by game destruction. *Zoologica Africana*, 5, 135–141.

Bursell, E. (1973a) Development of mitochondrial and contractile components of the flight muscle in adult tsetse flies, *Glossina morsitans*. *Journal of Insect Physiology*, 19, 1079–1086.

Bursell, E. (1973b) Entomological aspects of the epidemiology of sleeping sickness. *Central African Journal of Medicine*, 19, 201–204.

Bursell, E. (1975a) Substrates of oxidative metabolism in dipteran flight muscle. *Comparative Biochemistry and Physiology*, 52B, 235–238.

Bursell, E. (1975b) Glutamic dehydrogenase from sarcosomes of the tsetse fly (*Glossina morsitans*) and the blowfly (*Sarcophaga nodosa*). *Insect Biochemistry*, 5, 289–297.

Bursell, E. (1977a) Synthesis of proline by the fat body of the tsetse fly (*Glossina morsitans*): metabolic pathways. *Insect Biochemistry*, 7, 427–434.

Bursell, E. (1977b) Chemosterilisation of tsetse flies using a pressurised Metepa aerosol. *Transactions of the Rhodesian Scientific Association*, 58, 43–47.

Bursell, E. (1980) The future of tsetse biology. In: Locke, M. and Smith, D. (eds) *Insect Biology in the Future*. Academic Press, London, pp. 905–924.

Bursell, E. (1981) Energetics of haematophagous arthropods: influence of parasites. *Parasitology*, 82, 107–116.

Bursell, E. (1984a) Observations on the orientation of tsetse flies (*Glossina pallidipes*) to wind-borne odours. *Physiological Entomology*, 9, 133–137.

Bursell, E. (1984b) Effects of host odour on the behaviour of tsetse. *Insect Science and its Application*, 5, 345–350.

Bursell, E. (1987) The effect of wind-borne odours on the direction of flight in tsetse flies, *Glossina* spp. *Physiological Entomology*, 12, 149–156.

Bursell, E. (1990) The effect of host odour on the landing responses of tsetse flies (*Glossina morsitans morsitans*) in a wind tunnel with and without visual targets. *Physiological Entomology*, 15, 369–376.

Bursell, R. and Berridge, M.J. (1962) The invasion of polymorphic trypanosomes of the ectoperitrophic space of tsetse flies in relation to some properties of the rectal fluid. *Annals of the Society of Experimental Biology, January Meeting*, 1962.

Bursell, E. and Jackson, C.H.N. (1957) Notes on the choriothete and milk gland of *Glossina* and *Hippobosca* (Diptera). *Proceedings of the Royal Entomological Society (A)*, 32, 30–34.

Bursell, E. and Kuwengwa, T. (1972) The effect of flight on the development of flight musculature in the tsetse fly (*Glossina morsitans*). *Entomologia Experimentalis et Applicata*, 15, 229–237.

Bursell. E. and Slack, E. (1968) Indications concerning the flight activity of tsetse flies (*Glossina morsitans* Westw.) in the field. *Bulletin of Entomological Research*, 58, 575–579.

Bursell, E. and Slack, E. (1976) Oxidation of proline by sarcosomes of the tsetse fly, *Glossina morsitans*. *Insect Biochemistry*, 6, 159–167.

Bursell, E. and Taylor, P. (1980) An energy budget for *Glossina*. *Bulletin of Entomological Research*, 70, 187–197.

Bursell, E., Slack, E. and Kuwengwa, T. (1972) Aspects of the development of flight musculature in the tsetse fly (*Glossina morsitans*). *Transactions of the Royal Society of Tropical Medicine and Hygiene*, 66, 319–320.

Bursell, E., Billing, K.C., Hargrove, J.W., McCabe, C.T. and Slack, E. (1974) Metabolism of the bloodmeal in tsetse flies (A review). *Acta Tropica Sepparatum*, 31, 297–320.

Bursell, E., Gough, A.J.E., Beevor, P.S., Cork, A., Hall, D.R. and Vale, G.A. (1988) Identification of components of cattle urine attractive to tsetse flies, *Glossina* spp. (Diptera: Glossinidae). *Bulletin of Entomological Research*, 78, 281–291.

Burtt, E. (1942) Observations on the high proportion of polymorphic trypanosome infections found in the salivary glands of *Glossina brevipalpis* near Amani, Tanganyika Territory, with a note on the appearance of the infected glands. *Annals of Tropical Medicine and Parasitology*, 36, 170–176.

Burtt, E. (1945) Hypertrophied salivary glands in *Glossina*: evidence that *G. pallidipes* with this abnormality is peculiarly suited to trypanosome infection. *Annals of Tropical Medicine and Parasitology*, 39, 11–13.

Burtt, E. (1946a) The sex ratio of infected flies found in transmission-experiments with *Glossina morsitans* and *Trypanosoma rhodesiense*. *Annals of Tropical Medicine and Parasitology*, 40, 74–79.

Burtt, E. (1946b) Incubation of tsetse pupae: increased transmission rate of *Trypanosoma rhodesiense* in *Glossina morsitans*. *Annals of Tropical Medicine and Parasitology*, 40, 18–28.

Burtt, E. (1946c) Salivation by *Glossina morsitans* on to glass slides: a technique for isolating infected flies. *Annals of Tropical Medicine and Parasitology*, 40, 141–144.

Burtt, E. (1952) The occurrence in nature of tsetse pupae (*Glossina swynnertoni* Austen). *Acta Tropica*, 9, 304–344.

Burtt, E. and Fairbairn, H. (1946) Preparations of *Trypanosoma rhodesiense*. I. Metacyclic trypanosomes ejected by the tsetse-flies in the act of biting. *Transactions of the Royal Society of Tropical Medicine and Hygiene*, 39, 2–3.

Bushrod, F.M. (1984) Variations in the mitotic chromosomes of *Glossina morsitans centralis* in Zambia. *Transactions of the Royal Society of Tropical Medicine and Hygiene*, 78, 259.

Busvine, J.R. (1993) *Disease Transmission by Insects: its Discovery and 90 Years of Effort to Prevent it*. Springer-Verlag, Berlin, 361 pp.

Buxton, P.A. (1936) Studies on soils in relation to the biology of *Glossina submorsitans* and *tachinoides* in the north of Nigeria. *Bulletin of Entomological Research*, 27, 281–286.

Buxton, P.A. (1955a) Tsetse and climate: a consideration of the growth of knowledge. *Proceedings of the Royal Entomological Society of London (C)*, 19, 71–78.

Buxton, P.A. (1955b) The natural history of tsetse flies. *Memoirs of the London School of Hygiene and Tropical Medicine*, 10, Lewis, London, 816 pp.

Buyst, H. (1974) The epidemiology, clinical features, treatment and history of sleeping sickness on the northern edge of the Luangwa flybelt. *Medical Journal of Zambia*, 8, 2–12.

Buyst, H. (1977) The epidemiology of sleeping sickness in the historical Luangwa valley. *Annales de la Société Belge de Médecine Tropicale*, 57, 4–5 349–359.

Campbell, W.C. and Benz, G.W. (1984) Ivermectin: a review of efficacy and safety. *Journal of Veterinary Pharmacology and Therapeutics*, 7, 1–16.

Carlson, D.A., Langley, P.A. and Huyton, P. (1978) Sex pheromone of the tsetse fly: isolation, identification, and synthesis of contact aphrodisiacs. *Science*, 201, 750–753.

Carlson, D.A., Nelson, D.R., Langley, P.A., Coates, T.W., Davis, T.L. and Leegwater-van der Linden, M.E. (1984) Contact sex pheromone in the tsetse fly *Glossina pallidipes* (Austen) identification and synthesis. *Journal of Chemical Ecology*, 10, 429–449.

Carlson, D.A., Milstrey, S.K. and Narang, S.K. (1993) Classification of tsetse flies *Glossina* spp. (Diptera: Glossinidae) by gas chromatographic analysis of cuticular components. *Bulletin of Entomological Research*, 83, 507–515.

Carmichael, J. (1933) The virus of Rinderpest and its relation to *Glossina morsitans*, Westw. *Bulletin of Entomological Research*, 24, 337–342.

Carmichael, J. (1934) Trypanosomes pathogenic to domestic stock and their effect in certain species of wild fauna in Uganda. *Annals of Tropical Medicine and Parasitology*, 28, 41–45.

Carmichael, J. (1938) Rinderpest in African game. *Journal of Comparative Pathology and Therapeutics*, 51, 264–268.

Carpenter, G.D.H. (1912) Progress report on investigations into the bionomics of *Glossina palpalis* July 27 to August 5 1911. *Report of the Sleeping Sickness Commission of the Royal Society*, 12, 79–111.

Carr, W.R., MacLeod, J., Woolf, B. and Spooner, R.L. (1974) A survey of the relationship of genetic markers, tick infestation level and parasitic diseases in Zebu cattle in Zambia. *Tropical Animal Health and Production*, 6, 203–214.

Carson, R. (1962) *Silent Spring*. Houghton Mifflin, Cambridge, Massachusetts.

Carter, R.M. (1906) Tsetse fly in Arabia. *British Medical Journal*, 2, 1393–1394.

Castellani, A. (1903a) On the discovery of a species of *Trypanosoma* in the cerebrospinal fluid of cases of sleeping sickness. *Proceedings of the Royal Society London*, 71, 501–508.

Castellani, A. (1903b) *Trypanosoma* in sleeping sickness. *British Medical Journal*, 1218, May 23, 1903.

Castellani, A. (1903c) Presence of *Trypanosoma* in sleeping sickness. *Reports of the Sleeping Sickness Commission of the Royal Society*, 3, 2–10.

Cattand, P. (1994) World Health Organisation Press release. WHO/73, 7 October 1994.

Cattand, P. and de Raadt, P. (1991) Laboratory diagnosis of trypanosomiasis. *Clinics in Laboratory Medicine*, 11, 899–908.

Cavalli-Sforza, L.L. and Edwards, A.W.F. (1967) Phylogenetic analysis: models and estimation procedures. *Evolution*, 21, 550–570.

Cawdery, M.J.H. and Simmons, D.J.C. (1964) A review of bovine trypanocidal drug trials of the Uganda Veterinary Department. *ISCTRC/CCTA 10th meeting, Kampala*, Publication No. 97, 47–50.

Centurion-Lara, A., Barrett, L. and Van Voorhis, W.C. (1994) Quantitation of parasitaemia by competitive polymerase chain reaction amplification of parasite minicircles during chronic infection with *Trypanosoma cruzi*. *Journal of Infectious Diseases*, 170, 1334–1339.

Chadwick, P.R. (1964a) A study of the resting sites of *Glossina swynnertoni* Aust. in northern Tanganyika. *Bulletin of Entomological Research*, 55, 23–28.

Chadwick, P.R. (1964b) Effect of two chemosterilants on *Glossina morsitans*. *Nature*, 204, 299–300.

Chadwick, P.R., Beesley, J.S.S., White, P.J. and Matechi, H.T. (1964) An experiment on the eradication of *Glossina swynnertoni* Aust. by insecticidal treatment of its resting sites. *Bulletin of Entomological Research*, 55, 411–419.

Challier, A. (1965) Amélioration de la méthode de détermination de l'âge physiologique des Glossines. Etudes faites sur *Glossina palpalis gambiensis* Vanderplank, 1949. *Bulletin de la Société de Pathologie Exotique*, 58, 250–259.

Challier, A. (1973) Ecologie de *Glossina palpalis gambiensis* Vanderplank, 1949 (Diptera – Muscidae) en savane d'Afrique Occidentale. *Mémoires ORSTOM* No. 64, Paris. pp. 274.

Challier, A. (1982) The ecology of tsetse (*Glossina* spp.) (Diptera, Glossinidae): a review (1970–1981). *Insect Science and its Application*, 3, 97–143.

Challier, A. and Dejardin, J. (1987) Variations morphologiques chez les mâles de *Glossina palpalis palpalis* (Rob.-Desv.) et *G. p. gambiensis* Vanderplank. Leurs implications taxinomiques. *Cahiers ORSTOM Série Entomologie Médecine et Parasitologie*, No. Spécial, 83–99.

Challier, A. and Gouteux, J.-P. (1980) Ecology and epidemiological importance of *Glossina palpalis* in the Ivory Coast forest zone. *Insect Science and its Application*, 1, 77–83.

Challier, A. and Laveissière, C. (1973) Un nouveau piège pour la capture des Glossines (*Glossina*: Diptera, Muscidae): description et essais sur le terrain. *Cahiers ORSTOM Série Entomologie Médecine et Parasitologie*, 11, 251–262.

Challier, A. and Turner, D. (1985) Methods to calculate survival rate in tsetse fly populations. *Annales de la Société Belge de Médecine Tropicale*, 65, 191–197.

Chandler, R.L. (1952) Comparative tolerance of West African N'Dama cattle to trypanosomiasis. *Annals of Tropical Medicine and Parasitology*, 46, 127–134.

Chandler, R.L. (1958) Studies on the tolerance of N'Dama cattle to trypanosomiasis. *Journal of Comparative Pathology*, 68, 253–260.

Chapman, N.G. (1976) Aerial spraying of tsetse flies (*Glossina* spp.) in Rhodesia with ultra low volumes of endosulphan. *Transactions of the Rhodesian Scientific Association*, 57, 12–21.

Chapman, R.F. (1960) A note on *Glossina medicorum* Aust. (Diptera) in Ghana. *Bulletin of Entomological Research*, 51, 435–440.

Chapman, R.F. (1961) Some experiments to determine the methods used in host-finding by the tsetse fly, *Glossina medicorum* Austen. *Bulletin of Entomological Research*, 52, 83–97.

Chaudhury, M.F.B. and Dhadialla, T.S. (1976) Evidence of hormonal control of ovulation in tsetse flies. *Nature*, 260, 243–244.

Chaudhury, M.F.B., Dhadialla, T.S. and Kunyiha, R. (1980) Evidence of neuro-endocrine relationships between mating and ovulation in the tsetse fly *Glossina morsitans morsitans*. *Insect Science and its Application*, 1, 161–166.

Cheeseman, M.T. and Gooding, R.H. (1985) Proteolytic enzymes from tsetse flies, *Glossina morsitans* and *G. palpalis*. *Insect Biochemistry*, 15, 677–680.

Chege, G.M.M. and Beier, J.C. (1990) Effect of *Plasmodium falciparum* on the survival of naturally infected Afrotropical *Anopheles* (Diptera: Culicidae). *Journal of Medical Entomology*, 27, 454–458.

Cheke, R.A. and Garms, R. (1988) Trials of compounds to enhance trap catches of *Glossina palpalis palpalis* in Liberia. *Medical and Veterinary Entomology*, 2, 199–200.

Child, G. and Wilson, V.J. (1964a) Delayed effects of tsetse control hunting on a Duiker population. *Journal of Wildlife Management*, 28, 866–868.

Child, G. and Wilson, V.J. (1964b) Observations on ecology and behaviour of Roan and sable in three tsetse control areas. *Arnoldia*, No 16, 1.

Chorley, J.K. (1929) The bionomics of *Glossina morsitans* in the Umniati fly belt, southern Rhodesia, 1922–23. *Bulletin of Entomological Research*, 20, 279–301.

Chorley, J.K. (1947) Tsetse fly operations in southern Rhodesia. *Rhodesian Agricultural Journal*, 44, 520–531.

Chorley, T.W. (1948) *Glossina pallidipes* Austen attracted by the scent of cattle-dung and urine (Diptera). *Proceedings of the Royal Entomological Society of London (A)*, 23, 9–11.

Chorley, T.W. and Hopkins, G.H.E. (1942) Activity of *Glossina pallidipes* at night (Diptera). *Proceedings of the Royal Entomological Society of London (A)*, 17, 93–97.

Christie, J.R. (1936) Life history of *Agameris decaudata*, a nematode parasite of grasshoppers and other insects. *Journal of Agricultural Research*, 52, 161-000.

Christy, C. (1903a) The epidemiology and aetiology of the sleeping sickness in East Equatorial Africa with clinical observations. *Reports of the Sleeping Sickness Commission of the Royal Society*, 3, 1–42.

Christy, C. (1903b) The distribution of sleeping sickness, *Filaria perstans*, etc. in East Equatorial Africa. *Reports of the Sleeping Sickness Commission of the Royal Society*, 2, 2–7.

Cibulskis, R.E. (1992) Genetic variation in *Trypanosoma brucei* and the epidemiology of sleeping sickness in the Lambwe valley, Kenya. *Parasitology*, 104, 99–109.

Clarke, J.E. (1969) Trypanosome infection rates in the mouthparts of Zambian tsetse flies. *Annals of Tropical Medicine and Parasitology*, 63, 15–34.

Clarke, L. (1980) Insect growth regulators: their possible role for the control of *Glossina*. *Transactions of the Royal Society of Tropical Medicine and Hygiene*, 74, 279–280.

Clarke, L. (1982) Factors affecting uptake and loss of diflubenzuron in the tsetse fly *Glossina morsitans morsitans* Westwood (Diptera: Glossinidae). *Bulletin of Entomological Research*, 72, 511–522.

Clarkson, M.J. and McCabe, W. (1970) The transmission of monomorphic trypanosomes by laboratory-reared flies. *First Tsetse Fly breeding Symposium Lisbon 1969*, 403

Claxton, J.R., Faye, J.A. and Rawlings, P. (1992a) Trypanosome infections in warthogs (*Phacochoerus aethiopicus*) in The Gambia. *Veterinary Parasitology*, 41, 179–187.

Claxton, J.R., Leperre, P., Rawlings, P., Snow, W.F. and Dwinger, R.H. (1992b) Trypanosomiasis in cattle in Gambia: incidence, prevalence and tsetse challenge. *Acta Tropica*, 50, 219–225.

Clausen, P.-H., Sidibe, I., Kaboré, I. and Bauer, B. (1992b) Development of multiple drug resistance of *Trypanosoma congolense* in Zebu cattle under high natural tsetse fly challenge in the pastoral zone of Samorogouan, Burkina Faso. *Acta Tropica*, 51, 229–236.

Cmelik, S.H.W., Bursell, E. and Slack, E. (1969) Composition of the gut contents of third-instar tsetse larvae (*Glossina morsitans* Westwood). *Comparative Biochemistry and Physiology*, 29, 447–453.

Coates, T.W. and Langley, P.A. (1982a) The causes of mating abstention in male tsetse flies *Glossina morsitans*. *Physiological Entomology*, 7, 235–242.

Coates, T.W. and Langley, P.A. (1982b) Laboratory evaluation of contact sex pheromone and bisazir for autosterilization of *Glossina morsitans*. *Entomologia Experimentalis et Applicata*, 31, 276–284.

Cockbill, G.F. (1967) Recent developments in tsetse and trypanosomiasis control. The history and significance of trypanosomiasis problems in Rhodesia. *Transactions of the Rhodesian Scientific Association*, 52, 7–15.

Cockbill, G.F. (1971) The control of tsetse and trypanosomiasis in Rhodesia. *Bulletin de l'Office international des Epizooties*, 76, 347–352.

Cockbill, G.F., Lovemore, D.F. and Phelps, R.J. (1963) The control of tsetse flies (*Glossina*: Diptera, Muscidae) in a heavily infested area of southern Rhodesia by means of insecticide discharged from aircraft, followed by the settlement of indigenous people. *Bulletin of Entomological Research*, 54, 93–106.

Cockerell, T.D.A. (1907) A fossil tsetse fly in Colorado. *Nature*, 76, 414.

Cockerell, T.D.A. (1909) Another fossil tsetse fly. *Nature*, 80, 128.

Cockerell, T.D.A. (1919) New species of north American fossil beetles, cockroaches, and tsetse flies. *Proceedings of the US National Museum*, 54, 301–313.

Codjia, V., Woudyalew, M., Authié, E., Leak, S.G.A., Rowlands, G.J. and Peregrine, A.S. (1993) Epidemiology of cattle trypanosomiasis in the Ghibe valley, southwest Ethiopia. 3. Occurrence of populations of *Trypanosoma congolense* resistant to diminazene, isometamidium and homidium. *Acta Tropica*, 53, 151–163.

Colvin, J. and Gibson, G. (1992) Host seeking behaviour and management of tsetse. *Annual Review of Entomology*, 37, 21–40.

Colvin, J., Brady, J. and Gibson, G. (1989) Visually guided, upwind turning behaviour of free-flying tsetse flies in odour-laden wind: a wind tunnel study. *Physiological Entomology*, 14, 31–39.

Connor, R.J. (1989) *Final Report of the Trypanosomiasis Expert*. FGU-Kronberg Consulting and Engineering GmbH, West Germany, 123 pp.

Conrad, P.A., Iams, K., Brown, W.C., Sohanpal, B. and Ole-MoiYoi, O.K. (1987) DNA probes detect genomic diversity in *Theileria parva* stocks. *Molecular Biochemistry and Parasitology*, 25, 213–226.

Cook, J.H. (1901) Notes on cases of sleeping sickness occurring in the Uganda protectorate. *Journal of Tropical Medicine*, 4, 236–239.

Cooke, W.E., Gregg, A.L. and Manson-Bahr, P.H. (1937) Recent experiences of mild or symptomless infections with *Trypanosoma gambiense* from the Gold Coast and Nigeria. *Transactions of the Royal Society of Tropical Medicine and Hygiene*, 30, 461–466.

Cooper, J. and Dobson, H. (1994) *Aerial Spraying for Tsetse Fly Control. A handbook of aerial spray calibration and monitoring for the sequential aerosol technique*. Natural Resources Institute, Chatham, UK, 24 pp.

Coosemans, M. and Mouchet, J. (1990) Consequences of rural development on vectors and their control. *Annales de la Société Belge de Médecine Tropicale*, 70, 5–23.

Corson, J.F. (1932a) A note on the susceptibility of some birds and wild animals to infection with *Trypanosoma rhodesiense*. *Journal of Tropical Medicine and Hygiene*, 35, 123–124.

Corson, J.F. (1932b) The results of successive bites of an infected tsetse fly. *Journal of Tropical Medicine and Hygiene*, 35, 136–137.

Corson, J.F. (1936) A second note on a high rate of infection of the salivary glands of *Glossina morsitans* after feeding on a reedbuck infected with *Trypanosoma rhodesiense*. *Transactions of the Royal Society of Tropical Medicine and Hygiene*, 30, 207–212.

Corson, J.F. (1938) A third note on a strain of *Trypanosoma gambiense* transmitted by *Glossina morsitans*. *Annals of Tropical Medicine and Parasitology*, 32, 245–248.

Coulibaly, L., Rowlands, G.J., Authié, E., Hecker, P.A., D'Ieteren, G.D.M., Krebs, H., Leak, S.G.A. and Rarieya, J.M. (1995) Effect of tsetse control with insecticide impregnated traps on trypanosome prevalence and productivity of cattle and sheep in northern Côte d'Ivoire. In: *Proceedings of the OAU/ISCTRC 22nd meeting, Kampala, Uganda, 1993*, pp. 244–250.

Cranefield, P.F. (1991) *Science and Empire: East Coast Fever in Rhodesia and the Transvaal*. Cambridge History of Medicine, Cambridge University Press, Cambridge, UK, 385 pp.

Croft, S., East, J.S. and Molyneux, D.H. (1982) Anti-trypanosomal factor in the haemolymph of *Glossina*. *Acta Tropica*, 39, 293–302.

Crump, A.J. and Brady, J. (1979) Circadian activity patterns in three species of tsetse fly: *Glossina palpalis, G. austeni* and *G. morsitans*. *Physiological Entomology*, 4, 311–318

Cuisance, D. (1989) Le piégeage des tsetse. *Etudes et Synthèses de l'IEMVT*, 32, IEMVT, Paris, 172 pp.

Cuisance, D., Politzar, H., Mérot, P. and Tamboura, I. (1984) Les lâchers de mâles irradiés dans la campagne de lutte intégrée contre les glossines dans la zone pastorale de Sidéradougou (Burkina Faso). *Revue d'Élevage et de Médecine Vétérinaire des Pays Tropicaux*, 37, 449–467.

Cuisance, D., Février, J., Dejardin, J. and Filledier, J. (1985) Dispersion linéaire de *Glossina palpalis gambiensis* et de *Glossina tachinoides* dans une galerie forestière en zone Soudano-Guinéenne (Burkina Faso). *Revue d'Élevage et de Médecine Vétérinaire des Pays Tropicaux*, 38, 153–172.

Cunningham, I. and Slater, J.S. (1974) Amino-acid analyses of haemolymph of *Glossina morsitans morsitans* (Westwood). *Acta Tropica*, 31, 85–88.

Cunningham, M.P. and Van Hoeve, K. (1964) Diagnosis of trypanosomiasis in cattle. *ISCTRC/CCTA 10th meeting, Kampala*, Publication No. 97, 51–53.

Curson, H.H. (1924) Notes on *Glossina pallidipes* in Zululand. *Bulletin of Entomological Research*, 14, 445–453.

Curtis, C.F. (1968a) A possible genetic method for the control of insect pests with special reference to tsetse flies (*Glossina* spp.). *Bulletin of Entomological Research*, 57, 509–523.

Curtis, C.F. (1968b) Possible use of translocations to fix desirable genes in insect pest populations. *Nature*, 218, 368–369.

Curtis, C.F. (1968c) Some observations on reproduction and the effects of radiation on *Glossina austeni. Transactions of the Royal Society of Tropical Medicine and Hygiene*, 62, 124.

Curtis, C.F. (1971) Experiments on breeding translocation homozygotes in tsetse flies. *IAEA STI/PUB/265 Vienna* pp. 425–433.

Curtis, C.F. (1972) Sterility from crosses between sub-species of the tsetse fly *Glossina morsitans. Acta Tropica*, 29, 250–268.

Curtis, C.F. (1972) Sterility from crosses between subspecies of the tsetse fly *Glossina morsitans. Transactions of the Royal Society of Tropical Medicine and Hygiene*, 66, 310.

D'Amico, F., Geoffroy, B., Cuisance, D. and Bossy, J.P. (1991) Acquisition de nouvelles données sur l'équipement sensoriel des glossines (Diptera, Glossinidae). *Revue d'Élevage et de Médecine Vétérinaire des Pays Tropicaux*, 44, 75–79.

D'Amico, F., Geoffroy, B., Cuisance, D. and Bossy, J.P. (1992) Sites and abundance of chemoreceptors on the legs of tsetse, *Glossina tachinoides* (Diptera: Glossinidae) *Insect Science and its Application*, 13, 781–786.

D'Costa, M.A. and Rutesasira, A. (1973a) Phospholipid composition of flight muscle sarcosomes from the tsetse fly, *Glossina morsitans. Comparative Biochemistry and Physiology*, 45B, 491–498.

D'Costa, M.A. and Rutesasira, A. (1973b) Variations in lipids during the development of the tsetse fly, *Glossina morsitans. International Journal of Biochemistry*, 4, 467–478.

D'Costa, M.A., Rice, M.J. and Latif, A. (1973) Glycogen in the proventriculus of the tsetse fly. *Journal of Insect Physiology*, 19, 427–433.

D'Haeseleer, F., Van den Abbeele, J., Gooding, R.H., Rolseth, B.M. and Van der Vloedt, A. (1987) An eye colour mutant, *tan*, in the tsetse fly *Glossina palpalis palpalis* (Diptera: Glossinidae). *Genome*, 29, 828–833.

Dagnogo, M. and Gouteux, J.P. (1983) Essai sur le terrain de différents insecticides contre *Glossina palpalis* (Robineau-Desvoidy) et *Glossina tachinoides* Westwood. 1. Effet répulsif de OMDS 1998, OMS 2002, PMS 2000, OMS 18 et OMS 570. *Cahiers ORSTOM Série Entomologie Médecine et Parasitologie*, 21, 29–34.

Dagnogo, M., Lohuirignon, K. and Gouteux, J.P. (1985a) Comportement alimentaire des populations peridomestiques de *Glossina palpalis* (Robineau-Desvoidy) et de *Glossina tachinoides* Westwood du Domaine Guinéen de Côte d'Ivoire. *Cahiers ORSTOM Série Entomologie Médecine et Parasitologie*, 23, 3–8.

Dagnogo, M., Eouzan, J.P. and Lohuirignon, K. (1985b) Comparaison de différents pièges à tsetse (Diptera, Glossinidae) en Côte d'Ivoire et au Congo. *Revue d'Élevage et de Médecine Vétérinaire des Pays Tropicaux*, 38, 371–378.

Dagnogo, M., Eouzan, J.P. and Lohuirignon, K. (1985c) Données préliminaires sur l'efficacité comparée de trois supports attractifs toxiques pour les Glossines: le piège monoconique, le piège biconique et l'écran bleu-noir dans la région de Daloa (Côte d'Ivoire). *Revue d'Élevage et de Médecine Vétérinaire des Pays Tropicaux*, 38, 379–385.

Dagnogo, M., Lohuirignon, K. and Traore, G. (1997) Diversity of *Glossina* in the forest belt of Côte d'Ivoire. *Acta Tropica* 65, 149–153.

Dale, C., Welburn, S.C., Maudlin, I. and Milligan, P.J.M. (1995) The kinetics of maturation of trypanosome infections in tsetse. *Parasitology*, 111, 187–191.

Dame, D.A. and Ford, H.R. (1966) Effect of the chemosterilant Tepa on *Glossina morsitans* Westw. *Bulletin of Entomological Research*, 56, 649–658.

Dame, D.A. and Ford, H.R. (1968) Multiple mating of *Glossina morsitans* Westw. and its potential effect on the sterile male technique. *Bulletin of Entomological Research*, 58, 213–219.

Dame, D.A. and Mackenzie, P.K.I. (1968) Transmission of *Trypanosoma congolense* by chemosterilised male *Glossina morsitans*. *Annals of Tropical Medicine and Parasitology*, 62, 372–374.

Dame, D.A. and Schmidt, C.H. (1970) The sterile male technique against tsetse flies, *Glossina* spp. *Bulletin of the Entomological Society of America*, 16, 24–30.

Dame, D.A., Birkenmeyer, D.R. and Bursell, E. (1968) Development of the thoracic muscle and flight behaviour of *Glossina morsitans orientalis* Vanderplank. *Bulletin of Entomological Research*, 59, 345–350.

Dame, D.A., Birkenmeyer, D.R., Nash, T.A.M. and Jordan, A.M. (1975) The dispersal and survival of laboratory-bred and native *Glossina morsitans morsitans* Westw. (Diptera: Glossinidae) in the field. *Bulletin of Entomological Research*, 65, 453–457.

Davey, J.B. (1910) Notes on the habits of *Glossina fusca*. *Bulletin of Entomological Research*, 1, 143–160.

Davies, E.D.G. and Southern, D.I. (1976) Giemsa C-banding with the genus *Glossina* (Diptera, Glossinidae). *Genetica*, 46, 413–418.

Davies, H. (1977) *Tsetse Flies in Nigeria*, 3rd ed. Oxford University Press, Ibadan.

Davies, H. and Blasdale, P. (1960) The eradication of *Glossina m. submorsitans* and *G. tachinoides* in part of a river flood plain in northern Nigeria by chemical means. *Bulletin of Entomological Research*, 51, 265–270.

Davies, J.B. (1993) Description of a computer model of forest onchocerciasis transmission and its application to field scenarios of vector control and chemotherapy. *Annals of Tropical Medicine and Parasitology*, 87, 41–63.

Davies, J.E. (1978) The use of ageing techniques to evaluate the effects of aerial spraying against *Glossina morsitans centralis* Machado (Diptera: Glossinidae) in northern Botswana. *Bulletin of Entomological Research*, 68, 373–383.

Davies, J.N.P. (1962a) The cause of sleeping-sickness? Entebbe 1902–03 Part I. *East African Medical Journal*, 39, 81–99.

Davies, J.N.P. (1962b) The cause of sleeping sickness? Entebbe 1902–03 Part II. *East African Medical Journal*, 39, 145–160.

Davis, J.C. and Gooding, R.H. (1983) Spectral sensitivity and flicker fusion frequencies of the compound eye of salmon and wild-type tsetse flies, *Glossina morsitans*. *Physiological Entomology*, 8, 15–23.

De Greef, C., Imberects, H., Matthysens, G., Van Meirvenne, N. and Hamers, R. (1989) A gene expressed only in serum-resistant variants of *Trypanosoma brucei rhodesiense*. *Molecular Biochemistry and Parasitology*, 36, 169–176.

De Leeuw, P.N. and Rey, B. (1995) An analysis of current trends in the distribution patterns of ruminant livestock in sub-Saharan Africa. *World Animal Review*, 83, 47–59.

Dean, G.J.W. and Clements, S.A. (1969) Effect of gamma-irradiation on *Glossina pallidipes* Aust. *Bulletin of Entomological Research*, 58, 775–762.

Dean, G.J.W. and Wortham, S.M. (1969) Effect of gamma-irradiation on the tsetse fly *Glossina morsitans* Westw. *Bulletin of Entomological Research*, 58, 505–515.

Dean, G.J., Wilson, F. and Wortham, S. (1968) Some factors affecting eclosion of *Glossina morsitans* Westw. from pupae. *Bulletin of Entomological Research*, 57, 367–377.

Dean, G.J.W., Williamson, B.R. and Phelps, R.J. (1969a) Behavioural studies of *Glossina morsitans* Westw. using Tantalum-182. *Bulletin of Entomological Research*, 58, 763–771.

Dean, G.J.W., Dame, D.A. and Birkenmeyer, D. (1969b) Field cage evaluation of the competitiveness of male *Glossina morsitans orientalis* Vanderplank sterilised with Tepa or Gamma irradiation. *Bulletin of Entomological Research*, 59, 339–344.

Dehoux, J.P. (1990) Chimioprophylaxie anti-trypanosomienne de bovins N'Dama importes de Sénegambie et du Zaïre au Gabon. *Revue d'Élevage et de Médecine Vétérinaire des Pays Tropicaux*, 43, 337–341.

Delafosse, A., Bengaly, Z. and Duvallet, G. (1995) Absence d'interaction des infections à *Trypanosoma theileri* avec le diagnostic des trypanosomoses animales par détection des antigènes circulants. *Revue d'Élevage et de Médecine Vétérinaire des Pays Tropicaux*, 48, 18–20.

Den Otter, C.J. and Saini, R.K. (1985) Pheromone perception in the tsetse fly *Glossina morsitans morsitans*. *Entomologia Experimentalis et Applicata*, 39, 155–161.

Den Otter, C.J., Tchicaya, T. and Schutte, A. (1991) Effects of age, sex and hunger on the antennal olfactory sensitivity of tsetse flies. *Physiological Entomology*, 16, 173–182.

Denecke, K. (1941) Menschen pathogene Trypanosomen des Hundes auf Fernando Po. Ein Beitrag zur Epidemiologie der Schlafkrankheit. *Archiv für Hygiene und Bakteriologie*, 126, 38–42. (In: Soltys, 1971.)

Denlinger, D.L. (1975) Insect hormones as tsetse abortifacients. *Nature*, 253, 347–348.

Denlinger, D.L. and Ma, W.-C. (1974) Dynamics of the pregnancy cycle in the tsetse *Glossina morsitans*. *Journal of Insect Physiology*, 20, 1015–1026.

Denlinger, D.L. and Zdárek, J. (1992) Rhythmic pulses of haemolymph pressure associated with parturition and ovulation in the tsetse fly, *Glossina morsitans*. *Physiological Entomology*, 17, 127–130.

Denlinger, D.L. and Zdárek, J. (1994) Metamorphosis behavior of flies. *Annual Review of Entomology*, 39, 243–266.

Denlinger, D.L., Gnagey, A.L., Rockey, S.J., Chaudhury, M.F.B. and Fertel, R.H. (1984) Change in levels of cyclic AMP and cyclic GMP during pregnancy and larval development of the tsetse fly, *Glossina morsitans*. *Comparative Biochemistry and Physiology*, 77C, 233–236.

Deportes, I., Geoffroy, B., Cuisance, D., Den Otter, C.J., Carlson, D.A. and Ravallec, M. (1994) Les chimiorécepteurs des ailes chez la glossine (*Diptera: Glossinidae*). Approche structurale et électro-physiologique chez *Glossina fuscipes fuscipes*. *Revue d'Élevage et de Médecine Vétérinaire des Pays Tropicaux*, 47, 81–88.

Desowitz, R.S. and Fairbairn, H. (1955) The influence of temperature on the length of developmental cycle of *Trypanosoma vivax* in *Glossina palpalis*. *Annals of Tropical Medicine and Parasitology*, 49, 161–163.

Desquesnes, M. (1997) Evaluation of a simple PCR technique for the diagnosis of *Trypanosoma vivax* infection in the serum of cattle in comparison to parasitological techniques and antigen–enzyme-linked immuno-sorbent assay. *Acta Tropica*, 65, 139–148.

Desquesnes, M. and de La Rocque, S. (1995) Comparaison de la sensibilité du test Woo et d'un test de détection des antigènes de *Trypanosoma vivax* chez deux moutons expérimentalement infectés avec une souche guyanaise du parasite. *Revue d'Élevage et de Médecine Vétérinaire des Pays Tropicaux*, 48, 247–253.

Dethier, V.G. (1954) Notes on the biting response of tsetse flies. *American Journal of Tropical Medicine and Hygiene*, 3, 160–171.

Diack, A., Moloo, S.K. and Peregrine, A.S. (1997) Effect of diminazene aceturate on the infectivity and transmissibility of drug-resistant *Trypanosoma congolense* in *Glossina morsitans centralis*. *Veterinary Parasitology*, 70, 13–23.

Diall, O., Touré, O.B., Diarra, B. and Sanogo, Y. (1992) Trypanosomose et traitements trypanocides chez le veau Ndama en milieu fortement infesté de glossines (ranch de Madina-Diassa au Mali). *Revue d'Élevage et de Médecine Vétérinaire des Pays Tropicaux*, 45, 155–161.

Diallo, A. (1981) *Glossina morsitans submorsitans* Newstead 1910, (Diptera muscidae) en zone de savane soudano-guinéenne au Mali 1. Ecodistribution et fluctuations saisonnières. *Revue d'Élevage et de Médecine Vétérinaire des Pays Tropicaux*, 34, Part 2, 179–185.

Diallo, B.P., Truc, P. and Laveissière, C. (1997) A new method for identifying blood meals of human origin in tsetse flies. *Acta Tropica*, 63, 61–64.

Dietz, K. (1988) Density-dependence in parasite transmission dynamics. *Parasitology Today*, 4, 91–97.

Dillman, J.S.S. and Townsend, A.J. (1979) A trypanosomiasis survey of wild animals in the Luangwa valley, Zambia. *Acta Tropica*, 36, 349–356.

Dipeolu, O.O. (1975) Studies on the development of *Trypanosoma congolense* in tsetse flies (*Glossina*: Diptera) and the factors affecting it. *Acta Protozoologica*, 14, 241–251.

Dipeolu, O.O. and Adam, K.M.G. (1974) On the use of membrane feeding to study the development of *Trypanosoma brucei* in *Glossina*. *Acta Tropica*, 31, 185–201.

Dirie, M.F., Murphy, N.B. and Gardiner, P.R. (1993) DNA fingerprinting of *Trypanosoma vivax* isolates rapidly identifies intraspecific relationships. *Journal of Eukaryotic Microbiology*, 40, 132–134.

Distelmans, W., D'Haeseleer, F., Kaufman, L. and Rousseeuw, P. (1982) The susceptibility of *Glossina palpalis palpalis* at different ages to infection with *Trypanosoma congolense*. *Annales de la Société Belge de Médecine Tropicale*, 62, 41–47.

Distelmans, W., D'Haeseleer, F. and Mortelmans, J. (1983) Efficacy of systemic administration of Ivermectin against tsetse flies. *Annales de la Société Belge de Médecine Tropicale*, 63, 119–125.

Distelmans, W., Makumyaviri, A.M., D'Haeseleer, F., Claes, Y., Le Ray, D. and Gooding, R.H. (1985) Influence of the Salmon mutant of *Glossina morsitans* on the susceptibility to infection with *T. congolense*. *Acta Tropica*, 42, 143–148.

Dixon, J.B., Cull, R.S., Dunbar, I.F., Greenhill, B.J., Landeg, F.J. and Miller, W.M. (1971) Non-cyclical transmission of trypanosomiasis in Uganda: 1. Abundance and biting behaviour of Tabanidae and *Stomoxys*. *Veterinary Record*, 89, 228–233.

Dodd, C.W.H. (1971) Factors regulating ovarian cycle in tsetse flies. *Transactions of the Royal Society of Tropical Medicine and Hygiene*, 65, 223.

Doku, C. and Brady, J. (1989) Landing site preferences of *Glossina morsitans morsitans* Westwood (Diptera: Glossinidae) in the laboratory: avoidance of horizontal features? *Bulletin of Entomological Research*, 79, 521–528.

Dolan, R.B. (1987) Genetics and trypanotolerance. *Parasitology Today*, 3, 137–143.

Dolan, R.B., Sayer, P.D. and Heath, B.R. (1988) Pyrethroid impregnated ear-tags in trypanosomiasis control. *Tropical Animal Health and Production*, 20, 267–268.

Dorward, D.C. and Payne, A.I. (1975) Deforestation, the decline of the horse, and the spread of the tsetse fly and trypanosomiasis (*Nagana*) in nineteenth century Sierra Leone. *Journal of African History*, 16, 239–256.

Douati, A., Küpper, W., Kotia, K. and Badou, K. (1986) Contrôle des Glossines (*Glossina*: Diptera, Muscidae) a l'aide d'écrans et de pièges (méthodes statiques): bilan de deux années de lutte à Sirasso dans le nord de la Côte d'Ivoire. *Revue d'Élevage et de Médecine Vétérinaire des Pays Tropicaux*, 39, 213–219.

Douthwaite, R.J. (1986) Effects of drift sprays of endosulfan applied for tsetse-fly control on breeding of little bee-eaters in Somalia. *Environmental Pollution (Series A)*, 41, 11–22.

Douthwaite, R.J. (1992a) Effects of DDT treatments applied for tsetse fly control on white-browed sparrow-weaver (*Plocepasser mahali*) populations in NW Zimbabwe. *African Journal of Ecology*, 30, 233–244.

Douthwaite, R.J. (1992b) Effects of DDT treatments applied for tsetse fly control on white-headed black chat (*Thamnolaea arnoti*) populations in Zimbabwe. I. Population changes. *Ecotoxicology*, 1, 17–30.

Douthwaite, R.J. (1992c) Effects of DDT on the fish eagle *Haliaeetus vocifer* population of Lake Kariba in Zimbabwe. *Ibis*, 134, 250–258.

Douthwaite, R.J. (1993) Effect on birds of DDT applied for tsetse fly control in Zimbabwe. *Proceedings of the 8th Pan African Ornithological Congress, Burundi*, 268, 608–610.

Douthwaite, R.J. (1995) Occurrence and consequences of DDT residues in woodland birds following tsetse fly spraying operations in NW Zimbabwe. *Journal of Applied Ecology*, 32, 727–738.

Douthwaite, R.J. and Fry, C.H. (1982) Food and feeding behaviour of the little bee-eater *Merops pusillus* in relation to tsetse fly control by insecticides. *Biological Conservation*, 23, 71–78.

Douthwaite, R.J. and Tingle, C.C.D. (1992) Effects of DDT treatments applied for tsetse fly control on white-headed black-chat (*Thamnolaea arnoti*) populations in Zimbabwe. 2. Cause of decline. *Ecotoxicology*, 1, 101–115.

Douthwaite, R.J., Fox, P.J. Mattheissen, P. and Russell-Smith, A. (1981) Environmental impact of aerial spraying operations against tsetse flies in Botswana. *ISCTR 17th meeting, Arusha*, pp. 626–633.

Douthwaite, R.J., Hustler, C.W., Kruger, J. and Renzoni, A. (1992) DDT residues and mercury levels in Reed Cormorants on Lake Kariba: a hazard assessment. *Ostrich*, 63, 123–127.

Doutt, J. (1958) Vice, virtue and the vedalia. *Entomological Society of America Bulletin*, 4, 119–123.

Dowler, M.E., Schillinger, D. and Connor, R. (1989) Notes on the routine intravenous use of Isometamidium in the control of bovine trypanosomiasis on the Kenya coast. *Tropical Animal Health and Production*, 21, 4–10.

Drager, N. and Mehlitz, D. (1978) Investigations on the prevalence of trypanosome carriers and the antibody response in wildlife in northern Botswana. *Tropenmedizin und Parasitologie*, 29, 223–233.

Dransfield, R.D. (1984) The range of attraction of the biconical trap for *Glossina pallidipes* and *Glossina brevipalpis*. *Insect Science and its Application*, 5, 363–368.

Dransfield, R.D. and Brightwell, R. (1989) Problems of field testing theoretical models: A case study. *Annales de la Société Belge de Médecine Tropicale*, 69, Suppl. 1, 147–154.

Dransfield, R.D., Brightwell, R., Chaudhury, M.F., Golder, T.K. and Tarimo, S.A.R. (1986a) The use of odour attractants for sampling *Glossina pallidipes* Austen (Diptera: Glossinidae) at Nguruman, south-western Kenya. *Bulletin of Entomological Research*, 76, 607–619.

Dransfield, R.D., Chaudhury, M.F., Tarimo, S.A.R., Golder, T.K. and Brightwell, R. (1986b) Population dynamics of *Glossina pallidipes* under drought conditions at Nguruman, Kenya. *18th Meeting of the International Scientific Council for Trypanosomiasis Research and Control, Harare, Zimbabwe, 1985.* OAU/STRC, Nairobi, pp. 284–292.

Dransfield, R.D., Brightwell, R., Kiilu, J., Chaudhury, M.F. and Adabie, D.A. (1989) Size and mortality rates of *Glossina pallidipes* in the semi-arid zone of south-western Kenya. *Medical and Veterinary Entomology*, 3, 83–95.

Dransfield, R.D., Brightwell, R., Kyorku, C. and Williams, B. (1990) Control of tsetse fly (Diptera: Glossinidae) populations using traps at Nguruman, south-west Kenya. *Bulletin of Entomological Research*, 80, 265–276.

Dransfield, R.D., Williams, B. and Brightwell, R. (1991) Control of tsetse flies and trypanosomiasis: myth or reality? *Parasitology Today*, 7, 287–291.

Drummond, R.O. (1987) Economic aspects of ectoparasites of cattle in North America. In: Leaning, W.D.H. and Guerrero, J. (eds) *The Economic Impact of Parasitism in Cattle*. 23rd World Veterinary Congress, 19 August, Montreal, Quebec, Canada, pp. 9–24.

Du Toit, R. (1954) Trypanosomiasis control in Zululand and the control of tsetse flies by chemical means. *Ondestepoort Journal of Veterinary Research*, 26, 317–387.

Duggan, A.J. (1962) A survey of sleeping sickness in northern Nigeria from the earliest times to the present day. *Transactions of the Royal Society of Tropical Medicine and Hygiene*, 55, 439–481.

Duke, A.L. (1928) On the effect on the longevity of *G. palpalis* of trypanosome infections. *Annals of Tropical Medicine and Parasitology*, 22, 25–32.

Duke, H.L. (1919a) Some observations on the bionomics of *Glossina palpalis* on the islands of Victoria Nyanza. *Bulletin of Entomological Research*, 9, 263–272.

Duke, H.L. (1919b) Tsetse flies and trypanosomiasis. Some questions suggested by the later history of the sleeping sickness epidemic in Uganda protectorate. *Parasitology*, 11, 415–422.

Duke, H.L. (1923) An inquiry into an outbreak of human trypanosomiasis in a '*Glossina morsitans*' belt to the east of Mwanza, Tanganyika territory. *Proceedings of the Royal Society of London (B)*, 94, 250–265.

Duke, H.L. (1930) The discovery of *Trypanosoma rhodesiense* in man in the Uganda protectorate. *Transactions of the Royal Society of Tropical Medicine and Hygiene*, 24, 201–218.

Duke, H.L. (1933a) Studies on the factors that may influence the transmission of the polymorphic trypanosomes by tsetse. I. A review of existing knowledge on this subject, with some general observations. *Annals of Tropical Medicine and Parasitology*, 27, 99–118.

Duke, H.L. (1933b) Studies on the factors that may influence the transmission of the polymorphic trypanosomes by tsetse. II. On the transmitting power of different races of *Glossina. Annals of Tropical Medicine and Parasitology*, 27, 119–121.

Duke, H.L. (1933c) Studies on the factors that may influence the transmission of the polymorphic trypanosomes by tsetse. V. On the effects of temperature. *Annals of Tropical Medicine and Parasitology*, 27, 437–450.

Duke, H.L. (1933d) Studies on the factors that may influence the transmission of the polymorphic trypanosomes by tsetse. VI. On the duration of the biological cycle in *Glossina*. *Annals of Tropical Medicine and Parasitology*, 27, 451–467.

Duke, H.L. (1933e) Studies on the factors that may influence the transmission of the polymorphic trypanosomes by tsetse. VII. *T. rhodesiense* versus *T. gambiense*: a comparison of their power to develop cyclically in *Glossina*. *Annals of Tropical Medicine and Parasitology*, 27, 569–584.

Duke, H.L. (1933f) Relative susceptibility of the sexes of *Glossina* to infection with trypanosomes. *Annals of Tropical Medicine and Parasitology*, 27, 355–356.

Duke, H.L. (1935a) On the factors that may determine the infectivity of a trypanosome to tsetse. *Transactions of the Royal Society of Tropical Medicine and Hygiene*, 29, 203–206.

Duke, H.L. (1935b) Further studies of the behaviour of *Trypanosoma rhodesiense* recently isolated from man, in antelope and other African game animals. *Parasitology*, 27, 68–92.

Duke, H.L. (1935c) Studies on factors that may influence the transmission of the polymorphic trypanosomes by tsetse. IX. On the infectivity to *Glossina* of the trypanosome in the blood of the mammal. *Annals of Tropical Medicine and Parasitology*, 29, 131–143.

Duke, H.L. (1936a) On the power of *Glossina morsitans* and *Glossina palpalis* to transmit the trypanosomes of the *brucei* group. *Annals of Tropical Medicine and Parasitology*, 30, 37–38.

Duke, H.L. (1936b) Antelopes as reservoirs of *Trypanosoma gambiense*. *Transactions of the Royal Society of Tropical Medicine and Hygiene*, 29, 129–131.

Duke, H.L. (1936c) Biology of the trypanosomes of man in Africa. *Transactions of the Royal Society of Tropical Medicine and Hygiene*, 29, 275–308.

Duke, H.L. (1937) Studies on the effect on *T. gambiense* and *T. rhodesiense* of prolonged maintenance in mammals other than man; with special reference to the power of these trypanosomes to infect man. V. The effect of prolonged maintenance away from man on the infectivity of *T. rhodesiense* for man. *Parasitology*, 29, 12–34.

Duke, H.L. (1944) Rhodesian Sleeping sickness. *Transactions of the Royal Society of Tropical Medicine and Hygiene*, 38, 163–165.

Dukes, P., Kaukas, A., Hudson, K.M., Asonganyi, T. and Gashumba, J.K. (1989) A new method for isolating *Trypanosoma brucei gambiense* from sleeping sickness patients. *Transactions of the Royal Society of Tropical Medicine and Hygiene*, 83, 636–639.

Dukes, P., Gibson, W.C., Gashumba, J.K., Hudson, K.M., Bromidge, T.J., Kaukus, A., Asonganyi, T. and Magnus, E. (1992) Absence of the LiTAT 1.3 (CATT antigen) gene in *Trypanosoma brucei gambiense* stocks from Cameroon. *Acta Tropica*, 51, 123–134.

Dutton, J.E. (1902) Preliminary note upon a trypanosome occurring in the blood of man. *Thomson Yates Laboratory Reports*, 4, 455–467.

Dutton, J.E. and Todd J.L. (1906) The distribution and spread of sleeping sickness in the Congo Free State with suggestions on prophylaxis. Reports on the Expedition to the Congo, 1903–1905. *Liverpool School of Tropical Medicine Memoir*, 11, 23–38.

Dutton, J.E., Todd, J.L. and Tobey, E.N. (1906) Concerning certain parasitic protozoa observed in Africa. Part I. *Memoirs of the Liverpool School of Tropical Medicine*, 21, 87–97.

Dwinger, R.H., Rawlings, P., Jeannin, P. and Grieve, A.S. (1990) Experimental infection of N'Dama cattle with trypanosomes using *Glossina palpalis gambiensis* caught in the wild. *Tropical Animal Health and Production*, 22, 37–43.

Dwinger, R.H., Grieve, A.S., Snow, W.F., Rawlings, P., Jabang, B. and Williams, D.J.L. (1992) Maternal antibodies in N'Dama cattle kept under natural trypanosomiasis risk in The Gambia. *Parasite Immunology*, 14, 351–354.

Dwinger, R.H., Agyemang, K., Snow, W.F., Rawlings, P., Leperre, P. and Bah, M.L. (1994) Productivity of trypanotolerant cattle kept under traditional management conditions in The Gambia. *Veterinary Quarterly*, 16, 81–86.

Dye, C. (1986) Vectorial capacity: must we measure all its components? *Parasitology Today*, 2, 203–209.

Dye, C. (1992) The analysis of parasite transmission by bloodsucking insects. *Annual Review of Entomology*, 37, 1–19.

Dye, C. (1994a) Approaches to vector control: new and trusted. 5. The epidemiological context of vector control. *Transactions of the Royal Society of Tropical Medicine and Parasitology*, 88, 147–149.

Dye, C. (1994b) Models for Leishmania transmission. In: Perry, B.D. and Hansen, J.W. (eds) *Modelling Vector-borne and Other Parasitic Diseases*. Proceedings of a workshop organised by ILRAD in collaboration with FAO, 23–27 November 1992. ILRAD, Nairobi, pp. 95–104.

Dye, W.H. (1927) The relative importance of man and beast in human trypanosomiasis. *Transactions of the Royal Society of Tropical Medicine and Hygiene*, 21, 187–198.

East, J., Molyneux, D.H. and Hillen, N. (1980) Haemocytes of *Glossina*. *Annals of Tropical Medicine and Parasitology*, 74, 471–474.

East, J., Molyneux, D.H., Maudlin, I. and Dukes, P. (1983) Effect of *Glossina* haemolymph on salivarian trypanosomes *in vitro*. *Annals of Tropical Medicine and Parasitology*, 77, 97–99.

Ebert, D. and Herre, E.A. (1996) The evolution of parasitic diseases. *Parasitology Today*, 12, 96–101.

Echessah, P.N., Swallow, B.M., Kamara, D. and Curry, J.J. (1997) Willingness to contribute labor and money to tsetse control: application of contingent valuation in Busia district, Kenya. *World Development*, 25, 239–253.

Edeghere, H., Olise, P.O. and Olatunde, D.S. (1989) Human African trypanosomiasis (sleeping sickness): new endemic foci in Bendel state, Nigeria. *Tropical Medicine and Parasitology*, 40, 16–20.

Edney, E.B. and Barrass, R. (1962) The body temperature of the tsetse fly, *Glossina morsitans* Westwood (Diptera, Muscidae). *Journal of Insect Physiology*, 8, 469–481.

Eisler, M.C., Gault, E.A., Smith, H.V., Peregrine, A.S. and Holmes, P.H. (1993) Evaluation and improvement of an enzyme-linked immunosorbent assay (ELISA) for the detection of isometamidium in bovine serum. *Therapeutic Drug Monitor*, 15, 236–242.

Eisler, M., Lessard, P., Masake, R.A., Moloo, S.K. and Peregrine, A.S. (1998) Sensitivity and specificity of antigen-capture ELISAs for diagnosis of *Trypanosoma congolense* and *Trypanosoma vivax* infections in cattle. *Veterinary Parasitology*, in press.

Ejezie, G.C. (1983) Hormones and reproduction in tsetse flies. *Zeitschrift für Angewandte Zoologie*, 70, 1–12.

Ejezie, G.C. and Davey, K.G. (1974) Changes in the neurosecretory cells, corpus cardiacum and corpus allatum during pregnancy in *Glossina austeni* Newst. (Diptera, Glossinidae). *Bulletin of Entomological Research*, 64, 247–256.

Ejezie, G.C. and Davey, K.G. (1977) Some effects of mating in female tsetse, *Glossina morsitans* Newst. *Journal of Experimental Zoology*, 200, 303–310.

Ejezie, G.C. and Davey, K.G. (1982) Some effects of prolonged virginity on feeding and milk-gland activity in females of the tsetse *Glossina austeni* (Newst.). *Insect Science and its Application*, 3, 53–57.

Elger, D. and Liebisch, A. (1982) Recherches de terrain sur l'effet de Permethrine pour la lutte contre les mouches sur les bovins de prairie au Nord de l'Allemagne. *Tierarztl Umschau*, 37, 437–442.

Elhassan, E., Ikede, B.O. and Adeyemo, O. (1994) Trypanosomiasis and reproduction: effect of *Trypanosoma vivax* infection on the oestrous cycle in the ewe. *Tropical Animal Health and Production*, 26, 213–218.

Ellis, D.S. and Evans, D.A. (1976) Electron microscope studies of the passage through the anterior midgut cells of *Glossina morsitans morsitans* by *Trypanosoma brucei rhodesiense*. *Transactions of the Royal Society of Tropical Medicine and Hygiene*, 70, 20.

Ellis, D.S. and Evans, D.A. (1977) Passage of *Trypanosoma brucei rhodesiense* through the peritrophic membrane of *Glossina morsitans morsitans*. *Nature*, 267, 834–835.

Ellis, D.S. and Maudlin, I. (1987) Salivary gland hyperplasia in wild caught tsetse from Zimbabwe. *Entomologia Experimentalis et Applicata*, 45, 167–173.

Ellis, D.S., Evans, D.A. and Stamford, S. (1982) Studies by electron microscopy of the giant forms of some African and South American trypanosomes found other than within their mammalian hosts. *Folia Parasitologia (Praha)*, 29, 5–11.

Elsen, P., de Lil, E. and Roelants, P. (1989) Première mise en évidence de chromosomes polytenes chez les *Glossina* adultes. *Annales de la Société Belge de Médecine Tropicale*, 69, 245–250.

Elsen, P., Amoudi, M.A. and Leclercq, M. (1990) First record of *Glossina fuscipes* Newstead, 1910 and *Glossina morsitans submorsitans* Newstead, 1910 in southwestern Saudi Arabia. *Annales de la Société Belge de Médecine Tropicale*, 70, 281–287.

Eminson, R.A.F. (1915) Observations on *Glossina morsitans* in Northern Rhodesia. *Bulletin of Entomological Research*, 5, 381–382.

Emslie, V.W. and Steinberg, E.A. (1973) The production in the bovine of antibody specific for closely related species and the use of such antisera in the identification of tsetse blood meals. *Annals of Tropical Medicine and Parasitology*, 67, 213–217.

Endege, W.O., Lonsdale-Eccles, J.D., Olembo, N.K., Moloo, S.K. and Ole-MoiYoi, O.K. (1989) Purification and characterisation of two fibrinolysins from the midgut of adult female *Glossina morsitans centralis*. *Comparative Biochemistry and Physiology*, 92B, 25–34.

England, E.C. and Baldry, D.A.T. (1972) The hosts and trypanosome infection rates of *Glossina pallidipes* in the Lambwe and Roo valleys. *Bulletin of the World Health Organisation*, 47, 785–788.

Enyaru, J.C.K., Odiit, M., Gashumba, J.K., Carasco, J.F. and Rwendeire, A.J.J. (1992) Characterization by isoenzyme electrophoresis of *Trypanozoon* stocks from sleeping sickness endemic areas of south-eastern Uganda. *Bulletin of the World Health Organisation*, 70, 631–636.

Enyaru, J.C.K., Stevens, J.R., Odiit, M., Okuna, N.M. and Carasco, J.F. (1993) Isoenzyme comparison of *Trypanozoon* isolates from two sleeping sickness areas of south-eastern Uganda. *Acta Tropica*, 55, 97–115.

Eouzan, J.-P., Frézil, J.-L. and Lancien, J. (1981) Epidemiologie de la trypanosomiase humaine au Congo: Les déplacements des Glossines dans le foyer du 'Couloir'. *Cahiers ORSTOM Série Entomologie Médecine et Parasitologie*, 19, 81–85.

Epstein, H. (1971) *The Origin of Domestic Animals of Africa*. Africana Publishing Corporation, New York.

Evans, A.M. (1919) On the genital armature of the female tsetse flies. *Annals of Tropical Medicine and Parasitology*, 13, 31–56.

Evans, D.A. and Brown, R.C. (1972) The utilisation of glucose and proline by culture forms of *Trypanosoma brucei*. *Journal of Protozoology*, 19, 686–690.

Evans, D.A. and Ellis, D.S. (1975) Penetration of midgut cells of *Glossina morsitans morsitans* by *Trypanosoma brucei rhodesiense*. *Nature*, 258, 231–233.

Evans, D.A. and Ellis, D.S. (1983) Recent observations on the behaviour of certain trypanosomes within their insect hosts. *Advances in Parasitology*, 22, 1–42.

Everts, J.W., Van Frankenhuyzen, K., Roman, B., Cullen, J., Copplestone, J., Koeman, J.H. and Van Frankenhuytzen, K. (1983) Observations on side effects of endosulfan used to control tsetse in a settlement area in connection with a campaign against human sleeping sickness in Ivory Coast. *Tropical Pest Management*, 29, 177–182.

Fairbairn, H. (1943) The agricultural problems posed by sleeping sickness settlements. *East African Forestry Journal*, 19, 17–22.

Fairbairn, H. (1948) Sleeping sickness in Tanganyika territory, 1922–1946. *Tropical Diseases Bulletin*, 45, 1–17.

Fairbairn, H. (1954) The animal reservoirs of *Trypanosoma rhodesiense*. *Annales de la Société Belge de Médecine Tropicale*, 34, 663–669.

Fairbairn, H. (1957) The penetration of *Trypanosoma rhodesiense* through the peritrophic membrane of *Glossina palpalis*. *Annals of Tropical Medicine and Parasitology*, 51, 18–19.

Fairbairn, H. and Burtt, E. (1946) The infectivity to man of a strain of *Trypanosoma rhodesiense* transmitted cyclically by *Glossina morsitans* through sheep and antelope: evidence that man requires a minimum infective dose of metacyclic trypanosomes. *Annals of Tropical Medicine and Parasitology*, 40, 270–313.

Fairbairn, H. and Culwick, A.T. (1950) The transmission of the polymorphic trypanosomes. *Acta Tropica*, 7, 19–47.

Fairbairn, H. and Godfrey, D.G. (1957) The local reaction in man at the site of infection with *Trypanosoma rhodesiense*. *Annals of Tropical Medicine and Parasitology*, 51, 464–470.

Fairbairn, H. and Watson, H.J.C. (1955) The transmission of *Trypanosoma vivax* by *Glossina palpalis*. *Annals of Tropical Medicine and Parasitology*, 49, 250–259.

Fairclough, R. (1962) A summary of the use of Berenil in Kenya. *ISCTRC/CCTA 9th meeting, Conakry*, Publication No. 88, 81–86.

Fairclough, R. and Thomson, W.E.F. (1958) The effect of insecticidal spraying against *Glossina palpalis fuscipes*, Newstead in the Nyando river basin of Kenya. *East African Agricultural Journal*, 23, 186–189.

FAO (1982) Tsetse Control Training Manual, Vol. 1, *Tsetse Biology, Systematics and Distribution; Techniques*. Pollock, J.N. (ed.). Food and Agriculture Organisation of the United Nations, Rome, 280 pp.

FAO (1982a) Tsetse Control Training Manual, Vol. 2, *Ecology and Behaviour of Tsetse*. Pollock, J.N. (ed.). Food and Agriculture Organisation of the United Nations, Rome, 101 pp.

FAO (1982b) Tsetse Control Training Manual, Vol. 3, *Control Methods and Side-effects*. Pollock, J.N. (ed.) Food and Agriculture Organisation of the United Nations, Rome, 128 pp.

FAO (1992a) Tsetse Control Training Manual, Vol. 4, *Use of Attractive Devices for Tsetse Survey and Control*. Dransfield, R.D. and Brightwell, R. (eds). Food and Agriculture Organisation of the United Nations, Rome, 196 pp.

FAO (1992b) Tsetse Control Training Manual, Vol. 5, *Insecticides for Tsetse and Trypanosomiasis Control Using Attractive Bait Technology*. Alsop, N. (ed.). Food and Agriculture Organisation of the United Nations, Rome, 88 pp.

FAO (1992c) *The New World Screwworm Eradication Programme; North Africa 1988–1992*. Food and Agriculture Organisation of the United Nations, Rome, 192 pp.

FAO/WHO/OIE (1963) The economic losses caused by animal diseases. In: *Animal Health Yearbook 1962*. Food and Agricultural Organisation of the United Nations, Rome, pp. 284–313.

Fenn, R. (1992) Effect of physiological age and pregnancy stage on the tolerance of *Glossina* species to aerosol and topical application of endosulfan and the consequences for aerial control. *Tropical Pest Management*, 38, 453–458.

Ferenc, S.A., Raymond, H.L. and Lancelot, R. (1988) Essai de transmission mécanique de *Trypanosoma vivax* Ziemann (Kinetoplastida: Trypanosomatidae) par le taon neotropical *Cryptotylus unicolor* (Wiedemann) (Diptera: Tabanidae). *Proceedings of the 18th International Congress of Entomology*, p. 295.

Ferriere, C. (1935) Les hymenopteres parasites des mouches tsetse. *Mittelungen der Schweizerischen Entomologischen Gesellschaft*, 16, 328–340.

Fiedler, O.G.H. (1954) The parasites of tsetse flies in Zululand with special reference to the influence of the hosts upon them. *Onderstepoort Journal of Veterinary Research*, 26, 399–404.

Fiedler, O.G.H., du Toit, R. and Kluge, E.B. (1954) The influence of the tsetse fly eradication campaign on the breeding activity of *Glossinae* and their parasites in Zululand. *Ondestepoort Journal of Veterinary Research*, 26, 389–397.

Fiennes, R.W.T.-W. (1970) Pathogenesis and pathology of animal trypanosomiases. In: Mulligan, H.W. (ed.) *The African Trypanosomiases*. Allen and Unwin, London, pp. 729–750.

Filledier, J. and Mérot, P. (1989a) Etude de l'attractivité de solutions isolées par fractionnement de l'urine de bovin Baoulé pour *Glossina tachinoides* Westwood, 1850 au Burkina Faso. *Revue d'Élevage et de Médecine Vétérinaire des Pays Tropicaux*, 42, 453–455.

Filledier, J. and Mérot, P. (1989b) Pouvoir attractif de l'association M-cresol 1-octen-3-ol dans un type de diffuseur pratique et de longue durée d'action pour *Glossina tachinoides* Westwood 1850 dans le sud-ouest du Burkina Faso. *Revue d'Élevage et de Médecine Vétérinaire des Pays Tropicaux*, 42, 541–544.

Filledier, J., Duvallet, G. and Mérot, P. (1988) Comparaison du pouvoir attractif des bovins Zébu et Baoulé pour *Glossina tachinoides* Westwood, 1850 et *Glossina morsitans submorsitans* Newstead, 1910 en savane soudano-guinéenne (Burkina Faso). *Revue d'Élevage et de Médecine Vétérinaire des Pays Tropicaux*, 41, 191–196.

Fimmen, H.O., Mehlitz, D., Horchaer, F. and Karbe, E. (1982a) Colostral antibodies and *Trypanosoma congolense* infections in calves. In: Karbe, E. and Freitas, E.K. (eds) *Trypanotolerance Recherche and Applications*. GTZ, No. 116, Germany, pp. 173–187.

Fimmen, H.O., Mehlitz, D., Horchner, F., Karbe, E. and Freitas, E.K. (1982b) The effect of maternal colostral antibodies on the evolution of trypanosomiasis infection in N'Dama calves. *ISCTR Arusha Publication* No. 112, 352–360.

Findlay, G.M., Hardwicke, J. and Phelps, R.J. (1946) Tsetse fly repellents. *Transactions of the Royal Society of Tropical Medicine and Hygiene*, 40, 341–344.

Finlayson, L.H. (1967) Behaviour and regulation of puparium formation in the larva of the tsetse fly *Glossina morsitans orientalis* Vanderplank in relation to humidity, light and mechanical stimuli. *Bulletin of Entomological Research*, 57, 301–313.

Finlayson, L.H. (1972) Chemoreceptors, cuticular mechanoreceptors, and peripheral multiterminal neurones in the larva of the tsetse fly (*Glossina*). *Journal of Insect Physiology*, 18, 2265–2275.

Finlayson, L.H. and Rice, M.J. (1972) Sensory and neurosecretory innervation of the tsetse fly proventricular complex. *Transactions of the Royal Society of Tropical Medicine and Hygiene*, 66, 317.

Fiske, W.F. (1913) The bionomics of *Glossina*; a review with hypothetical conclusions. *Bulletin of Entomological Research*, 4, 95–111.

Fiske, W.F. (1920) Investigations into the bionomics of *Glossina palpalis*. *Bulletin of Entomological Research*, 10, 347–463.

Fleming, A.M. (1913) Trypanosomiasis in southern Rhodesia. *Transactions of the Royal Society of Tropical Medicine and Hygiene*, 6, 298–311.

Flint, S. (1985) A comparison of various traps for *Glossina* spp. (Glossinidae) and other Diptera. *Bulletin of Entomological Research*, 75, 529–534.

Flynn, J.N. and Sileghem, M. (1994) Involvement of γδ T cells in immunity to trypanosomiasis. *Immunology*, 83, 86–92.

Foil, L.D. (1989) Tabanids as vectors of disease agents. *Parasitology Today*, 5, 88–96.

Folkers, C. and Jones-Davies, W.J. (1966) The incidence of trypanosomes in blood smears of cattle presented for trypanosomiasis treatment in Northern Nigeria. *Bulletin de l'Office International des Epizooties*, 14, 409–421.

Folkers, C. and Mohammed, A.N. (1965) The importance of biting flies other than *Glossina* in the epidemiology of trypanosomiasis in cattle in Shika, northern Nigeria. *Bulletin de l'Office International des Epizooties*, 13, 331–339.

Ford, J. (1960) The influence of tsetse flies on the distribution of African cattle. *Proceedings of the 1st Federal Scientific Congress, Salisbury*, 357–365.

Ford, J. (1962) Microclimates of tsetse fly resting sites in the Zambesi valley, southern Rhodesia. *ISCTRC 9th meeting Conakry 1962*, 165–170.

Ford, J. (1963) The distribution of the vectors of African pathogenic trypanosomes. *Bulletin of the World Health Organisation*, 28, 653–669.

Ford, J. (1964) The geographical distribution of trypanosome infections in African cattle populations. *Bulletin de l'Office International des Epizooties*, 12, 307–320.

Ford, J. (1966) The role of elephants in controlling the distribution of tsetse flies. *Bulletin of the International Journal of Conservation*, 19, 6.

Ford, J. (1969) Control of the African trypanosomiases with special reference to land use. *Bulletin of the World Health Organisation*, 40, 879–892.

Ford, J. (1970a) Recent information on changes in distribution of tsetse flies. *Joint WHO/FAO African trypanosomiasis information service*. TRYP/INF/70.41, 1–8.

Ford, J. (1970b) The geographical distribution of *Glossina*. In: Mulligan, H.W. (ed.) *The African Trypanosomiases*. George Allen and Unwin, London, pp. 274–297.

Ford, J. (1971) *The Role of the Trypanosomiases in African Ecology*. Clarendon Press, Oxford, 698 pp.

Ford, J. and Katondo, K.M. (1977) Maps of tsetse fly (*Glossina*) distribution in Africa 1973, according to sub-generic groups on scale of 1:5,000,000. *Bulletin of Animal Health and Production in Africa*, 15, 188–194.

Ford, J. and Leggate, B.M. (1961) The geographical and climatic distribution of trypanosome infection rates in *G. morsitans* group of tsetse flies (*Glossina* Wied., Diptera). *Transactions of the Royal Society of Tropical Medicine and Hygiene*, 55, 383–397.

Ford, J., Glasgow, J.P., Johns, D.L. and Welch, J. (1959) Transect fly-rounds in field studies of *Glossina*. *Bulletin of Entomological Research*, 50, 275–284.

Ford, J., Maudlin, I. and Humphryes, K.C. (1972) Comparisons between three small collections of *Glossina morsitans morsitans* (Machado) (Diptera: Glossinidae) from the Kilombero River valley, Tanzania. Part 1. Characteristics of flies exhibiting different patterns of behaviour. *Acta Tropica*, 29, 231–249.

Forde, R.M. (1902a) Some clinical notes on a European patient in whose blood a trypanosome was observed. *Journal of Tropical Medicine and Hygiene*, 5, 261–263.

Forde, R.M. (1902b) The discovery of the human *Trypanosoma*. *British Medical Journal*, November 29.

Foster, R. (1957) Observations on laboratory colonies of the tsetse flies *Glossina morsitans* Westw. and *Glossina austeni* Newstead. *Parasitology*, 47, 361–374.

Foster, R. (1963a) Contributions to the epidemiology of human Sleeping sickness in Liberia. Bionomics of the vector *G. palpalis* R.-D. in a savanna habitat in a focus of the disease. *Transactions of the Royal Society of Tropical Medicine and Hygiene*, 57, 465–475.

Foster, R. (1963b) Infestation of *Glossina palpalis* R.-D. 1830 (Diptera) by larval Mermithidae Braun 1883 (Nematoda) in west Africa, with some comments on the parasitization of man by the worms. *Annals of Tropical Medicine and Parasitology*, 57, 347–358.

Foster, R. (1964) Contributions to the epidemiology of human sleeping sickness in Liberia: bionomics of the vector *Glossina palpalis* (R.-D.) in a forest habitat. *Bulletin of Entomological Research*, 54, 727–744.

Foster, W.A. (1972) Influence of medial neurosecretory cells on reproduction in female *Glossina austeni*. *Transactions of the Royal Society of Tropical Medicine and Hygiene*, 66, 322.

Foster, W.A. (1974) Surgical inhibition of ovulation and gestation in the tsetse fly *Glossina austeni* Newst. (Dipt., Glossinidae). *Bulletin of Entomological Research*, 63, 483–493.

Foster, W.A. (1976) Male sexual maturation of the tsetse flies *Glossina morsitans* Westwood and *G. austeni* Newstead (*Dipt., Glossinidae*) in relation to blood feeding. *Bulletin of Entomological Research*, 66, 389–399.

Fox, R.G.R., Mbando, S.O., Fox, M.S. and Wilson, A. (1993) Effect on herd health and productivity of controlling tsetse and trypanosomiasis by applying deltamethrin to cattle. *Tropical Animal Health and Production*, 25, 203–214.

Fraser, A.D. and Duke, H.L. (1912) Duration of the infectivity of the *Glossina palpalis* after the removal of the lake-shore population. *Report of the Sleeping Sickness Commission of the Royal Society*, 12, 63–75.

Freeman, J.C. (1973) The penetration of the peritrophic membrane of the tsetse flies by trypanosomes. *Acta Tropica*, 30, 347–355.

Freidheim, E.A.H. (1949) Mel B in the treatment of human trypanosomiasis. *American Journal of Tropical Medicine*, 29, 173.

French, F.E. and Kline, D.L. (1989) 1-Octen-3-ol, an effective attractant for Tabanidae (Diptera). *Journal of Medical Entomology*, 26, 459–461.

Fuller, C. and Mossop, M.C. (1929) Entomological notes on *Glossina pallidipes. Bulletin 67*, Union of South Africa Department of Agricultural Science, South Africa.

Fuller, G.K. (1975) *Hippopotamus amphibius* along the Omo river. *Walia*, 6, 4–7.

Fuller, G.K. (1978) Distribution of *Glossina* in south western Ethiopia. *Bulletin of Entomological Research*, 68, 299–302.

Galey, J.B., Mérot, P., Mitteault, A., Filledier, J. and Politzar, H. (1986) Efficacité du dioxyde de carbone comme attractif pour *Glossina tachinoides* en savane humide d'Afrique de l'ouest. *Revue d'Elevage et de Médecine Vétérinaire des Pays Tropicale*, 39, 351–354.

Galun, R., Ben-Eliahu, M.N., Ben-Tamar, D. and Simkin, J. (1980) Long-term protection of animals from tsetse bites through controlled release repellents. In: *Isotope and Radiation Research on Animal Diseases and their Vectors.* IAEA, Vienna, pp. 207–217.

Garcia, E.S., Mello, C.B., Azambuja, P. and Ribeiro, J.M.C. (1994) *Rhodnius prolixus*: salivary antihemostatic components decrease with *Trypanosoma rangeli* infection. *Experimental Parasitology*, 78, 287–293.

Gardiner, P.R. (1989) Recent studies of the biology of *Trypanosoma vivax. Advances in Parasitology*, 28, 229–316.

Garnham, P.C.C. (1960) Blood parasites of hippopotamus in Uganda. *Africa Medical Journal*, 37, No 7.

Gashumba, J.K. (1981) Sleeping sickness in Iganga district, Uganda: application of a new diagnostic technique. *East African Medical Journal*, 58, 699–702.

Gashumba, J.K., Komba, E.K., Truc, P., Allingham, R.M., Ferris, V. and Godfrey, D.G. (1994) The persistence of genetic homogeneity among *Trypanosoma brucei rhodesiense* isolates from patients in north-west Tanzania. *Acta Tropica*, 56, 341–348.

Gass, R.F. and Yeates, R.A. (1979) *In vitro* damage of cultured ookinetes of *Plasmodium gallinaceum* by digestive proteinases from susceptible *Aedes aegypti. Acta Tropica*, 36, 243–252.

Gaston, K.A. and Randolph, S.E. (1993) Reproductive under-performance of tsetse flies in the laboratory, related to feeding frequency. *Physiological Entomology*, 18, 130–136.

Gates, D.B., Cobb, P.E., Williamson, D.L., Bakuli, B. and Dame, D.A. (1983) Integration of insect sterility and insecticides for control of *Glossina morsitans morsitans* Westwood (Diptera: Glossinidae) in Tanzania. III. Test site characteristics and the natural distribution of tsetse flies. *Bulletin of Entomological Research*, 73, 373–381.

Gee, J.D. (1975a) Diuresis in the tsetse fly *Glossina austeni. Journal of Experimental Biology*, 63, 381–390.

Gee, J.D. (1975b) The control of diuresis in the tsetse fly *Glossina austeni*: a preliminary investigation of the diuretic hormone. *Journal of Experimental Biology*, 63, 391–401.

Gee, J.D. (1977) The effects of dietary sodium and potassium on rapid diuresis in the tsetse fly *Glossina morsitans. Journal of Insect Physiology*, 23, 137–143.

Geigy, R., Mwambu, P.M. and Kauffmann, M. (1971) Sleeping sickness survey in Musoma district Tanzania. IV. Examination of wild mammals as a potential reservoir for *Trypanosoma rhodesiense*. *Acta Tropica*, 28, 211–220.

Geigy, R., Jenni, L., Kauffmann, M., Onyango, R.J. and Weiss, N. (1975) Identification of *T. brucei*-subgroup strains isolated from game. *Acta Tropica*, 32, 190–205.

Gelfland, M. (1966) The early clinical features of Rhodesian trypanosomiasis with special reference to the 'Chancre' (local reaction). *Transactions of the Royal Society of Tropical Medicine and Hygiene*, 60, 376–379.

Gettinby, G. (1989) Understanding infectious diseases: modelling approaches for the trypanosomiases. *Annales de la Société Belge de Médecine Tropicale*, 69, Suppl. 1, 21–30.

Gibson, G.A. (1992) Do tsetse flies 'see' zebras? A field study of the visual response of tsetse to striped targets. *Physiological Entomology*, 17, 141–147.

Gibson, G. and Brady, J. (1985) Anemotactic flight paths of tsetse flies in relation to host odour: a preliminary video study in nature of the response to loss of odour. *Physiological Entomology*, 10, 395–406.

Gibson, G. and Brady, J. (1988) Flight behaviour of tsetse flies in host odour plumes: the initial response to leaving or entering odour. *Physiological Entomology*, 13, 29–42.

Gibson, G. and Young, S. (1991) The optics of tsetse fly eyes in relation to their behaviour and ecology. *Physiological Entomology*, 16, 273–282.

Gibson, G., Packer, M.J., Steullet, P. and Brady, J. (1991) Orientation of tsetse flies to wind, within and outside host odour plumes in the field. *Physiological Entomology*, 16, 47–56.

Gibson, W. (1996) More on sleeping sickness in Uganda. *Parasitology Today*, 12, 40.

Gibson, W. and Ferris, V. (1992) Sequential infection of tsetse flies with *Trypanosoma congolense* and *Trypanosoma brucei*. *Acta Tropica*, 50, 345–352.

Gibson, W.C. and Gashumba, J.K. (1983) Isoenzyme characterisation of some Trypanozoon stocks from a recent trypanosomiasis epidemic in Uganda. *Transactions of the Royal Society of Tropical Medicine and Hygiene*, 77, 114–118.

Gibson, W.C., Mehlitz, D., Lanham, S.M. and Godfrey, D.G. (1978) The identification of *Trypanosoma brucei gambiense* in Liberian pigs and dogs by isoenzymes and by resistance to human plasma. *Tropenmedizin und Parasitologie*, 29, 335–345.

Gibson, W.C., de C. Marshall, T.F. and Godfrey, D.G. (1980) Numerical analysis of enzyme polymorphism: a new approach to the epidemiology and taxonomy of the subgenus *Trypanozoon*. *Advances in Parasitology*, 18, 175–246.

Gibson, W.C., Dukes, P. and Gashumba, J.K. (1988) Species-specific DNA probes for the identification of African trypanosomes in tsetse flies. *Parasitology*, 97, 63–73.

Gibson, W., Kanmogne, G. and Bailey, M. (1995) A successful backcross in *Trypanosoma brucei*. *Molecular and Biochemical Parasitology*, 69, 101–110.

Gidudu, A.M., Cuisance, D., Reifenberg, J.M. and Frézil, J.L. (1995) Amélioration de la technique de salivation des glossines pour la détection des métatrypanosomes infectants: étude de quelques facteurs biologiques et non biologiques sur le comportement de sondage des glossines. *Revue d'Elevage et de Médecine Vétérinaire des Pays Tropicale*, 48, 153–160.

Gillman, C. (1936) A population map of Tanganyika territory. *The Geographical Review*, 26, 353–375.

Gillott, C. and Langley, P.A. (1981) The control of receptivity and ovulation in the tsetse fly, *Glossina morsitans*. *Physiological Entomology*, 6, 269–281.

Gingrich, J.B., Ward, R.A., Macken, L.M. and Schoenbechler, M.J. (1982) *Trypanosoma brucei rhodesiense* (Trypanosomatidae): factors influencing infection rates of a recent human isolate in the tsetse. *Journal of Medical Entomology*, 19, 268–274.

Gitatha, S.K. (1979) *T. congolense* (Shimba Hills) resistant to various trypanocidal drugs. In: *Proceedings of the 16th Meeting of the International Scientific Council for Trypanosomiasis Research and Control, Yaounde, Cameroon.* Publication No. 111, pp. 257–263.

Glasgow, J.P. (1960) The variability of fly-round catches in field studies of *Glossina. Bulletin of Entomological Research*, 51, 781–788.

Glasgow, J.P. (1961a) The feeding habits of *Glossina swynnertoni* Austen. *Journal of Animal Ecology*, 30, 77–85.

Glasgow, J.P. (1961b) Seasonal changes in the breeding places of *Glossina morsitans morsitans* Westwood. *Acta Tropica*, 18, 252–254.

Glasgow, J.P. (1961c) Selection for size in tsetse flies. *Journal of Animal Ecology*, 30, 87–94.

Glasgow, J.P. (1963) The distribution and abundance of tsetse. *International Series of Monographs on Pure and Applied Biology*, 20, Pergamon Press, Oxford, 241 pp.

Glasgow, J.P. (1967) Recent fundamental work on tsetse flies. *Annual Review of Entomology*, 12, 421–438.

Glasgow, J.P. (1970) The *Glossina* Community. In: Mulligan, H.W. (ed.) *The African Trypanosomiases*. George Allen and Unwin, London, pp. 348–381.

Glasgow, J.P. and Bursell, E. (1961) Seasonal variations in the fat content and size of *Glossina swynnertoni* Austen. *Bulletin of Entomological Research*, 51, 705–713.

Glasgow, J.P. and Duffy, B.J. (1947) The extermination of *Glossina palpalis fuscipes*, Newstead, by hand catching. *Bulletin of Entomological Research*, 38, 465–477.

Glasgow, J.P. and Duffy, B.J. (1951) Further observations on the extermination of *Glossina palpalis fuscipes* Newstead by hand-catching. *Bulletin of Entomological Research*, 42, 55–63.

Glasgow, J.P. and Welch (1962) Long-term fluctuation in numbers of the tsetse fly *Glossina swynnertoni. Bulletin of Entomological Research*, 53, 129–137.

Glasgow, J.P. and Wilson, F. (1953) A census of the tsetse-fly *Glossina pallidipes* Austen and of its host animals. *Journal of Animal Ecology*, 22, 47–56.

Glasgow, J.P., Isherwood, F., Lee-Jones, F. and Weitz, B. (1958) Factors influencing the staple food of tsetse flies. *Journal of Animal Ecology*, 27, 59–69.

Godfrey, D.G. (1966) Diagnosis of trypanosome infections in tsetse flies. In: *Proceedings of the 1st International Congress of Parasitology, Rome, 1964.* Pergamon Press, Oxford, pp. 990–991.

Godfrey, D.G. (1979) The zymodemes of trypanosomes. *Symposia of The British Society for Parasitology*, 17, 31–53.

Godfrey, D.G. and Fairbairn, H. (1958) Sections cut through a chancre developing in a human volunteer previously exposed to the bite of an infected tsetse. *Transactions of the Royal Society of Tropical Medicine and Hygiene*, 52, 21–22.

Godfrey, D.G. and Killick-Kendrick, R. (1961) Bovine trypanosomiasis in Nigeria. I. The inoculation of blood into rats as a method of survey in the Donga valley, Benue Province. *Annals of Tropical Medicine and Parasitology*, 55, 287–297.

Godfrey, D.G., Leach, T.M. and Killick-Kendrick, R. (1964) Bovine trypanosomiasis in Nigeria. III. A high incidence in a group of west African humpless cattle. *Annals of Tropical Medicine and Parasitology*, 58, 204–215.

Golder, T.K. and Patel, N.Y. (1980) Some effects of trypanosome development on the saliva and salivary glands of the tsetse fly, *Glossina morsitans*. *European Journal of Cell Biology*, 22, 511.

Golder, T.K. and Patel, N.Y. (1982) Localisation and characterisation of salivary gland cholinesterase in the tsetse, *Glossina morsitans*. *Insect Science and its Application*, 3, 167–171.

Golder, T.K., Otieno, L.H., Patel, N.Y. and Onyango, P. (1982) Increased sensitivity to endosulfan of *Trypanosoma*-infected *Glossina morsitans*. *Annals of Tropical Medicine and Parasitology*, 76, 483–484.

Golder, T.K., Otieno, L.H., Patel, N.Y. and Onyango, P. (1984) Increased sensitivity to a natural pyrethrum extract of *Trypanosoma*-infected *Glossina morsitans*. *Acta Tropica*, 41, 77–79.

Gooding, R.H. (1974) Digestive processes of haematophagous insects: control of trypsin secretion in *Glossina morsitans*. *Journal of Insect Physiology*, 20, 957–964.

Gooding, R.H. (1977a) Digestive processes of haematophagous insects. XII. Secretion of trypsin and carboxypeptidase B by *Glossina morsitans morsitans* Westwood (Diptera: Glossinidae). *Canadian Journal of Zoology*, 55, 215–222.

Gooding, R.H. (1977b) Digestive processes of haematophagous insects. XIII. Evidence for the digestive function of midgut proteinases of *Glossina morsitans morsitans* Westwood (Diptera: Glossinidae). *Canadian Journal of Zoology*, 55, 1557–1562.

Gooding, R.H. (1977c) Digestive processes of haematophagous insects. XIV. Haemolytic activity in the midgut of *Glossina morsitans morsitans* Westwood (Diptera: Glossinidae). *Canadian Journal of Zoology*, 55, 1899–1905.

Gooding, R.H. (1979) Genetics of *Glossina morsitans morsitans* (Diptera: Glossinidae). III. *Salmon*, a sex-linked maternally influenced, semi-lethal eye colour mutant. *Canadian Entomologist*, 111, 557–560.

Gooding, R.H. (1981) Genetic polymorphism in three species of tsetse flies (Diptera: Glossinidae) in Upper Volta. *Acta Tropica*, 38, 149–161.

Gooding, R.H. (1982) Classification of nine species and subspecies of tsetse flies (Diptera: Glossinidae: *Glossina* Wiedemann) based on molecular genetics and breeding data. *Canadian Journal of Zoology*, 60, 2737–2744.

Gooding, R.H. (1984a) Tsetse genetics: a review. *Quaestiones Entomologicae*, 20, 89–128.

Gooding, R.H. (1984b) Genetics of *Glossina morsitans morsitans* (Diptera: Glossinidae). X. A mutant (sabr) having long scutellar apical bristles in females. *Canadian Journal of Genetic Cytology*, 26, 770–775.

Gooding, R.H. (1988a) Preliminary analysis of genetics of hybrid sterility in crosses of *Glossina palpalis gambiensis* Vanderplank. *Canadian Entomologist*, 120, 997–1001.

Gooding, R.H. (1988b) Héritabilité de la capacité vectorielle chez les insectes haematophages. *Annales de Médecine Véterinar*, 132, 521–532.

Gooding, R.H. (1989) Genetics of two populations of *Glossina morsitans centralis* (Diptera: Glossinidae) from Zambia. *Acta Tropica*, 46, 17–22.

Gooding, R.H. (1992) Genetic variation in tsetse flies and implications for trypanosomiasis. *Parasitology Today*, 8, No. 3.

Gooding, R.H. and Hollebone, J.E. (1976) Heritability of adult weight in the tsetse fly *Glossina morsitans morsitans* Westw. (Diptera: Glossinidae). *Entomologia Experimentalis et Applicata*, 32, 1507–1509.

Gooding, R.H. and Rolseth, B.M. (1981) Genetics of *Glossina morsitans morsitans* (Diptera: Glossinidae). VI. Multilocus comparison of three laboratory populations. *Canadian Journal of Genetic Cytology*, 24, 109–115.

Gooding, R.H., Moloo, S.K. and Rolseth, B.M (1991) Genetic variation in *Glossina brevipalpis, G. longipennis* and *G. pallidipes*, and the phenetic relationships of *Glossina* species. *Medical and Veterinary Entomology*, 5, 165–173.

Gooding, R.H., Mbise, S., Macha, P. and Rolseth, B.M. (1993) Genetic variation in a Tanzanian population of *Glossina swynnertoni* (Diptera: Glossinidae). *Journal of Medical Entomology*, 30, 489–492.

Gordon, R.M. (1957) *Trypanosoma congolense* in its passage through the peritrophic membrane of *Glossina morsitans*. *Transactions of the Royal Society of Tropical Medicine and Hygiene, Laboratory meeting*, 51, 296.

Gordon, R.M. and Crewe, W. (1948) The bite of the tsetse. *Transactions of the Royal Society of Tropical Medicine and Hygiene*, 41, 439–440.

Gordon, R.M. and Davey, T.H. (1930) An account of trypanosomiasis at the Cape lighthouse peninsula, Sierra Leone. *Annals of Tropical Medicine and Parasitology*, 24, 289–318.

Gouteux, J.-P. (1982) Analyse des groupes d'âge physiologique des femelles de glossines calcul de la courbe de survie du taux de mortalité des âge max. *Cahiers Série Entomologie Médecine et Parasitologie*, 20, 189–197.

Gouteux, J.-P. (1985) Ecologie des Glossines en secteur pré-forestière de Côte d'Ivoire Relation avec la trypanosomiase humaine et possibilités de lutte. *Annales de Parasitologie Humaine et Comparée*, 60, 329–347.

Gouteux, J.-P. (1987a) Une nouvelle glossine du Congo: *Glossina (Austenina) frezili* sp. nov. (Diptera: Glossinidae). *Tropical Medicine and Parasitology*, 38, 97–100.

Gouteux, J.-P. (1987b) Prevalence of enlarged salivary glands in *Glossina palpalis, G. pallicera* and *G. nigrofusca* from the Vavoua area, Ivory Coast. *Journal of Medical Entomology*, 24, 121–140.

Gouteux, J.-P. (1990) Current considerations on the distribution of *Glossina* in west and central Africa. *Acta Tropica*, 47, 185–187.

Gouteux, J.-P. (1991) La lutte par piégeage contre *Glossina fuscipes fuscipes* pour la protection de l'élevage en République Centrafricaine. II. Caractéristiques du piège bipyramidal. *Revue d'Elevage et de Médecine Vétérinaire des Pays Tropicale*, 44, 295–299.

Gouteux, J.-P. and Artzrouni, M. (1996) Faut-il ou non un contrôle des vecteurs dans la lutte contre la maladie du sommeil ? Une approche bio-mathématique du problème. *Bulletin de la Société de Pathologie Exotique*, 89, 299–305.

Gouteux, J.-P. and Buckland, S.T. (1984) Ecologie des glossines en secteur pre-forestière de Côte d'Ivoire. 8. Dynamique des populations. *Cahiers Série Entomologie Médecine et Parasitologie*, 23, 19–34.

Gouteux, J.-P. and Dagnogo, M. (1985) Homogénéité morphologique des genitalia mâles de *Glossina palpalis palpalis* en Côte d'Ivoire. *Cahiers Série Entomologie Médecine et Parasitologie*, 23, 55–59.

Gouteux, J.-P. and Gibson, W.C. (1996) Detection of infections of *Trypanosoma grayi* in *Glossina fuscipes fuscipes* in the Central African Republic. *Annals of Tropical Medicine and Parasitology*, 90, 555–557.

Gouteux, J.-P. and Lancien, J. (1986) Le piège pyramidal à tsétsé (Diptera: Glossinidae) pour la capture et la lutte: Essais comparatifs et description de nouveaux systèmes de capture. *Tropical Medicine and Parasitology*, 37, 61–66.

Gouteux, J.-P. and Laveissière, C. (1982) Ecologie des Glossines en secteur pre-forestière de Côte d'Ivoire. 4. Dynamique de l'écodistribution en terroir villageois. *Cahiers ORSTOM Série Entomologie Médecine et Parasitologie*, 20, 199–229.

Gouteux, J.-P. and Le Gall, F. (1992) Piège bipyramidal à tsetse pour la protection de l'élevage en République Centrafricaine. *World Animal Review*, 70/71, 37–43.

Gouteux, J.-P. and Noireau, F. (1986) Un nouvel écran–piège pour la lutte anti-tsetse. *Entomologia Experimentalis et Applicata*, 4, 291–297.

Gouteux, J.-P. and Sinda, D. (1990) Community participation in the control of tsetse flies. Large scale trials using the Pyramid trap in the Congo. *Tropical Medicine and Parasitology*, 41, 49–55.

Gouteux, J.-P., Challier, C., Laveissière, C. and Couret, D. (1982a) L'utilisation des écrans dans la lutte anti-tsetse en zone forestière, *Tropenmedizin und Parasitologie*, 33, 163–168.

Gouteux, J.-P., Laveissière, C. and Boreham, P. (1982b) Ecologie des Glossines en secteur pre-forestière de Côte d'Ivoire. 3. Les préférences trophiques de *Glossina pallicera* et *G. nigrofusca*. *Cahiers ORSTOM Série Entomologie Médecine et Parasitologie*, 20, 109–124.

Gouteux, J.-P., Bois, J.F., Laveissière, C., Couret, D. and Mustapha, A. (1984) Ecologie des Glossines en secteur pre-forestière de Côte d'Ivoire. 9. Les lieux de repos. *Cahiers ORSTOM Série Entomologie Médecine et Parasitologie*, 22, 159–174.

Gouteux, J.-P., Noireau, F., Sinda, D. and Frézil, J.-L. (1986) Essais du piège pyramidal contre *Glossina palpalis palpalis* (Rob.- Desv.) dans le foyer du Niari. *Cahiers ORSTOM Série Entomologie Médecine et Parasitologie*, 24, 181–190.

Gouteux, J.-P., Cuisance, D., Demba, D., N'Dokue, F. and Le Gall, F. (1991) La lutte par piégeage contre *Glossina fuscipes fuscipes* pour la protection de l'élevage en République Centrafricaine. I. Mise au point d'un piège adapté à un milieu d'éleveurs semi-nomades. *Revue d'Elevage et de Médecine Vétérinaire des Pays Tropicale*, 44, 287–294.

Gouteux, J.-P., Kounda Gboumi, J.C., Noutoua, L., D'Amico, F., Bailly, C. and Roungou, J.B. (1993) Man–fly contact in the Gambian trypanosomiasis focus of Nola-Bilolo (Central African Republic). *Tropical Medicine and Parasitology*, 44, 213–218.

Gouteux, J.-P., Blanc, F., Pounekrozou, E., Cuisance, D., Mainguet, M., D'Amico, F. and Le Gall, F. (1994) Tsetse and livestock in the Central African Republic: retreat of *Glossina morsitans submorsitans* (Diptera: Glossinidae). *Bulletin de la Société de Pathologie Exotique*, 87, 52–56.

Graber, M. (1968) Note sur la résistance au Bérénil d'une souche tchadienne de *Trypanosoma vivax*. *Revue d'Elevage et de Médecine Vétérinaire des Pays Tropicale*, 21, 463–466.

Graham, P. (1964) Destruction of birds and other wildlife, by dieldrex spraying against tsetse fly in Bechuanaland. *Arnoldia*, 1, 1–4.

Gray, A.C.H. and Tulloch, F.M.G. (1905) The multiplication of *T. gambiense* in the alimentary canal of *Glossina palpalis*. *Reports of the Sleeping Sickness Commission of the Royal Society*, 6, 282–287.

Gray, A.R. and Luckins, A.G. (1980) Features of epidemiological importance in the development of cyclically transmitted stocks of *Trypanosoma congolense* in vertebrates. *Insect Science and its Application*, 1, 69–72.

Gray, A.R. and Nixon, J. (1967) Observations on the incidence and importance of *Trypanosoma theileri* in Nigeria. *Annals of Tropical Medicine and Parasitology*, 61, 251–160.

Gray, A.R. and Roberts, C.J. (1968) A summary of progress in experiments on the cyclical transmission of drug resistant strains of trypanosomes. *ISCTR 12th meeting Bangui*, 103–108.

Gray, A.R. and Roberts, C.J. (1971) The cyclical transmission of strains of *Trypanosoma congolense* and *T. vivax* resistant to normal therapeutic doses of trypanocidal drugs. *Parasitology*, 63, 67–89.

Green, C.H. (1984) A comparison of phototactic responses to red and green light in *Glossina morsitans morsitans* and *Musca domestica*. *Physiological Entomology*, 16, 165–172.

Green, C.H. (1986) Effects of colours and synthetic odours on the attraction of *Glossina pallidipes* and *G. morsitans morsitans* (Diptera: Glossinidae) to coloured traps and screens. *Physiological Entomology*, 11, 411–421.

Green, C.H. (1988) The effect of colour on trap and screen-orientated responses in *Glossina palpalis palpalis* (Robineau-Desvoidy) (Diptera: Glossinidae). *Bulletin of Entomological Research*, 78, 591–604.

Green, C.H. (1989) The use of two-coloured screens for catching *Glossina palpalis palpalis* (Robineau-Desvoidy) (Diptera: Glossinidae). *Bulletin of Entomological Research*, 79, 81–93.

Green, C.H. (1990) The effect of colour on the numbers, age and nutritional status of *Glossina tachinoides* (Diptera: Glossinidae) attracted to targets. *Physiological Entomology*, 15, 317–329.

Green, C.H. (1993) The effects of odours and target colour on landing responses of *Glossina morsitans morsitans* and *G. pallidipes* (Diptera: Glossinidae). *Bulletin of Entomological Research*, 83, 553–562.

Green, C.H. (1994) Bait methods for tsetse fly control. *Advances in Parasitology*, 34, 229–291.

Green, C.H. and Cosens, D. (1983) Spectral responses of the tsetse fly, *Glossina morsitans morsitans*. *Journal of Insect Physiology*, 29, 795–800.

Green, C.H. and Flint, S. (1986) An analysis of colour effects in the performance of the F2 trap against *Glossina pallidipes* Austen and *G. morsitans morsitans* Westwood (Diptera: Glossinidae). *Bulletin of Entomological Research*, 76, 409–418.

Green, C.H. and Maudlin, I. (1982) Evolution of insecticide resistance in tsetse. In: *Sterile Insect Technique and Radiation in Insect Control*. Proceedings of a Symposium IAEA/FAO, 1981, Vienna, pp. 401–410.

Greig, S. and Ashall, F. (1987) Detection of south American trypanosomes using total parasite DNA probes. *Parasitology Today*, 3, 375–376.

Griffiths, G.C.D. (1976) Comments on some recent studies of tsetse-fly phylogeny and structure. *Systematic Entomology*, 1, 15–18.

Griffiths, N. and Brady, J. (1994) Analysis of the components of 'electric nets' that affect their sampling efficiency for tsetse flies (Diptera: Glossinidae). *Bulletin of Entomological Research*, 84, 325–330.

Grigson, C. (1991) An African origin for African cattle? – some archaeological evidence. *African Archaeological Review*, 9, 119–144.

Groenendijk, C.A. (1996) The responses of tsetse flies to artificial baits in relation to age, nutritional and reproductive state. *Entomologia Experimentalis et Applicata*, 78, 335–340.

Grubhoffer, L., Muska, M. and Volf, P. (1994) Midgut hemagglutinins in five species of tsetse flies (*Glossina* spp.): two different lectin systems in the midgut of *Glossina tachinoides*. *Folia Parasitologia (Praha)*, 41, 229–232.

Gruvel, J. (1980) Considérations générales sur la signification de la transmission mécanique des trypanosomes chez le bétail. *Insect Science and its Application*, 1, 55–57.

Gruvel, J. and Tazé, Y. (1978) Essais d'un nouveau pyrethrinoide: La décamethrine contre *Glossina tachinoides* au Tchad. *Revue d'Élevage et de Médecine Vétérinaire des Pays Tropicaux*, 31, 193–203.

Guidot, G. and Roelants, G.E. (1982) Sensibilité de taurins Baoulé et de Zébus à *Trypanosoma* (*Duttonella*) *vivax* et *T.* (*Nannomonas*) *congolense*. *Revue d'Élevage et de Médecine Vétérinaire des Pays Tropicaux*, 35, 233–244.

Habbema, J.D.F., Alley, E.S., Plaisier, A.P., van Oortmarssen, G.J. and Remme, J.H.F. (1992) Epidemiological modelling for onchocerciasis control. *Parasitology Today*, 8, 99–103.

Habtemariam, T. (1987a) Epidemiologic modelling to select the best method for the control of African trypanosomiasis. *Tropical Veterinarian*, 5, 1–6.

Habtemariam, T. (1987b) Computer modelling of African trypanosomiasis. *Tropical Veterinarian*, 5, 7–12.

Habtemariam, T., Howitt, R.E., Ruppanner, R. and Riemann, H.P. (1982a) The benefit/cost analysis of alternative strategies for the control of bovine trypanosomiasis in Ethiopia. *Preventive Veterinary Medicine*, 1, 157–168.

Habtemariam, T., Ruppaner, R., Riemann, H.P. and Theis, J.H. (1982b) Epidemic and endemic characteristics of trypanosomiasis in cattle. A simulation model. *Preventive Veterinary Medicine*, 1, 137–145.

Habtemariam, T., Ruppanner, R., Riemann, H.P. and Theis, J.H. (1986) Estimating the probability of effective transmission of trypanosomes using the Poisson distribution. *Preventive Veterinary Medicine*, 4, 57–68.

Hadaway, A.B. (1972) Toxicity of insecticides to tsetse flies. *Bulletin of the World Health Organisation*, 46, 353–362.

Hadaway, A.B. (1978) *Post-treatment Temperature and the Toxicity of some Insecticides to Tsetse Flies*. WHO/VBC/78.693. World Health Organisation, Geneva.

Hale Carpenter, G.D. (1923) Report on a test of a method of attacking *Glossina* by artificial breeding places. *Bulletin of Entomological Research*, 13, 443–445.

Hall, D.R., Beevor, P.S., Cork, A., Nesbitt, B.F. and Vale, G.A. (1984) 1-Octen-3-ol: a potent olfactory stimulant and attractant for tsetse isolated from cattle odours. *Insect Science and its Application*, 5, 335–339.

Hall, M.J.R. (1987) The orientation of males of *Glossina morsitans* Westwood (Diptera: Glossinidae) to pheromone-baited decoy 'females' in the field. *Bulletin of Entomological Research*, 77, 487–495.

Hall, M.J.R. (1988) Characterisation of the sexual responses of male tsetse flies, *Glossina morsitans morsitans*, to pheromone-baited decoy 'females' in the field. *Physiological Entomology*, 13, 49–58.

Hall, M.J.R. and Langley, P.A. (1987) Development of a system for sterilising tsetse flies, *Glossina* spp., in the field. *Medical and Veterinary Entomology*, 1, 201–210.

Hamann, H.J. and Iwannek, K.H. (1981) Sterilisation of *Glossina palpalis palpalis* by beta irradiation. *Bulletin of Entomological Research*, 71, 513–519.

Hargrove, J.W. (1975a) Some changes in the flight and apparatus of tsetse flies during maturation. *Journal of Insect Physiology*, 21, 1485–1489.

Hargrove, J.W. (1975b) The flight performance of tsetse flies. *Journal of Insect Physiology*, 21, 1385–1395.

Hargrove, J.W. (1976a) Amino acid metabolism during flight in tsetse flies. *Journal of Insect Physiology*, 22, 309–313.

Hargrove, J.W. (1976b) The effect of human presence on the behaviour of tsetse (*Glossina* spp.) (Diptera, Glossinidae) near a stationary ox. *Bulletin of Entomological Research*, 66, 173–178.

Hargrove, J.W. (1977) Some advances in the trapping of tsetse (*Glossina* spp.) and other flies. *Ecological Entomology*, 2, 123–137.

Hargrove, J.W. (1980) The effect of ambient temperature on the flight performance of the mature male tsetse fly, *Glossina morsitans. Physiological Entomology*, 5, 397–400.

Hargrove, J.W. (1981) Discrepancies between estimates of tsetse fly populations using mark–recapture and removal trapping techniques. *Journal of Applied Ecology*, 18, 737–748.

Hargrove, J.W. (1988) Tsetse: the limits to population growth. *Medical and Veterinary Entomology*, 2, 203–217.

Hargrove, J.W. (1990) Age-dependent changes in the probabilities of survival and capture of the tsetse, *Glossina morsitans morsitans* Westwood. *Insect Science and its Application*, 11, 323–330.

Hargrove, J.W. (1991) Ovarian ages of tsetse flies (Diptera: Glossinidae) caught from mobile and stationary baits in the presence and absence of humans. *Bulletin of Entomological Research*, 81, 43–50.

Hargrove, J.W. (1993) Target barriers for tsetse flies (*Glossina* spp.) (Diptera: Glossinidae): quick estimates of optimal target densities and barrier widths. *Bulletin of Entomological Research*, 83, 197–200.

Hargrove, J.W. (1994) Reproductive rates of tsetse flies in the field in Zimbabwe. *Physiological Entomology*, 19, 307–318.

Hargrove, J.W. and Borland, C.H. (1994) Pooled population parameter estimates from mark–release–recapture data. *Biometrics*, 50, 1129–1141.

Hargrove, J.W. and Brady, J. (1992) Activity rhythms of tsetse flies (*Glossina* spp.) (Diptera: Glossinidae) at low and high temperatures in nature. *Bulletin of Entomological Research*, 82, 321–326.

Hargrove, J.W. and Coates, T.W. (1990) Metabolic rates of tsetse flies in the field as measured by the excretion of injected caesium. *Physiological Entomology*, 15, 157–166.

Hargrove, J.W. and Langley, P.A. (1990) Sterilising tsetse (Diptera: Glossinidae) in the field: a successful trial. *Bulletin of Entomological Research*, 80, 397–403.

Hargrove, J.W. and Langley, P.A. (1993) A field trial of pyriproxyfen-treated targets as an alternative method for controlling tsetse (Diptera: Glossinidae). *Bulletin of Entomological Research*, 83, 361–368.

Hargrove, J.W. and Packer, M.J. (1992) Fat and haematin contents of male tsetse flies *Glossina pallidipes* and *G. m. morsitans* (Diptera: Glossinidae) caught in odour-baited traps and artificial refuges in Zimbabwe. In: *Tsetse Control, Diagnosis and Chemotherapy Using Nuclear Techniques.* Proceedings of a seminar jointly organised by the International Atomic Energy Agency and the Food and Agriculture Organisation of the United Nations. Kenya 11–15 February 1991. IAEA, Vienna, pp. 65–89.

Hargrove, J.W. and Packer, M.J. (1993) Nutritional states of male tsetse flies (*Glossina* spp.) (Diptera: Glossinidae) caught in odour-baited traps and artificial refuges: models for feeding and digestion. *Bulletin of Entomological Research*, 83, 29–46.

Hargrove, J.W. and Vale, G.A. (1978) The effect of host odour concentration on catches of tsetse flies (Glossinidae) and other Diptera in the field. *Bulletin of Entomological Research*, 68, 607–612.

Hargrove, J.W. and Vale, G.A. (1979) Aspects of the feasibility of employing odour baited traps for controlling tsetse flies. *Bulletin of Entomological Research*, 69, 283–291.

Hargrove, J.W. and Williams, B.G. (1995) A cost–benefit analysis of feeding in female tsetse. *Medical and Veterinary Entomology*, 9, 109–119.

Harley, J.M.B. (1954) The breeding sites of the tsetse fly *Glossina morsitans*. *Acta Tropica*, 11, 379–401.

Harley, J.M.B. (1958) The availability of *Glossina morsitans* in Ankole Uganda. *Bulletin of Entomological Research*, 49, 225–228.

Harley, J.M.B. (1965) Activity cycles of *Glossina pallidipes* Aust., *G. palpalis fuscipes* Newst. and *G. brevipalpis* Newst. *Bulletin of Entomological Research*, 56, 141–161.

Harley, J.M.B. (1966a) Studies on age and trypanosome infection rates in females of *Glossina pallidipes* Aust., *G. palpalis fuscipes* Newst. and *G. brevipalpis* Newst. in Uganda. *Bulletin of Entomological Research*, 57, 23–37.

Harley, J.M.B. (1966b) Seasonal and diurnal variation in physiological age and trypanosome infection rates of female *Glossina pallidipes* Aust., *G. palpalis fuscipes* Newst. and *G. brevipalpis* Newst. *Bulletin of Entomological Research*, 56, 595–614.

Harley, J.M.B. (1967a) The influence of age in *Glossina morsitans* at the time of the infected meal on infection with *Trypanosoma rhodesiense*. *Report of the East African Trypanosomiasis Research Organisation*, 1966, 51.

Harley, J.M.B. (1967b) Further studies on age and trypanosome infection rates in *Glossina pallidipes* Aust., *G. palpalis fuscipes* Newst. and *G. brevipalpis* Newst. in Uganda. *Bulletin of Entomological Research*, 57, 459–477.

Harley, J.M.B. (1967c) The influence of sampling method on the trypanosome infection rates of catches of *G. pallidipes* and *G. fuscipes*. *Entomologia Experimentalis et Applicata*, 10, 240–252.

Harley, J.M.B. (1970) The influence of the age of the fly at the time of the infecting feed on infection of *Glossina fuscipes* with *Trypanosoma rhodesiense*. *Annals of Tropical Medicine and Parasitology*, 65, 191–196.

Harley, J.M.B. (1971) Comparison of the susceptibility to infection with *Trypanosoma rhodesiense* of *Glossina pallidipes, G. morsitans, G. fuscipes* and *G. brevipalpis*. *Annals of Tropical Medicine and Parasitology*, 65, 185–189.

Harley, J.M.B. and Pilson, R.D. (1961) An experiment in the use of discriminative clearing for the control of *Glossina morsitans* Westw. in Ankole, Uganda. *Bulletin of Entomological Research*, 52, 561–576.

Harley, J.M.B. and Wilson, A.J. (1968) Comparison between *Glossina morsitans, G. pallidipes* and *G. fuscipes* as vectors of trypanosomes of the *Trypanosoma congolense* group: the proportions infected experimentally and the numbers of infective organisms extruded during feeding. *Annals of Tropical Medicine and Parasitology*, 62, 178–187.

Harley, J.M.B., Cunningham, M.P. and Van Hoeve, K. (1966) The numbers of infective *Trypanosoma rhodesiense* extruded by *Glossina morsitans* during feeding. *Annals of Tropical Medicine and Parasitology*, 60, 455–460.

Harmsen, R. (1970) Pteridines in flies of the genus *Glossina*. *Acta Tropica*, 27, 165–172.

Harmsen, R. (1973) The nature of the establishment barrier for *Trypanosoma brucei* in the gut of *Glossina pallidipes*. *Transactions of the Royal Society of Tropical Medicine and Hygiene*, 67, 364–373.

Harris, R.H.T.P. (1938) The control and possible extermination of the tsetse fly by trapping. *Acta Conv Ter Trop Malar Morb Amsterdam*, 1, 663–677.

Harrison, H. (1936) The Shinyanga game experiment: a few of the early observations. *Journal of Animal Ecology*, 5, 271–293.

Harrison, P. (1987) *The Greening of Africa*. Penguin Books, Harmondsworth, UK, 380 pp.

Hassanali, A., McDowell, P.G., Owaga, M.L.A. and Saini, R.K. (1986) Identification of tsetse attractants from excretory products of a wild host animal, *Syncerus caffer*. *Insect Science and its Application*, 7, 5–9.

Hawking, F. (1963) Action of drugs upon *Trypanosoma congolense, T. vivax* and *T. rhodesiense* in tsetse flies and in culture. *Annals of Tropical Medicine and Parasitology*, 57, 255–261.

Hawking, F. (1976a) The resistance of human plasma of *Trypanosoma brucei, T. rhodesiense* and *T. gambiense*. I. Analysis of the composition of trypanosome strains. *Transactions of the Royal Society of Tropical Medicine and Hygiene*, 70, 504–512.

Hawking, F. (1976b) The resistance of human plasma of *Trypanosoma brucei, T. rhodesiense* and *T. gambiense*. II. Survey of strains from East Africa and Nigeria. *Transactions of the Royal Society of Tropical Medicine and Hygiene*, 70, 513–320.

Heaversedge, R.C. (1968) Variation in the size of insect parasites of puparia of *Glossina* spp. *Bulletin of Entomological Research*, 58, 153–159.

Heckenroth, F. and Blanchard, M. (1913) Transmission du *Trypanosoma gambiense* par des moustiques (*Mansonia uniformis*). *Bulletin de la Société de Pathologie Exotique*, 6, 442–443.

Hecker, H. and Moloo, S.K. (1981) Influence of *Trypanosoma* (*Trypanozoon*) *brucei* infections on the fine structure of midgut cells of *Glossina morsitans morsitans*. *Parasitology*, 82, 106–107.

Heisch, R.B., McMahon, J.P. and Manson-Bahr, P.E.C. (1958) The isolation of *Trypanosoma rhodesiense* from a bushbuck. *British Medical Journal*, 2, 1202–1204.

Hendrickx, G. and Napala, A. (1997) Le contrôle de la trypanosomose 'à la carte': une approche intégrée basée sur un Système d'Information Géographique. *Académie Royale des Sciences d'Outre Mer. Mémoires Classe Sciences Naturelles and Médicales*. In: 8th Nouvelle Série, Tome 24–2, Bruxelles, 1997, 100 pp.

Hendrickx, G., Slingenbergh, J.H.W., Dao, B., Bastiansen, P. and Napala, A. (1997) Geographical information systems (GIS), powerful tools in decision-making. In: *Proceedings of the 24th meeting of the International Scientific Council for Trypanosomosis Research and Control Maputo*, (in press).

Hendry, K.A.K. and Vickerman, K. (1988) The requirement for epimastigote attachment during division and metacyclogenesis in *Trypanosoma congolense*. *Parasitology Research*, 74, 403–408.

Hervouët, J.-P. and Laveissière, C. (1987) Ecologie humaine et maladie du sommeil en Côte d'Ivoire forestière. *Cahiers ORSTOM Série Entomologie Médecine et Parasitologie*, No. Spécial, 101–111.

Hide, G. (1997) The molecular epidemiology of Trypanosomatids. In: Hide, G., Mottram, J.C., Coombs, G.H. and Holmes, P.H. (eds) *Trypanosomiasis and Leishmaniasis*. CAB International, Wallingford, pp. 289–303.

Hide, G., Welburn, S.C., Tait, A. and Maudlin, I. (1994) Epidemiological relationships of *Trypanosoma brucei* stocks from South East Uganda: evidence for different population structures in human infective and non-human infective isolates. *Parasitology*, 109, 95–111.

Hide, G., Tait, A., Maudlin, I. and Welburn, S.C. (1996) The origins, dynamics and generation of *Trypanosoma brucei rhodesiense* epidemics in East Africa. *Parasitology Today*, 12, 50–55.

Hill, S.M. and Crampton, J.M. (1994) DNA-based methods for the identification of insect vectors. *Annals of Tropical Medicine and Parasitology*, 88, 227–250.

Hoare, C.A. (1929) Studies on *Trypanosoma grayi*. 2. Experimental transmission to the crocodile. *Transactions of the Royal Society of Tropical Medicine and Hygiene*, 23, 39–56.

Hoare, C.A. (1931) Studies on *Trypanosoma grayi*. 3. Life-cycle in the tsetse-fly and in the crocodile. *Parasitology*, 23, 449–479.

Hoare, C.A. (1936a) Morphological and taxonomic studies on mammalian trypanosomes. II. *Trypanosoma simiae* and acute porcine trypanosomiasis in tropical Africa. *Transactions of the Royal Society of Tropical Medicine and Hygiene*, 29, 619–645.

Hoare, C.A. (1936b) Note on *Trypanosoma simiae* from an outbreak amongst pigs in the Gold Coast. *Transactions of the Royal Society of Tropical Medicine and Hygiene*, 30, 315–316.

Hoare, C.A. (1947) Tsetse-borne trypanosomiases outside their natural boundaries. *Annales de la Société Belge de Médecine Tropicale*, 27, supplément, 267–277.

Hoare, C.A. (1962) Reservoir hosts and natural foci of human protozoal infections. *Acta Tropica*, 19, 281–315.

Hoare, C.A. (1965) Vampire bats as vectors and hosts of equine and bovine trypanosomes. *Acta Tropica*, 22, 204–216.

Hoare, C.A. (1970) The mammalian trypanosomes of Africa. In: Mulligan, H.W. (ed.) *The African Trypanosomiases*. George Allen and Unwin, London, pp. 3–23.

Hocking, K.S. (1961) Discriminative application of insecticide against *Glossina morsitans* Westw. *Bulletin of Entomological Research*, 52, 17–22.

Hocking, K.S., Burnett, G.F. and Sell, R.C. (1954a) Aircraft applications of insecticides in east Africa. VII. An experiment against the tsetse flies *Glossina morsitans* Westw. and *G. swynnertoni* Aust., in the rainy season. *Bulletin of Entomological Research*, 45, 605–612.

Hocking, K.S., Burnett, G.F. and Sell, R.C. (1954b) Aircraft applications of insecticides in east Africa. VIII. An experiment against the tsetse fly *Glossina swynnertoni* Aust., in an isolated area of thornbush and thicket. *Bulletin of Entomological Research*, 45, 613–622.

Hocking, K.S., Lee, C.W., Beesley, J.S.S. and Matechi, H.T. (1966) Aircraft applications of insecticides in east Africa. XVI. Airspray experiment with endosulfan against *Glossina morsitans* Westw., *G. swynnertoni* Aust. *Bulletin of Entomological Research*, 56, 737–744.

Hodges, A.D.P. (1911) Sleeping sickness news. Uganda. *Sleeping Sickness Bureau Bulletin*, 3, 426–430.

Hoek, J.B., Pearson, D.J. and Olembo, N.K. (1976) Nicotinamide–adenine dinucleotide-linked 'malic' enzyme in flight muscle of the tsetse fly (*Glossina*) and other insects. *Biochemical Journal*, 160, 253–262.

Hoeppli, R. and Lucasse, C. (1964) Old ideas regarding cause and treatment of sleeping sickness held in west Africa. *Journal of Tropical Medicine and Hygiene*, 67, 60–68.

Hoffmann, R. (1954) Zur Fortpflanzungsbiologie und zur intra-uterinen Entwicklung von *Glossina palpalis*. *Acta Tropica*, 11, 1–57.

Hogsette, J.A. and Ruff, J.P. (1987) Control of stable flies and horn flies (Diptera: Muscidae) with Permethrin tapes applied to tails of beef and dairy cattle. *Journal of Economic Entomology*, 80, 417–426.

Holden, J.R. and Findlay, G.M. (1944) Pyrethrum as a tsetse fly repellent: Human experiments. *Transactions of the Royal Society of Tropical Medicine and Hygiene*, 38, 199–203.

Holloway, M.T.P. (1990) Alternatives to DDT for use in ground spraying control operations against tsetse flies (Diptera: Glossinidae). *Transactions of the Zimbabwe Science Association*, 64, 33–40.

Hornby, H.E. and French, M.H. (1943) Introduction to the study of tsetse fly repellents in the field of veterinary science. *Transactions of the Royal Society of Tropical Medicine and Hygiene*, 37, 41–54.

House, A.P.R. (1982) Chemosterilisation of *Glossina morsitans morsitans* Westwood and *G. pallidipes* Austen (Diptera: Glossinidae) in the field. *Bulletin of Entomological Research*, 48, 561–579.

Houseman, J.G. (1980) Anterior proteinase inhibitor from *Glossina morsitans morsitans* Westwood (Diptera: Glossinidae) and its effects upon tsetse digestive enzymes. *Canadian Journal of Zoology*, 58, 79–87.

Howe, M.A. and Lehane, M.J. (1986) Post-feed buzzing in the tsetse, *Glossina morsitans morsitans* is an endothermic mechanism. *Physiological Entomology*, 11, 279–286.

Huber, M., Cabib, E. and Miller, L.H. (1991) Malaria parasite chitinase and penetration of the mosquito peritrophic membrane. *Proceedings of the National Academy of Sciences, USA*, 88, 2807–2810.

Huebner, E. and Davey, K.G. (1974) Bacteroids in the ovaries of a tsetse fly. *Nature*, 249, 260–261.

Huebner, E., Tobe, S.S. and Davey, K.G. (1975) Structural and functional dynamics of oogenesis in *Glossina austeni*: vitellogenesis with special reference to the follicular epithelium. *Tissue and Cell*, 7, 535–558.

Hugh-Jones, M. (1989) Applications of remote sensing to the identification of the habitats of parasites and disease vectors. *Parasitology Today*, 5, 244–251.

Hulley, P.E. (1968) Mitotic chromosomes of *Glossina pallidipes* Austen. *Nature*, 217, 977–979.

Hunter, F.F. and Bayly, R. (1991) ELISA for identification of blood meal source in black flies (Diptera: Simuliidae). *Journal of Medical Entomology*, 28, 527–532.

Hussain, M., Han, L.-F. and Rathor, M.N. (1994) Evaluation of oil formulations of deltamethrin for use on cotton targets for tsetse fly control. *Pesticide Science*, 40, 299–306.

Hutchinson, M.P. (1971) Human trypanosomiasis in south-west Ethiopia (March 1967–March 1970). *Ethiopian Medical Journal*, 9, 3–69.

Hutchinson, M.P. and Watson, H.J.C. (1962) Berenil in the treatment of *Trypanosoma gambiense* infection in man. *Transactions of the Royal Society of Tropical Medicine and Hygiene*, 56, 227–230.

Huyton, P.M. and Brady, J. (1975) Some effects of light and heat on the feeding and resting behaviour of tsetse flies, *Glossina morsitans* Westwood. *Journal of Entomology*, Series A 50, 23–30.

Huyton, P.M., Langley, P.A., Carlson, D.A. and Schwarz, M. (1980a) Specificity of contact sex pheromones in tsetse flies, *Glossina* spp. *Physiological Entomology*, 5, 253–264.

Huyton, P.M., Langley, P.A., Carlson, D.A. and Coates, T.W. (1980b) The role of sex pheromones in initiation of copulatory behaviour by male tsetse flies, *Glossina morsitans morsitans*. *Physiological Entomology*, 5, 243–252.

IAEA (1997) Expert group confirms: *Tsetse fly eradicated on Zanzibar*. Press release PR 97/38, International Atomic Energy Agency.

Ibrahim, E.A. and Molyneux, D.H. (1987) Pathogenicity of *Crithidia fasciculata* in the haemocoele of *Glossina*. *Acta Tropica*, 44, 13–22.

Ibrahim, E.A., Ingram, G.A. and Molyneux, D.H. (1984) Haemagglutinins and parasite agglutinins in haemolymph and gut of *Glossina*. *Tropenmedizin und Parasitologie*, 35, 151–156.

ICIPE (1992) *1992 Annual Report*. ICIPE Science Press, International Centre for Insect Physiology and Ecology, Nairobi, Kenya, 150 pp.

Ikede, B.O. (1986) Trypanosomiasis and livestock production in Africa: is current emphasis misplaced? *Tropical Veterinarian*, 4, 1–4.

Ikede, B.O., Reynolds, L., Ogunsanmi, A.O., Fawumi, N.K., Ekwuruke, J.O. and Taiwo, V.O. (1987) The epizootiology of bovine trypanosomiasis in the derived savanna zone of Nigeria – a preliminary report. In: *19th Meeting of the International Scientific Council for Trypanosomiasis Research and Control, Lomé, Togo*.

Ilemobade, A.A. (1979) Drug sensitivity of mouse infective *Trypanosoma vivax* isolates in cattle and sheep. In: *Proceedings of the 16th Meeting of the International Scientific Council for Trypanosomiasis Research and Control, Yaounde, Cameroun*. Publication no. 111, pp. 251–253.

ILRI (no date) *The Genetics of Disease Resistance in Tropical Livestock*. International Livestock Research Institute, Nairobi, Kenya, pp. 1–4.

Imbuga, M.O., Osir, E.O. and Labongo, V.L. (1992) Inhibitory effect of *Trypanosoma brucei brucei* on *Glossina morsitans* midgut trypsin *in vitro*. *Parasitology Research*, 78, 273–276.

Ingram, G.A. and Molyneux, D.H. (1988) Sugar specificities of anti-human ABO(H) blood group erythrocyte agglutinins (lectins) and haemolytic activity in the haemolymph and gut extracts of three *Glossina* species. *Insect Biochemistry*, 18, 269–279.

Ingram, G.A. and Molyneux, D.H. (1990) Lectins (haemagglutinins) in the haemolymph of *Glossina fuscipes fuscipes*: Isolation, partial characterisation, selected physico-chemical properties and carbohydrate binding specificities. *Insect Biochemistry*, 20, 13–27.

Irving, N.S. (1968) The absorption and storage of insecticide by the *in utero* larva of the tsetse fly *Glossina pallidipes* Aust. *Bulletin of Entomological Research*, 58, 221–226.

Isherwood, F. (1957) The resting sites of *Glossina swynnertoni* Aust. in the wet season. *Bulletin of Entomological Research*, 48, 601–605.

Itard, J. (1966a) Chromosomes de Glossines (Diptera – Muscidae). *Comptes Rendus de l'Académie des Sciences, Paris*, 263, Séries D, 1395–1397.

Itard, J. (1966b) Cycle de l'oogénèse chez les femelles de *Glossina tachinoides* West. et détermination de l'âge physiologique. *Revue d'Élevage et de Médecine Vétérinaire des Pays Tropicaux*, 19, 331–350.

Itard, J. (1970) L'appareil reproducteur mâle des Glossines (Diptera – Muscidae). Les étapes de sa formation chez la pupe. La spermatogenèse. *Revue d'Élevage et de Médecine Vétérinaire des Pays Tropicaux*, 23, 57–81.

Itard, J. (1973) Revue des connaissances actuelles sur la cytogénétique des Glossines. *Revue d'Élevage et de Médecine Vétérinaire des Pays Tropicaux*, 26, 151–167.

Itard, J. (1974) Caryotype de *Glossina palpalis gambiensis* Vanderplank, 1949. Comparaison avec d'autres espèces de groupe *palpalis* et du groupe *morsitans*. *Revue d'Élevage et de Médecine Vétérinaire des Pays Tropicaux*, 27, 431–436.

Itty, P. (1992) *Economics of Village Cattle Production in Tsetse-affected Areas of Africa: A study of trypanosomiasis control using trypanotolerant cattle and chemotherapy in Ethiopia, Kenya, the Gambia, Côte d'Ivoire, Zaire and Togo.* Hartung Gore Verlag, Constance, Germany.

Itty, P., Rowlands, G.J., Minengu, M., Ngamuna, S., Van Winkel, F. and D'Ieteren, G.D.M. (1995) The economics of recently introduced village cattle production in a tsetse affected area (I): trypanotolerant N'Dama cattle in Zaire. *Agricultural Systems*, 47, 347–366.

Jack, R.W. (1914) Tsetse fly and big game in southern Rhodesia. *Bulletin of Entomological Research*, 5, 97–110.

Jack, R.W. (1939) Studies in the physiology and behaviour of *Glossina morsitans* Westwood. *Memoirs of the Department of Agriculture, Southern Rhodesia*, No 1, 203 pp.

Jack, R.W. (1941) Notes on the behaviour of *Glossina pallidipes* and *G. brevipalpis* and some comparisons with *G. morsitans. Bulletin of Entomological Research*, 31, 407–430.

Jack, R.W. and Williams, W.L. (1937) The effect of temperature on the reaction of *Glossina morsitans* Westw. to light. *Bulletin of Entomological Research*, 28, 499–503.

Jackson, C.H.N. (1930) Contribution to the bionomics of *Glossina morsitans. Bulletin of Entomological Research*, 21, 491–527.

Jackson, C.H.N. (1933) Notes on a method of marking tsetse flies. *Journal of Animal Ecology*, 2, 289–290.

Jackson, C.H.N. (1937) Some new methods in the study of *Glossina morsitans. Proceedings of the Zoological Society of London*, 1936, 811–896.

Jackson, C.H.N. (1939) The analysis of an animal population. *Journal of Animal Ecology*, 8, 238–246.

Jackson, C.H.N. (1945) Comparative studies of the habitat requirements of tsetse fly species. *Journal of Animal Ecology*, 4, 46–51.

Jackson, C.H.N. (1946) An artificially isolated generation of tsetse-flies (Diptera). *Bulletin of Entomological Research*, 32, 291–299.

Jackson, C.H.N. (1948a) Some further isolated generations of tsetse flies. *Bulletin of Entomological Research*, 39, 441–451.

Jackson, C.H.N. (1948b) The analysis of a tsetse-fly population. III. *Annals of Eugenics, Cambridge*, 14, 91–108.

Jackson, C.H.N. (1950a) Pairing of *Glossina morsitans* Westwood with *G. swynnertoni* Austen (Diptera). *Proceedings of the Royal Entomological Society of London (A)*, 25, 106.

Jackson, C.H.N. (1950b) Wet season fraying of wings of tsetse, *Glossina morsitans. Bulletin of Entomological Research*, 41, 159–160.

Jackson, C.H.N. (1952) Seasonal variations in the mean size of tsetse flies. *Bulletin of Entomological Research*, 43, 703–706.

Jackson, C.H.N. (1953) A mixed population of *Glossina morsitans* and *G. swynnertoni. Journal of Animal Ecology*, 22, 78–86.

Jackson, C.H.N. (1954a) The hunger-cycles of *Glossina morsitans* Westwood and *G. swynnertoni* Austen. *Journal of Animal Ecology*, 23, 368–371.

Jackson, C.H.N. (1954b) The availability of tsetse flies. *ISCTRC Fifth meeting Pretoria*, Publication No. 206, 97.

Jackson, C.H.N. (1955) The natural reservoir of *Trypanosoma rhodesiense*. *Transactions of the Royal Society of Tropical Medicine and Hygiene*, 49, 582–587.

Jacobs-Lorena, M. and Lemos, F.J.A. (1995) Immunological strategies for control of insect disease vectors: a critical assessment. *Parasitology Today*, 11, 144–147.

Jacobson, R.L. and Doyle, R.J. (1996) Lectin–parasite interactions. *Parasitology Today*, 12, 55–61.

Jaenson, T.G.T. (1978) Virus-like rods associated with salivary gland hyperplasia in tsetse, *Glossina pallidipes*. *Transactions of the Royal Society of Tropical Medicine and Hygiene*, 72, 234–238.

Jaenson, T.G.T. (1980) Mating behaviour of females of *Glossina pallidipes*. *Bulletin of Entomological Research*, 70, 49–60.

Jaenson, T.G.T. (1986) Sex ratio distortion and reduced lifespan of *Glossina pallidipes* infected with the virus causing salivary gland hyperplasia. *Entomologia Experimentalis et Applicata*, 41, 265–271.

Jaenson, T.G.T., Rui, C., Dos Santos, B. and Hall, D.R. (1991) Attraction of *Glossina longipalpis* (Diptera: Glossinidae) in Guinea-Bissau to odor-baited biconical traps. *Journal of Medical Entomology*, 28, 284–286.

Jahnke, H.E. (1976) *Tsetse Flies and Livestock Development in East Africa*. Afrika-Studien Weltforum. Verlag, Munchen 1–169.

Jahnke, H.E. (1982) *Livestock Production Systems in Livestock Development in Tropical Africa*. Kieler Wissenschaftsverlag Vauk, Kiel, FRG.

Jahnke, H.E., Tacher, G., Keil, P. and Rojat, D. (1988) Livestock production in tropical Africa, with special reference to the tsetse-affected zone. In: *Livestock Production in Tsetse Affected Areas of Africa. Proceedings of a meeting held in Nairobi, 23–27 November 1987*. English Press, Nairobi, pp. 3–21.

Janssen, J.A.H.A. and Wijers, D.J.B. (1974) *Trypanosoma simiae* at the Kenya coast. A correlation between virulence and the transmitting species of *Glossina*. *Annals of Tropical Medicine and Parasitology*, 68, 5–19.

Jefferies, D. and Jenni, L. (1987) The effect of trypanocidal drugs on the transmission of *Trypanosoma brucei brucei* by *Glossina morsitans centralis*. *Acta Tropica*, 44, 23–28.

Jefferies, D., Helfrich, M.P. and Molyneux, D.H. (1987) Cibarial infections of *Trypanosoma vivax* and *T. congolense* in *Glossina*. *Parasitology Research*, 73, 289–292.

Jemal, A. and Hugh-Jones, M. (1995) Association of tsetse control with health and productivity of cattle in the Didessa Valley, western Ethiopia. *Preventive Veterinary Medicine*, 22, 29–40.

Jenni, L. (1973) Virus-like particles in a strain of *G. morsitans centralis* Machado 1970. *Transactions of the Royal Society of Tropical Medicine and Hygiene*, 67, 295.

Jenni, L. and Böhringer, S. (1976) Nuclear coat and viruslike particles in the midgut epithelium of *Glossina morsitans* sspp. *Acta Tropica Separatum*, 33, 380–389.

Jenni, L. and Steiger, R. (1974a) Viruslike particles of *Glossina fuscipes fuscipes* Newst. 1910. *Acta Tropica*, 31, 177–180.

Jenni, L. and Steiger, R.F. (1974b) Viruslike particles in the tsetse fly *Glossina morsitans* sspp. Preliminary results. *Revue Suisse de Zoologie*, 81, 663–666.

Jenni, L., Molyneux, D.H. and Livesey, J.L. (1980) Feeding behaviour of tsetse flies infected with salivarian trypanosomes. *Nature*, 283, 383–385.

Jenni, L., Marti, S., Schweizer, J., Betschart, B., Le Page, R.W.F., Wells, J.M., Tait, A., Paindavoine, P., Pays, E. and Steinert, M. (1986) Hybrid formation between African trypanosomes during cyclical transmission. *Nature*, 322, 173–175.

Jewell, G.R. (1956) Marking of tsetse flies for their detection at night. *Nature*, 178, 750.

Jewell, G.R. (1958) Detection of tsetse fly at night. *Nature*, 181, 1354.

Johnson, W.B. and Lloyd, L. (1923) First report of the tsetse-fly investigation in the northern provinces of Nigeria. *Bulletin of Entomological Research*, 15, 373–396.

Johnstone, D.R. and Cooper, J.F. (1986) Forecasting the efficiency of the sequential aerosol technique for tsetse fly control. *Pesticide Science*, 17, 675–685.

Jones, R.N. and Rees, H. (1982) *B-Chromosomes*. Academic Press, London. (In: Warnes and Maudlin, 1992.)

Jones-Davies, W.J. (1967a) A Berenil-resistant train of *Trypanosoma vivax* in cattle. *Veterinary Record*, 80, 567–568.

Jones-Davies, W.J. (1967b) The discovery of berenil-resistant *Trypanosoma vivax* in northern Nigeria. *Veterinary Record*, 80, 531–532.

Jones-Davies, W.J. (1968) Diminazene aceturate and Homidium chloride resistance in tsetse-transmitted trypanosomes of cattle in Northern Nigeria. *Veterinary Record*, 83, 433–437.

Jones-Davies, W.J. and Folkers, C. (1966a) Some observations on cross-resistance to Samorin and Berenil of homidium-resistant field strains of *Trypanosoma congolense* in northern Nigerian cattle. In: *11th meeting of the International Scientific Council for Trypanosomiasis Research and Control, Nairobi*. Publication no. 100, pp. 35–40.

Jones-Davies, W.J. and Folkers, C. (1966b) The prevalence of homidium-resistant strains of trypanosomes in cattle in northern Nigeria. *Bulletin of Epizootic Diseases in Africa*, 14, 65–72.

Jordan, A.M. (1958) The mating behaviour of females of *Glossina palpalis* (R.-D.) in captivity. *Bulletin of Entomological Research*, 49, 35–43.

Jordan, A.M. (1962a) The pregnancy rate in *Glossina palpalis* (R.-D.) in southern Nigeria. *Bulletin of Entomological Research*, 53, 387–393.

Jordan, A.M. (1962b) The ecology of the *Fusca* group of tsetse flies (*Glossina*) in southern Nigeria. *Bulletin of Entomological Research*, 53, 355–385.

Jordan, A.M. (1963) The distribution of the *Fusca* group of tsetse flies in Nigeria and West Cameroun. *Bulletin of Entomological Research*, 54, 307–323.

Jordan, A.M. (1964a) Long-term fluctuations in numbers of a population of *Glossina palpalis palpalis* (R-D.): sixteen years observations. In: *10th meeting of the International Scientific Council for Trypanosomiasis Research and Control, Kampala*. Publication no. 97, pp. 85–90.

Jordan, A.M. (1964b) Trypanosome infection rates in *Glossina morsitans submorsitans* Newst. in northern Nigeria. *Bulletin of Entomological Research*, 55, 219–231.

Jordan, A.M. (1965a) Bovine trypanosomiasis in Nigeria. V. The tsetse fly challenge to a herd of cattle trekked along a trade cattle route. *Annals of Tropical Medicine and Parasitology*, 59, 270–276.

Jordan, A.M. (1965b) Observations on the ecology of *G. morsitans submorsitans* in the northern Guinea savanna of northern Nigeria. *Bulletin of Entomological Research*, 56, 1–17.

Jordan, A.M. (1965c) The hosts of *Glossina* as the main factor affecting trypanosome infection rates of tsetse flies in Nigeria. *Transactions of the Royal Society of Tropical Medicine and Hygiene*, 59, 423–431.

Jordan, A.M. (1972) Extracellular ducts within the wall of the spermatheca of tsetse flies (*Glossina* spp) (Diptera, Glossinidae). *Bulletin of Entomological Research*, 61, 669–672.

Jordan, A.M. (1974) Recent developments in the ecology and methods of control of tsetse flies (*Glossina* spp.) (Diptera, Glossinidae) – a review. *Bulletin of Entomological Research*, 63, 361–399.

Jordan, A.M. (1976) Tsetse flies as vectors of trypanosomes. *Veterinary Parasitology*, 2, 143–152.

Jordan, A.M. (1986) *Trypanosomiasis Control and African Rural Development.* Longman, London, 357 pp.

Jordan, A.M. (1989a) Man and changing patterns of the African trypanosomiases. In: Service, M.W. (ed.) *Demography and Vector-borne Diseases.* CRC Press, Boca Raton, Florida, pp. 47–58.

Jordan, A.M. (1989b) Importance of land use on the incidence of *Trypanosoma brucei gambiense* sleeping sickness (Summary). *Annales de la Société Belge de Médecine Tropicale*, 69 (Suppl. 1), 254.

Jordan, A.M. (1993) Tsetse-flies (Glossinidae). In: Lane, R.P. and Crosskey, R.W. (eds) *Medical Insects and Arachnids.* Chapman & Hall, London, pp. 333–388.

Jordan, A.M. and Curtis, C.F. (1972) Productivity of *Glossina morsitans* Westwood maintained in the laboratory, with particular reference to the sterile-insect release method. *Bulletin of the World Health Organisation*, 46, 33–38.

Jordan, A.M. and Green, C. (1984) Visual responses of tsetse to stationary targets. *Insect Science and its Application*, 5, 331–334.

Jordan, A.M. and Okoth, J.O. (1990) A record of *Glossina medicorum* Austen (Diptera: Glossinidae) from Uganda. *Annals of Tropical Medicine and Parasitology*, 84, 423–426.

Jordan, A.M. and Trewern, M.A. (1978) Larvicidal effect of diflubenzuron in the tsetse fly. *Nature*, 272, 719–720.

Jordan, A.M., Page, W.A. and McDonald, W.A. (1958) Progress made in ascertaining the natural hosts favoured by different species of tsetse. *ISCTR/CCTA 7th meeting Bruxelles 1958*, 315–317.

Jordan, A.M., Lee-Jones, F. and Weitz, B. (1961) The natural hosts of tsetse flies in the forest belt of Nigeria and the Southern Cameroons. *Annals of Tropical Medicine and Parasitology*, 55, 167–179.

Jordan, A.M., Lee-Jones, F. and Weitz, B. (1962) The natural hosts of tsetse flies in Nigeria. *Annals of Tropical Medicine and Parasitology*, 56, 430–442.

Jordan, A.M., Trewern, M.A., Southern, D.I., Pell, P.E. and Davies, E.D.G. (1977) Differences in laboratory performance between strains of *Glossina morsitans morsitans* Westwood from Rhodesia and Tanzania and associated chromosome diversity. *Bulletin of Entomological Research*, 67, 35–48.

Jordan, A.M., Trewern, M.A., Borkovec, A.B. and DeMilo, A.B. (1979) Laboratory studies on the potential of three insect growth regulators for control of the tsetse *Glossina morsitans morsitans* Westwood (Diptera: Glossinidae). *Bulletin of Entomological Research*, 69, 55–65.

Jordt, T. and Lorenzini, E. (1990) Multiple superovulations in N'Dama heifers. *Tropical Animal Health and Production*, 22, 178–184.

Joshua, R.A. (1988) Drug resistance in recent isolates of *Trypanosoma brucei* and *Trypanosoma congolense. Revue d'Élevage et de Médecine Vétérinaire des Pays Tropicaux*, 41, 359–364.

Kaaya, G. (1989a) A review of the progress made in recent years on research and understanding of immunity in insect vectors of human and animal diseases. *Insect Science and its Application*, 10, 751–769.

Kaaya, G.P. (1989b) *Glossina morsitans morsitans:* mortalities caused in adults by experimental infection with entomopathogenic fungi. *Acta Tropica*, 46, 107–114.

Kaaya, G.P., Otieno, L.H., Darji, N. and Alemu, P. (1986) Defence reactions of *Glossina morsitans morsitans* against different species of bacteria and *Trypanosoma brucei brucei. Acta Tropica*, 43, 31–42.

Kaaya, G.P., Darji, N. and Otieno, L.H. (1987) Effects of bacteria, antibacterial compounds and trypanosomes on tsetse reproduction and longevity. *Insect Science and its Application*, 8, 217–220.

Kageruka, P., Colaert, J. and Nkuku-Pela, N. (1977) Strain of *Trypanosoma (Trypanozoon) brucei* isolated from pigs in Bas-Zaire. *Annales de la Société Belge de Médecine Tropicale*, 57, 85–88.

Kageruka, P., Mangus, E., Bajyana Songa, E., Nantulya, V., Jochems, M., Hamers, R. and Mortelmans, J. (1991) Infectivity of *Trypanosoma (Trypanozoon) brucei gambiense* for baboons (*Papio hamadryas, Papio papio). Annales de la Société Belge de Médecine Tropicale*, 71, 39–45.

Kaminsky, R. (1984) Breeding sites of *Glossina palpalis gambiensis* and *G. pallicera* in the leaf axils of oilpalms. *Zeitschrift für Angewandte Entomologie*, 98, 508–511.

Kaminsky, R. (1990) *In vitro* techniques for assessment of drug resistance in trypanosomes. *AgBiotech News and Information*, 2, 205–210.

Kaminsky, R., Chuma, F., Zweygarth, E., Kitosi, D.S. and Moloo, S.K. (1991) The effect of Isometamidium chloride on insect forms of drug-sensitive and drug-resistant stocks of *Trypanosoma vivax:* studies *in vitro* and in tsetse flies. *Parasitology Research*, 77, 13–17.

Kangwagye, T.N. (1975) Control of *Glossina fuscipes* in the (1971) Rhodesian Sleeping sickness outbreak at Busesa, south Busoga, Uganda. *ISCTR 14th meeting, Dakar*, 365–370.

Kanwe, A.B., Bengaly, Z., Saulnier, D. and Duvallet, G. (1992) Évaluation du test de détection des antigènes circulants de trypanosomes à l'aide d'anticorps monoclonaux. Infections expérimentales et naturelles. *Revue d'Élevage et de Médecine Vétérinaire des Pays Tropicaux*, 45, 265–271.

Kashiwazaki, Y., Snowden, K., Smith, D.H. and Hommel, M. (1994) A multiple antigen detection dipstick colloidal dye immunoassay for the field diagnosis of trypanosome infections in cattle. *Veterinary Parasitology*, 55, 57–69.

Katondo, K.M. (1984) Revision of second edition of tsetse distribution maps. An interim report. *Insect Science and its Application*, 5, 381–388.

Kayang, B.B., Bosompem, K.M., Assoku, R.K. and Awumbila, B. (1997) Detection of *Trypanosoma brucei, T. congolense* and *T. vivax* infections in cattle, sheep and goats using latex agglutination. *International Journal of Parasitology*, 1, 83–87.

Kazadi, J.M.L., Van Hees, J., Jochems, M. and Kageruka, P. (1991) Etude de la capacité vectorielle de *Glossina palpalis gambiensis* (Bobo-Dioulasso) vis-à-vis de *Trypanosoma brucei brucei* EATRO 1125. *Revue d'Élevage et de Médecine Vétérinaire des Pays Tropicaux*, 44, 437–442.

Kazyumba, G.L. (1989) Dépistage de la maladie du sommeil. Expérience du Zaïre (Résume). *Annales de la Société Belge de Médecine Tropicale*, 69 (Suppl. 1), 207–208.

Keay, R.W.J. (1959) *Vegetation Map of Africa, Explanatory Notes.* Oxford University Press, London.

Kemp, S.J. and Teale, A.J. (1994) Randomly primed PCR amplification of pooled DNA reveals polymorphism in a ruminant repetitive DNA sequence which differentiates *Bos indicus* and *B. taurus. Animal Genetics*, 25, 83–88.

Kence, A., Otieno, L.H., Darji, N. and Mahamat, H. (1995) Genetic polymorphisms in natural populations of tsetse fly, *Glossina pallidipes* Austen in Kenya. *Insect Science and its Application*, 16, 369–373.

Kennedy, J.S. (1983) Zigzagging and casting as a programmed response to windborne odour: a review. *Physiological Entomology*, 8, 109–120.

Kerrich, G.J. (1961) The forms of *Syntomosphyrum* (Hym., Eulophidae) parasitic on tsetse flies. *Bulletin of Entomological Research*, 51, 21–23.

Keymer, I.F. (1969) A survey of trypanosome infections in wild ungulates in the Luangwa valley, Zambia. *Annals of Tropical Medicine and Parasitology*, 63, 195–200.

Khonde, N., Pépin, J., Niyonsenga, T., Milord, F. and de Wals, P. (1995) Epidemiological evidence for immunity following *Trypanosoma brucei gambiense* sleeping sickness. *Transactions of the Royal Society of Tropical Medicine and Hygiene*, 89, 607–611.

Kilgour, V. and Godfrey, D.G. (1978) The influence of lorry transport on the *Trypanosoma vivax* infection rate in Nigerian trade cattle. *Tropical Animal Health and Production*, 10, 145–148.

Killick-Kendrick, R. and Godfrey, D.G. (1963a) Observations on a close association between *Glossina tachinoides* and domestic pigs near Nsukka, eastern Nigeria 1. *Trypanosoma congolense* and *T. brucei* infections in the pigs. *Annals of Tropical Medicine and Parasitology*, 57, 225–231.

Killick-Kendrick, R. and Godfrey, D.G. (1963b) Bovine trypanosomiasis in Nigeria. II. Incidence in migrating cattle. *Annals of Tropical Medicine and Parasitology*, 57, 117–126.

Killick-Kendrick, R., Leaney, A.J., Ready, P.D. and Molyneux, D.H. (1977) Leishmania in phlebotomid sandflies. IV. The transmission of *Leishmania mexicana amazonensis* to hamsters by the bite of experimentally infected *Lutzomyia longipalpis. Proceedings of the Royal Society of London, Series B*, 196, 105–115.

Kinghorn, A. (1925) Human trypanosomiasis in the Luangwa valley, Northern Rhodesia. *Annals of Tropical Medicine and Parasitology*, 19, 281–300.

Kinghorn, A. and Yorke, W. (1912a) Trypanosomes obtained by feeding wild *Glossina morsitans* on monkeys in Luangwa valley, Northern Rhodesia. *Annals of Tropical Medicine and Parasitology*, 6, 317–325.

Kinghorn, A. and Yorke, W. (1912b) Trypanosomes infecting game and domestic animals in the Luangwa valley, N.E. Rhodesia. *Annals of Tropical Medicine and Parasitology*, 6, 301–315.

Kinghorn, A. and Yorke, W. (1912c) Further observations on the trypanosomes of game and domestic stock in north eastern Rhodesia. *Annals of Tropical Medicine and Parasitology*, 6, 483–493.

Kinghorn, A. and Yorke, W. (1912d) On the influence of meteorological conditions on the development of *Trypanosoma rhodesiense* in *Glossina morsitans. Annals of Tropical Medicine and Parasitology*, 6, 405–413.

Kinghorn, A. and Yorke, W. (1912e) On the transmission of human trypanosomes by *Glossina morsitans* West. and on the occurrence of human trypanosomes in game. *Annals of Tropical Medicine and Parasitology*, 6, 1–23.

Kinghorn, A., Yorke, W. and Lloyd, Ll. (1913) Final report of the Luangwa valley sleeping sickness Commission of the British South Africa Company 1911–1912. *Annals of Tropical Medicine and Parasitology*, 7, 183–302.

Kitron, U., Otieno, L.H., Hungerford, L.L., Odulaja, A., Brigham, W.U., Okello, O.O., Joselyn, M., Mohamed-Ahmed, M.M. and Cook, E. (1996) Spatial analysis of the distribution of tsetse flies in the Lambwe Valley, Kenya, using Landsat TM satellite imagery and GIS. *Journal of Animal Ecology*, 65, 371–380.

Kleine, F. (1909) Positive infektinversuche mit *Trypanosoma brucei* durch *Glossina palpalis*. *Deutsche medizinische Wochenschift*, 35, 409–470, 1257–1260 (In: Wells, E.A. 1972).

Kleine, F. and Eckard, B. (1913) Zur Epidemiologie des Schlafkrankheit. *Archiv für Schiffs und Trop Hygiene*, 17, 325.

Kleine, F. and Taute, M. (1911) Erganzungen zu unseren Trypanosomenstudien. *Arbeiten aus dem Kaiserlichen Gesundheitsamte*, 31, 321–376.

Knight, R.H. and Southon, H.A.W. (1963) A simple method for marking haematophagous insects during the act of feeding. *Bulletin of Entomological Research*, 54, 379–382.

Knipling, E.F. (1959) Sterile male method of population control. *Science*, 130, 902.

Knipling, E.F. (1964) *The Potential Role of the Sterility Method for Insect Population Control with Special Reference to Combining this Method with Conventional Methods*. A.R.S. 33–98, US Department of Agriculture, Washington, DC.

Knipling, E.F. (1979) *The Basic Principles of Insect Population Suppression and Management*. US Department of Agriculture, Washington, DC.

Knottenbelt, D.C. (1974) An investigation into the incidence and pathology of natural trypanosomiasis of bushbuck (*Tragelaphus scriptus*) and Kudu (*T. strepsiceros*). *Tropical Animal Health and Production*, 6, 131–143.

Knudsen, A.B. and Slooff, R. (1992) Vector-borne disease problems in rapid urbanisation: New approaches to vector control. *Bulletin of the World Health Organisation*, 70, 1–6.

Koella, J.C. and Packer, M.J. (1996) Malaria parasites enhance blood-feeding of their naturally infected vector *Anopheles punctulatus*. *Parasitology*, 113, 105–109.

Koeman, J.H., Den Boer, W.M.J., Feith, A.F., de Iongh, H.H., Spliethoff, P.C., Na'isa, B.K. and Spielberger, U. (1978) Three years observations on side effects of helicopter application of insecticides used to exterminate *Glossina* species in Nigeria. *Environmental Pollution*, 15, 31–59.

Koen, C. (1990) A linear programming model of trypanosomiasis control reconsidered. *Preventive Veterinary Medicine*, 9, 37–44.

Koerner, T., de Raadt, P. and Maudlin. I. (1995) The 1901 Uganda sleeping sickness epidemic revisited: a case of mistaken identity? *Parasitology Today*, 11, 303–306.

Kokwaro, E.D. and Odhiambo, T.R. (1980) Spermatophore of the tsetse, *Glossina morsitans morsitans* Westwood: an ultrastructural study. *Insect Science and its Application*, 1, 185–190.

Kokwaro, E.D., Odhiambo, T.R. and Murithii, J.K. (1981) Ultrastructural and histochemical study of the spermathecae of the tsetse, *Glossina morsitans morsitans* Westwood. *Insect Science and its Application*, 2, 135–143.

Kokwaro, E.D., Okot-Kotber, B.M., Odhiambo, T.R. and Murithi, J.K. (1986) Biochemical and immunochemical evidence for the origin of the spermatophore material in *Glossina morsitans morsitans* Westwood. *Entomologia Experimentalis et Applicata*, 43, 448–451.

Kokwaro, E.D., Otieno, L.H. and Chimtawi, M. (1991) Salivary glands of the tsetse *Glossina pallidipes* Austen infected with *Trypanosoma brucei* and virus particles: ultrastructural study. *Insect Science and its Application*, 12, 661–669.

Komba, E., Odiit, M., Mbulamberi, D.B., Chimtwembe, E.C. and Nantulya, V.M. (1992) Multicentre evaluation of an antigen-detection ELISA for the diagnosis of *Trypanosoma brucei rhodesiense* sleeping sickness. *Bulletin of the World Health Organisation*, 70, 57–61.

Komoin-Oka, C., Truc, P., Bengaly, Z., Formenty, P., Duvallet, G., Lauginie, F., Raath, J.P., N'Depo, A.E. and Leforban, Y. (1994) Etude de la prévalance des infections à trypanosomes chez différentes espèces d'animaux sauvages du parc national de la Comoé en Côte d'Ivoire: résultats préliminaires sur la comparaison de trois méthodes de diagnostic. *Revue d'Elevage et de Médecine Vétérinaire des Pays Tropicale*, 47, 189–194.

Konji, V.N., Olembo, N.K. and Pearson, D.J. (1984) Enzyme activities in the fat body of the tsetse fly *Glossina morsitans* and the fleshfly *Sarcophaga tibialis* in relation to proline. *Insect Biochemistry*, 14, 685–690.

Konji, V.N., Olembo, N.K. and Pearson, D.J. (1988) Proline synthesis in the fat body of the tsetse fly *Glossina morsitans*, and its stimulation by isocitrate. *Insect Biochemistry*, 18, 449–452.

Krafsur, E.S., Griffiths, N., Brockhouse, C.L. and Brady, J. (1997) Breeding structure of *Glossina pallidipes* (Diptera: Glossinidae) populations in East and southern Africa. *Bulletin of Entomological Research*, 87, 67–73.

Krampitz, H.E. and Persoons, C. (1967) Ectoparasitic mites on tsetse flies. *EATRO Annual Report 1966*, 55.

Krinsky, W.L. (1976) Animal disease agents transmitted by horse flies and deer flies (Diptera: Tabanidae). *Journal of Medical Entomology*, 13, 225–275.

Kukla, B.A., Majiwa, P.A.O., Young, J.R., Moloo, S.K. and Ole-Moiyoi, O. (1987) Use of species-specific DNA probes for detection and identification of trypanosome infection in tsetse flies. *Parasitology*, 95, 1–16.

Küpper, W. and Harbers, F. (1987) Cypermethrin-impregnated ear tags for the prevention of *Glossina* transmitted trypanosomiasis. *Animal Research and Development*, 21, 52–58.

Küpper, W., Wolters, M. and Tscharf, I. (1983) Observations on Kob antelopes (*Kobus kob*) in northern Ivory Coast and their epizootiological role in trypanosomiasis transmission. *Zeitschrift Für Angewandte Zoologie*, 70, 277–283.

Küpper, W., Manno, A., Douati, A. and Koulibali, S. (1984) Impact des pièges biconiques imprégnés sur les populations de *Glossina palpalis gambiensis* et *Glossina tachinoides*. Résultat d'une campagne de lutte à grande échelle contre la trypanosomose animale au Nord de la Côte d'Ivoire. *Revue d'Élevage et de Médecine Vétérinaire des Pays Tropicaux*, 37, No. Spécial, 176–185.

Küpper, W., Manno, A., Clair, M. and Kotia, K. (1985) The large scale control of *Glossina palpalis* s.l., *G. fusca fusca*, *G. medicorum* and *G. longipalpis* in the southern Guinea zone of the Ivory Coast by deltamethrin impregnated biconical traps. *Cahiers ORSTOM Série Entomologie Médecine et Parasitologie*, 23, 9–16.

Küpper, W., Staak, C., Krober, T. and Späth, J. (1990) Natural hosts of *Glossina tachinoides* (Diptera: Glossinidae) in northern Côte d'Ivoire. *Tropical Medicine and Parasitology*, 41, 217–218.

Küpper, W., Späth, J. and Krober, T. (1991) Attractiveness of chemicals to *Glossina tachinoides* Westwood (Diptera, Glossinidae) in Côte d'Ivoire. *Tropical Pest Management*, 37, 436–438.

Kutuza, S.B. and Okoth, J.O. (1981) A tsetse survey of parts of Uganda during a sleeping sickness outbreak. *Bulletin of Animal Health and Production in Africa*, 29, 55–58.

Kuzoe, F.A.S. (1989) Current knowledge on epidemiology and control of sleeping sickness. *Annales de la Société Belge de Médecine Tropicale*, 69, Suppl. 1, 217–220.

Kuzoe, F.A.S., Baldry, D.A.T., Van der Vloedt, A. and Cullen, J.R. (1985) Observations on an apparent population extension of *Glossina tachinoides* Westwood in southern Ivory Coast. *Insect Science and its Application*, 6, 55–58.

Kwan, W.H., Gatehouse, A.G. and Kerridge, E. (1982) Effects of endosulfan on pregnant females of *Glossina morsitans morsitans* Westwood (Diptera: Glossinidae) and their offspring. *Bulletin of Entomological Research*, 72, 391–401.

Kyorku, C., Brightwell, R. and Dransfield, R.D. (1990) Traps and odour baits for the tsetse fly, *Glossina longipennis* (Diptera: Glossinidae). *Bulletin of Entomological Research*, 80, 405–416.

La Rocque, A. de, Geoffroy, B. and Cuisance, D. (1996) Nouvelle approche pour l'estimation de l'âge des glossines par analyse d'image de l'aile. *Revue d'Élevage et de Médecine Vétérinaire des Pays Tropicaux*, 49, 46–48.

Ladikpo, E. and Seureau, C. (1988) *Trypanosoma (Nannomonas) congolense* Broden, 1904 (Kinetoplastida, Trypanosomatidae) dans les cellules epitheliales du segment antérieur. *Revue d'Élevage et de Médecine Vétérinaire des Pays Tropicaux*, 41, 165–167.

Lambert, M.R.K. (1993) Effects of DDT ground-spraying against tsetse flies on lizards in NW Zimbabwe. *Environmental Pollution*, 82, 231–237.

Lamborn, W.A. (1916) Third report on *Glossina* investigations in Nyasaland. *Bulletin of Entomological Research*, 7, 29–50.

Lamborn, W.A. (1925) An attempt to control *Glossina morsitans* by means of *Syntomosphyrum glossinae,* Waterston. *Bulletin of Entomological Research*, 15, 303–310.

Lamborn, W.A. and Howat, C.H. (1936) A possible reservoir host of *Trypanosoma rhodesiense*. *British Medical Journal*, June 6 1936, 1153–1155.

Lambrecht, F.L. (1964) Aspects of evolution and ecology of tsetse flies and trypanosomiasis in prehistoric African environments. *Journal of African History*, 5, 1–24.

Lambrecht, F.L. (1972) Field studies of *Glossina morsitans* Westw. (Diptera: Glossinidae) in relation to Rhodesian sleeping sickness in N'Gamiland, Botswana. *Bulletin of Entomological Research*, 62, 183–193.

Lambrecht, F.L. (1980) Ecological and physiological factors in the cyclic transmission of African trypanosomiasis. *Insect Science and its Application*, 1, 47–54.

Lamprey, H.F., Glasgow, J.P., Lee-Jones, F. and Weitz, B.A. (1962) A simultaneous census of the potential and actual food sources of the tsetse fly *Glossina swynnertoni* Austen. *Journal of Animal Ecology*, 31, 151–156.

Lancien, J. (1981) Description du piège monoconique utilisé pour l'élimination des glossines en République Populaire du Congo. *Cahiers ORSTOM Série Entomologie Médecine et Parasitologie,* 19, 235–238.

Lancien, J. (1991) Lutte contre la maladie du sommeil dans le sud-est Ouganda par piégeage des Glossines. *Annales de la Société Belge de Médecine Tropicale,* 71, Suppl. 1, 35–47.

Lane, R.P. and Crosskey, R.W. (1993) (eds) *Medical Insects and Arachnids.* British Museum (Natural History), Chapman & Hall, London, UK, 723 pp.

Lang, J.T., Schreck, C.E. and Pamintuan, H. (1981) Permethrin for biting-fly (*Diptera: Muscidae; Tabanidae*) control on horses in Central Luzon, Philippines. *Journal of Medical Entomology,* 18, 522–529.

Langley, P.A. (1965) The neuroendocrine system and stomogastric nervous system of the adult tsetse fly *Glossina morsitans. Proceedings of the Zoological Society of London (B),* 144, 415–424.

Langley, P.A. (1966) The control of digestion in the tsetse fly, *Glossina morsitans.* Enzyme activity in relation to the size and nature of the meal. *Journal of Insect Physiology,* 12, 439–448.

Langley, P.A. (1967a) Experimental evidence for a hormonal control of digestion in the tsetse fly, *Glossina morsitans* Westwood: a study of the larva, pupa, and teneral adult fly. *Journal of Insect Physiology,* 13, 1921–1931.

Langley, P.A. (1967b) The control of digestion in the tsetse fly, *Glossina morsitans:* a comparison between field flies and flies reared in captivity. *Journal of Insect Physiology,* 13, 477–486.

Langley, P.A. (1967c) Effect of ligaturing on puparium formation in the larva of the tsetse fly, *Glossina morsitans* Westwood. *Nature,* 214, 389–390.

Langley, P.A. (1970) Post-teneral development of thoracic flight musculature in the tsetse-flies *Glossina austeni* and *G. morsitans. Entomologia Experimentalis et Applicata,* 13, 133–140.

Langley, P.A. (1971) The respiratory metabolism of tsetse fly puparia in relation to fat consumption. *Bulletin of Entomological Research,* 60, 351–358.

Langley, P.A. (1977) Physiology of tsetse flies, a review. *Bulletin of Entomological Research,* 67, 523–574.

Langley, P.A. (1979) Sex pheromones of *Glossina:* possible aids to control. *Transactions of the Royal Society of Tropical Medicine and Hygiene,* 74, 280–281.

Langley, P.A. (1998) Use of triflumeron for the autosterilization of tsetse in the field. In: *24th Meeting of the International Scientific Council for Trypanosomosis Research and Control, Maputo,* Publication no. 119 (in press).

Langley, P.A. and Abasa, R.O. (1970) Blood meal utilisation and flight muscle development in the tsetse fly *Glossina austeni,* following sterilising doses of gamma irradiation. *Entomologia Experimentalis et Applicata,* 13, 141–152.

Langley, P.A. and Bursell, E. (1980) Role of fat body and uterine gland in milk synthesis by adult female *Glossina morsitans. Acta Tropica,* 10, 11–17.

Langley, P.A. and Carlson, D.A. (1983) Biosynthesis of contact sex pheromone in the female tsetse fly, *Glossina morsitans morsitans* Westwood. *Insect Physiology,* 29, 825–831.

Langley, P.A. and Carlson, D.A. (1986) Laboratory evaluation of bisazir as a practical chemosterilant for the control of tsetse, *Glossina* spp. (Diptera: Glossinidae). *Bulletin of Entomological Research,* 76, 583–592.

Langley, P.A. and Hall, M.J.R. (1986) Tsetse control by sterilisation. *Parasitology Today*, 2, 125–126.

Langley, P.A. and Pimley, R.W. (1973) Influence of diet composition on feeding and water excretion by the tsetse fly, *Glossina morsitans. Journal of Insect Physiology*, 19, 1097–1109.

Langley, P.A. and Pimley, R.W. (1974) Utilisation of U-^{14}C amino acids or U-^{14}C protein by adult *Glossina morsitans* during *in utero* development of larva. *Journal of Insect Physiology*, 20, 2157–2170.

Langley, P.A. and Pimley, R.W. (1979a) Storage and mobilisation of nutriment for uterine milk synthesis by *Glossina morsitans. Journal of Insect Physiology*, 25, 193–197.

Langley, P.A. and Pimley, R.W. (1979b) Influence of diet on synthesis and utilisation of lipids for reproduction by the tsetse fly *Glossina morsitans. Journal of Insect Physiology*, 25, 79–85.

Langley, P.A. and Pimley, R.W. (1986) A role for juvenile hormone and the effects of so-called anti-juvenile hormones in *Glossina morsitans. Journal of Insect Physiology*, 32, 727–734.

Langley, P.A. and Roe, J.M. (1984) Ivermectin as a possible control agent for the tsetse fly, *Glossina morsitans. Entomologia Experimentalis et Applicata*, 3, 137–143.

Langley, P.A. and Stafford, K. (1990) Feeding frequency in relation to reproduction in *Glossina morsitans morsitans* and *G. pallidipes. Physiological Entomology*, 15, 415–421.

Langley, P.A. and Wall, R. (1990) The implications of hunger in the tsetse fly, *Glossina pallidipes,* in relation to its availability to trapping techniques. *Journal of Insect Physiology*, 36, 903–908.

Langley, P.A. and Weidhaas, D. (1986) Trapping as a means of controlling tsetse, *Glossina* spp. (Diptera: Glossinidae): The relative merits of killing and of sterilisation. *Bulletin of Entomological Research*, 76, 89–95.

Langley, P.A., Curtis, C.F. and Brady, J. (1974) The viability, fertility and behaviour of tsetse flies (*Glossina morsitans*) sterilized by irradiation under various conditions. *Entomologia Experimentalis et Applicata*, 17, 97–111.

Langley, P.A., Pimley, R.W. and Carlson, D.A. (1975) Sex recognition pheromone in tsetse fly, *Glossina morsitans. Nature*, 254, 51–53.

Langley, P.A., Huyton, P.M., Carlson, D.A. and Schwarz, M. (1981a) Effects of *Glossina morsitans morsitans* Westwood (Diptera: Glossinidae) sex pheromone on behaviour of males in field and laboratory in the UK. *Bulletin of Entomological Research*, 71, 57–65.

Langley, P.A., Bursell, E., Kabayo, J., Pimley, R.W., Trewern, M.A. and Marshall, J. (1981b) Haemolymph lipid transport from fat body to uterine gland in pregnant females of *Glossina morsitans. Insect Biochemistry*, 11, 225–231.

Langley, P.A., Coates, T.W., Carlson, D.A., Vale, G.A. and Marshall, J. (1982a) Prospects for chemosterilisation of tsetse flies, *Glossina* spp. (Diptera: Glossinidae) using sex-pheromone and Bisazir in the field. *Bulletin of Entomological Research*, 72, 319–329.

Langley, P.A., Trewern, M.A. and Jurd, L. (1982b) Sterilising effects of benzyl-1,3-benzodioxoles on the tsetse fly *Glossina morsitans morsitans* Westwood (Diptera: Glossinidae). *Bulletin of Entomological Research*, 72, 473–481.

Langley, P.A., Maudlin, I. and Leedham, M.P. (1984) Genetic and behavioural differences between *Glossina pallidipes* from Uganda and Zimbabwe. *Entomologia Experimentalis et Applicata*, 35, 55–60.

Langley, P.A., Huyton, P.M. and Carlson, D.A. (1987) Sex pheromone perception by males of the tsetse fly, *Glossina morsitans morsitans*. *Physiological Entomology*, 12, 425–433.

Langley, P.A., Felton, T. and Oouchi, H. (1988a) Juvenile hormone mimics as effective sterilants for the tsetse fly *Glossina morsitans morsitans*. *Medical and Veterinary Entomology*, 2, 29–35.

Langley, P.A., Hall, M.J.R. and Felton, T. (1988b) Determining the age of tsetse flies, *Glossina* spp. (Diptera: Glossinidae): an appraisal of the pteridine fluorescence technique. *Bulletin of Entomological Research*, 78, 387–395.

Langley, P.A., Felton, T., Stafford, K. and Oouchi, H. (1990a) Formulation of pyriproxyfen, a juvenile hormone mimic, for tsetse control. *Medical and Veterinary Entomology*, 4, 127–133.

Langley, P.A., Hargrove, J.W. and Wall, R.L. (1990b) Maturation of the tsetse fly *Glossina pallidipes* (Diptera: Glossinidae) in relation to trap-oriented behaviour. *Physiological Entomology*, 15, 179–186.

Langley, P.A., Perschke, H. and Hussain, M. (1992) Oil formulation of pyrethroids for contamination of tsetse flies (*Glossina* spp.) through tarsal contact with treated targets. *Pesticide Science*, 35, 309–313.

Langridge, W.P. (1960) Scent attractants for tsetse flies. *International Council for Trypanosomiasis Research and Control 8th Meeting, Jos, Nigeria, 1960.* Publication no. 62, 235–241.

Langridge, W.P. (1975) Design and operation of the 'Langridge' tsetse fly trap. *International Council for Trypanosomiasis Research and Control 14th Meeting, Dakar 1975.* Publication no. 109, 277–281.

Langridge, W.P. (1976) *A Tsetse and Trypanosomiasis Survey of Ethiopia.* Ministry of Overseas Development Report. MOD, London, 100 pp.

Langridge, W.P. and Mugutu, S.P. (1968) Some observations on the destruction of wildlife and insects after spraying with organochlorine pesticides for tsetse fly control measures. *International Council for Trypanosomiasis Research and Control 12th Meeting, Bangui 1968.* Publication no. 102, 195–201.

Lanham, S. and Godfrey, D. (1970) Isolation of salivarian trypanosomes from man and other animals using DEAE-cellulose. *Experimental Parasitology*, 28, 521–534.

Laufer, I. (1955) Aspects of medical control of Rhodesian sleeping sickness in Tanganyika. *East African Medical Journal*, 32, 465–480.

Laveissière, C. and Boreham, P.F.L. (1976) Ecologie de *Glossina tachinoides* Westwood, 1850, en savane humide d'Afrique de l'ouest. I. Préférences trophiques. *Cahiers ORSTOM Série Entomologie Médecine et Parasitologie*, 14, 187–200.

Laveissière, C. and Couret, D. (1981a) Trapping as a means of controlling human trypanosome vectors. *ISCTRC 17th meeting Arusha*, 609–617.

Laveissière, C. and Couret, D. (1981b) Essai de lutte contre les Glossines riveraines à l'aide d'écrans imprégnés d'insecticide. *Cahiers ORSTOM Série Entomologie Médecine et Parasitologie*, 19, 271–283.

Laveissière, C. and Couret, D. (1983) Dieldrine et écrans pour la lutte contre les Glossines riveraines. *Cahiers ORSTOM Série Entomologie Médecine et Parasitologie*, 21, 57–62.

Laveissière, C. and Couret, D. (1986) La Campagne pilote de lutte contre la trypanosomiase humaine dans le foyer de Vavoua (Côte d'Ivoire). 5. Bilan financier. *Cahiers ORSTOM Série Entomologie Médecine et Parasitologie*, 24, 149–153.

Laveissière, C. and Grébaut, P. (1990) Recherches sur les pièges à Glossines (Diptera : Glossinidae). Mise au point d'un modèle économique: le piège 'Vavoua'. *Tropical Medicine and Parasitology*, 41, 185–192.

Laveissière, C. and Méda, H.H. (1992) La lutte par piégeage contre la maladie du sommeil: Pas aussi simple que l'on croit. *Annales de la Société Belge de Médecine Tropicale*, 72, Suppl. 1, 57–68.

Laveissière, C. and Sané, B. (1994) Régulateur de croissance et piégeage pour la lutte contre *Glossina palpalis palpalis* en Côte d'Ivoire: Essai sur le terrain de l'OMS 3019 (Pyriproxyfen® Sumitomo). *Insect Science and its Application*, 15, 105–110.

Laveissière, C., Couret, D. and Kienou, J.P. (1981) Lutte contre les Glossines riveraines à l'aide de pièges biconiques imprégnés d'insecticide en zone de savane humide. 4. Expérimentation à grande échelle. *Cahiers ORSTOM Série Entomologie Médecine et Parasitologie*, 19, 41–48.

Laveissière, C., Kienou, J.P. and Traoré, T. (1984a) Ecologie de *Glossina tachinoides* Westwood, 1850, en savane humide d'Afrique de l'ouest X. Durée du stade pupal. Importance de ce paramètre dans la dynamique des populations. *Cahiers ORSTOM Série Entomologie Médecine et Parasitologie*, 22, 219–230.

Laveissière, C., Couret, D. and Traoré, T. (1984b) Tests d'efficacité et de rémanence d'insecticides utilises en imprégnation sur tissus pour la lutte par piégeage contre les Glossines. I. Protocole expérimental. L'effet 'knock-down' des pyrethrinoides. *Cahiers ORSTOM Série Entomologie Médecine et Parasitologie*, 23, 61–67.

Laveissière, C., Traoré, T. and Kienon, J.-P. (1984c) Ecologie de *Glossina tachinoides* Westwood, 1850, en savane humide d'Afrique de l'ouest. *Cahiers ORSTOM Série Entomologie Médecine et Parasitologie*, 22, 231–243.

Laveissière, C., Couret, D., Staak, C. and Hervouët, J.-P. (1985a) *Glossina palpalis* et ses hôtes en secteur forestière de Côte d'Ivoire. Relations avec l'épidémiologie de la trypanosomiase humaine. *Cahiers ORSTOM Série Entomologie Médecine et Parasitologie*, 23, 297–303.

Laveissière, C., Hervouët, J.-P., Couret, D., Eouzan, J.-P. and Merouze, F. (1985b) La campagne pilote de lutte contre la trypanosomiase humaine dans le foyer de Vavoua (Côte d'Ivoire). 2. La mobilisation des communautés rurales et l'application du piégeage. *Cahiers ORSTOM Série Entomologie Médecine et Parasitologie*, 23, 167–185.

Laveissière, C., Couret, D. and Eouzan, J.-P. (1986a) La campagne pilote de lutte contre la trypanosomiase humaine dans le foyer de Vavoua (Côte d'Ivoire). 3. Résultats des évaluations entomologiques. *Cahiers ORSTOM série Entomologie Médicale et Parasitologie*, 24, 7–20.

Laveissière, C., Couret, D. and Hervouët, J.-P. (1986b) Localisation et fréquence du contact Homme/Glossine en secteur forestière de Côte d'Ivoire. 1. Recherche des points épidémiologique. *Cahiers ORSTOM Série Entomologie Médecine et Parasitologie*, 24, 21–35.

Laveissière, C., Hervouët, J.-P. and Couret, D. (1986c) Localisation et fréquence du contact Homme/Glossine en secteur forestier de Côte d'Ivoire 2. Le facteur humain et la transmission de la trypanosomiase. *Cahiers ORSTOM Série Entomologie Médecine et Parasitologie*, 24, 45–57.

Laveissière, C., Couret, D. and Manno, A. (1987) Importance de la nature des tissus dans la lutte par piégeage contre les Glossines. *Cahiers ORSTOM Série Entomologie Médecine et Parasitologie*, 25, 133–143.

Laveissière, C., Couret, D. and Grébaut, P. (1988) Recherche sur les écrans pour la lutte contre les Glossines en région forestière de Côte d'Ivoire. Mise au point d'un nouvel écran. *Rapport Final. Final report No 03/IPR/RAP/88*, OCCGE ORSTOM, Institut Pierre Richet, Bouaké, Côte d'Ivoire.

Laveissière, C., Vale, G.A. and Gouteux, J.-P. (1990) Bait methods for tsetse control. In: Curtis, C.F. (ed.) *Appropriate Technology in Vector Control.* CRC Press, Boca Raton, Florida, pp. 47–74.

Laveissière, C., Sané, B. and Méda, H.A. (1994) Measurement of risk in endemic areas of human African trypanosomiasis in Côte d'Ivoire. *Transactions of the Royal Society of Tropical Medicine and Hygiene*, 88, 645–648.

Laveran, A. (1905) Trypanosomiases et tsetse dans la Guinée Française. *Comptes Rendus de l'Académie des Sciences, Paris*, 139, p. 658.

Le Ray, D. (1989) Vector susceptibility to African trypanosomes. *Annales de la Société Belge de Médecine Tropicale*, 69, Supplément No. 1, 165–171.

Le Roux, J.G. and Platt, D.C. (1968) Application of a dieldrin invert emulsion by helicopter for tsetse control. *ISCTR 12th Meeting Bangui*, 219–229.

Leach, T.M. and Roberts, C.J. (1981) Present status of chemotherapy and chemoprophylaxis of animal trypanosomiasis in the Eastern Hemisphere. *Pharmacology and Therapeutics*, 13, 91–147.

Leak, S.G.A. and Mulatu, W. (1993) Advance of *Glossina morsitans submorsitans* and *G. pallidipes* along the Ghibe-river system in southwest Ethiopia. *Acta Tropica*, 55, 91–95.

Leak, S.G.A. and Rowlands, G.J. (1997) The dynamics of trypanosome infections in natural populations of tsetse (Diptera: Glossinidae); studied using wing-fray and ovarian ageing techniques. *Bulletin of Entomological Research*, 87, 273–282.

Leak, S.G.A., Awuomé, K., Colardelle, C., Duffera, W., Féron, A., Mahamet, B., Mawuena, K., Minengu, M., Mulungo, M., Nankodaba, G., Ordner, G., Pelo, M., Sheria, M., Tikubet, G., Touré, M. and Yangari, G. (1988) Determination of tsetse challenge and its relationship with trypanosome prevalence in trypanotolerant livestock at sites of the African Trypanotolerant Livestock Network. In: *Livestock Production in Tsetse Affected Areas of Africa.* Proceedings of a meeting, November 1987, Nairobi, Kenya. English Press, Nairobi, pp. 43–54.

Leak, S.G.A., Colardelle, C., Coulibaly, L., Dumont, P., Féron, A., Hecker, P., D'Ieteren, G.D., Jeannin, P., Minengu, M., Minja, S., Mulatu, W., Nankodaba, G., Ordner, G., Rowlands, G.J., Sauveroche, B., Tikubet, G. and Trail, J.C.M. (1990) Relationships between tsetse challenge and trypanosome prevalence in trypanotolerant and susceptible cattle. *Insect Science and its Application*, 11, 293–299.

Leak, S.G.A., Colardelle, C., D'Ieteren, G., Dumont, P., Féron, A., Jeannin, P., Minengu, M., Mulungu, M., Ngamuna, S., Ordner, G., Sauveroche, B., Trail, J.C.M. and Yangari, G. (1991) *Glossina fusca* group tsetse as vectors of cattle trypanosomiasis in Gabon and Zaire. *Medical and Veterinary Entomology*, 5, 111–120.

Leak, S.G.A., Woudyalew Mulatu, Rowlands, G.J., and D'Ieteren, G.D.M. (1995) A trial of a cypermethrin 'pour-on' insecticide to control *Glossina pallidipes*, *G. fuscipes fuscipes* and *G. morsitans submorsitans* (Diptera: Glossinidae) in southwest Ethiopia. *Bulletin of Entomological Research*, 85, 241–251.

Leak, S.G.A., Peregrine, A.S., Woudyalew Mulatu and Rowlands, G.J. (1996) Use of insecticide-impregnated targets for the control of tsetse flies (*Glossina* spp.) and trypanosomiasis occurring in cattle in an area of southwest Ethiopia with a high prevalence of drug-resistant trypanosomes. *Tropical Medicine and International Health*, 1, 599–609.

Leeflang, P. (1975) The predominance of *Trypanosoma vivax* infections of cattle at a distance from savanna tsetse concentration. *Tropical Animal Health and Production*, 7, 201–204.

Leger, M. and Vienne, M. (1919) Epizootie à trypanosomes chez les Bovidés de la Guyane française. *Bulletin Société Pathologie Exotique*, 12, 258–266.

Leggate, B.M. and Pilson, R.D. (1961) The diurnal feeding activity of *Glossina pallidipes* Aust. in relation to trypanosome challenge. *Bulletin of Entomological Research*, 51, 697–704.

Lehane, M.J. (1976) Formation and histochemical structure of the peritrophic membrane in the stablefly, *Stomoxys calcitrans*. *Journal of Insect Physiology*, 22, 1551–1557.

Lehane, M.J. and Mail, T.S. (1985) Determining the age of adult male and female *Glossina morsitans morsitans* using a new technique. *Ecological Entomology*, 10, 219–224.

Lehane, M.J. and Msangi, A.R. (1991) Lectin and peritrophic membrane development in the gut of *Glossina m. morsitans* and a discussion of their role in protecting the fly against trypanosome infection. *Medical and Veterinary Entomology*, 5, 495–501.

Lejeune, E. (1923) La prophylaxie de la maladie du sommeil: son organisation au Congo Belge. *Revista Medica de Angola*, 4, 180–199.

Lenoble, B.J. and Denlinger, D.L. (1982) The milk gland of the sheep ked, *Melophagus ovinus*: a comparison with *Glossina*. *Journal of Insect Physiology*, 28, 165–172.

Leonard, D.E. and Saini, R.K. (1993) Semiochemicals from anal exudate of larvae of tsetse flies *Glossina morsitans morsitans* Westwood and *G. morsitans centralis* Machado attract gravid females. *Journal of Chemical Ecology*, 19, 2039–2046.

Leperre, P. and Claxton, J.R. (1994) Comparative study of trypanosomiasis in Zebu and N'Dama cattle in The Gambia. *Tropical Animal Health and Production*, 26, 139–145.

Leslie, P.H. (1952) The estimation of population parameters from data obtained by means of the capture–recapture method. II. The estimation of total numbers. *Biometrika*, 39, 363–388.

Leslie, P.H. and Chitty, D. (1951) The estimation of population parameters from data obtained by means of the capture–recapture method. I. The maximum likelihood equations for estimating the death-rate. *Biometrika*, 38, 269–292.

Lessard, P., L'Eplattenier, R., Norval, R.A.I., Kundert, K., Dolan, T.T., Croze, H., Walker, J.B., Irvin, A.D. and Perry, B.D. (1990) Geographical information systems for studying the epidemiology of cattle diseases caused by *Theileria parva*. *Veterinary Record*, 126, 255–262.

Lester, H.M.O. (1933) Sleeping sickness in northern Nigeria, a review of events leading to adoption of present methods. *West African Medical Journal*, 6, 50–53.

Lester, H.M.O. (1939) Certain aspects of trypanosomiasis in some African dependencies. *Transactions of the Royal Society of Tropical Medicine and Hygiene*, 33, 11–36.

Lester, H.M.O. and Lloyd, L. (1928) Notes on the process of digestion in tsetse-flies. *Bulletin of Entomological Research*, 19, 39–60.

Lewis, A.R. and Thomson, J.W. (1974) Observations on an isometamidium resistant strain of *Trypanosoma congolense* in Rhodesia. *Rhodesian Veterinary Journal*, 4, 62–67.

Lewis, D.J. (1934) The behaviour of the larvae of tsetse-flies before pupation. *Bulletin of Entomological Research*, 25, 195–199.

Lewis, D.J. (1953a) *Simulium damnosum* and its relation to onchocerciasis in the Anglo-Egyptian Sudan. *Bulletin of Entomological Research*, 43, 597–644.

Lewis, D.J. (1953b) The tabanidae of the Anglo-Egyptian Sudan. *Bulletin of Entomological Research*, 44, 175–216.

Lewis, E.A. (1939) Observations on *Glossina fuscipleuris* and other tsetses in the Oyani valley, Kenya colony. *Bulletin of Entomological Research*, 30, 345–360.

Lewis, E.A. (1942) Notes on *Glossina longipennis* and its breeding places. *Bulletin of Entomological Research*, 32, 303–307.

Lewis, E.A. (1947) *Second Progress Report of Tsetse Fly and Trypanosomiasis Survey and Control in Kenya Colony*. Office of the Member for Agriculture and Natural Resources, Kenya, Nairobi.

Lewis, E.A. (1953) Land use and tsetse control. *East African Agricultural Journal*, 18, 160–168.

Liddel, J.S. and Clayton, R. (1982) Long duration fly control on cattle using cyper-methrin impregnated eartags. *Veterinary Record*, 110, 502.

Lincoln, F.C. (1930) Calculating wildfowl abundances on the basis of banding returns. *US Department of Agriculture Circular*, No. 118, May 1930, pp. 67–79.

Lindquist, K.J., Gathuma, J.M. and Kaburia, H.F.A. (1982) Analysis of bloodmeals of haematophagous insects by haemagglutination inhibition and enzyme immunoassay. Tukei, P. and Njogu, A.R. (eds) *Proceedings of the 3rd Annual Medical Science Conference, Nairobi, Kenya*, 77 pp.

Linear, M. (1981) Zapping Africa's flies. *Vole*, 9, 14–18.

Linear, M. (1982) Gift of poison – the unacceptable face of development aid. *Ambio*, 11, 2–8.

Linear, M. (1985a) The phoney tsetse war. *Africa Now*, April 1985, 51–52.

Linear, M. (1985b) The tsetse war. *The Ecologist*, 15, 27–35.

Livesey, J.L., Molyneux, D.H. and Jenni, L. (1980) Mechanoreceptor–trypanosome interactions in the labrum of *Glossina*: fluid mechanics. *Acta Tropica* 37, 151–161.

Lloyd, L. (1913) Notes on *Glossina morsitans* Westw. in the Luangwa valley, Northern Rhodesia. *Bulletin of Entomological Research*, 3, 233–239.

Lloyd, L. (1914) Further notes on the bionomics of *Glossina morsitans* in Northern Rhodesia. *Bulletin of Entomological Research*, 5, 49–60.

Lloyd, L. (1916) Report on the investigation into the bionomics of *Glossina morsitans* in Northern Rhodesia, 1915. *Bulletin of Entomological Research*, 7, 67–79.

Lloyd, L. (1930) Some factors influencing the trypanosome infection rate in tsetse flies. *Transactions of the Royal Society of Tropical Medicine and Hygiene*, 23, 533–542.

Lloyd, L. (1935) Notes on the bionomics of *Glossina swynnertoni* Austen. *Bulletin of Entomological Research*, 26, 439–468.

Lloyd, L. and Johnson, W.B. (1924) The trypanosome infections of tsetse flies in northern Nigeria and a new method of estimation. *Bulletin of Entomological Research*, 14, 265–288.

Lloyd, L., Johnson, W.B., Young, W.A. and Morrison, H. (1924) Second report of the tsetse fly investigation in the Northern Provinces of Nigeria. *Bulletin of Entomological Research*, 15, 1–27.

Lloyd, L., Johnson, W.B. and Rawson, P. (1927) Experiments in the control of tsetse fly. *Bulletin of Entomological Research*, 17, 423–457.

Lloyd, L., Lester, H.M.O., Taylor, A.W. and Thornewill, A.S. (1933) Experiments in the control of tsetse fly. Part II. *Bulletin of Entomological Research*, 24, 233–251.

Loder, P.M.J. (1997) Size of blood meals taken by tsetse flies (*Glossina* spp.) (Diptera: Glossinidae) correlates with fat reserves. *Bulletin of Entomological Research*, 87, 547–549.

Loftus, R.T., MacHugh, D.E., Bradley, D.G., Sharp, P.M. and Cunningham, P. (1994) Evidence for two independent domestications of cattle. *Proceedings of the National Academy of Sciences, USA*, 91, 2757–2761.

Löhr, K.-F., Omukuba, J.N., Njogu, A.R., Maloo, S.H., Gisemba, F., Okedi, T. and Mwongela, S. (1991) Investigation of the efficacy of flumethrin pour-on for the control of high tsetse and trypanosomiasis challenge in Kenya. *Tropical Medicine and Parasitology*, 42, 131–134.

Lourens, J.H.M. (1980) Inheritance of organochlorine resistance in the cattle tick *Rhipicephalus appendiculatus* Neumann (Acari: Ixodidae) in East Africa. *Bulletin of Entomological Research*, 70, 1–10.

Lucas, J.M.S. (1955) Transmission of *T. congolense* in cattle under field conditions in the absence of tsetse flies. *Veterinary Record*, 67, 403–407.

Luckins, A.G. and Mehlitz, D. (1978) Evaluation of an indirect fluorescent antibody test, enzyme-linked immunosorbent assay and quantification of immuno-globulins in the diagnosis of bovine trypanosomiasis. *Tropical Animal Health and Production*, 10, 149–159.

Luckins, A.G., Sutherland, D., Mwangi, D. and Hopkins, J. (1994) Early stages of infection with *Trypanosoma congolense*: parasite kinetics and expression of metacyclic variable antigen types. *Acta Tropica*, 58, 199–206.

Lyons, M. (1992) *The Colonial Disease. A Social History of Sleeping Sickness in Northern Zaire, 1900–1940*. Cambridge University Press, Cambridge, 335 pp.

Ma, W.-C. and Denlinger, D.L. (1974) Secretory discharge and microflora of milk gland in tsetse flies. *Nature*, 247, 301–303.

Ma, W.-C., Denlinger, D.L., Jarlfors, U. and Smith, D.S. (1975) Structural modulations in the tsetse fly milk gland during a pregnancy cycle. *Tissue and Cell*, 7, 319–330.

MacDonald, G. (1957) *The Epidemiology and Control of Malaria*. Oxford University Press, London.

MacDonald, G. (1965) On the scientific basis of tropical hygiene. *Transactions of the Royal Society of Tropical Medicine and Hygiene*, 59, 611–620.

Macfie, J.W.S. (1913a) The distribution of *Glossina* in the Ilorin province of northern Nigeria. *Bulletin of Entomological Research*, 4, 1–28.

Macfie, J.W.S. (1913b) Trypanosomiasis of domestic animals in northern Nigeria. *Annals of Tropical Medicine and Parasitology*, 7, 1–27.

Machado, A. de Barros (1954) Révision systématique des glossines du groupe *palpalis* (Diptera). *Subsidios para o Estudo da Biologia na Lunda Publication Culturais Co Diamantés Angola*, no. 22. Museo do Dundo, Angola, 189 pp.

Machado, A. de Barros (1959) Nouvelles contributions à l'étude systématique et biogéographique des Glossines (Diptera). *Companhia de Diamantés de Angola Museo de Dundo Publicaes Culturais*, No. 46.

MacKichan, I.W. (1944) Rhodesian sleeping sickness in Eastern Uganda. *Transactions of the Royal Society of Tropical Medicine and Hygiene*, 38, 49–60.

Maclean, G. (1926) History of an outbreak of Rhodesian sleeping sickness in the Ufipa district of Tanganyika territory with short notes on cases and treatment. *Annals of Tropical Medicine and Parasitology*, 20, 329–339.

Maclean, G. (1929) The relationship between economic development and Rhodesian sleeping sickness in Tanganyika territory. *Annals of Tropical Medicine and Parasitology*, 23, 37–46.

MacLennan, K.J.R. (1963) Cattle trypanosomiasis in northern Nigeria. The problem in the field. *Bulletin of Epizootic Diseases in Africa*, 11, 381–390.

MacLennan, K.J.R. (1970) The epizootiology of trypanosomiasis in livestock in West Africa. In: Mulligan, H.W. (ed.) *The African Trypanosomiases*. ODA, George Allen and Unwin, London, pp. 751–765.

MacLennan, K.J.R. (1973) A consideration of environmental consequences following anti-tsetse operations in Nigeria. *Tropical Animal Health and Production*, 5, 40–45.

MacLennan, K.J.R. (1974) The epizootiology of tsetse transmitted trypanosomiasis in relation to livestock development and control measures. In: *Les Moyens de Lutte Contre les Trypanosomes et leurs Vecteurs*. IEMVT, Paris, pp. 259–268.

MacLennan, K.J.R. (1977) Tsetse habitats, distribution and reclamation prospects. In: *Final Report, Mali Range Survey*. Earth Satellite Corporation, Washington, DC, pp. 54–60.

MacLennan, K.J.R. and Cooke, M.G. (1972) The resting behaviour of *Glossina morsitans submorsitans* in the northern Guinea zone. *Entomologist*, 105, 589.

MacLennan, K.J.R. and Jones-Davies, W.J. (1967) The occurrence of a Berenil-resistant *Trypanosoma congolense* strain in northern Nigeria. *Veterinary Record*, 80, 389–390.

Macleod, J. and Donnelly, J. (1958) Individual and group marking methods for fly-population studies. *Bulletin of Entomological Research*, 49, 585–592.

Madubunyi, L.C. (1975) A technique for detecting abortions in wild populations of *Glossina* species. In: *Sterility Principle for Insect Control 1974*. Publication No. IAEA-SM-186/32, International Atomic Energy Agency, Vienna, pp. 477–485.

Madubunyi, L.C. (1978) Relative frequency of reproductive abnormalities in a natural population of *Glossina morsitans* Westwood (Diptera: Glossinidae) in Zambia. *Bulletin of Entomological Research*, 68, 437–442.

Magnus, E., Vervoort, T. and Van Meirvenne, N. (1978) A card-agglutination test with stained trypanosomes (C.A.T.T.) for the serological diagnosis of *T. b. gambiense* trypanosomiasis. *Annales de la Société Belge de Médecine Tropicale*, 58, 169–176.

Mahamat, H., Okech, M.A. and Maniania, N. (1998) The lethal insect technique (LIT): a new concept for the control of *Glossina* spp. in the laboratory and field. In: *24th Meeting of the International Scientific Council for Trypanosomosis Research and Control, Maputo*, Publication no. 119 (in press).

Mahood, A.R. (1962) A note on the ecology of *Glossina morsitans submorsitans* Newst. in the Guinea savannah zone of northern Nigeria. *ISCTR 9th Meeting Conakry, 1962*, 181–185.

Mail, R.S., Chadwick, J. and Lehane, M.J. (1983) Determining the age of adults of *Stomoxys calcitrans*. *Bulletin of Entomological Research*, 73, 501–525.

Maillard, J.C. (1974) Recherches sur les bacilles présumés pathogènes pour les glossines. Etude sur *Glossina tachinoides* en République du Tchad. *Revue d'Elevage et de Médecine Vétérinaire des Pays Tropicale*, 27, 67–73.

Maillard, J.C. and Provost, A. (1975) Recherche du pouvoir pathogène de *Bacillus thuringiensis* sur les glossines (Diptera-Muscidae). Etude sur *Glossina tachinoides* en République du Tchad. *Revue d'Elevage et de Médecine Vétérinaire des Pays Tropicale*, 28, 61–65.

Majiwa, P.A.O. and Otieno, L.H. (1990) Recombinant DNA probes reveal simultaneous infection of tsetse flies with different trypanosome species. *Molecular Biochemistry and Parasitology*, 40, 245–254.

Majiwa, P.A.O. and Webster, P. (1987) A repetitive deoxyribonucleic acid sequence distinguishes *Trypanosoma simiae* from *T. congolense. Parasitology*, 95, 543–598.

Majiwa, P.A.O., Thatthi, R., Moloo, S.K., Nyeko, J.H.P., Otieno, L.H. and Maloo, S. (1994) Detection of trypanosome infections in the saliva of tsetse flies and buffy-coat samples from antigenaemic but aparasitaemic cattle. *Parasitology*, 108, 313–322.

Makumi, J.N. and Moloo, S.K. (1991) *Trypanosoma vivax* in *Glossina palpalis gambiensis* do not appear to affect feeding behaviour, longevity or reproductive performance of the vector. *Medical and Veterinary Entomology*, 5, 35–42.

Makumyaviri, A.M., Distelmans, W., Claes, Y., D'Haeseleer, F., Le Ray, D. and Gooding, R.H. (1984a) Capacité vectorielle du type sauvage et du mutant salmon de *Glossina morsitans morsitans* Westwood, 1850 (Diptera: Glossinidae) dans la transmission de *Trypanosoma brucei* Plimmer et Bradford, 1899. *Cahiers ORSTOM Série Entomologie Médecine et Parasitologie*, 22, 283–288.

Makumyaviri, A.M., Demey, F., Claes, Y., Verhulst, A. and Le Ray, D. (1984b) Caractérisation de la capacité vectorielle de *Glossina morsitans morsitans* (Diptera: Glossinidae) vis-à-vis de *Trypanosoma brucei brucei* EATRO 1125 (AnTAR 1). *Annales de la Société Belge de Médecine Tropicale*, 64, 365–372.

Mansfield-Aders, W. (1923) Trypanosomiasis of stock in Zanzibar. *Transactions of the Royal Society for Tropical Medicine and Hygiene*, 17, 192–200.

Mant, M.J. and Parker, K.R. (1981) Two platelet aggregation inhibitors in tsetse (*Glossina*) saliva with studies of roles of thrombin and citrate in *in vitro* platelet aggregation. *British Journal of Haematology*, 48, 601–608.

Margalit, J., Galun, R. and Rice, M.J. (1972) Mouthpart sensilla of the tsetse fly and their function. I. Feeding patterns. *Annals of Tropical Medicine and Parasitology*, 66, 525–536.

Masake, R.A., Majiwa, P.A.O., Moloo, S.K., Makau, J.M., Njuguna, J.T., Maina, M., Kabata, J., Ole-Moi-Yoi, O.K. and Nantulya, V.M. (1997) Sensitive and specific detection of *Trypanosoma vivax* using the polymerase chain reaction. *Experimental Parasitology*, 85, 193–205.

Masiga, D.K., Smyth, A.J., Hayes, P., Bromidge, T.J. and Gibson, W.C. (1992) Sensitive detection of trypanosomes in tsetse flies by DNA amplification. *International Journal for Parasitology*, 22, 909–918.

Masiga, D.K., McNamara, J.J., Laveissière, C., Truc, P. and Gibson, W. (1996a) A high prevalence of mixed trypanosome infections in tsetse flies in Sinfra, Côte d'Ivoire, detected by DNA amplification. *Parasitology*, 112, 75–80.

Masiga, D.K., McNamara, J.J. and Gibson, W.C. (1996b) A repetitive DNA sequence specific for *Trypanosoma (Nannomonas) godfreyi. Veterinary Parasitology*, 62, 27–33.

Mathieu-Daudé, F., Bicart-See, A., Bosseno, M.F., Breniere, S.F. and Tibayrenc, M. (1994) Identification of *Trypanosoma brucei gambiense* group I by a specific kinetoplast DNA probe. *American Journal of Tropical Medicine and Hygiene*, 50, 13–19.

Matovu, F.S. (1982) Rhodesian sleeping sickness in south-eastern Uganda: the present problems. *East African Medical Journal*, 59, 390–393.

Matthysse, J.G. (1946) DDT to control hornflies and Gulf Coast ticks on range cattle in Florida. *Journal of Economic Entomology*, 39, 62–65.

Mattioli, R.C., Bah, M., Faye, J., Kora, S. and Cassama, M. (1993) A comparison of field tick infestation on N'Dama, Zebu and N'Dama × Zebu crossbred cattle. *Veterinary Parasitology*, 47, 139–148.

Mattioli, R.C., Bah, M., Kora, S., Cassama, M. and Clifford, D.J. (1995) Susceptibility to different tick genera in Gambian N'Dama and Gobra Zebu cattle exposed to naturally occurring tick infestations. *Tropical Animal Health and Production*, 27, 95–103.

Matzke, G. (1983) A reassessment of the expected development consequences of tsetse control efforts in Africa. *Social Science and Medicine*, 17, 531–537.

Maudlin, I. (1970) Preliminary studies on the karyotypes of five species of *Glossina*. *Parasitology*, 61, 71–74.

Maudlin, I. (1979) Chromosome polymorphism and sex determination in a wild population of tsetse. *Nature, London*, 277, 300–301.

Maudlin, I. (1980) Population genetics of tsetse flies and its relevance to trypanosomiasis research. *Insect Science and its Application*, 1, 35–38.

Maudlin, I. (1982) Inheritance of susceptibility to *Trypanosoma congolense* infection in *Glossina morsitans*. *Annals of Tropical Medicine and Parasitology*, 76, 225–227.

Maudlin, I. (1985) Inheritance of susceptibility to trypanosomes in tsetse flies. *Parasitology Today*, 1, 59–60.

Maudlin, I. (1991) Transmission of African trypanosomiasis: interactions among tsetse immune system, symbionts, and parasites. *Advances in Disease Vector Research*, 7, 117–148.

Maudlin, I. and Dukes, P. (1985) Extrachromosomal inheritance of susceptibility to trypanosome infection in tsetse flies. I. Selection of susceptible and refractory lines of *Glossina morsitans morsitans*. *Annals of Tropical Medicine and Parasitology*, 79, 317–324.

Maudlin, I. and Ellis, D.S. (1985) Association between intracellular rickettsia like infections of midgut cells and susceptibility to trypanosome infection in tsetse. *Zeitschrift für Parasitenkunde*, 71, 683–687.

Maudlin, I. and Welburn, S.C. (1987) Lectin mediated establishment of midgut infections of *Trypanosoma congolense* and *Trypanosoma brucei* in *Glossina morsitans*. *Tropical Medicine and Parasitology*, 38, 167–170.

Maudlin, I. and Welburn, S.C. (1988) The role of lectins and trypanosome genotype in the maturation of midgut infections in *Glossina morsitans*. *Tropical Medicine and Parasitology*, 39, 56–58.

Maudlin, I. and Welburn, S.C. (1989) A single trypanosome is sufficient to infect a tsetse fly. *Annals of Tropical Medicine and Parasitology*, 83, 431–433.

Maudlin, I. and Welburn, S.C. (1994) Maturation of trypanosome infections in tsetse. *Experimental Parasitology*, 79, 202–205.

Maudlin, I., Green, C.H. and Barlow, F. (1981) The potential for insecticide resistance in *Glossina* (Diptera: Glossinidae) – an investigation by computer simulation and chemical analysis. *Bulletin of Entomological Research*, 71, 691–702.

Maudlin, I., Turner, M.J., Dukes, P. and Miller, N. (1984) Maintenance of *Glossina morsitans* on antiserum to procyclic trypanosomes reduces infection rates with homologous and heterologous *Trypanosoma congolense* stocks. *Acta Tropica*, 41, 253–257.

Maudlin, I., Dukes, P., Luckins, A.G. and Hudson, K.M. (1986) Extra-chromosomal inheritance of susceptibility to trypanosome infection in tsetse flies. II. Susceptibility of selected lines of *Glossina morsitans morsitans* to different stocks and species of trypanosomes. *Annals of Tropical Medicine and Parasitology*, 80, 97–105.

Maudlin, I., Welburn, S.C. and Mehlitz, D. (1990) The relationship between Rickettsia-like organisms and trypanosome infections in natural populations of tsetse in Liberia. *Tropical Medicine and Parasitology*, 41, 265–267.

Maudlin, I., Welburn, S.C. and Milligan, P. (1991) Salivary gland infection: a sex-linked recessive character in tsetse? *Acta Tropica*, 48, 9–15.

Maudlin, I., Welburn, S.C. and Milligan, P.J.M. (1998) Trypanosome infections and survival in tsetse. *Parasitology*, 116, S23–S28.

Mawuena, K. (1987) Haut degré de tolérance à la trypanosomose des moutons et des chèvres de race Naine Djallonké des régions sud-guinéennes du Togo. Comparaison avec des bovins trypanotolérants. *Revue d'Élevage et de Médecine Vétérinaire des Pays Tropicaux*, 40, 55–58.

Mayer, J. and Denoulet, W. (1984) Résultats d'utilisation de boucles d'oreille imprégnées de pyréthrinoides dans la lutte contre les Glossines (perméthrine). *Revue d'Élevage et de Médecine Vétérinaire des Pays Tropicaux*, 37, 290–292.

Mbulamberi, D.L. (1989) Possible causes leading to an epidemic of sleeping sickness. Facts and hypotheses. *Annales de la Société Belge de Médecine Tropicale*, 69, 173–179.

Mbwambo, H.A., Mella, P.N.P. and Lekaki, K.A. (1988) Berenil (diminazene aceturate)-resistant *Trypanosoma congolense* in cattle under natural tsetse challenge at Kibaha, Tanzania. *Acta Tropica*, 45, 239–244.

MacDonald, G. (1952) The analysis of equilibrium in malaria. *Tropical Diseases Bulletin*, 49, 813–828.

MacLennan, K.J.R. (1967) Recent advances in techniques for tsetse-fly control. *Bulletin of the World Health Organisation*, 37, 615–628.

McAlpine, J.F. (1989) Phylogeny and classification of the Muscomorpha. In: McAlpine, J.F. (ed.) *Manual of Nearctic Diptera, 2*. Research Branch, Agriculture Canada (Monograph No. 28), pp. 1397–1518.

McCabe, C.T. and Bursell, E. (1975a) Metabolism of digestive products in the tsetse fly *Glossina morsitans*. *Insect Biochemistry*, 5, 769–779.

McCabe, C.T. and Bursell, E. (1975b) Interrelationships between amino acid and lipid metabolism in the tsetse fly, *Glossina morsitans*. *Insect Biochemistry*, 5, 781–789.

McCall, P.J., Wilson, M.D., Dueben, B.D., de Clare Bronsvoort, B.M. and Heath, R.R. (1997) Similarity in oviposition aggregation pheromone composition within the *Simulium damnosum* (Diptera: Simuliidae) species complex. *Bulletin of Entomological Research*, 87, 609–616.

McCalla, A.F. (1994) *Agriculture and Food Needs to 2025: Why we should be concerned*. Sir John Crawford Memorial Lecture, CGIAR, Washington, 29 pp.

McClelland, G.A.H. and Weitz, B. (1963) Serological identification of the natural hosts of *Aedes aegypti* (L.) and some other mosquitoes (Diptera: Culicidae) caught resting in vegetation in Kenya and Uganda. *Annals of Tropical Medicine and Parasitology*, 57, 214–224.

McConnell, E., Hutchinson, M.P. and Baker, J.R. (1970) Human trypanosomiasis in Ethiopia: The Gilo river area. *Transactions of the Royal Society of Tropical Medicine and Hygiene*, 64, 683–691.

McDonald, W.A. (1957) A calliphorid host of *Thyridanthrax abruptus* (LW.) in Nigeria (Diptera, Bombyliidae). *Bulletin of Entomological Research*, 48, 533.

McDonald, W.A. (1960) Nocturnal detection of tsetse flies in Nigeria with UV light. *Nature*, 185, 867–868.

McDowell, P.G., Whitehead, D.L. and Chaudhury, M.F.B. (1981) The isolation and identification of the cuticular sex-stimulant pheromone of the tsetse *Glossina pallidipes* Austen (Diptera: Glossinidae). *Insect Science and its Application*, 2, 181–187.

McEwan Jenkinson, D., Hutchinson, G., Jackson, D. and McQueen, L. (1986) Route of passage of cypermethrin across the surface of sheep skin. *Research in Veterinary Science*, 41, 237–241.

McIntyre, G.S. and Gooding, R.H. (1995) Pteridine accumulation in *Musca domestica*. *Journal of Insect Physiology*, 41, 357–368.

McIntyre, G.S. and Gooding, R.H. (1996) Variation in the pteridine content in the heads of tsetse flies (Diptera: Glossinidae: *Glossina* Wiedemann): evidence for genetic control. *Canadian Journal of Zoology*, 74, 621–626.

McLelland, G.A.H. and Weitz, B. (1963) Serological identification of the natural hosts of *Aedes aegypti* (L.) and some other mosquitoes (Diptera: Culicidae) caught resting in vegetation in Kenya and Uganda. *Annals of Tropical Medicine and Parasitology*, 57, 214–224.

McLetchie, J.L. (1953) Sleeping sickness activities in Nigeria, 1931–1952 (Part 1). *West African Medical Journal*, 2, 70–78.

McNamara, J.J. and Snow, W.F. (1991) Improved identification of *Nannomonas* infections in tsetse flies from The Gambia. *Acta Tropica*, 48, 127–136.

McNamara, J.J., Dukes, P., Snow, W.F. and Gibson, W.C. (1989) Use of DNA probes to identify *Trypanosoma congolense* and *T. simiae* in tsetse flies from The Gambia. *Acta Tropica*, 46, 55–61.

McNamara, J.J., Mohammed, G. and Gibson, W.C. (1994) *Trypanosoma (Nannomonas) godfreyi* sp. nov. from tsetse flies in The Gambia: biological and biochemical characterization. *Parasitology*, 109, 497–509.

McNamara, J.J., Laveissière, C. and Masiga, D.K. (1995) Multiple trypanosome infections in wild tsetse in Côte d'Ivoire detected by PCR analysis and DNA probes. *Acta Tropica*, 59, 85–92.

Mehlitz, D. (1977) The behaviour in the blood incubation infectivity test of four Trypanozoon strains isolated from pigs in Liberia. *Transactions of the Royal Society of Tropical Medicine and Hygiene*, 71, 86.

Mehlitz, D. (1979) Trypanosome infections in domestic animals in Liberia. *Tropenmedizin und Parasitologie*, 30, 212–219.

Mehlitz, D. (1986) Le réservoir animal de la maladie du sommeil à *Trypanosoma brucei gambiense*. *Etudes de Synthèse*, 18, IEMVT, pp. 156.

Mehlitz, D., Zillmann, U., Scott, C.M. and Godfrey, D.G. (1982) Epidemiological studies on the animal reservoir of *Gambiense* sleeping sickness. Part III. Characterization of *Trypanozoon* stocks by isoenzymes and sensitivity to human serum. *Tropenmedizin und Parasitologie*, 33, 113–118.

Meidell, E.-M. (1982) Effects of a synthetic juvenile hormone mimic on the reproduction of the tsetse fly *Glossina morsitans*. *Insect Science and its Application*, 3, 263–266.

Mellanby, K. (1936) Experimental work with the tsetse-fly *Glossina palpalis* in Uganda. *Bulletin of Entomological Research*, 27, 611–633.

Mellanby, H. (1937) Experimental work on reproduction in the tsetse fly, *Glossina palpalis*. *Parasitology*, 29, 131–141.

Mérot, P. and Filledier, J. (1985) Efficacité contre *Glossina morsitans submorsitans* d'écrans de différentes couleurs, avec ou sans adjonction de panneaux en moustiquaire noire. *Revue d'Élevage et de Médecine Vétérinaire des Pays Tropicaux*, 38, 64–71.

Mérot, P., Galey, J.B., Politzar, H., Tamboura, I. and Cuisance, D. (1984) Résultats d'une campagne de lutte contre les glossines riveraines en Burkina par l'emploi d'écrans imprégnés de deltaméthrine. *Revue d'Élevage et de Médecine Vétérinaire des Pays Tropicaux*, 37, 175–184.

Mérot, P., Galey, J.B., Politzar, H., Filledier, J. and Mitteault, A. (1986) Pouvoir attractif de l'odeur des hôtes nourriciers pour *Glossina tachinoides* en zone Soudano-Guinéene (Burkina Faso). *Revue d'Élevage et de Médecine Vétérinaire des Pays Tropicaux*, 39, 345–350.

Mesnil, F. and Blanchard, M. (1912) Infections des poules deux aux *Trypanosoma gambiense* et *T. rhodesiense*. *Proceedings of the Royal Society, (B)*, 72, 938–940.

Mettam, R.W.M. (1940) *Annual Report of the Veterinary Pathologist Report of the Veterinary Department of Nigeria*, 1938, p. 20. (In: Killick-Kendrick and Godfrey, 1963.)

Miezan, T., Doua, F., Cattand, P. and Raadt, P. de (1991) Evaluation du Testryp CATT appliqué au sang prélevé sur papier filtre et au sang dilué, dans le foyer de trypanosomiase à *Trypanosoma brucei gambiense* en Côte d'Ivoire. *Bulletin of the World Health Organisation*, 69, 603–606.

Mihok, S., Otieno, L.H., Darji, N. and Munyinyi, D. (1992a) Influence of D+ glucosamine on infection rates and parasite loads in tsetse flies (*Glossina* spp.) infected with *Trypanosoma brucei*. *Acta Tropica*, 51, 217–228.

Mihok, S., Otieno, L.H. and Tarimo, C.S. (1992b) Trypanosome infection rates in tsetse flies (Diptera: Glossinidae) and cattle during control operations in the Kagera river region in Rwanda. *Bulletin of Entomological Research*, 82, 361–367.

Mihok, S., Stiles, J.K., Mpanga, E. and Olubayo, R.O. (1994a) Relationships between protease activity, host blood and infection rates in *Glossina morsitans* spp. infected with *Trypanosoma congolense*, *T. brucei* and *T. simiae*. *Medical and Veterinary Entomology*, 8, 47–50.

Mihok, S., Zweygarth, E., Munyoki, E.N., Wambua, J. and Kock, R. (1994b) *Trypanosoma simiae* in the white rhinoceros (*Ceratotherium simum*) and the dromedary camel (*Camelus dromedarius*). *Veterinary Parasitology*, 53, 191–196.

Mihok, S., Machika, C., Darji, N., Kang'ethe, E.K. and Otieno, L.H. (1995a) Relationships between host blood factors and proteases in *Glossina morsitans* subspecies infected with *Trypanosoma congolense*. *Medical and Veterinary Entomology*, 9, 155–160.

Mihok, S., Maramba, O., Munyoki, E. and Kagoiya, J. (1995b) Mechanical transmission of *Trypanosoma* spp. by African Stomoxyinae (Diptera: Muscidae). *Tropical Medicine and Parasitology*, 46, 103–105.

Milligan, P.J.M. (1990) Modelling trypanosomiasis transmission. *Insect Science and its Application* 11, 301–307.

Milligan, P.J.M. and Baker, R.D. (1988) A model of tsetse-transmitted animal trypanosomiasis. *Parasitology*, 96, 211–139.

Milligan, P.J.M., Maudlin, I. and Welburn, S.C. (1995) *Trypanozoon*: Infectivity to humans is linked to reduced transmissibility in tsetse II. Genetic mechanisms. *Experimental Parasitology*, 81, 409–415.

Miltgen, F. and Landau, I. (1982) *Culicoides nubeculosus*, vecteur expérimental d'un nouveau trypanosome de Psittaciforme: *Trypanosoma bakeri* n. sp. *Annales de Parasitologie Humaine et Comparée*, 57, 423–428.

Minchin, E.A. (1905) Report on the anatomy of the tsetse-fly (*Glossina palpalis*). *Proceedings of the Royal Society of London (B)*, 76, 531–547.

Minter, D.M. and Goedbloed, E. (1971) The preservation in liquid nitrogen of tsetse flies and phlebotomine sandflies naturally infected with trypanosomatid flagellates. *Transactions of the Royal Society of Tropical Medicine and Hygiene*, 65, 175–181.

Minter-Goedbloed, E. and Minter, D.M. (1989) Salivary gland hyperplasia and trypanosome infection of *Glossina* in two areas of Kenya. *Transactions of the Royal Society of Tropical Medicine and Hygiene*, 83, 640–641.

Mitchell, B.K. (1976) ATP reception by the tsetse fly, *Glossina morsitans* West. *Experientia*, 32, 192–194.

Moggridge, J.Y. (1936a) Experiments on the crossing of open spaces by *Glossina swynnertoni*. *Bulletin of Entomological Research*, 27, 435–448.

Moggridge, J.Y. (1936b) Some observations on the seasonal spread of *Glossina pallidipes* in Italian Somaliland with notes on *G. brevipalpis* and *G. austeni*. *Bulletin of Entomological Research*, 27, 449–466.

Moggridge, J.Y. (1948) Night activity of *Glossina* on the Kenya coast. *Proceedings of the Royal Entomological Society of London (A)*, 23, 87–92.

Moggridge, J.Y. (1949a) Observations on the control of Kenya coast *Glossina*. *Bulletin of Entomological Research*, 40, 345–349.

Moggridge, J.Y. (1949b) Climate and the activity of the Kenya coastal *Glossina*. *Bulletin of Entomological Research*, 40, 307–321.

Mohamed-Ahmed, M. and Odulaja, A. (1997) Diel activity patterns and host preferences of *Glossina fuscipes fuscipes* (Diptera: Glossinidae) along the shores of Lake Victoria, Kenya. *Bulletin of Entomological Research*, 87, 179–186.

Mohamed-Ahmed, M.M., Rahman, A.H. and Abdel Karim, E.I. (1992) Multiple drug-resistant bovine trypanosomes in South Darfur province, Sudan. *Tropical Animal Health and Production*, 24, 179–181.

Moloo, S.K. (1971) Oocyte differentiation and vitellogenesis in *G. morsitans*. *Acta Tropica*, 28, 334–339.

Moloo, S.K. (1973) A new trap for *Glossina pallidipes* Aust. and *G. fuscipes* Newst. (Diptera: Glossinidae). *Bulletin of Entomological Research*, 63, 231–236.

Moloo, S.K. (1976a) Nutrition of *Glossina morsitans*: metabolism of U-^{14}C glucose during pregnancy. *Journal of Insect Physiology*, 22, 195–200.

Moloo, S.K. (1976b) Aspects of the nutrition of adult female *Glossina morsitans* during pregnancy. *Journal of Insect Physiology*, 22, 563–567.

Moloo, S.K. (1976c) Storage of nutriments by adult female *Glossina morsitans* and their transfer to intra-uterine larva. *Journal of Insect Physiology*, 22, 1111–1115.

Moloo, S.K. (1978) Excretion of uric acid and amino acids during diuresis in the adult female *Glossina morsitans*. *Acta Tropica*, 35, 247–252.

Moloo, S.K. (1980) Interacting factors in the epidemiology of trypanosomiasis in an endemic/enzootic region of Uganda and its contiguous area of Kenya. *Insect Science and its Application*, 1, 117–121.

Moloo, S.K. (1981) Effects of maintaining *Glossina morsitans morsitans* on different hosts upon the vector's subsequent infection rates with pathogenic trypanosomes. *Acta Tropica*, 38, 125–136.

Moloo, S.K. (1982a) Studies on the infection rates of a West African stock of *Trypanosoma vivax* in *Glossina morsitans morsitans* and *G. m. centralis*. *Annals of Tropical Medicine and Parasitology*, 76, 355–359.

Moloo, S.K. (1982b) Cyclical transmission of pathogenic *Trypanosoma* species by irradiated *Glossina morsitans morsitans*. *Parasitology*, 84, 289–296.

Moloo, S.K. (1983) Feeding behaviour of *Glossina morsitans morsitans* infected with *Trypanosoma vivax, T. congolense* or *T. brucei. Parasitology*, 86, 51–56.

Moloo, S.K. (1985) Distribution of *Glossina* species in Africa. *Acta Tropica*, 42, 275–281.

Moloo, S.K. (1993a) The distribution of *Glossina* species in Africa and their natural hosts. *Insect Science and its Application*, 14, 511–527.

Moloo, S.K. (1993b) A comparison of the susceptibility of two allopatric populations of *Glossina pallidipes* for stocks of *Trypanosoma congolense*. *Medical and Veterinary Entomology*, 7, 369–372.

Moloo, S.K. and Gooding, R.H. (1995) A comparison of *Glossina morsitans centralis* originating from Tanzania and Zambia, with respect to vectorial competence for pathogenic *Trypanosoma* species, genetic variation and inter-colony fertility. *Medical and Veterinary Entomology*, 9, 365–371.

Moloo, S.K. and Gray, M.A. (1989) New observations on cyclical development of *Trypanosoma vivax* in *Glossina. Acta Tropica*, 46, 167–172.

Moloo, S.K. and Kamunya, G.W. (1987) Suppressive action of Samorin on the cyclical development of pathogenic trypanosomes in *Glossina morsitans centralis. Medical and Veterinary Entomology*, 1, 285–287.

Moloo, S.K. and Kutuza, S.B. (1970) Feeding and crop emptying in *Glossina brevipalpis* Newstead. *Acta Tropica Sepparatum*, 27, 355–377.

Moloo, S.K. and Kutuza, S.B. (1984) Vectorial capacity of gamma-irradiated sterile male *Glossina morsitans centralis, G. austeni* and *G. tachinoides* for pathogenic *Trypanosoma* species. *Insect Science and its Application*, 5, 411–414.

Moloo, S.K. and Kutuza, S.B. (1985) Survival and reproductive performance of female *Glossina morsitans* when maintained on livestock infected with salivarian trypanosomes. *Annals of Tropical Medicine and Parasitology*, 79, 223–224.

Moloo, S.K. and Kutuza, S.B. (1986) Effect of Samorin administered to a bovine host on the survival and reproductive performance of female *Glossina morsitans centralis. Annals of Tropical Medicine and Parasitology*, 81, 743–744.

Moloo, S.K. and Kutuza, S.B. (1988) Comparative study on the susceptibility of different *Glossina* species to *Trypanosoma brucei brucei* infection. *Tropical Medicine and Parasitology*, 39, 211–213.

Moloo, S.K. and Okumu, I.O. (1995) A comparison of susceptibility to stocks of *Trypanosoma vivax* of *Glossina pallidipes* from allopatric populations in Kenya. *Medical and Veterinary Entomology*, 9, 202–204.

Moloo, S.K. and Shaw, M.K. (1989) Rickettsial infections of midgut cells are not associated with susceptibility of *Glossina morsitans centralis* to *Trypanosoma congolense* infection. *Acta Tropica*, 46, 223–227.

Moloo, S.K., Steiger, R.F. and Hecker, H. (1970) Ultrastructure of the peritrophic membrane formation in *Glossina* Wiedemann. *Acta Tropica*, 27, 378–383.

Moloo, S.K., Steiger, R.F., Brun, R. and Boreham, P.F.L. (1971) Sleeping sickness survey in the Musoma district Tanzania. *Acta Tropica*, 28, 189–205.

Moloo, S.K., Kutuza, S.B. and Boreham, P.F. (1980) Studies on *Glossina pallidipes, G. fuscipes fuscipes* and *G. brevipalpis* in terms of the epidemiology and epizootiology of trypanosomiases in south-eastern Uganda. *Annals of Tropical Medicine and Parasitology*, 74, 219–237.

Moloo, S.K., Dar, F. and Kamunya, G.W. (1982) The transmission of mixed infections of pathogenic trypanosome species to susceptible hosts by *G. morsitans morsitans*. *Acta Tropica*, 39, 303–306.

Moloo, S.K., Asonganyi, T. and Jenni, L. (1986) Cyclical development of *Trypanosoma brucei gambiense* from cattle and goats in *Glossina*. *Acta Tropica*, 43, 407–408.

Moloo, S.K., Kutuza, S.B. and Desai, J. (1987) Comparative study on the infection rates of different *Glossina* species for east and west African *Trypanosoma vivax* stocks. *Parasitology*, 95, 537–542.

Moloo, S.K., Grootenhuis, J.G., Kar, K. and Karstad, L. (1988) Survival and reproductive performance of female *Glossina morsitans morsitans* when maintained on the blood of different species of wild animals. *Medical and Veterinary Entomology*, 2, 347–350.

Moloo, S.K., Sabwa, C.L. and Kabata, J.M. (1992a) Vector competence of *Glossina pallidipes* and *G. morsitans centralis* for *Trypanosoma vivax*, *T. congolense* and *T. b. brucei*. *Acta Tropica*, 51, 271–280.

Moloo, S.K., Olubayo, R.O., Kabata, J.M. and Okumu, I.O. (1992b) A comparison of African buffalo, N'Dama and Boran cattle as reservoirs of *Trypanosoma congolense* for different *Glossina* species. *Medical and Veterinary Entomology*, 6, 225–230.

Moloo, S.K., Gettinby, G., Olubayo, R.O., Kabata, J.M. and Okumu, I.O. (1993) A comparison of African buffalo, N'Dama and Boran cattle as reservoirs of *Trypanosoma vivax* for different *Glossina* species. *Parasitology*, 106, 277–282.

Moloo, S.K., Kabata, J.M. and Sabwa, C.L. (1994a) A study on the maturation of procyclic *Trypanosoma brucei brucei* in *Glossina morsitans centralis* and *G. brevipalpis*. *Medical and Veterinary Entomology*, 8, 369–374.

Moloo, S.K., Zweygarth, E. and Sabwa, C.L. (1994b) Comparative study on the susceptibility of different laboratory strains of *Glossina* species to *Trypanosoma simiae*. *Medical and Veterinary Entomology*, 8, 225–230.

Molyneux, D.H. (1973a) *Trypanosoma everetti* sp. nov. a trypanosome from the black-rumped waxbill *Estrilda t. troglodytes* Lichtenstein. *Annals of Tropical Medicine and Parasitology*, 67, 219–222.

Molyneux, D.H. (1973b) Experimental infections of avian trypanosomes in *Glossina*. *Annals of Tropical Medicine and Parasitology*, 67, 223–228.

Molyneux, D.H. (1976) Vector relationships in the Trypanosomatidae. *Advances in Parasitology*, 15, 1–82.

Molyneux, D.H. (1977) The attachment of *Trypanosoma grayi* in the hindgut of *Glossina*. *Protozoology*, 111, 83–86.

Molyneux, D.H. (1980a) Host–trypanosome interactions in *Glossina*. *Insect Science and its Application*, 1, 39–46.

Molyneux, D.H. (1980b) Patterns of development of trypanosomes and related parasites in insect hosts. In: *Isotope and Radiation Research on Animal Diseases and their Vectors*. IAEA, Vienna, pp. 179–190.

Molyneux, D.H. (1980c) Animal reservoirs and residual foci of *Trypanosoma brucei gambiense* sleeping sickness in west Africa. *Insect Science and its Application*, 1, 59–63.

Molyneux, D.H. (1983) Selective primary health care: Strategies for control of disease in the developing World. VIII. African trypanosomiasis. *Review of Infectious Diseases*, 5, No. 5, 945–956.

Molyneux, D.H. and Jefferies, D. (1986) Feeding behaviour of pathogen-infected vectors. *Parasitology*, 92, 721–736.

Molyneux, D.H. and Jenni, L. (1981) Mechanoreceptors, feeding behaviour and trypanosome transmission in *Glossina*. *Transactions of the Royal Society of Tropical Medicine and Hygiene*, 75, 160–163.

Molyneux, D.H. and Stiles, J.K. (1991) Trypanosomatid – Vector interactions. *Annales de la Société Belge de Médecine Tropicale*, 71 (Suppl. 1), 151–166.

Molyneux, D.H., Baldry, D.A.T., Van Wettere, P., Takken, W. and de Raadt, P. (1978a) The experimental application of insecticides from a helicopter for the control of riverine populations of *G. tachinoides* in west Africa 1. Objectives, experimental area and insecticides evaluated. *PANS*, 24, 391–403.

Molyneux, D.H., Baldry, D.A.T., de Raadt, P., Lee, C.W. and Hamon, J. (1978b) Helicopter application of insecticides for the control of riverine *Glossina* vectors of African human trypanosomiasis in the moist savanna zones. *Annales de la Société Belge de Médecine Tropicale*, 58, 185–203.

Molyneux, D.H., Baldry, D.A.T. and Fairhurst, C. (1979a) Tsetse movement in wind fields: Possible epidemiological and entomological implications for trypanosomiasis and its control. *Acta Tropica*, 36, 53.

Molyneux, D.H., Lavin, D.R. and Elce, B. (1979b) A possible relationship between salivarian trypanosomes and *Glossina* labrum mechano-receptors. *Annals of Tropical Medicine and Parasitology*, 73, 288–290.

Montgomery, R.E. and Kinghorn, A. (1908) A report on trypanosomiasis of domestic stock in north-western Rhodesia. *Annals of Tropical Medicine and Parasitology*, 2, 97–132.

Morris, K.R.S. (1934) The bionomics and importance of *Glossina longipalpis*, Wied. in the Gold Coast. *Bulletin of Entomological Research*, 26, 309–335.

Morris, K.R.S. (1949) Planning the control of sleeping sickness. *Transactions of the Royal Society of Tropical Medicine and Hygiene*, 43, 165–198.

Morris, K.R.S. (1951) The ecology of epidemic sleeping sickness. I. The significance of location. *Bulletin of Entomological Research*, 42, 427–433.

Morris, K.R.S. (1952) The ecology of epidemic sleeping sickness. II. The effects of an epidemic. *Bulletin of Entomological Research*, 43, 375–397.

Morris, K.R.S. (1960) Studies on the epidemiology of sleeping sickness in East Africa. II. Sleeping sickness in Kenya. *Transactions of the Royal Society of Tropical Medicine and Hygiene*, 54, No. 1, 71–86.

Morris, K.R.S. (1962a) The food of *Glossina palpalis* (R-D) and its bearing on the control of sleeping sickness in forest country. *Journal of Tropical Medicine and Hygiene*, 65, 12–23.

Morris, K.R.S. (1962b) The epidemiology of sleeping sickness in East Africa. V. Epidemics on the Albert Nile. *Transactions of the Royal Society of Tropical Medicine and Hygiene*, 56, 316–338.

Morris, K.R.S. (1963) The movement of Sleeping sickness across central Africa. *Journal of Tropical Medicine and Hygiene*, 66, 59–76.

Morris, K.R.S. and Morris, M.G. (1949) The use of traps against tsetse in west Africa. *Bulletin of Entomological Research*, 39, 491–523.

Mortelmans, J. (1984) Socio-economic problems related to animal trypanosomiasis in Africa. *Social Sciences and Medicine*, 19, 1105–1107.

Morzaria, S.P., Latif, A.A., Jongejan, F. and Walker, A.R. (1986) Transmission of a *Trypanosoma* sp. to cattle by the tick *Hyalomma anatolicum anatolicum*. *Veterinary Parasitology*, 19, 13–21.

Moser, D.R., Cook, G.A., Ochs, D.E., Bailey, C.P., McKane, M.R. and Donelson J.E. (1989) Detection of *Trypanosoma congolense* and *Trypanosoma brucei* subspecies by DNA amplification using the polymerase-catalyzed chain reaction. *Parasitology*, 99, 57–66.

Mott, K.E., Nuttall, I., Desjeux, P. and Cattand, P. (1995) New geographical approaches to control of some parasitic zoonoses. *Bulletin of the World Health Organisation*, 73, 247–257.

Msangi, A. and Lehane, M.J. (1991) A method for determining the age of very young tsetse flies (Diptera: Glossinidae) and an investigation of the factors determining head fluorescent levels in newly emerged adults. *Bulletin of Entomological Research*, 81, 185–188.

Mshelbwala, A.S. (1972) *Trypanosoma brucei* infections in the haemocoel of tsetse flies. *Transactions of the Royal Society of Tropical Medicine and Hygiene*, 66, 637–643.

Mulatu, W., Swallow, B.M., Rowlands, G.J., Leak, S.G.A., D'Ieteren, G. and Nagda, S.M. (1997) Economic benefits to farmers of six years of application of an insecticidal 'pour-on' to control tsetse in Ghibe, southwest Ethiopia. In: *24th Meeting of the International Scientific Council for Trypanosomosis Research and Control, Maputo* (in press).

Mulla, A.F. and Rickman, L.R. (1988) The isolation of human serum-resistant *Trypanosoma* (Trypanozoon) species from Zebra and Impala in Luangwa valley, Zambia. *Transactions of the Royal Society of Tropical Medicine and Hygiene*, 82, 718.

Mulligan, H.W. (ed.) (1970) *The African Trypanosomiases*. Ministry of Overseas Development and George Allen and Unwin, London, 950 pp.

Mungomba, L.M., Molyneux, D.H. and Wallbanks, K.R. (1987) A record of trypanosomes from *Ixodes ricinus* in Britain. *Medical and Veterinary Entomology*, 1, 435–437.

Munsterman, S., Mbura, R.J., Maloo, S.H. and Löhr, C. (1992) Trypanosomiasis control in Boran cattle in Kenya: a comparison between chemoprophylaxis and a parasite detection and treatment. *Tropical Animal Health and Production*, 24, 17–27.

Muranjan, M., Wang, Q., Li, Y.L., Hamilton, E., Otieno-Omondi, F.P., Wang, J., Van Praagh, A., Grootenhuis, J.G. and Black, S.J. (1997) The trypanocidal Cape Buffalo serum protein is xanthine oxidase. *Infection and Immunity*, 65, 3806–3814.

Murlis, J., Elkington, J.S. and Cardé, R.T. (1992) Odor plumes and how insects use them. *Annual Review of Entomology*, 37, 505–532.

Murray, M. and Black, S.J. (1985) African trypanosomiasis in cattle: Working with nature's solution. *Veterinary Parasitology*, 18, 167–182.

Murray, M. and Dexter, T.M. (1988) Anaemia in bovine African trypanosomiasis. *Acta Tropica*, 45, 389–432.

Murray, M. and Gray, A.R. (1984) The current situation on animal trypanosomiasis in Africa. *Preventive Veterinary Medicine*, 2, 23–30.

Murray, M., Clifford, D.J., Gettinby, G., Snow, W.F. and McIntyre, W.I.M. (1981a) Susceptibility to African trypanosomiasis of N'Dama and Zebu cattle in an area of *Glossina morsitans submorsitans* challenge. *The Veterinary Record*, 109, 503–510.

Murray, M., Grootenhuis, J.G., Akol, G.W.O., Emery, D.L., Shapiro, S.Z., Moloo, S.K., Dar, F., Bovell, D.L. and Paris, J. (1981b) Potential application of research on African trypanosomiases in wildlife and preliminary studies on animals exposed to tsetse infected with *Trypanosoma congolense*. *ISCTRC 16th Meeting Nairobi*, 265–270.

Murray, M., Trail, J.C.M., Turner, D.A. and Wissocq, Y. (1983) *Livestock Productivity and Trypanotolerance. Network Training Manual.* International Livestock Centre for Africa, Addis Ababa, 198 pp.

Murray, M., Hirumi, H. and Moloo, S.K. (1985) Suppression of *Trypanosoma congolense, T. vivax* and *T. brucei* infection rates in tsetse flies maintained on goats immunized with uncoated forms of trypanosomes grown *in vitro. Parasitology,* 91, 53–66.

Murray, M., Trail, J.C.M. and D'Ieteren, G. (1990) Trypanotolerance in cattle and prospects for the control of trypanosomiasis by selective breeding. *Revue Scientifique et Technique Office International des Epizooties,* 9, 369–386.

Murray, M., Stear, M.J., Trail, J.C.M., D'Ieteren, G.D.M., Agyemang, K. and Dwinger, R.H. (1991) Trypanosomiasis in cattle: prospects for control. In: Owen, J.B. and Axford, R.F.E. (eds) *Breeding for Disease Resistance in Farm Animals.* CAB International, Wallingford, pp. 203–223.

Murray, W.A. (1921) History of the introduction and spread of human trypanosomiasis (sleeping sickness) in British Nyasaland in 1908 and following years. *Transactions of the Royal Society of Tropical Medicine and Hygiene,* 15, 121–128.

Mutayoba, B.M., Gombe, S., Waindi, E.N. and Kaaya, G.P. (1989) Comparative trypanotolerance of the small east African breed of goats from different localities to *Trypanosoma congolense* infection. *Veterinary Parasitology,* 31, 95–105.

Mwambo, H.A., Mella, P.N.P. and Lekaki, K.A. (1988) Berenil (diminazene aceturate)-resistant *Trypanosoma congolense* in cattle under natural tsetse challenge at Kibaha, Tanzania. *Acta Tropica,* 45, 239–244.

Mwambu, P.M. and Mayende, J.S.P. (1971) Berenil resistant *Trypanosoma vivax,* isolated from naturally infected cattle in Teso District, eastern Uganda. In: *Proceedings of the International Scientific Council for Trypanosomiasis Research and Control (ISCTRC) 13th Meeting, Lagos, Nigeria,* Publication No. 105, pp. 133–138.

Mwambu, P.M. and Woodford, M.H. (1972) Trypanosomes from game animals of the Queen Elizabeth National park, western Uganda. *Tropical Animal Health and Production,* 4, 152–155.

Mwangelwa, M.I., Otieno, L.H. and Reid, G. (1987) Some barriers to *Trypanosoma congolense* development in *Glossina morsitans morsitans. Insect Science and its Application,* 8, 33–37.

Mwangelwa, M.I., Dransfield, R.D., Otieno, L.H. and Mbata, K.J. (1990) Distribution and diel activity patterns of *Glossina fuscipes* Newst. on Rusinga island and mainland in Mbita, Kenya. *Insect Science and its Application,* 11, 315–322.

Mwangi, E., Stevenson, P., Gettinby, G. and Murray, M. (1993) Variation in susceptibility to tsetse-borne trypanosomiasis among *Bos indicus* cattle breeds in east Africa. In: Sones, K. (ed.) *Proceedings of the 22nd Meeting of the International Scientific Council for Trypanosomiasis Research and Control.* Publication no. 117, pp. 125–128.

Na'isa, B.K. (1967) Follow-up survey on the prevalence of homidium-resistant strains of trypanosomes in cattle in northern Nigeria and drug cross-resistance tests on the strains with Samorin and Berenil. *Bulletin of Epizootic Diseases in Africa,* 15, 231–241.

Na'isa, B.K. (1969) The protection of 60 Nigerian trade cattle from trypanosomiasis using Samorin. *Bulletin of Epizootic Diseases in Africa,* 17, 45–54.

Nantulya, V.M. (1989) An antigen detection enzyme immunoassay for the diagnosis of *T. rhodesiense* Sleeping sickness. *Parasite Immunology,* 11, 69–75.

Nantulya, V.M. (1990) Trypanosomiasis in domestic animals: the problems of diagnosis. *Revue Scientifique et Technique de l'Office International des Epizooties*, 9, 357–367.

Nantulya, V.M. (1993) The development of latex-agglutination antigen tests for diagnosis of African trypanosomiasis. In: *ISCTRC/OAU 22nd meeting Kampala, Nov. 1993*, p. 37.

Nantulya, V.M. (1994) Suratex: A simple latex agglutination antigen test for diagnosis of *Trypanosoma evansi* infections (Surra). *Tropical Medicine and Parasitology*, 45, 9–12.

Nantulya, V.M. and Moloo, S.K. (1988) Suppression of cyclical development of *Trypanosoma brucei brucei* in *Glossina morsitans centralis* by an anti-procyclics monoclonal anti-bodies. *Acta Tropica*, 45, 137–144.

Nantulya, V.M., Bajana Songa, E. and Hamers, R. (1989) Detection of circulating trypanosomal antigens in *Trypanosoma evansi*-infected animals using a *T. brucei* group-specific monoclonal. *Tropical Medicine and Parasitology*, 40, 263–266.

Nash, T.A.M. (1930) A contribution to our knowledge of the bionomics of *Glossina morsitans*. *Bulletin of Entomological Research*, 21, 201–256.

Nash, T.A.M. (1931) The relationship between *Glossina morsitans* and the evaporation rate. *Bulletin of Entomological Research*, 22, 383–384.

Nash, T.A.M. (1933a) A statistical analysis of the climatic factors influencing the density of tsetse flies, *Glossina morsitans* Westw. *Journal of Animal Ecology*, 2, 197–203.

Nash, T.A.M. (1933) The ecology of *Glossina morsitans* Westw. and two possible methods for its destruction. Part II. *Bulletin of Entomological Research*, 24, 107–157.

Nash, T.A.M. (1933c) The ecology of *Glossina morsitans* Westw. and two possible methods for its destruction. Part III. *Bulletin of Entomological Research*, 24, 163–195.

Nash, T.A.M. (1936) The relationship between the maximum temperature and the seasonal longevity of *Glossina submorsitans* Westw., and *G. tachinoides* in Northern Nigeria. *Bulletin of Entomological Research*, 27, 273–279.

Nash, T.A.M. (1937) Climate: the vital factor in the ecology of *Glossina. Bulletin of Entomological Research*, 28, 75–131.

Nash, T.A.M. (1939) The ecology of the puparium of *Glossina* in northern Nigeria. *Bulletin of Entomological Research*, 30, 259–284.

Nash, T.A.M. (1940) The effect upon *Glossina* of changing the climate in the true habitat by partial clearing of vegetation. *Bulletin of Entomological Research*, 31, 69–84.

Nash, T.A.M. (1941) Bats as a source of food for *Glossina morsitans* and *G. tachinoides. Bulletin of Entomological Research*, 32, 249.

Nash, T.A.M. (1944a) The control of sleeping sickness in the *Raphia* pole trade. *Bulletin of Entomological Research*, 35, 49.

Nash, T.A.M. (1944b) A low density of tsetse flies associated with a high incidence of sleeping sickness. *Bulletin of Entomological Research*, 35, 51–59.

Nash, T.A.M. (1947) A record of *Syntomosphyrum glossinae* from Nigeria. *Bulletin of Entomological Research*, 38, 525.

Nash, T.A.M. (1948) *Tsetse Flies in British West Africa.* Colonial Office, 1–75, HMSO, London.

Nash, T.A.M. (1952) Some observations on resting tsetse-fly populations, and evidence that *Glossina medicorum* is a carrier of trypanosomes. *Bulletin of Entomological Research*, 43, 33–42.

Nash, T.A.M. (1960) A review of the African trypanosomiasis problem. *Tropical Diseases Bulletin*, 57, 973–1003.

Nash, T.A.M. and Davey, J.T. (1950) The resting habits of *Glossina medicorum*, *G. fusca* and *G. longipalpis*. *Bulletin of Entomological Research*, 41, 153–157.

Nash, T.A.M. and Jordan, A.M. (1958) A guide to the identification of the west African species of the *fusca* group of tsetse-flies, by dissection of the genitalia. *Annals of Tropical Medicine and Parasitology*, 53, 72–88.

Nash, T.A.M. and Page, W.A. (1953) On the ecology of *Glossina palpalis* in northern Nigeria. *Transactions of the Royal Entomological Society of London*, 104, 71–170.

Nash, T.A.M. and Trewern, M.A. (1972) Diurnal rhythm of larviposition by *Glossina morsitans orientalis* Vanderplank and *G. austeni* Newst. *Transactions of the Royal Society of Tropical Medicine and Hygiene*, 66, 309.

Nash, T.A.M., Page, W.A., Jordan, A.M. and Petana, W. (1958) The rearing of *Glossina palpalis* in the laboratory for experimental work. *ISCTRC 7th Meeting*, Publication no. 41, pp. 343–350.

Nash, T.A.M., Trewern, M.A. and Moloo, S.K. (1976) Observations on the free larval stage of *Glossina morsitans morsitans* West. (Diptera, Glossinidae): the possibility of a larval pheromone. *Bulletin of Entomological Research*, 66, 17–24.

Ndegwa, P., Irungu, L. and Moloo, S.K. (1992) Effect of puparia incubation temperature: increased infection rates of *Trypanosoma congolense* in *Glossina morsitans centralis*, *G. fuscipes fuscipes*. *Medical and Veterinary Entomology*, 6, 127–130.

Ndoutamia, G., Moloo, S.K., Murphy, N.B. and Peregrine, A.S. (1993) Derivation and characterization of a quinapyramine-resistant clone of *Trypanosoma congolense*. *Antimicrobial Agents and their Chemotherapy*, 37, 1163–1166.

Neave, S.A. (1912) Notes on the blood sucking insects of Eastern tropical Africa. *Bulletin of Entomological Research*, 3, 275–323.

Négrin, M. and MacLennan, K.J.R. (1977) Economic assessment of tsetse control in Botswana. *FAO Mission report* No. 1, AGDP/RAF/75/001.

Nehili, M., Ilk, C., Melhorn, H., Ruhnau, K., Dick, W. and Njayou, M. (1994) Experiments on the possible role of leeches as vectors of animal and human pathogens: a light and electron microscope study. *Parasitology Research*, 80, 277–290.

Nelson, D.R. and Carlson, D.A. (1986) Cuticular hydrocarbons of the tsetse flies *Glossina morsitans morsitans*, *G. austeni*, and *G. pallidipes*. *Insect Biochemistry*, 16, 403–416.

Newstead, R. (1911a) On the genital armature of the males of *Glossina medicorum* Austen and *Glossina tabaniformis* Westwood. *Bulletin of Entomological Research*, 2, 107–110.

Newstead, R. (1911b) A revision of the tsetse-flies (*Glossina*) based on a study of the male genital armature. *Bulletin of Entomological Research*, 2, 9–36.

Newstead, R. (1912) A new tsetse-fly from British East Africa. *Annals of Tropical Medicine and Parasitology*, 6, 129–130.

Newstead, R., Evans, A.M. and Potts, W.H. (1924) *Guide to the Study of Tsetse Flies*. Memoirs of the Liverpool School of Tropical Medicine, University of Liverpool Press, Liverpool, UK, pp. 1–332.

Newton, B.A., Cross, G.A.M. and Baker, J.R. (1973a) Differentiation in trypanosomatidae. *Symposium of the Society for General Microbiology*, 23, 339–373.

Newton, B.A., Steinert, M. and Borst, P. (1973b) Differentiation of haemoflagellate species by hybridisation of complementary RNA and kinetoplast DNA. *Transactions of the Royal Society of Tropical Medicine and Hygiene*, 67, 259.

Ngeranwa, J.J.N., Mutiga, E.R., Agumboli, G.J.O., Gathumbi, P.K. and Munyua, W.K. (1991) The effects of experimental *Trypanosoma* (*Trypanozoon*) (*brucei*) *evansi* infection on the fertility of male goats. *Veterinary Research Communications*, 15, 301–308.

Nitcheman, S. (1988) Comparaison des longévités des Glossines (*Glossina morsitans morsitans* Westwood, 1850) infectées par les trypanosomes (*Trypanosoma* (*Nannomonas*) *congolense* Broden, 1904) et des glossines saines. *Annales de Parasitologie Humaine et Comparée*, 63, 163–164.

Nitcheman, S. (1990) Comparison of the susceptibility to deltamethrin of female *Glossina morsitans morsitans* Westwood, 1850 (Diptera: Glossinidae) uninfected and infected with *Trypanosoma* (*Nannomonas*) *congolense* Broden, 1904 (Kinetoplastida, Trypanosomatidae). *Annals of Tropical Medicine and Parasitology*, 84, 483–491.

Njagi, E.N.M., Olembo, N.K. and Pearson, D.J. (1992) Proline transport by tsetse fly *Glossina morsitans* flight muscle mitochondria. *Comparative Biochemistry and Physiology*, 102B, 579–584.

Njau, B.C., Mkonyi, P.M. and Kundy, D.J. (1981) Berenil resistant *Trypanosoma congolense* isolated from naturally infected goats in Tanga Region, Tanzania. In: *Proceedings of the International Scientific Council for Trypanosomiasis Research and Control (ISCTRC) 17th Meeting, Arusha, Tanzania*. Publication no. 112, pp. 289–298.

Njogu, A.R., Dolan, R.B., Wilson, A.J. and Sayer, P.D. (1985) Trypanotolerance in east African Orma Boran cattle. *Veterinary Record*, 117, 632–636.

Nogge, G. (1976) Sterility in tsetse flies caused by loss of symbionts. *Experientia*, 32, 995.

Nogge, G. (1978) Aposymbiotic tsetse flies, *Glossina morsitans morsitans* obtained by feeding on rabbits immunized specifically with symbionts. *Journal of Insect Physiology*, 24, 299–304.

Nogge, G. (1981) Significance of symbionts for the maintenance of an optimal nutritional state of successful reproduction in haematophagous arthropods. *Parasitology*, 82, 299–304.

Nogge, G. and Giannetti, M. (1980) Specific antibodies: a potential insecticide. *Science*, 209, 1028–1029.

Nogge, G. and Ritz, R. (1982) Number of symbionts and its regulation in tsetse flies, *Glossina* spp. *Entomologia Experimentalis et Applicata*, 31, 249–254.

Noireau, F., Gouteux, J.P., Toudic, A., Samba, F. and Frézil, J.L. (1986) Importance épidémiologique du réservoir animal à *Trypanosoma brucei gambiense* au Congo. 1. Prévalance des trypanosomoses animales dans les foyers de maladie du sommeil. *Tropical Medicine and Parasitology*, 37, 393–398.

Noireau, F., Gouteux, J.P., Paindavoine, P., Lemesre, J.L., Toudic, A., Pays, E., Steinert, M. and Frézil, J.L. (1989) The epidemiological importance of the animal reservoir of *Trypanosoma brucei gambiense* in the Congo. *Tropical Medicine and Parasitology*, 40, 9–11.

Nolan, R.A. (1977) Pathogens of *Glossina*. *Bulletin of the World Health Organisation*, 55 (Suppl.), 265–270.

Noller, W. (1925) Der Nachweis des Ubertragers des gemeiner Rindertrypanosomas *Trypanosoma theileri* mit hilpe der Kurturverfahrens. *Zentralblatt für Bakteriologie Parasitenkunde*, 79, 133–142.

Norden, D.A. and Matanganyidze, C. (1977) Some properties of a mitochondrial malic enzyme from the flight muscle of the tsetse fly (*Glossina*). *Insect Biochemistry*, 7, 215–222.

Norden, D.A. and Paterson, D.J. (1969) Carbohydrate metabolism in flight muscle of the tsetse fly (*Glossina*) and the blowfly (*Sarcophaga*). *Comparative Biochemistry and Physiology*, 31, 819–827.

Norden, D.A. and Venturas, D.J. (1972) Substances affecting the activity of proline dehydrogenase in the sarcosomes of the tsetse fly (*Glossina*) and a comparison with some other insects. *Insect Biochemistry*, 2, 226–234.

Nuttall, G.H.F. (1904) *Blood Immunity and Blood Relationships*. Cambridge University Press, Cambridge.

Nyeko, J.H.P., Ole-MoiYoi, O.K., Majiwa, P.A.O., Otieno, L.H. and Ociba, P.M. (1990) Characterisation of trypanosome isolates from cattle in Uganda using species-specific DNA probes reveals predominance of mixed infections. *Insect Science and its Application*, 11, 271–280.

Odhiambo, T.R. (1971) The regulation of ovulation in the tsetse-fly *Glossina pallidipes* Austen. *Journal of Experimental Zoology*, 177, 447–454.

Odindo, M.O. and Hominick, W.M. (1985) *Hexameris glossinae* (Nematoda: Mermithidae) parasitizing tsetse (Diptera: Glossinidae) in Kenya. *Entomophaga*, 30, 199–205.

Odindo, M.O., Sabwa, D.M., Amutalla, P.A. and Otieno, W.A. (1981) Preliminary tests on the transmission of virus-like particles to the tsetse *Glossina pallidipes*. *Insect Science and its Application*, 2, 219–221.

Offor, M.I.I., Carlson, D.A., Gadzama, N.M. and Bozimo, H.T. (1981) Sex recognition pheromone in the west African tsetse fly *Glossina palpalis palpalis* (Robineau-Desvoidy). *Insect Science and its Application*, 1, 417–420.

Okiwelu, S.N. (1976) Resting sites of *Glossina morsitans morsitans* during the dry season in the Republic of Zambia. *Bulletin of Entomological Research*, 66, 413–419.

Okiwelu, S.N. (1977a) Insemination, pregnancy and suspected abortion rates in a natural population of *Glossina morsitans morsitans* (Diptera: Glossinidae) in the Republic of Zambia. *Journal of Medical Entomology*, 14, 19–23.

Okiwelu, S.N. (1977b) Host preference and trypanosome infection rates of *Glossina morsitans morsitans* Westwood in the Republic of Zambia. *Annals of Tropical Medicine and Parasitology*, 71, 101–107.

Okiwelu, S.N. (1981) Resting site preferences of the tsetse *Glossina palpalis gambiensis* Vanderplank (Diptera: Glossinidae) in Mali. *Insect Science and its Application*, 1, 289–294.

Okiwelu, S.N. Van Wettere, Maiga, S., Bouare, S. and Crans, W. (1981) Contribution to the distribution of *Glossina* (Diptera: Glossinidae) in Mali. *Bulletin of Entomological Research*, 71, 195–207.

Okoth, J.O. (1985) The use of indigenous plant materials for the construction of tsetse traps in Uganda. *Insect Science and its Application*, 6, 569–572.

Okoth, J.O. (1986) Peridomestic breeding sites of *Glossina fuscipes fuscipes* Newst. in Busoga, Uganda, and epidemiological implications for trypanosomiasis. *Acta Tropica*, 43, 283–286.

Okoth, J.O. (1991) Description of a mono-screen trap for *Glossina fuscipes fuscipes* Newstead in Uganda. *Annals of Tropical Medicine and Parasitology*, 85, 309–314.

Okoth, J.O. and Kapaata, R. (1986) Trypanosome infection rates in *Glossina fuscipes fuscipes* Newst. in the Busoga sleeping sickness focus Uganda. *Annals of Tropical Medicine and Parasitology*, 80, 459–461.

Okoth, J.O., Kirumira, E.K. and Kapaata, R. (1991) A new approach to community participation in tsetse control in the Busoga sleeping sickness focus, Uganda. A preliminary report. *Annals of Tropical Medicine and Parasitology*, 85, 315–322.

Okuna, N.M., Mayende, J.S.P. and Guloba, A. (1986) *Trypanosoma brucei* infection in domestic pigs in a sleeping sickness epidemic area of Uganda. *Acta Tropica*, 43, 183–184.

Oladunmade, M. and Balogun, R.A. (1979) Effects of Samorin on some aspects of the biology of *Glossina morsitans* Westwood. *Nigerian Journal of Entomology*, 3, 207–222.

Oladunmade, M.A., Feldmann, U., Takken, W., Tenabe, S.O., Hamann, H.J., Onah, J., Dengwat, L., Van der Vloedt, A.M.W. and Gingrich, R.E. (1990) Eradication of *Glossina palpalis palpalis* (Robineau-Desvoidy) (Diptera: Glossinidae) from agropastoral land in central Nigeria by means of the sterile insect technique. *IAEA Proceedings of a meeting in Nigeria*, 1988, 5–23.

Ole-Moi Yoi, O.K. (1987) Trypanosome species-specific DNA probes to detect infection in tsetse flies. *Parasitology Today*, 3, 371–374.

Omoogun, G.A. (1985) The resting behaviour of *G. longipalpis* and *G. fusca congolensis* at Oke-Ako, Nigeria. *Insect Science and its Application*, 6, 83–89.

Omoogun, G.A., Dipeolu, O.O. and Akinboade, O.A. (1991) The decline of a *Glossina morsitans submorsitans* belt in the Egbe area of the derived savanna zone, Kwara state, Nigeria. *Medical and Veterinary Entomology*, 5, 43–50.

O'Neill, S.L., Gooding, R.H. and Aksoy, S. (1993) Phylogenetically distant symbiotic micro-organisms reside in *Glossina* midgut and ovary tissues. *Medical and Veterinary Entomology*, 7, 377–383.

Onori, E. and Grab, B. (1980) Indicators for the forecasting of malaria epidemics. *Bulletin of the World Health Organisation*, 58, 91–98.

Onoviran, O., Hamann, H.J., Adegboye, D.S., Ajufo, J.C., Chima, J.C., Makinde, A.A., Pam, G. and Garba, A. (198) A bacterium pathogenic to tsetse fly (*Glossina palpalis*). *Tropical Veterinarian*, 3, 22–24.

Onyango, R.J., Van Hoeve, K. and de Raadt, P. (1966) The epidemiology of *Trypanosoma rhodesiense* sleeping sickness in Alego location, Central Nyanza, Kenya. I. Evidence that cattle may act as reservoir hosts of trypanosomes infective to man. *Transactions of the Royal Society of Tropical Medicine and Hygiene*, 60, 175–182.

Onyango, R.J., Geigy, R., Kauffmann, M., Jenni, L. and Steiger, R. (1973) New animal reservoirs of *T. rhodesiense* sleeping sickness. *Acta Tropica*, 30, 275.

Onyiah, J.A. (1980) Mechanisms of host selection by tsetse flies. *Insect Science and its Application*, 1, 31–34.

Onyiah, J.A. and Riordan, K. (1978) Field and laboratory observations on parasitization rates of *Glossina* puparia by *Syntomosphyrum* species in Nigeria. *Acta Tropica*, 35, 291–294.

Opdebeeck, J.P. (1994) Vaccines against blood-sucking arthropods. *Veterinary Parasitology*, 54, 205–222.

Ormerod, W.E. (1960) Cell inclusions and the epidemiology of Rhodesian sleeping sickness. *Transactions of the Royal Society of Tropical Medicine and Hygiene*, 54, 299–300.

Ormerod, W.E. (1961) The epidemic spread of Rhodesian sleeping sickness 1908–1960. *Transactions of the Royal Society of Tropical Medicine and Hygiene*, 55, 525–538.

Ormerod, W.E. (1967) Taxonomy of the sleeping sickness trypanosomes. *Journal of Parasitology*, 53, 824–830.

Ormerod, W.E. (1976) Ecological effect of control of African trypanosomiasis. *Science*, 191, No. 4229, 815–821.

Osaer, S., Goossens, B., Clifford, D.J., Kora, S. and Kassama, M. (1994) A comparison of the susceptibility of Djallonké sheep and West African Dwarf goats to experimental infection with two different strains of *Trypanosoma congolense*. *Veterinary Parasitology*, 51, 191–204.

Osir, E.O., Imbuga, M.O. and Onyango, P. (1993) Inhibition of *Glossina morsitans* midgut trypsin activity by D-glucosamine. *Parasitology Research*, 79, 93–97.

Osir, E.O., Abubakar, L. and Imbuga, M.O. (1995) Purification and characterization of a midgut lectin–trypsin complex from the tsetse fly *Glossina longipennis*. *Parasitology Research*, 81, 276–281.

Otieno, L.H. (1973) *Trypanosoma (Trypanozoon) brucei* in the haemolymph of experimentally infected young *Glossina morsitans*. *Transactions of the Royal Society of Tropical Medicine and Hygiene*, 67, 886–887.

Otieno, L.H. and Darji, N. (1979) The abundance of pathogenic African trypanosomes in salivary secretions of wild *Glossina pallidipes*. *Annals of Tropical Medicine and Parasitology*, 73, 53–58.

Otieno, L.H., Darji, N., Onyango, P. and Mpanga, E. (1983) Some observations on factors associated with the development of *Trypanosoma brucei brucei* infections in *Glossina morsitans morsitans*. *Acta Tropica*, 40, 113–120.

Otieno, L.H., Vundla, R.M.W. and Mongi, A. (1984) Observations on *Glossina morsitans morsitans* maintained on rabbits immunized with crude tsetse midgut proteases. *Insect Science and its Application*, 5, 297–302.

Otte, M.J. and Abuabara, J.Y. (1991) Transmission of south American *Trypanosoma vivax* by the neotropical horsefly *Tabanus nebulosus*. *Acta Tropica*, 49, 73–76.

Owaga, M.L.A. (1984) Preliminary observations on the efficacy of olfactory attractants derived from wild hosts of tsetse. *Insect Science and its Application*, 5, 87–90.

Owaga, M.L.A. (1985) Observations on the efficacy of buffalo urine as a potent olfactory attractant for *Glossina pallidipes* Austen. *Insect Science and its Application*, 6, 561–566.

Owaga, M.L.A. and Challier, A. (1985) Catch composition of the tsetse, *Glossina pallidipes* Austen in revolving and stationary traps, with respect to age and sex ratio. *Insect Science and its Application*, 6, 711–718.

Owaga, M.L.A., Hassanali, A. and McDowell, P.G. (1988) The role of 4-cresol and 3-*N*-propylphenol in the attraction of tsetse flies to buffalo urine. *Insect Science and its Application*, 9, 95–100.

Owaga, M.L.A., Okelo, R.O. and Chaudhury, M. (1993) Diel activity pattern of the tsetse fly *Glossina austeni* Newstead (Diptera: Glossinidae) in the field and in the laboratory. *Insect Science and its Application*, 14, 701–705.

Packer, M.J. and Brady, J. (1990) Efficiency of electric nets as sampling devices for tsetse flies (Diptera: Glossinidae). *Bulletin of Entomological Research*, 80, 43–47.

Packer, M.J. and Warnes, M.L. (1991) Responses of tsetse flies to ox sebum: a video study in the field. *Medical and Veterinary Entomology*, 5, 59–65.

Page, W.A. (1959a) The ecology of *Glossina longipalpis* Wied. in Southern Nigeria. *Bulletin of Entomological Research*, 50, 595–615.

Page, W.A. (1959b) Some observations on the *Fusca* group of tsetse flies (*Glossina*) in the south of Nigeria. *Bulletin of Entomological Research*, 50, 633–646.

Page, W.A. (1972) The infection of *Glossina morsitans* Weid. by *Trypanosoma brucei* in relation to the parasitaemia in the mouse host. *Tropical Animal Health and Production*, 4, 41–48.

Page, W.A. and McDonald, W.A. (1959) An assessment of the degree of man–fly contact exhibited by *Glossina palpalis* at water-holes in northern and southern Nigeria. *Annals of Tropical Medicine and Parasitology*, 53, 162–165.

Pagot, J., Coulomb, J. and Petit, J.P. (1972) Review and present situation of the use of trypanotolerance breeds. *Inter Regional Seminar FAO/WHO on African Trypanosomiasis Kinshasa*, 119–140.

Paling, R.W., Leak, S.G.A., Katende, J., Kamunya, G. and Moloo, S.K. (1987) Epidemiology of animal trypanosomiasis on a cattle ranch in Kilifi, Kenya. *Acta Tropica*, 44, 67–82.

Paling, R.W., Moloo, S.K., Logan, L. and Murray, M. (1991a) Susceptibility of N'Dama and Boran cattle to tsetse-transmitted primary and re-challenge infection with a homologous serodeme of *Trypanosoma congolense*. *Parasite Immunology*, 13, 413–425.

Paling, R.W., Moloo, S.K., Logan, L. and Murray, M. (1991b) Susceptibility of N'Dama and Boran cattle to sequential challenges with tsetse-transmitted clones of *Trypanosoma congolense*. *Parasite Immunology*, 13, 413–425.

Parashar, B.D., Gupta, G.P. and Rao, K.M. (1989) Control of haematophagous flies on equines with permethrin-impregnated eartags. *Medical and Veterinary Entomology*, 3, 137–140.

Paris, J., Murray, M. and McOdimba, F. (1982) A comparative evaluation of the parasitological techniques currently available for the diagnosis of African trypanosomiasis. *Acta Tropica*, 39, 307–316.

Parker, A.H. (1956a) Experiments on the behaviour of *Glossina palpalis* larvae, together with observations on the natural breeding-places of the species during the wet season. *Annals of Tropical Medicine and Parasitology*, 50, 69–74.

Parker, A.H. (1956b) Laboratory studies on the selection of the breeding-site by *Glossina palpalis*. *Annals of Tropical Medicine and Parasitology*, 50, 49–68.

Patel, N.Y., Otieno, L.H. and Golder, T.K. (1982) Effect of *Trypanosoma brucei* infection on the salivary gland secretions of the tsetse fly *Glossina morsitans morsitans* (Westwood). *Insect Science and its Application*, 3, 35–38.

Payne, W.J.A. (1970) *Cattle Production in the Tropics*. Longman, London.

Payne, W.J.A. (1990) *An Introduction to Animal Husbandry in the Tropics*, 4th Edn. Longman, Harlow, UK, and John Wiley & Sons, UK.

Paynter, Q. and Brady, J. (1992) Flight behaviour of tsetse flies in thick bush (*Glossina pallidipes* (Diptera: Glossinidae)). *Bulletin of Entomological Research*, 82, 513–516.

Peacock, A.J. (1986) Effect of anions Acetazolamide and copper on diuresis in the tsetse fly *Glossina morsitans morsitans*. *Journal of Insect Physiology*, 32, 157–160.

Peel, E. and Chardome, M. (1954) Etude expérimentale de souches de *Trypanosoma simiae* Bruce 1912, transmises par *G. brevipalpis* du Mosso (Urundi). *Annales de la Société Belge de Médecine Tropicale*, 34, 345.

Pell, P.E and Southern, D.I. (1975) Symbionts in the female tsetse fly *Glossina morsitans morsitans*. *Experientia*, 31, 650–651.

Penchenier, L., Jannin, J., Moulia-Pelat, J.P., Elfassi de la Baume, F., Fadat, G., Chanfreau, B. and Eozenou, P. (1991) Le problème de l'interprétation du CATT dans le dépistage de la trypanosomiase humaine à *Trypanosoma brucei gambiense*. *Annales de la Société Belge de Médecine Tropicale*, 71, 221–228.

Pépin, J., Guern, C., Mercier, D. and Moore, P. (1986) Utilisation du Testryp® CATT pour le dépistage de la trypanosomiase à Nioki, Zaïre. *Annales de la Société Belge de Médecine Tropicale*, 66, 213–224.

Pépin, J., Guern, C., Milord, F. and Schechter, P.J. (1987) Difluoromethylornithine for arseno-resistant *Trypanosoma brucei gambiense* sleeping sickness. *The Lancet*, ii, 1431–1433.

Pépin, J., Guern, C., Milord, F. and Mpia, Bokelo (1989) Intégration de la lutte contre la trypanosomiase humaine Africaine dans un réseau de centres de santé polyvalents. *Bulletin of the World Health Organisation*, 67, 301–308.

Peregrine, A.S. (1994) Chemotherapy and delivery systems: haemoparasites. *Veterinary Parasitology*, 54, 223–248.

Peregrine, A.S., Moloo, S.K. and Whitelaw, D.D. (1987) Therapeutic and prophylactic activity of Isometamidium chloride in Boran cattle against *Trypanosoma vivax* transmitted by *Glossina morsitans centralis*. *Research in Veterinary Science*, 43, 268–270.

Peregrine, A.S., Ogunyemi, O., Whitelaw, D.D., Holmes, P.H., Moloo, S.K., Hirumi, H., Urquhart, G.M. and Murray, M. (1988) Factors influencing the duration of Isometamidium chloride (Samorin) prophylaxis against experimental challenge with metacyclic forms of *Trypanosoma congolense*. *Veterinary Parasitology*, 28, 53–64.

Peregrine, A.S., Knowles, G., Ibitayo, A.I., Scott, J.R., Moloo, S.K. and Murphy, N.B. (1991a) Variation in resistance to Isometamidium chloride and diminazene aceturate by clones derived from a stock of *Trypanosoma congolense*. *Parasitology*, 102, 93–100.

Peregrine, A.S., Moloo, S.K. and Whitelaw, D.D. (1991b) Differences in sensitivity of Kenyan *Trypanosoma vivax* populations to the prophylactic and therapeutic actions of Isometamidium chloride in Boran cattle. *Tropical Animal Health and Production*, 23, 29–38.

Peregrine, A.S., Gray, M.A. and Moloo, S.K. (1997) Cross-resistance associated with development of resistance to isometamidium in a clone of *Trypanosoma congolense*. *Antimicrobial Agents and Chemotherapy*, 41, 1604–1606.

Pereira, M.E.A., Loures, M.A., Villalta, F. and Andrade, A.F.B. (1980) Lectin receptors as markers for *Trypanosoma cruzi*: developmental stages and a study of wheat germ agglutinin with sialic acid residues on epimastigote cells. *Journal of Experimental Medicine*, 152, 1375–1392.

Pereira, M.E.A., Andrade, A.F.B. and Ribeiro, (1981) Lectins of distinct specificity in *Rhodnius prolixus* interact selectively with *Trypanosoma cruzi*. *Science*, 211, 597–599.

Persoons, C.J. (1966) Trapping *Glossina pallidipes* and *G. palpalis fuscipes* in scented traps. *ISCTRC 11th meeting Nairobi, 1966*, Publication no. 100, 127–132.

Persoons, C.J. (1967) The mechanical transmission of *Trypanosoma congolense* group by *Glossina morsitans* and *Stomoxys*. *East African Trypanosomiasis Research Organisation Annual Report 1966*, 52–53.

Phelps, R.J. (1973) The effect of temperature on fat consumption during the puparial stages of *Glossina morsitans morsitans* Westw. (Dipt., Glossinidae) under laboratory conditions, and its implication in the field. *Bulletin of Entomological Research*, 62, 423–438.

Phelps, R.J. and Burrows, P.M. (1969a) Prediction of the pupal duration of *Glossina morsitans orientalis* Vanderplank under field conditions. *Journal of Applied Ecology*, 6, 323–337.

Phelps, R.J. and Burrows, P.M. (1969b) Lethal temperatures for puparia of *Glossina morsitans orientalis* Vanderplank. *Entomologia Experimentalis et Applicata*, 12, 23–32.

Phelps, R.J. and Clarke, G.P.Y. (1974) Seasonal elimination of some size classes in males of *Glossina morsitans morsitans* Westw. (Diptera, Glossinidae). *Bulletin of Entomological Research*, 64, 313–324.

Phelps, R.J. and Lovemore, D.F. (1994) Vectors: tsetse flies. In: Coetzer, J.A.W., Thomson, G.R. and Tustin, R.C. (eds) *Infectious Diseases of Livestock*. Oxford University Press, Cape Town, pp. 25–52.

Phelps, R.J. and Vale, G. (1978) Studies on populations of *Glossina morsitans* and *G. pallidipes* in Rhodesia. *Journal of Applied Ecology*, 15, 743–760.

Pierre, C. (1906) *L'élevage dans l'Afrique Occidentale Française*. Gouvernement Général de l'Afrique Occidentale Française, Paris.

Piessens, W.F., McReynolds, L.A. and Williams, S.A. (1987) Highly repeated DNA sequences as species-specific probes for *Brugia*. *Parasitology Today*, 3, 378–379.

Pilson, R.D. and Leggate, B.M. (1962a) A diurnal and seasonal study of the feeding activity of *Glossina pallidipes* Aust. *Bulletin of Entomological Research*, 53, 541–550.

Pilson, R.D. and Leggate, B.M. (1962b) A diurnal and seasonal study of the resting behaviour of *Glossina pallidipes* Aust. *Bulletin of Entomological Research*, 53, 551–562.

Pilson, R.D. and Pilson, B.M. (1967) Behaviour studies of *Glossina morsitans* Westw. in the field. *Bulletin of Entomological Research*, 57, 227–257.

Pilson, R.D., Boyt, W.P. and MacKenzie, P.K.I. (1978) The relative attractiveness of cattle, sheep and goats to *Glossina morsitans morsitans* Westwood and *G. pallidipes* Austen (Diptera: Glossinidae) in the Zambesi valley of Rhodesia. *Bulletin of Entomological Research*, 68, 489–495.

Pimenta, P.F.P., Modi, G.B., Pereira, S.T., Shahabuddin, M. and Sacks, D.L. (1997) A novel role for the peritrophic matrix in protecting *Leishmania* from the hydrolytic activities of the sand fly midgut. *Parasitology*, 115, 359–369.

Pimley, R.W. (1983) Neuroendocrine stimulation of uterine gland protein synthesis in the tsetse fly, *Glossina morsitans*. *Physiological Entomology*, 8, 429–437.

Pimley, R.W. (1985) Cyclic AMP and calcium mediate the regulation of fat cell activity by Octopamine and peptide hormones in *Glossina morsitans*. *Insect Biochemistry*, 15, 293–298.

Pimley, R.W. and Langley, P.A. (1981) Hormonal control of lipid synthesis in the fat body of the adult female tsetse-fly *Glossina morsitans*. *Journal of Insect Physiology*, 27, 839–847.

Pimley, R.W. and Langley, P.A. (1982) Hormone stimulated lipolysis and proline synthesis in the fat body of the adult tsetse-fly *Glossina morsitans*. *Journal of Insect Physiology*, 28, 781–789.

Pinder, M. and Authié, E. (1984) The appearance of isometamidium resistant *Trypanosoma congolense* in West Africa. *Acta Tropica*, 41, 247–252.

Pinder, M., Bauer, J., Fumoux, F. and Roelants, G.E. (1987) *Trypanosoma congolense*: Lack of correlation between the resistance of cattle subjected to experimental cyclic infection or to field challenge. *Experimental Parasitology*, 64, 410–417.

Pingali, P., Bigot, Y. and Binswanger, H.P. (1987) *Agricultural Mechanisation and the Evolution of Farming Systems in Sub-Saharan Africa*. World Bank, Washington, DC.

Pinnock, D.E. and Hess, R.T. (1974) The occurrence of intracellular rickettsia-like organisms in the tsetse flies, *Glossina morsitans*, *G. fuscipes*, *G. brevipalpis* and *G. pallidipes*. *Acta Tropica*, 31, 70–79.

Poinar, G.O. (1981) *Hexameris Glossinae* sp. nov.: a parasite of tsetse flies in west Africa. *Canadian Journal of Zoology*, 59, 858–861.

Poinar, G.O., Wassink, H.J.M. and Leegwater-van der Linden, M.E. (1979) *Serratia marcescens* as a pathogen of tsetse flies. *Acta Tropica*, 36, 223–227.

Politzar, H. and Cuisance, D. (1982) SIT in the control and eradication of *Glossina palpalis gambiensis*. In: *IAEA Proceedings of a Conference on Sterile Male Insect Release for Insect Pest control*, pp. 101–109.

Politzar, H. and Cuisance, D. (1983) A trap-barrier to block reinvasion of a river system by riverine tsetse species. *Revue d'Élevage et de Médecine Vétérinaire des Pays Tropicaux*, 36, 364–370.

Politzar, H. and Mérot, P. (1984) Attraction of the tsetse fly *Glossina morsitans submorsitans* to acetone, 1-octen-3-ol, and the combination of these compounds in West Africa. *Revue d'Élevage et de Médecine Vétérinaire des Pays Tropicaux*, 37, 468–473.

Politzar, H., Mérot, P., and Brandl, F.E. (1984) Experimental aerial release of sterile males of *Glossina palpalis gambiensis* and of *Glossina tachinoides* in a biological control operation. *Revue d'Élevage et de Médecine Vétérinaire des Pays Tropicaux*, 37, 198–202.

Pollard, J. (1912) Notes on the tsetse-flies of Muri province, northern Nigeria. *Bulletin of Entomological Research*, 3, 219–221.

Pollock, J.N. (1970) Sperm transfer by spermatophores in *Glossina austeni* Newstead. *Nature*, 225, 1063–1064.

Pollock, J.N. (1971) Origin of the tsetse flies: a new theory. *Journal of Entomology (B)*, 40, 101–109.

Pollock, J.N. (1973) A comparison of the male genitalia and abdominal segmentation in *Gasterophilus* and *Glossina* with notes on the Gasterophiloid origin. *Transactions of the Royal Entomological Society of London*, 125, 107–124.

Pollock, J.N. (1974) Functional morphology of the phallosome in *Glossina* and its evolutionary implications. *Zoologica Scripta*, 3, 185–192.

Popham, E.J. (1972) The effects of local agriculture on the distribution of the species of *Glossina* in northern Nigeria. *Transactions of the Royal Society of Tropical Medicine and Hygiene*, 66, 321.

Popham, E.J., Abdillahi, M. and Lavin, D. (1979) Tsetse fly feeding sites (Diptera: Glossinidae). *Experientia*, 35, 607–609.

Potts, W.H. (1930a) Observations on *Glossina morsitans*, Westw., in East Africa. *Bulletin of Entomological Research*, 24, 293–300.

Potts, W.H. (1930b) A contribution to the study of numbers of tsetse fly (*Glossina morsitans* Westw.) by quantitative methods. *South African Journal of Science*, 27, 491–497.

Potts, W.H. (1940) The tsetse position at Shinyanga with special regard to the Shinyanga Kahama fire exclusion experiment. *Tsetse Research Report Tanganyika Territory Dar Es Salaam*, 19–27.

Potts, W.H. (1944) Hybridization between *Glossina* species and suggested new method for control of certain species of tsetse. *Nature*, 154, 606–607.

Potts, W.H. (1950) *Tsetse and Trypanosomiasis Research: Annual Report of the East African Trypanosomiasis Research Organisation, 1950*. East African Trypanosomiasis Research Organisation, Uganda.

Potts, W.H. (1958) Sterilisation of tsetse flies (*Glossina*) by gamma irradiation. *Annals of Tropical Medicine and Parasitology*, 52, 484–499.

Potts, W.H. (1970) Systematics and identification of *Glossina*. In: Mulligan, H.W. (ed.) *The African Trypanosomiases*. ODA, George Allen and Unwin, London, pp. 243–273.

Potts, W.H. and Jackson, C.H.N. (1952) The Shinyanga game destruction experiment. *Bulletin of Entomological Research*, 43, 365–374.

Power, R.J.B. (1964) The activity pattern of *Glossina longipennis* Corti (Diptera: Muscidae). *Proceedings of the Royal Entomological Society of London (A)*, 39, 5–14.

Proverbs, M.D. (1982) Sterile insect technique in codling moth control. In: *Sterile Insect Technique and Radiation in Insect Control*. Proceedings of a Symposium IAEA/FAO, 1981. IAEA, Vienna, pp. 85–99.

Putt, S.N.H., Shaw, A.P.M., Matthewman, R.W., Bourn, D.M., Underwood, M., James, A.D., Hallam, M.J. and Ellis, P.R. (1980) *The Social and Economic Implications of Trypanosomiasis Control: a Study of its Impact on Livestock Production and Rural Development in Northern Nigeria*. Veterinary Epidemiology and Economics Research Unit, Department of Agriculture and Horticulture, University of Reading, 549 pp.

Putt, S.N.H., Leslie, J. and Willemse, L. (1988) The economics of trypanosomiasis control in western Zambia. *Acta Veterinaria Scandinavia*, 84 (Suppl.), 394–397.

Quinlan, R.J. and Gatehouse, A.G. (1981) Characteristics and implications of knockdown of the tsetse fly *Glossina morsitans morsitans* Westw. by deltamethrin. *Pesticide Science*, 12, 439–442.

Raadt, P. de (1975) *Sleeping Sickness Today*. WHO, Geneva.

Raadt, P. de (1989) Epidemiological models for Africa trypanosomiasis: a waste of time? *Annales de la Société Belge de Médecine Tropicale*, 69 (Suppl. 1), 7–10.

Radley D.E., Brown C.G.D., Burridge M.J., Cunningham, M.P., Kirimi, I.M., Purnell, R.E. and Young, A.S. (1975) East Coast Fever. 1. Chemoprophylactic immunisation of cattle against *Theileria parva* (Muguga) and five Theilerial strains. *Veterinary Parasitology*, 1, 35–41.

Rae, P.F. and Luckins, A.G. (1984) Detection of circulating trypanosomal antigens by enzyme immunoassay. *Annals of Tropical Medicine and Parasitology*, 78, 587–596.

Rajagopal, P.K. and Bursell, E. (1965) The effect of temperature on the oxygen consumption of tsetse pupae. *Bulletin of Entomological Research*, 56, 219–225.

Rajagopal, P.K. and Bursell, E. (1966) The respiratory metabolism of resting tsetse flies. *Journal of Insect Physiology*, 12, 287–297.

Randolph, S.E. and Rogers, D.J. (1978) Feeding cycles and flight activity in field populations of tsetse (Diptera: Glossinidae). *Bulletin of Entomological Research*, 68, 655–671.

Randolph, S.E. and Rogers, D.J. (1981) Physiological correlates of the availability of *Glossina morsitans centralis* Machado to different sampling methods. *Ecological Entomology*, 6, 63–77.

Randolph, S.E. and Rogers, D.J. (1984a) Local variation in the population dynamics of *Glossina palpalis palpalis* 2. The effect of insecticidal spray programmes. *Bulletin of Entomological Research*, 74, 425–438.

Randolph, S.E. and Rogers, D.J. (1984b) Movement patterns of the tsetse fly *Glossina palpalis palpalis* around villages in the pre-forest zone of Ivory Coast. *Bulletin of Entomological Research*, 74, 689–705.

Randolph, S.E. and Rogers, D.J. (1986) The use of discriminant analysis for the estimation of sampling biases for female tsetse. *Ecological Entomology*, 11, 205–220.

Randolph, S.E., Rogers, D.J. and Kiilu, J. (1991a) The feeding behaviour, activity and trappability of wild female *Glossina pallidipes* in relation to their pregnancy cycle. *Medical and Veterinary Entomology*, 5, 335–350.

Randolph, S.E., Rogers, D.J., Dransfield, R.D. and Brightwell, R. (1991b) Trap-catches, nutritional condition and the timing of activity of the tsetse fly *Glossina longipennis* (Diptera: Glossinidae). *Bulletin of Entomological Research*, 81, 455–464.

Randolph, S.E., Williams, B.G., Rogers, D.J. and Connor, H. (1992) Modelling the effect of feeding-related mortality on the feeding strategy of tsetse. *Medical and Veterinary Entomology*, 6, 231–240.

Rawlings, P. and Maudlin, I. (1984) Sex ratio distortion in *Glossina morsitans submorsitans* Newstead (Diptera: Glossinidae). *Bulletin of Entomological Research* 74, 311–315.

Rawlings, P., Dwinger, R.H. and Snow, W.F. (1991) An analysis of survey measurements of tsetse challenge to trypanotolerant cattle in relation to aspects of analytical models of trypanosomiasis. *Parasitology*, 102, 371–377.

Rawlings, P., Ceesay, M.L., Wacher, T.J. and Snow, W.F. (1993) The distribution of the tsetse flies *Glossina morsitans submorsitans* and *G. palpalis gambiensis* (Diptera: Glossinidae) in The Gambia and the application of survey results to tsetse and trypanosomiasis control. *Bulletin of Entomological Research*, 83, 625–632.

Rawlings, P., Wacher, T.J. and Snow, W.F. (1994) Cattle–tsetse contact in relation to the daily activity patterns of *Glossina morsitans submorsitans* in The Gambia. *Medical and Veterinary Entomology*, 8, 57–62.

Raynaud, J.P., Sones, K.R. and Friedheim, E.A.H. (1989) A review of Cymelarsan®, a new trypanocide proposed for the treatment of *T. evansi* infections. *20th meeting of the International Scientific Council for Trypanosomosis Research and Control, Mombasa*, pp. 334–338.

Reduth, D., Grootenhuis, J.G., Olubayo, R.O., Muranjan, M., Otieno-Omondi, F.P., Morgan, G.A., Brun, R., Williams, D.J. and Black, S. (1995) African Buffalo serum contains novel trypanocidal protein. *Journal of Eukaryotic Microbiology*, 41, 95–103.

Rehacek, J., Sixl, W. and Sebek, Z. (1974) Trypanosomen in der haemolymphe von Zecken. *Mittelungen Arbeitsgemeinschaft Zoologischen Landesmus Joanneum*, 33, 161.

Reid, R.S., Wilson, C.J., Kruska, R.L. and Woudyalew Mulatu (1997a) Impacts of tsetse control and land-use on vegetative structure and tree species composition in south-western Ethiopia. *Journal of Applied Ecology*, 34, 731–747.

Reid, R.S., Kruska, R.L., Deichmann, U., Thornton, P.K. and Leak, S.G.A. (1997b) Will human population growth and land-use change control tsetse during our lifetimes? In: *24th Meeting of the International Scientific Council for Trypanosomosis Research and Control, Maputo, Mozambique* (in press).

Reid, R.S., Kruska, R.L., Muthui, N., Wotton, S., Wilson, C.J., Andualam Taye and Woudyalew Mulatu (1997c) Distinguishing the impacts of change in trypanosomosis severity from other factors that cause land-use change over time in southwestern Ethiopia. In: *24th Meeting of the International Scientific Council for Trypanosomosis Research and Control, Maputo, Mozambique* (in press).

Reid, R.S., Kruska, R.L., Wilson, C.J. and Perry, B.D. (1999) The impacts of controlling the tsetse fly on land-use and the environment. In: Lynam, J., Carter, S. and Reid. R.S. (eds) *Spatial and Temporal Dynamics of African Farming Systems*, (in press).

Reinhardt, C., Steiger, R. and Hecker, H. (1972) Ultrastructural study of the midgut mycetome-bacteroids of the tsetse flies *Glossina morsitans morsitans, G. fuscipes* and *G. brevipalpis. Acta Tropica*, 29, 280–288.

Reinouts van Haga, H.A. and Mitchell, B.K. (1975) Temperature receptors on tarsi of the tsetse fly *Glossina morsitans* West. *Nature*, 255, 225–226.

Remme, J. and Zongo, J.B. (1989) Demographic aspects of the epidemiology and control of Onchocerciasis in West Africa. In: Service, M.W. (ed.) *Demography and Vector-borne Diseases*. CRC Press, Boca Raton, Florida, pp. 367–386.

Remme, J., Ba, K.Y., Dadzie, K.Y. and Karam, M. (1986) A force-of-infection model for onchocerciasis and its applications in the epidemiological evaluation of the Onchocerciasis Control Programme in the Volta River basin area. *Bulletin of the World Health Organisation*, 64, 667–681.

Rennison, B.D. and Robertson, D.H.H. (1960) The use of carbon dioxide as an attractant for catching tsetse. *EATRO Annual Report 1959*, p. 26.

Rice, M.J. (1970a) Supercontracting and non-supercontracting visceral muscles in the tsetse fly, *Glossina austeni. Journal of Insect Physiology*, 16, 1109–1122.

Rice, M.J. (1970b) Cibarial stretch receptors in the tsetse fly (*Glossina austeni*) and the blowfly (*Calliphora erythrocephala*). *Journal of Insect Physiology*, 16, 277–289.

Rice, M.J. (1970c) Function of resilin in tsetse fly feeding mechanism. *Nature*, 228, 1337–1338.

Rice, M.J. (1972) The nervous system in tsetse fly feeding. *Transactions of the Royal Society of Tropical Medicine and Hygiene*, 66, 317–318.

Rice, M.J., Kamugisha, C.K. and Sebugwayo, B.S. (1972) Bat's wing membrane for tsetse fly synthetic feeding. *Transactions of the Royal Society of Tropical Medicine and Hygiene*, 66, 328.

Rice, M.J., Galun, R. and Margalit, J. (1973a) Mouthpart sensilla of the tsetse fly and their function. II. Labial sensilla. *Annals of Tropical Medicine and Parasitology*, 67, 101–107.

Rice, M.J., Galun, R. and Margalit, J. (1973b) Mouthpart sensilla of the tsetse fly and their function. III. Labrocibarial sensilla. *Annals of Tropical Medicine and Parasitology*, 67, 109–116.

Rickman, L.R. and Robson, J. (1970a) The blood incubation infectivity test: a simple test which may serve to distinguish *Trypanosoma brucei* from *T. rhodesiense. Bulletin of the World Health Organisation*, 42, 650–651.

Rickman, L.R. and Robson, J. (1970b) The testing of proven *Trypanosoma brucei* and *T. rhodesiense* strains by the blood incubation infectivity. *Bulletin of the World Health Organisation*, 42, 911–916.

Riddiford, L.M. and Dhadialla, T.S. (1990) Protein synthesis by the milk gland and fat body of the tsetse fly, *Glossina pallidipes*. *Insect Biochemistry*, 20, 493–500.

Riordan, K. (1968) Chromosomes of the tsetse, *Glossina palpalis* R.-D. *Parasitology*, 58, 835–838.

Riordan, K. (1976) Rate of linear advance by *Glossina morsitans submorsitans* Newst. (Diptera, Glossinidae) on a trade cattle route in south-western Nigeria. *Bulletin of Entomological Research*, 66, 365–372.

Robert, A., Grillot, J.P., Guilleminot, J. and Raabe, M. (1984) Experimental and ultra-structural study of the control of ovulation and parturition in the tsetse fly *Glossina fuscipes* (Diptera). *Journal of Insect Physiology*, 30, 671–684.

Roberts, C.J. and Gray, A.R. (1972) A comparison of *Glossina morsitans submorsitans* and *G. tachinoides* collected and maintained under similar conditions as vectors of *Trypanosoma (Nannomonas) congolense, T. (N.) simiae* and *T. (Duttonella) vivax. Annals of Tropical Medicine and Parasitology*, 66, 41–53.

Roberts, C.J. and Gray, A.R. (1973a) Studies on trypanosome-resistant cattle. I. The breeding and growth performance of N'Dama, Muturu and Zebu cattle maintained under the same conditions of husbandry. *Tropical Animal Health and Production*, 5, 211–219.

Roberts, C.J. and Gray, A.R. (1973b) Studies on trypanosome resistant cattle. II. The effect of trypanosomiasis on N'Dama, Muturu and Zebu cattle. *Tropical Animal Health and Production*, 5, 220–233.

Roberts, L.W. (1981) Probing by *Glossina morsitans morsitans* and transmission of *Trypanosoma (Nannomonas) congolense. American Journal of Tropical Medicine and Hygiene*, 30, 948–951.

Roberts, L.W., Wellde, B.T., Reardon, M.J. and Onyango, F.K. (1989) Mechanical transmission of *Trypanosoma brucei rhodesiense* by *Glossina morsitans morsitans* (Diptera: Glossinidae). *Annals of Tropical Medicine and Parasitology*, 83, Supplement 1, 127–131.

Roberts, M.J. (1971) The functional anatomy of the head in the larvae of the tsetse fly, *Glossina austeni*, Newstead (Diptera, Glossinidae). *Entomologist*, 104, 190–203.

Roberts, M.J. (1972) The position of the ejaculatory duct of tsetse flies (Diptera: Glossinidae). *Entomologist*, 105, 2–5.

Roberts, M.J. (1973a) The control of fertilisation in tsetse flies. *Annals of Tropical Medicine and Parasitology*, 67, 117–123.

Roberts, M.J. (1973b) Observations on the function of the choriothete and on egg hatching in *Glossina* spp. (Dipt., Glossinidae). *Bulletin of Entomological Research*, 62, 371–374.

Robertson, A.G. (1983) The feeding habits of tsetse flies in Zimbabwe (formerly Rhodesia) and their relevance to some tsetse control measures. *Smithersia*, 1, 1–72.

Robertson, A.G. and Bernacca, J.P. (1958) Game elimination as a tsetse control measure in Uganda. *East African Agricultural and Forestry Journal*, 23, 254–261.

Robertson, D.H.H. (1963) Human trypanosomiasis in south-east Uganda: a further study of the epidemiology of the disease among fishermen and peasant cultivators. *Bulletin of the World Health Organisation*, 28, 627–643.

Robertson, D.H.H. and Baker, J.R. (1958) Human trypanosomiasis in south-east Uganda. I. A study of the epidemiology and present virulence of the disease. *Transactions of the Royal Society of Tropical Medicine and Hygiene*, 52, 337–348.

Robinson, G.G. (1965) A note on the nocturnal resting sites of *Glossina morsitans* Westw. in the Republic of Zambia. *Bulletin of Entomological Research*, 56, 351–355.

Robinson, M.W., Baker, P.S. and Finlayson, L.H. (1985) Influence of temperature changes on larviposition rhythm in the tsetse fly *Glossina morsitans*. *Physiological Entomology*, 10, 215–220.

Robson, J. and Rickman, L.R. (1973) Blood incubation infectivity test results for *Trypanosoma brucei* sub-group isolates tested in the Lambwe valley, south Nyanza, Kenya. *Tropical Animal Health and Production*, 5, 187–191.

Rodhain, J. (1926) Le recul de la tsetse; *Glossina morsitans* devant l'occupation Européenne au Katanga. *Bulletin de la Société Pathologie Exotique*, 19, 222–234.

Rodhain, J., Pons, C., Van den Branden, F. and Bequaert, J. (191) *Rapport sur les travaux de la mission scientifique du Katanga (Octobre 1910 à Septembre 1912)*. Hayez, Brussels.

Rogers, A., Kenyanjui, E.N. and Wiggwah, A. (1972) A high rate of infection of wild *Glossina fuscipes* with *Trypanosoma brucei* subgroup trypanosome. 12th Seminar on Trypanosomiasis. *Transactions of the Royal Society of Tropical Medicine and Parasitology*, 66, 328–329.

Rogers, D.J. (1974) Natural regulation and movement of tsetse fly populations. In: *Les Moyens de Lutte Contre les Trypanosomes et Leurs Vecteurs*. IEMVT Symposium, Paris, pp. 35–38.

Rogers, D.J. (1977) Study of a natural population of *Glossina fuscipes fuscipes* Newstead and a model of fly movement. *Journal of Animal Ecology*, 46, 309–330.

Rogers, D.J. (1979) Tsetse population dynamics and distribution. A new analytical approach. *Journal of Animal Ecology*, 48, 825–849.

Rogers, D.J. (1984) The estimation of sampling biases for male tsetse. *Insect Science and its Application*, 5, 369–373.

Rogers, D.J. (1985a) Population ecology of tsetse. *Annual Review of Entomology*, 30, 107–216.

Rogers, D.J. (1985b) Trypanosomiasis risk or challenge: a review. *Acta Tropica*, 42, 5–23.

Rogers, D.J. (1988a) The dynamics of vector-transmitted diseases in human communities. *Philosophical Transactions of the Royal Society London (B)*, 321, 513–539.

Rogers, D.J. (1988b) A general model for the African trypanosomiases. *Parasitology*, 97, 193–212.

Rogers, D.J. (1988c) Tsetse flies in Africa: bane or boon? *Conservation Biology*, 2, 57–63.

Rogers, D.J. (1989) The development of analytical models for human trypanosomiasis. *Annales de la Société Belge de Médecine Tropicale*, 69, Suppl. 1, 73–88.

Rogers, D.J. (1990) A general model for tsetse populations. *Insect Science and its Application*, 11, 331–346.

Rogers, D.J. (1991) Satellite imagery, tsetse and trypanosomiasis in Africa. *Preventive Veterinary Medicine*, 11, 201–220.

Rogers, D.J. and Boreham, P.F.L. (1973) Sleeping sickness survey in the Serengeti area (Tanzania) 1971. II. The vector role of *Glossina swynnertoni* Austen. *Acta Tropica*, 30, 24–35.

Rogers, D.J. and Muynck, A. de. (1988) Workshop on modelling sleeping sickness epidemiology. *Parasitology Today*, 4, 111–112.

Rogers, D.J. and Randolph S.E. (1978a) A comparison of electric traps and handnet catches of *G. palpalis* and *G. tachinoides* in the Sudan vegetation zone of northern Nigeria. *Bulletin of Entomological Research*, 68, 283–297.

Rogers, D.J. and Randolph, S.E. (1978b) Metabolic strategies of male and female tsetse in the field. *Bulletin of Entomological Research*, 68, 639–654.

Rogers, D.J. and Randolph, S.E. (1984a) A review of density-dependent processes in tsetse populations. *Insect Science and its Application*, 5, 397–402.

Rogers, D.J. and Randolph, S.E. (1984b) From a case study to a theoretical basis for tsetse control. *Insect Science and its Application*, 5, 419–423.

Rogers, D.J. and Randolph, S.E. (1985) Population ecology of tsetse. *Annual Review of Entomology*, 30, 197–216.

Rogers, D.J. and Randolph, S.E. (1986a) Detection of activity cycles of tsetse from capture recapture data. *Ecological Entomology*, 11, 95–109.

Rogers, D.J. and Randolph, S.E. (1986b) Distribution and abundance of tsetse flies (*Glossina* spp.). *Journal of Animal Ecology*, 55, 1007–1025.

Rogers, D.J. and Randolph, S.E. (1990) Estimation of rates of predation on tsetse. *Medical and Veterinary Entomology*, 4, 195–204.

Rogers, D.J. and Randolph, S.E. (1991) Mortality rates and population density of tsetse flies correlated with satellite imagery. *Nature*, 351, 739–741.

Rogers, D.J. and Smith, D.T. (1977) A new electric trap for tsetse flies. *Bulletin of Entomological Research*, 67, 153–159.

Rogers, D.J. and Williams, B.G. (1993) Monitoring trypanosomiasis in space and time. *Parasitology*, 106, S77-S92.

Rogers, D.J., Randolph, S.E. and Kuzoe, F.A.S. (1984) Local variation in the population dynamics of *Glossina palpalis palpalis* (Robineau-Desvoidy) (Diptera: Glossinidae). I. Natural population regulation. *Bulletin of Entomological Research*, 74, 403–423.

Rogers, D.J., Hargrove, J. and Jordan, A.M. (1986) Report on a visit to the International trypanotolerance Centre, The Gambia 12–28th February 1986. Unpublished report, 20 pp. + Appendices. ILCA/ILRAD, Trypanotolerance Network.

Rogers, D.J., Hendrickx, G. and Slingenbergh, J. (1994) Tsetse flies and their control. *Revue Scientifique et Technique de l'Office International des Epizooties*, 13, 1075–1124.

Rogers, D.J., Hay, S.I. and Packer, M.J. (1996) Predicting the distribution of tsetse flies in West Africa using temporal Fourier processed meteorological satellite data. *Annals of Tropical Medicine and Parasitology*, 90, 225–241.

Ross, G.R. and Blair, D.M. (1956) Cas de porteurs en bonne santé de trypanosomiase humaine en Rhodésie du sud. *ISCTRC Sixth meeting Salisbury, CCTA*, 9–29.

Röttcher, D. and Schillinger, D. (1985) Multiple drug resistance in *Trypanosoma vivax* in the Tana River District of Kenya. *Veterinary Record*, 117, 557–558.

Rowcliffe, C. and Finlayson, L.H. (1981) Factors influencing the selection of larviposition sites in the laboratory by *Glossina morsitans morsitans* Westwood (Diptera: Glossinidae). *Bulletin of Entomological Research*, 71, 81–97.

Rowcliffe, C. and Finlayson, L.H. (1982) Active and resting behaviour of virgin and pregnant females of *Glossina morsitans morsitans* Westwood (Diptera: Glossinidae) in the laboratory. *Bulletin of Entomological Research*, 72, 271–289.

Rudin, W. and Hecker, H. (1989) Lectin-binding sites in the midgut of the mosquitoes *Anopheles stephensi* Liston and *Aedes aegyptii* L. (Diptera: Culicidae). *Parasitology Research*, 75, 268–279.

Rudin, W., Schwarzenbach, M. and Hecker, H. (1989) Binding of lectins to culture and vector forms of *Trypanosoma rangeli* Tejera, 1920 (Protozoa: Kinetoplastida) and to structures of the vector gut. *Journal of Protozoology*, 36, 532–538.

Ruppol, J.F. and Kazyumba, L. (1977) Situation actuelle de la lutte contre la maladie du sommeil au Zaïre. *Annales de la Société Belge de Médecine Tropicale*, 57, 299–314.

Rurangirwa, F.R., Minja, S.H., Musoke, A.J., Nantulya, V.M., Grootenhuis, J.G. and Moloo, S.K. (1986) Production and evaluation of specific antisera against sera of various vertebrate species for identification of bloodmeals of *Glossina morsitans*. *Acta Tropica*, 43, 379–389.

Ruttledge, W. (1928) Tsetse-fly (*Glossina morsitans*) in the Koalib hills, Nuba mountains Province, Sudan. *Bulletin of Entomological Research*, 19, 309–316.

Ryan, L. (1981) *Glossina* (Diptera: Glossinidae) population growth rates. *Bulletin of Entomological Research*, 71, 519–533.

Ryan, L. (1984) The effect of trypanosome infection on a natural population of *Glossina longipalpis* in Ivory Coast. *Acta Tropica*, 41, 355–359.

Ryan, L. and Molyneux, D.H. (1981) Non-setting adhesives for insect traps. *Insect Science and its Application*, 1, 349–355.

Ryan, L. and Molyneux, D.H. (1982) Observations on and comparisons of various traps for the collection of Glossinidae and other Diptera in Africa. *Revue d'Élevage et de Médecine Vétérinaire des Pays Tropicaux*, 35, 165–172.

Ryan, L., Molyneux, D.H. and Kuzoe, F.A.S. (1980) Differences in rate of wing-fray between *Glossina* species. *Tropenmedizin und Parasitologie*, 31, 111–116.

Ryan, L., Molyneux, D.H., Kuzoe, F.A.S. and Baldry, D.A.T. (1981) Traps to control and estimate populations of *Glossina* species. *Tropenmedizin und Parasitologie*, 32, 145–148.

Ryan, L., Croft, S.L., East, J.S., Molyneux, D.H. and Baldry, D.A.T. (1982a) Naturally occurring cicatrices of *Glossina morsitans centralis* (Diptera: Glossinidae). *Zeitschrift für Angewandte Zoologie*, 60, 299–307.

Ryan, L., Küpper, W., Croft, S.L., Molyneux, D.H. and Clair, M. (1982b) Differences in rates of acquisition of trypanosome infections in the field. *Annales de la Société Belge de Médecine Tropicale*, 291–300.

Ryan, L., Küpper, W., Molyneux, D.H. and Clair, M. (1986) Relationships between geographical and dietary factors and trypanosome infection rates of tsetse flies in the field (Diptera: Glossinidae). *Entomologia Generalis*, 12, 77–81.

Saini, R.K. (1986) Antennal responses of *Glossina morsitans morsitans* to buffalo urine – a potent olfactory attractant of tsetse flies. *Insect Science and its Application*, 7, 771–775.

Saini, R.K. and Dransfield, R.D. (1987) A behavioural bioassay to identify attractive odours for Glossinidae. *Medical and Veterinary Entomology*, 1, 313–318.

Saini, R.K., Hassanali, A., Ahuya, P., Andoke, J. and Nyandat, E. (1993) Close range responses of tsetse flies *Glossina morsitans morsitans* Westwood (Diptera: Glossinidae) to host body kairomones. *Discovery and Innovation*, 5, 149–153.

Saleh, K.M., Vreysen, M., Kassim, S.S., Suleiman, F.W., Juma, K.G., Zhu, Z.-R., Pan, H. and Dyck, V.A (1998) The successful application of the sterile insect technique (SIT) for the eradication of *Glossina austeni* Newstead (Diptera: Glossinidae) from Unguja island (Zanzibar). *24th meeting of the International Scientific Council for Trypanosomosis Research and Control Maputo, Mozambique* (in press).

Samaranayaka-Ramasamy, M. (1981) Influence of feeding and hormonal factors on sexual maturation in male *Glossina morsitans morsitans* Westwood 1850. (Diptera: Glossinidae). *Insect Science and its Application*, 1, 273–280.

Sanford, W.W. and Isichei, A.O. (1986) Savanna. In: Lawson, G.W. (ed.) *Plant Ecology in West Africa*. Wiley and Sons, Chichester, UK, pp. 95–149.

Saunders, D.S. (1960) The ovulation cycle in *Glossina morsitans* Westwood (Diptera: Muscidae) and a possible method of age determination for female tsetse flies by the examination of their ovaries. *Transactions of the Royal Entomological Society of London*, 112, 221–238.

Saunders, D.S. (1961a) On the stages in the development of *Syntomosphyrum albiclavus* Kerrich (Hym., Eulophidae), a parasite of tsetse flies. *Bulletin of Entomological Research*, 51, 25–31.

Saunders, D.S. (1961b) Laboratory studies on the biology of *Syntomosphyrum albiclavus* Kerrich (Hym., Eulophidae), a parasite of tsetse flies. *Bulletin of Entomological Research*, 52, 413–429.

Saunders, D.S. (1961c) The 'White-clubbed' form of *Syntomosphyrum* (Hym: Eulophidae) parasitic on tsetse flies. *Bulletin of Entomological Research*, 51, 17–20.

Saunders, D.S. (1962) Age determination for female tsetse flies and the age compositions of samples of *Glossina pallidipes* Aust., *G. palpalis fuscipes* Newst., and *G. brevipalpis* Newst. *Bulletin of Entomological Research*, 53, 579–595.

Saunders, D.S. (1967) Survival and reproduction in a natural population of the tsetse fly, *Glossina palpalis palpalis* (Robineau-Desvoidy). *Proceedings of the Royal Entomological Society of London (A)*, 42, 129–137.

Saunders, D.S. (1972) The effect of starvation on the length of the interlarval period in the tsetse fly *Glossina morsitans orientalis* Vanderplank. *Journal of Entomology (A)*, 46, 197–202.

Saunders, D.S. and Dodd, C.W.H. (1972) Mating, insemination and ovulation in the tsetse fly, *Glossina morsitans*. *Journal of Insect Physiology*, 18, 187–198.

Schares, G. and Mehlitz, D. (1996) Sleeping sickness in Zaire: a nested polymerase chain reaction improves the identification of *Trypanosoma* (*Trypanozoon*) *brucei gambiense* by specific kinetoplast DNA probes. *Tropical Medicine and International Health*, 1, 59–70.

Schaub, G.A. (1994) Pathogenicity of trypanosomatids on insects. *Parasitology Today*, 10, 463–468.

Schilling, C. (1935) Immunisation against trypanosomiasis. *Journal of Tropical Medicine and Hygiene*, 38, 106.

Schlein, Y. (1979) Age grading of tsetse flies by the cuticular growth layers in the thoracic phragma. *Annals of Tropical Medicine and Parasitology*, 73, 297–298.

Schlein, Y. and Lewis, C.T. (1976) Lesions in haematophagous flies after feeding on rabbits immunised with fly tissues. *Physiological Entomology*, 1, 55–59.

Schlein, Y., Galun, R. and Ben-Eliahu, M.N. (1980) The legs of *Musca domestica* and *Glossina morsitans* females as the site of sex pheromone release. *Experientia*, 36, 1174–1176.

Schlein, Y., Jacobson, R.L. and Schlomai, J. (1991) Chitinase secreted by *Leishmania* functions in the sandfly vector. *Proceedings of the Royal Society of London (B)*, 245, 121–126.

Schlein, Y., Jacobson, R.L. and Messer, G. (1992) *Leishmania* infections damage the feeding mechanism of the sandfly vector and implement parasite transmission by bite. *Proceedings of the National Academy of Sciences, USA*, 89, 9944–9948.

Schmittner, S.M. and McGhee, R.B. (1970) Host specificity of various species of *Crithidia* Leger. *Journal of Parasitology*, 56, 684–693.

Schofield, C.J. (1991) Vector population responses to control interventions. *Annales de la Société Belge de Médecine Tropicale*, 71 (Suppl. 1), 201–217.

Scholdt, L.L., Grothaus, R.H., Schreck, C.E. and Gouck, H.K. (1975) Field studies using repellent treated wide-mesh net jackets against *Glossina morsitans* in Ethiopia. *East African Medical Journal*, 52, 277–283.

Scholdt, L.L., Schreck, C.E., Mwangelwa, M.I., Nondo, J. and Siachinji, V.J. (1989) Evaluations of permethrin-impregnated clothing and three topical repellent formulations of deet against tsetse flies in Zambia. *Medical and Veterinary Entomology*, 3, 153–158.

Scholz, E., Spielberger, U. and Ali, J. (1976) The night resting sites of the tsetse fly *Glossina palpalis palpalis* (Robineau-Desvoidy) (Diptera: Glossinidae) in northern Nigeria. *Bulletin of Entomological Research*, 66, 443–452.

Schönefeld, A. (1983) Essai de lutte contre *Glossina morsitans submorsitans* par utilisation d'écrans imprégnés de deltamethrine. *Revue d'Élevage et de Médecine Vétérinaire des Pays Tropicaux*, 36, 33–44.

Schönefeld, A. (1988) Tsetse control with insecticide treated cattle in Zanzibar. Joint meeting of the Panels of experts on ecological/technical and development aspects of the programme for the control of African animal trypanosomiasis and related development. FAO, Accra, Ghana, *7–9 November 1988*.

Schönefeld, A., Rottcher, D. and Moloo, S. (1987) The sensitivity to trypanocidal drugs of *Trypanosoma vivax* isolated in Kenya and Somalia. *Tropical Medicine and Parasitology*, 38, 177–180.

Schwetz, J. (1915) Preliminary note on the general distribution of *Glossina palpalis* in the Lomani district, Belgian Congo. *Annals of Tropical Medicine and Parasitology*, 9, 513–526.

Scott, C.M., Frézil, J.-L., Toudic, A. and Godfrey, D.G. (1983) The sheep as a potential reservoir of human trypanosomiasis in the Republic of the Congo. *Transactions of the Royal Society of Tropical Medicine and Hygiene*, 77, 397–401.

Scott, C.M. (1981) Mixed populations of *Trypanosoma brucei* in a naturally infected pig. *Tropenmedizin und Parasitologie*, 32, 221–222.

Scott, D. (1957) The epidemiology of human trypanosomiasis in Ashanti, Ghana. *Journal of Tropical Medicine and Hygiene*, 60, Sept. 1957, 205–217.

Scott, J.M. and Pegram, R. (1974) A high incidence of *Trypanosoma congolense* strains resistant to homidium bromide in Ethiopia. *Tropical Animal Health and Production*, 6, 215–221.

Sékétéli, A. and Kuzoe, F.A.S. (1984) Gîtes à pupes de *Glossina palpalis* dans une zone pré-forestière de Côte d'Ivoire. *Acta Tropica*, 41, 293–301.

Sékétéli, A. and Kuzoe, F.A.S. (1986a) Effet résiduel sur *Glossina palpalis palpalis* de l'alphaméthrine (pyréthrinoide de synthèse) en concentré emulsifiable 10% appliqué par épandage au sol en zone pré-forestière de Côte d'Ivoire. *Insect Science and its Application*, 7, 757–775.

Sékétéli, A. and Kuzoe, F.A.S. (1986b) Essais d'épandage au sol de trois pyréthrinoides de synthèse (OMS 2012 p.m. 10%, OMS 2013 p.m. 5%, OMS 3004 p.m. 5%), contre *Glossina palpalis palpalis* et *Glossina tachinoides* en zone pré-forestière de Côte d'Ivoire. *Insect Science and its Application*, 7, 763–769.

Sékétéli, A. and Kuzoe, F.A.S. (1994) Lieux de repos diurnes de *Glossina palpalis palpalis* (Robineau-Desvoidy) dans une zone pré-forestière de Côte d'Ivoire. *Insect Science and its Application*, 15, 75–85.

Sekoni, V.O. (1990) Effect of Novidium (Homidium chloride) chemotherapy on genital lesions induced by *Trypanosoma vivax* and *Trypanosoma congolense* infection. *British Medical Journal*, 146, 181–185.

Sekoni, V.O., Njoku, C.O., Kumni-Diaka, J. and Saror, D.I. (1990) Pathological changes in male genitalia of cattle infected with *Trypanosoma vivax* and *Trypanosoma congolense*. *British Medical Journal*, 146, 175–180.

Semple, J.L. and Forno, I.W. (1990) Susceptibility of the Salvinia biological control agent *Cyrtobagous salviniae* (Coleoptera: Curculionidae) to chemicals used to control tsetse fly *Glossina morsitans* (Diptera: Glossinidae) in Botswana. *Bulletin of Entomological Research*, 80, 233–234.

Sequeira, L.A.F. de (1935) *Rapport de la Mission Médicale à la Colonie de Guinée en 1932*. Ecole de Médecine Tropicale.

Service, M.W. (1993) Community participation in vector-borne disease control. *Annals of Tropical Medicine and Parasitology*, 87, 223–234.

Service, M.W., Voller, A. and Bidwell, D. (1986) The enzyme-linked immunosorbent assay (ELISA) test for the identification of blood-meals of haematophagous insects. *Bulletin of Entomological Research*, 76, 321–330.

Shannon, C.E. and Weaver, W. (1948) *The Mathematical Theory of Communication*. Urbana University Press, Illinois, pp. 117–127.

Shastri, U.V. and Deshpande, P.D. (1981) *Hyalomma anatolicum anatolicum* (Koch, 1844) as a possible vector for transmission of *Trypanosoma theileri*, Laveran, 1902 in cattle. *Veterinary Parasitology*, 9, 151–155.

Shaw, A.P.M. and Hoste, C. (1991a) Les échanges internationaux de bovins trypanotolerants. I. Historique et synthèse. *Revue d'Élevage et de Médecine Vétérinaire des Pays Tropicaux*, 44, 229–237.

Shaw, A.P.M. and Hoste, C. (1991b) Les échanges internationaux de bovins trypanotolerants. II. Tendances et perspectives. *Revue d'Élevage et de Médecine Vétérinaire des Pays Tropicaux*, 44, 229–237.

Shaw, M.K. and Moloo, S.K. (1991) Comparative study on Rickettsia-like organisms in the midgut epithelial cells of different *Glossina* species. *Parasitology*, 102, 193–199.

Shaw, M.K. and Moloo, S.K (1993) Virus-like particles in Rickettsia within the midgut epithelial cells of *Glossina morsitans centralis* and *Glossina brevipalpis*. *Journal of Invertebrate Pathology*, 61, 162–166.

Simmonds, A.M. and Leggate, B.M. (1962) A survey method of trypanosome infections in *Glossina*. *Nature*, 194, 1297–1298.

Simpson, H.R. (1958) The effect of sterilised males on a natural tsetse fly population. *Biometrics*, 14, 159–173.

Simpson, J.J. (1918) Bionomics of tsetse and other parasitological notes in the Gold Coast. *Bulletin of Entomological Research*, 8, 193–214.

Slingenbergh, J. (1988) Visual and olfactory attractive devices for tsetse flies. *AGA: TRYP/EDA/88/28/A September 1988; Joint meeting of the Panels of Experts on Ecological/Technical and Development Aspects of the Programme for the Control of African Animal Trypanosomiasis and Related Development, Accra, Ghana, 7–9 November 1988.*

Smith, I.M. and Rennison, B.D. (1958) Some factors concerned in trypanosome challenge. *ISCTRC/CCTA, 7th Meeting Bruxelles 1958*, pp. 63–66.

Smith, I.M. and Rennison, B.D. (1961a) Studies on the sampling of *Glossina pallidipes* Aust. I. The numbers caught daily on cattle, in Morris traps and on a fly round. *Bulletin of Entomological Research*, 52, 165–182.

Smith, I.M. and Rennison, B.D. (1961b) Studies on the sampling of *Glossina pallidipes* Aust. II The daily pattern of flies caught on cattle, in Morris traps and on a fly-round. *Bulletin of Entomological Research*, 52, 183–189.

Smith, I.M. and Scott, W.N. (1961) Chemoprophylaxis against bovine trypanosomiasis. III. The cure of infected cattle removed from a high tsetse density. *Journal of Comparative Pathology*, 71, 325–342.

Smith, K.G.V. and Baldry, D.A.T. (1968) Some dipterous puparia resembling, and found among, those of tsetse flies. *Bulletin of Entomological Research*, 59, 367–370.

Smith, S.C., Harris, E.G. and Wilson, K. (1994) Effect of temperature regime on the toxicity of endosulfan and deltamethrin to tsetse flies, *Glossina morsitans morsitans. Tropical Science*, 34, 391–400.

Smith, T., Charlwood, J.D., Takken, W., Tanner, M. and Spiegelhalter, D.J. (1995) Mapping the densities of malaria vectors within a single village. *Acta Tropica*, 59, 1–18.

Snow, W.F. (1980) Host location and feeding patterns in tsetse. *Insect Science and its Application*, 1, 23–30.

Snow, W.F. (1984) Tsetse feeding habits in an area of endemic sleeping sickness in southern Sudan. *Transactions of the Royal Society of Tropical Medicine and Hygiene*, 78, 413–414.

Snow, W.F. and Tarimo, S.A. (1983) A quantification of the risk of trypanosomiasis infection to cattle on the south Kenya coast. *Acta Tropica*, 40, 331–340.

Sokal, R.R. and Rohlf, F.J. (1981) *Biometry. The principles and practice of statistics in biological research*, 2nd Edn. W.H. Freeman, San Francisco, 859 pp.

Solano, P. and Amsler-Delafosse, S. (1995) *Trypanosoma congolense* chez différentes espèces de taons (Diptera: Tabanidae) au Burkina Faso. *Revue d'Elevage et de Médecine Vétérinaire des Pays Tropicale*, 48, 145–146.

Solano, P., Agriro, L., Reifenberg, J.M., Yao Yao and Duvallet, G. (1995) Field application of the polymerase chain reaction (PCR) to the detection and characterization of trypanosomes in *Glossina longipalpis* (Diptera: Glossinidae) in Côte d'Ivoire. *Molecular Ecology*, 4, 781–785.

Solano, P., Duvallet, G., Dumas, V., Cuisance, D. and Cuny, G. (1997) Microsatellite markers for genetic population studies in *Glossina palpalis* (Diptera: Glossinidae). *Acta Tropica*, 65, 175–180.

Soltys, M.A. (1971) Epidemiology of trypanosomiasis in man. *Tropenmedizin und Parasitologie*, 22, 120–133.

Soltys, M.A. and Woo, P. (1968) Leeches as possible vectors for mammalian trypanosomes. *Transactions of the Royal Society of Tropical Medicine and Hygiene*, 62, 154–156.

Soltys, M.A. and Woo, P. (1970) Further studies on forms of *T. brucei* in a vertebrate host. *Transactions of the Royal Society of Tropical Medicine and Hygiene*, 64, 692–694.

Soltys, M.A. and Woo, P.T.K. (1977) Trypanosomes producing disease in livestock in Africa. *Parasitic Protozoa*, 1, 239–268.

Sonenshine, D.E., Allan, S.A., Norval. R.A.I. and Burridge, M.J. (1996) A self-medicating applicator for control of ticks on deer. *Medical and Veterinary Entomology*, 10, 149–154.

Southern, D.I. (1980) Chromosome polymorphism and aneuploidism in tsetse flies. *Transactions of the Royal Society of Tropical Medicine and Hygiene*, 74, 278–279.

Southern, D.I. and Pell, P.E. (1974) Comparative analysis of the polytene chromosomes of *Glossina austeni* and *Glossina morsitans*. *Chromosoma (Berlin)*, 47, 213–226.

Southern, D.I. and Pell, P.E. (1981) Cytogenetical aspects of *morsitans* tsetse flies with particular reference to *Glossina pallidipes* (Diptera, Glossinidae). *Cytobios*, 30, 135–52.

Southern, D.I., Craig-Cameron, T.A. and Pell, P.E. (1972) The meiotic sequence in *Glossina morsitans morsitans*. *Transactions of the Royal Society of Tropical Medicine and Hygiene*, 66, 145–149.

Southern, D.I., Craig-Cameron, T.A. and Pell, P.E. (1973) Polytene chromosomes of the tsetse fly, *Glossina morsitans*. *Chromosoma*, 40, 107–120.

Southon, H.A.W. (1958) Night observations on *Glossina swynnertoni* Aust. *ISCTR/CCTA 7th meeting Bruxelles*, 219–221.

Southon, H.A.W. (1959) Studies in predation on *Glossina*. *East African Trypanosomiasis Research Organisation Annual Report, 1958 Nairobi*. 56–58.

Southon, H.A.W. and Cunningham, M.P. (1966) Infectivity of trypanosomes derived from individual *Glossina morsitans* Westw. *Nature*, 212, 1477–1478.

Southon, H.A.W. and Robertson, D.H.H. (1961) Isolation of *Trypanosoma rhodesiense* from wild *Glossina palpalis*. *Nature*, 189, 411–412.

Southwood, T.R.E. (1978) *Ecological Methods, with particular reference to the study of insect populations*. Chapman and Hall, London.

Späth, J. (1995a) Trap-orientated behaviour of the tsetse fly species *Glossina tachinoides* (Diptera, Glossinidae). *Entomologia Generalis*, 19, 209–224.

Späth, J. (1995b) Olfactory attractants for West African tsetse flies, *Glossina* spp. (Diptera: Glossinidae). *Tropical Medicine and Parasitology*, 46, 253–257.

Späth, J. (1997) Natural host odours as possible attractants for *Glossina tachinoides* and *G. longipalpis* (Diptera: Glossinidae). *Acta Tropica*, 68, 149–158.

Späth, J. and Küpper, W. (1991) The trap orientated behaviour of *Glossina tachinoides* (Diptera, Glossinidae). *ISCTRC/OAU 21st Meeting, Yamoussoukro, Côte d'Ivoire*, Poster.

Spielberger, U. (1971) Lutte contre la trypanosomiase animales au Niger. *OAU/STRC 13th Meeting, Lagos, Nigeria*. OAU/STRC Publication No. 105, pp. 289–293.

Spielberger, U. and Barwinek, F. (1978) The night resting sites of *Glossina tachinoides* Westwood (Diptera, Glossinidae) in northern Nigeria. *Bulletin of Entomological Research*, 68, 137–144.

Spielberger, U. and Na'isa, B.K. (1977) Aerial application of insecticides by helicopters against *Glossina* spp. Field trials with tetrachlorvinphos, bromphos, dieldrin and Hosalthion in northern Nigeria. *ISCTRC 14th Meeting Dakar, Senegal 1975*, 341–395. Nairobi, OAU/STRC Publication No. 109.

Spielberger, U., Na'isa, B.K. and Abdurrahim, U. (1977) Tsetse (Diptera: Glossinidae) eradication by aerial (helicopter) spraying of persistent insecticides in Nigeria. *Bulletin of Entomological Research*, 67, 589–598.

Spielman, A. (1994) Why entomological antimalaria research should not focus on transgenic mosquitoes. *Parasitology Today*, 10, 374–376.

Staak, C., Allmang, B., Kämpe, U. and Mehlitz, D. (1981) The complement fixation test for the species identification of bloodmeals from tsetse flies. *Tropenmedizin und Parasitologie*, 32, 97–98.

Stafford, W.L. (1973) Changes in the content of glycogen, sugars, nucleic acids, protein and free amino-acids in the pupa of *Glossina morsitans* during its development. *Comparative Biochemistry and Physiology*, 45B, 763–768.

Steelman, C.D. (1976) Effects of external and internal arthropod parasites on domestic livestock production. *Annual Review of Entomology*, 21, 155–178.

Stephen L.E. (1966a) *Pig Trypanosomiasis in Tropical Africa.* Review Series No. 8, Commonwealth Agricultural Bureaux, Farnham Royal, UK, 65 pp.

Stephen, L.E. (1966b) Observations on the resistance of west African N'Dama and Zebu cattle to trypanosomiasis following challenge by *Glossina morsitans*. *Annals of Tropical Medicine and Parasitology*, 60, 230–246.

Stephen, L.E. (1986) *Trypanosomiasis – a Veterinary Perspective.* Pergamon Press, Oxford, 551 pp.

Stephen, L.E. and Gray, A.R. (1960) The trypanocidal activity of nucleocidin against *Trypanosoma vivax* in West African Zebu cattle. *Journal of Parasitology*, 46, 509.

Stephens, J.W.W. and Fantham, H.B. (1910) On the peculiar morphology of a trypanosome from a case of sleeping sickness and the possibility of its being a new species (*T. rhodesiense*). *Annals of Tropical Medicine and Parasitology*, 4, 343–350.

Stevens, J.R. and Welburn, S.C. (1993) Genetic processes within an epidemic of sleeping sickness in Uganda. *Parasitology Research*, 79, 421–427.

Stevens, J.R., Mathieu-Daudé, F., McNamara, J.J., Mizen, V.H. and Nzila, A. (1994) Mixed populations of *Trypanosoma brucei* in wild *Glossina palpalis palpalis*. *Tropical Medicine and Parasitology*, 45, 313–318.

Stewart, J.L. (1947) Porcine trypanosomiasis. *Veterinary Record*, 59, 648.

Stiles, J.K., Molyneux, D.H. and Wallbanks, K.R. (1988) Viruslike particles in *Glossina palpalis gambiensis* (Diptera: Glossinidae). *Annales de la Société Belge de Médecine Tropicale*, 68, 161–163.

Stiles, J.K., Ingram, G.A., Wallbanks, K.R., Molyneux, D.H., Maudlin, I. and Welburn, S.C. (1990) Identification of midgut trypanolysin and trypanoagglutinin of *Glossina palpalis* sspp. (Diptera: Glossinidae). *Parasitology*, 101, 369–376.

Stiles, J.K., Wallbanks, K.R. and Molyneux, D.H. (1991) The use of casein substrate gels for determining trypsin-like activity in the midgut of *Glossina palpalis* spp. (Diptera: Glossinidae). *Journal of Insect Physiology*, 37, 247–254.

Stone, B.F. (1981) A review of the genetics of resistance to acaricidal organochlorine and organophosphorus compounds with particular reference to the cattle tick *Boophilus microplus*. In: Whitehead, G.B. and Gibson, J.D. (eds) *Tick Biology and Control*. Proceedings of an International conference at the Tick Research Unit, Rhodes University, South Africa, pp. 95–102.

Strong, I. and Brown, T.A. (1987) Avermectins in insect control and biology: a review. *Bulletin of Entomological Research*, 77, 357–389.

Stuhlmann, F. (1907) Beiträge zur Kentniss der Tsetsefliege (*Glossina fusca* und *Gl. tachinoides*). *Arbeiten aus dem Kaiserlichen Gesundheitsamte*, 26, 301–308. (In: Evans and Ellis, 1983).

Sutherland, I.A., Moloo, S.K., Holmes, P.H. and Peregrine, A.S. (1991) Therapeutic and prophylactic activity of Isometamidium chloride against a tsetse-transmitted drug-resistant clone of *Trypanosoma congolense* in Boran cattle. *Acta Tropica*, 49, 57–64.

Swallow, B.M. and Mulatu, W. (1994) Evaluating willingness to contribute to a local public good: application of contingent valuation to tsetse control in Ethiopia. *Ecological Economics*, 11, 153–161.

Swallow, B.M., Mulatu, W. and Leak, S.G.A. (1995) Potential demand for a mixed public–private animal health input: evaluation of a pour-on insecticide for controlling tsetse-transmitted trypanosomiasis in Ethiopia. *Preventive Veterinary Medicine*, 24, 265–275.

Swynnerton, C.F.M. (1923a) Tsetse flies breeding in open ground. *Bulletin of Entomological Research*, 14, 119.

Swynnerton, C.F.M. (1923b) The entomological aspects of an outbreak of sleeping sickness near Mwanza, Tanganyika territory. *Bulletin of Entomological Research*, 13, 317–372.

Swynnerton, C.F.M. (1925) The tsetse-fly problem in the Nzega sub-district Tanganyika Territory. *Bulletin of Entomological Research*, 16, 99–110.

Swynnerton, C.F.M. (1933) Some traps for tsetse flies. *Bulletin of Entomological Research*, 24, 69–106.

Swynnerton, C.F.M. (1936) The tsetse flies of East Africa. *Transactions of the Royal Entomological Society of London*, 84, 1–579.

Symes, C.B. and McMahon, J.P. (1937) The food of tsetse-flies (*Glossina swynnertoni* and *G. palpalis*) as determined by the precipitin test. *Bulletin of Entomological Research*, 28, 31–42.

Tacher, G., Jahnke, H.E., Rojat, D. and Kell, P. (1988) Livestock development and economic production in tsetse-infested Africa. In: *Livestock Production in Tsetse Affected Areas of Africa*. Proceedings of a meeting, November 1987, Nairobi, Kenya. English Press, Nairobi, pp. 329–349.

Tait, A. (1980) Evidence for diploidy and mating in trypanosomes. *Nature*, 237, 536–538.

Tait, A., Buchanan, N., Hide, G. and Turner, C.M.R. (1996) Self-fertilisation in *Trypanosoma brucei*. *Molecular and Biochemical Parasitology*, 76, 31–42.

Taiwo, V.O., Nantulya, V.M., Moloo, S.K. and Ikede, B.O. (1990) Role of the chancre in induction of immunity to tsetse-transmitted *Trypanosoma* (*Nannomonas*) *congolense* in goats. *Veterinary Immunology and Immunopathology*, 26, 59–70.

Takken, W. (1984) Studies on the biconical trap as a sampling device for tsetse in Mozambique. *Insect Science and its Application*, 5, 357–361.

Takken, W., Balk, F., Jansen, R.C. and Koeman, J.H. (1978) The experimental application of insecticides from a helicopter for the control of riverine populations of *Glossina tachinoides* in West Africa. VI. Observations on side effects. *PANS*, 24, 455–466.

Takken, W., Oladunmade, M.A., Dengwat, L., Feldman, H.U., Onah, J.A., Tenabe, S.O. and Hammann, H.J. (1986) The eradication of *Glossina palpalis palpalis* using traps, insecticide-impregnated targets and the sterile insect technique in Central Nigeria. *Bulletin of Entomological Research*, 76, 275–286.

Targett, G.A.T. (1980) Malnutrition and immunity in to protozoan parasites. In: *The Impact of Malnutrition on Immune Defense in Parasitic Infestation.* Nestlé Foundation Workshop, Lausanne, 1980. Nestlé Foundation Publication Series, no. 2, pp. 158–179.

Tarimo, C.S. (1980) Contribution to the epidemiology of *Trypanosoma rhodesiense* Sleeping sickness in lower Kitete, northern Tanzania, by cultural. *Insect Science and its Application*, 1, 73–76.

Tarimo, C.S., Lee, C.W., Parker, J.D. and Matechi, H.T. (1970) Aircraft application of insecticides in East Africa. XIX. A comparison of two sampling techniques for assessing the effectiveness of pyrethrum applications on *Glossina pallidipes* Aust. *Bulletin of Entomological Research*, 60, 221–223.

Tarimo, S.R.A., Snow, W.F. and Butler, L. (1984) Trypanosome infections in wild tsetse, *Glossina pallidipes* Austen on the Kenya coast. *Insect Science and its Application*, 5, 415–418.

Tarimo, S.R.A., Snow, W.F., Butler, L. and Dransfield, R.D. (1985) The probability of tsetse acquiring trypanosome infection from a single blood meal in different localities in Kenya. *Acta Tropica*, 42, 199–207.

Tarimo Nesbitt, S., Njau, B.C. and Otieno, L.H. (1991) Epizootiology of trypanosomiasis in Lambwe valley, Kenya, east Africa. *Insect Science and its Application*, 12, 379–384.

Taute, M. (1911) Experimentelle studien uber die Beziehungen der *Glossina morsitans* zur Schlafkrankheit. *Zeitschrift für Hygiene und Infektionskrankheiten*, 69, 553–558.

Taute, M. (1913) Untersungen über die Bedeutung der Grosswildes und der Haustiere für die Verbreitung der Schlafkrankheit. *Arbeiten aus dem Kaiserlichen Gesundheitsamte*, 45, 102.

Taylor, A.W. (1932a) The development of West African strains of *Trypanosoma gambiense* in *Glossina tachinoides* under normal laboratory conditions, and at raised temperatures. *Parasitology*, 24, 401–418.

Taylor, A.W. (1932b) Pupal parasitism in *Glossina morsitans* and *Glossina tachinoides* at Gadau, Northern Nigeria. *Bulletin of Entomological Research*, 23, 463–467.

Taylor, P. (1976) Blood-meal size of *Glossina morsitans* Westw. and *G. pallidipes* Austen (Diptera: Glossinidae) under field conditions. *Transactions of the Rhodesian Scientific Association*, 57, 29–34.

Taylor, P. (1979) The construction of a life-table for *Glossina morsitans morsitans* Westwood (Diptera: Glossinidae) from seasonal age-measurements of a wild population. *Bulletin of Entomological Research*, 69, 553–560.

Taylor, P., Govere, J. and Crees, M.J. (1986) A field trial of microencapsulated deltamethrin, a synthetic pyrethroid, for malaria control. *Transactions of the Royal Society for Tropical Medicine and Hygiene*, 80, 537–545.

Taylor, S.M., Elliott, C.T. and Blanchflower, J. (1985) Cypermethrin concentrations in hair of cattle after application of impregnated ear-tags. *Veterinary Record*, 116, 620.

Teale, A.J. (1993) The trypanosomiasis research programme at ILRAD. In: McKeever, D.J. (ed.) *Novel Immunization Strategies against Protozoan Parasites.* Proceedings of a workshop held at ILRAD, November 1993, Nairobi, Kenya, pp. 3–5.

Teale, A.J., Wambugu, J., Gwakisa, P.S., Stranzinger, G., Bradley, D. and Kemp, S.J. (1995) A polymorphism in randomly amplified DNA that differentiates the Y chromosomes of *Bos indicus* and *Bos taurus. Animal Genetics*, 26, 243–248.

Teesdale, C. (1940) Fertilisation in the tsetse fly, *Glossina palpalis*, in a population of low density. *Journal of Animal Ecology*, 9, 24–26.

Thévenaz, P. and Hecker, H. (1980) Distribution and attachment of *Trypanosoma* (*Nannomonas*) *congolense* in the proximal part of the proboscis of *Glossina morsitans morsitans*. *Acta Tropica*, 37, 163–175.

Thomas, R.E., McDonough, K.A. and Schwan, T.G. (1989) Use of DNA hybridisation's probes for detection of the plague Bacillus (*Yersinia pestis*) in fleas (Siphonaptera: Pulicidae and Ceratophyllidae). *Journal of Medical Entomology*, 26, 342–348.

Thomson, J.W. and Wilson, A. (1992a) A review of developments in tsetse fly (*Glossina* spp.) control by application of insecticides to cattle. *Bulletin of Animal Health and Production in Africa*, 40, 1–4.

Thomson, J.W. and Wilson, A. (1992b) The control of tsetse flies and trypanosomiasis by the application of deltamethrin to cattle. *Bulletin of Animal Health and Production in Africa*, 40, 5–8.

Thomson, J.W., Mitchell, M., Rees, R.B., Shereni, W., Schönefeld, A.H. and Wilson, A. (1991) Studies on the efficacy of deltamethrin applied to cattle for the control of tsetse flies (*Glossina* spp.) in southern Africa. *Tropical Animal Health and Production*, 23, 221–226.

Thomson, M. (1987) The effect on tsetse flies (*Glossina* spp.) of Deltamethrin applied to cattle either as a spray or incorporated into eartags. *Pest Management*, 33, 329–335.

Thomson, W.E.F. (1947) Nematodes in tsetse. *Annals of the Society of Tropical Medicine and Parasitology*, 41, 164.

Thomson, W.E.F., Glover, P.E. and Trump, E.C. (1960) The extermination of *Glossina pallidipes* from an isolated area on Lake Victoria with the use of insecticides. *OAU/ISCTRC 8th Meeting, Jos*, Publication No. 62, 303–308.

Thrusfield, M. (1995) *Veterinary Epidemiology*, 2nd Edn. Blackwell Science, London, 479 pp.

Tibayrenc, R., Itard, J. and Cuisance, D. (1971) Marquage des glossines par des poudres fluorescentes. *Revue d'Elevage et de Médecine Vétérinaire des Pays Tropicale*, 56, 430–442.

Tibayrenc, M., Kjellberg, F. and Ayala, F.J. (1990) A clonal theory of parasitic protozoa: the population structures of *Entamoeba, Giardia, Leishmania, Naegleria, Plasmodium, Trichomonas* and *Trypanosoma* and their medical and taxonomical consequences. *Proceedings of the National Academy of Sciences, USA*, 87, 2414–2418.

Tikubet, G. and Gemetchu, T. (1984) Altitudinal distribution of tsetse in the Finchaa river valley (western part of Ethiopia). *Insect Science and its Application*, 5, 389–395.

Tingle, C.C.D. (1993) Bait location by ground foraging ants (Hymenoptera: Formicidae) in Mopane woodland selectively sprayed to control tsetse flies (Diptera: Glossinidae) in Zimbabwe. *Bulletin of Entomological Research*, 83, 259–265.

Titus, R.G. and Ribeiro, J.M.C. (1990) The role of vector saliva in transmission of arthropod-borne disease. *Parasitology Today*, 6, 157–160.

Tobe, S.S. and Davey, K.G. (1971) The choriothete of *Glossina austeni. Bulletin of Entomological Research*, 61, 363–368.

Tobe, S.S. and Davey, K.G. (1972) Volume relationships during the larviposition cycle of *Glossina austeni. Canadian Journal of Zoology*, 50, 999–1010.

Tobe, S.S. and Davey, K.G. (1974a) Nucleic acid synthesis during the reproductive cycle of *Glossina austeni*. *Insect Biochemistry*, 4, 215–223.

Tobe, S. and Davey, K.G. (1974b) Autoradiographic study of protein synthesis in abdominal tissues of *Glossina austeni*. *Tissue and Cell*, 6, 255–268.

Tobe, S. and Langley, P.A. (1978) Reproductive physiology of *Glossina*. *Annual Review of Entomology*, 23, 283–307.

Tobe, S.S., Davey, K.G. and Huebner, E. (1973) Nutrient transfer during the reproductive cycle in *Glossina austeni* Newst.: histology and histochemistry of the milk gland, fat body, and oenocytes. *Tissue and Cell*, 5, 633–650.

Torr, S.J. (1985) The susceptibility of *Glossina pallidipes* Austen (Diptera: Glossinidae) to insecticide deposits on targets. *Bulletin of Entomological Research*, 75, 451–458.

Torr, S.J. (1987) The host-orientation behaviour in tsetse flies (Diptera: Glossinidae). PhD thesis, London.

Torr, S.J. (1988a) The activation of resting tsetse flies (*Glossina*) in response to visual and olfactory stimuli in the field. *Physiological Entomology*, 13, 315–325.

Torr, S.J. (1988b) Behaviour of tsetse flies (*Glossina*) in host odour plumes in the field. *Physiological Entomology*, 13, 467–478.

Torr, S.J. (1989) The host-orientated behaviour of tsetse flies (*Glossina*): the interaction of visual and olfactory stimuli. *Physiological Entomology*, 14, 325–340.

Torr, S.J. (1990) Dose responses of tsetse flies (*Glossina*) to carbon dioxide, acetone and octenol in the field. *Physiological Entomology*, 15, 93–103.

Torr, S.J. (1994) Responses of tsetse flies (Diptera: Glossinidae) to warthog (*Phacochoerus aethiopicus* Pallas). *Bulletin of Entomological Research*, 84, 411–419.

Torr, S.J., Parker, A.G. and Leigh-Browne, G. (1989) The responses of *Glossina pallidipes* Austen (Diptera: Glossinidae) to odour baited traps and targets in southern Somalia. *Bulletin of Entomological Research*, 79, 99–108.

Torr, S.J., Holloway, M.T.P. and Vale, G.A. (1992) Improved persistence of insecticide deposits on targets for controlling *Glossina pallidipes* (Diptera: Glossinidae). *Bulletin of Entomological Research*, 82, 525–533.

Torr, S.J., Mangwiro, T.N.C. and Hall, D.R. (1996) Responses of *Glossina pallidipes* (Diptera: Glossinidae) to synthetic repellents in the field. *Bulletin of Entomological Research*, 86, 609–616.

Torr, S.J., Hall, D.R., Phelps, R.J. and Vale, G.A. (1997) Methods for dispensing odour attractants for tsetse flies (Diptera: Glossinidae) *Bulletin of Entomological Research*, 87, 299–311.

Touré, S.M., Seye, M., Gueye, E. and Diaite, M. (1981) Etudes comparatives sur les bovins N'Dama de haute Casamance pour évaluer leur trypanotolérance en fonction de la couleur de leur robe. *Revue d'Élevage et de Médecine Vétérinaire des Pays Tropicale*, 34, 281–287.

Trail, J.C.M., Murray, M., Sones, K., Jibbo, J.M.C., Durkin, J. and Light, D. (1985) Boran cattle maintained by chemoprophylaxis under trypanosomiasis risk. *Journal of Agricultural Science Cambridge*, 105, 147–166.

Trail, J.C.M., D'Ieteren, G.D.M., Colardelle, C., Maille, J.C., Ordner, G., Sauveroche, B. and Yangari, G. (1991) Evaluation of a field test for trypanotolerance in young N'Dama cattle. *Acta Tropica*, 48, 47–57.

Travassos Santos Dias, J.A. (1987) Contribução para o estudo da sistemática do género *Glossina* Wiedemann, 1830 (Insecta, Brachycera, Cyclorrhapa, Glossinidae) Proposta para a criação de um novo subgénero? *Garcia de Orta, Series Zoologia, Lisboa*, 14, 67–78.

Truc, P., Mathieu-Daudé, F. and Tibayrenc, M. (1991) Multilocus isozyme identification of *Trypanosoma brucei* stocks isolated in Central Africa: evidence for an animal reservoir of sleeping sickness. *Acta Tropica,* 49, 127–135.

Truc, P., Aerts, D., McNamara, J.J., Claes, Y., Allingham, R., Le Ray, D. and Godfrey, D.G. (1992) Direct isolation *in vitro* of *Trypanosoma brucei* from man and other animals, and its potential value for the diagnosis of Gambian trypanosomiasis. *Transactions of the Royal Society of Tropical Medicine and Hygiene,* 86, 627–629.

Tucker, C.J., Dregne, H. and Newcomb, W. (1991) Expansion and contraction of the Sahara desert from 1980 to 1990. *Science,* 253, 299–301.

Turner, D.A. (1980a) A novel marking–release recapture method for possible use in determining aspects of tsetse fly behaviour. *Insect Science and its Application,* 1, 1–7.

Turner, D.A. (1980b) Tsetse ecological studies in Niger and Mozambique. 1. Population sampling. *Insect Science and its Application,* 1, 9–13.

Turner, D.A. (1980c) Tsetse ecological studies in Niger and Mozambique. II. Resting behaviour. *Insect Science and its Application,* 1, 15–21.

Turner, D.A. and Brightwell, R. (1986) An evaluation of a sequential aerial spraying operation against *Glossina pallidipes* Austen (Diptera: Glossinidae) in the Lambwe valley of Kenya: aspects of post-spray recovery and evidence of natural population regulation. *Bulletin of Entomological Research,* 76, 331–349.

Turner, D.A. and Golder, T.K. (1986) Susceptibility to topical applications of endosulfan and dieldrin in sprayed and unsprayed populations of *Glossina pallidipes* Austen (*Diptera: Glossinidae*) in Kenya. *Bulletin of Entomological Research,* 76, 519–527.

Turner, R.E. (1915) A new species of *Mutilla* parasitic on *Glossina morsitans. Bulletin of Entomological Research,* 5, 383–386.

Turner, R.E. (1917) On a Braconid parasite of *Glossina. Bulletin of Entomological Research,* 8, 177.

Turton, J.D. (1953) An outbreak of *Trypanosoma brucei* infection in polo ponies with spread apparently by biting flies other than *Glossina. Veterinary Record,* 65, 11–12.

Tyndale-Biscoe, M. (1984) Age grading methods in adult insects: a review. *Bulletin of Entomological Research,* 74, 341–377.

Umali, D.L., Feder, G. and de Haan, C. (1994) Animal health services: finding the balance between public and private delivery. *The World Bank Research Observer,* 9, 71–96.

Unnasch, T.R. (1987) DNA probes to identify *Onchocerca volvulus. Parasitology Today,* 3, 377–378.

Urquhart, G.M. (1980) The pathogenesis and immunology of African trypanosomiasis in domestic animals. *Transactions of the Royal Society of Tropical Medicine and Hygiene,* 74, 726–729.

USAID (1976) *Feasibility Report on the Upper Didessa Project.* TAMS Agricultural Group, Washington, DC.

Vale, G.A. (1969) Mobile attractants for tsetse flies. *Arnoldia,* 33, 1–6.

Vale, G.A. (1971) Artificial refuges for tsetse flies (*Glossina* spp.). *Bulletin of Entomological Research,* 61, 331–350.

Vale, G.A. (1974a) New field methods for studying the responses of tsetse flies (Diptera, Glossinidae) to hosts. *Bulletin of Entomological Research,* 64, 199–208.

Vale, G.A. (1974b) The responses of tsetse flies (Diptera, Glossinidae) to mobile and stationary baits. *Bulletin of Entomological Research*, 64, 545–588.

Vale, G.A. (1974c) Direct observations on the responses of tsetse flies (Diptera, Glossinidae) to hosts. *Bulletin of Entomological Research*, 64, 589–594.

Vale, G.A. (1974d) Attractants for controlling and surveying tsetse populations. *Transactions of the Royal Society of Tropical Medicine and Parasitology*, 68, 11–12.

Vale, G.A. (1977a) The flight of tsetse flies (Diptera: Glossinidae) to and from a stationary ox. *Bulletin of Entomological Research*, 67, 297–303.

Vale, G.A. (1977b) Feeding responses of tsetse flies (Diptera: Glossinidae) to stationary hosts. *Bulletin of Entomological Research*, 67, 635–649.

Vale, G.A. (1978) Changes in our understanding of the behaviour and distribution of tsetse flies. *Rhodesian Science News*, 12, 144–146.

Vale, G.A. (1982) The interaction of men and traps as baits for tsetse. *Zimbabwe Journal of Agricultural Research*, 20, 179–183.

Vale, G.A. (1984) The responses of *Glossina* (Glossinidae) and other Diptera to odour plumes in the field. *Bulletin of Entomological Research*, 74, 143–152.

Vale, G.A. (1986) Prospects for tsetse control. *Proceedings of the 6th International Congress of Protozoology*, 665–670.

Vale, G.A. (1988) *The Recent Implementation and Investigations of Spraying Measures and Bait Techniques for Tsetse Control in Zimbabwe*. Meeting of the FAO panel of experts on ecological/technical development aspects of the programme for the control of African Animal Trypanosomiasis and related development, Accra, Ghana, 7–9 November 1988.

Vale, G.A. (1991) Responses of tsetse flies (Diptera: Glossinidae) to odour-baited trees. *Bulletin of Entomological Research*, 81, 323–331.

Vale, G.A. (1993a) Development of baits for tsetse flies (Diptera: Glossinidae) in Zimbabwe. *Journal of Medical Entomology*, 30, 831–842.

Vale, G.A. (1993b) Visual responses of tsetse flies (Diptera: Glossinidae) to odour-baited targets. *Bulletin of Entomological Research*, 83, 277–289.

Vale, G.A. and Cumming, D.H.M. (1976) The effects of selective elimination of hosts on a population of tsetse flies *Glossina morsitans morsitans* Westwood (Diptera: Glossinidae). *Bulletin of Entomological Research*, 66, 713–729.

Vale, G.A. and Flint, S. (1986) The field responses of tsetse flies, *Glossina* spp. (Diptera: Glossinidae), to odours of host residues. *Bulletin of Entomological Research*, 76, 685–693.

Vale, G.A. and Hall, D.R. (1985) The use of 1-octen-3-ol and carbon dioxide to improve baits for tsetse flies *Glossina* spp. (Diptera: Glossinidae). *Bulletin of Entomological Research*, 75, 219–231.

Vale, G.A. and Hargrove, J.W. (1975) Field attraction of tsetse flies (Diptera: Glossinidae) to ox odour; the effects of dose. *Transactions of the Rhodesian Scientific Association*, 56, 46–50.

Vale, G.A. and Hargrove, J.W. (1979) A method of studying the efficiency of traps for tsetse flies (Diptera: Glossinidae) and other insects. *Bulletin of Entomological Research*, 69, 183–193.

Vale, G.A. and Phelps, R.J. (1978) Sampling problems with tsetse flies (Diptera: Glossinidae). *Journal of Applied Ecology*, 15, 715–726.

Vale, G.A., Bursell, E. and Hargrove, J.W. (1985) Catching-out the tsetse. *Parasitology Today*, 1, 106–110.

Vale, G.A., Hargrove, J.W., Cockbill, G.F., and Phelps, R.J. (1986) Field trials of baits to control populations of *Glossina morsitans morsitans* Westwood and *G. pallidipes* Austen (Diptera: Glossinidae). *Bulletin of Entomological Research*, 76, 179–193.

Vale, G.A., Hall, D.R. and Gough, A.J.E. (1988a) The olfactory responses of tsetse flies, *Glossina* spp. (Diptera: Glossinidae), to phenols and urine in the field. *Bulletin of Entomological Research*, 78, 293–300.

Vale, G.A., Lovemore, D.F., Flint, S. and Cockbill, G.F. (1988b) Odour-baited targets to control tsetse flies, *Glossina* spp. (Diptera: Glossinidae), in Zimbabwe. *Bulletin of Entomological Research*, 78, 31–49.

Vale, G.A., Wilcox, J. and Abson, J. (1994) Prospects for using odour-baited trees to control tsetse flies (Diptera: Glossinidae). *Bulletin of Entomological Research*, 84, 123–130.

Van den Abbeele, J., D'Haeseleer, F. and Goosens, M. (1986) Efficacy of Ivermectin on the reproductive biology of *Glossina palpalis palpalis* (Rob. Desv.) (*Glossinidae: Diptera*). *Annales de la Société Belge de Médecine Tropicale*, 66, 167–172.

Van den Berghe, L. and Lambrecht, F.L. (1963) The epidemiology and control of human trypanosomiasis in *Glossina morsitans* flybelts. *American Journal of Tropical Medicine and Hygiene*, 21, 129.

Van den Berghe, L. and Zaghi, A.J. (1963) Wild pigs as hosts of *Glossina vanhoofi* Henrard and *Trypanosoma suis* Ochmann in the Central African forest. *Nature*, 197, 1126–1127.

Van den Bosch, R. and Messenger, P.S. (1973) *Biological Control*. International Textbook Company, Aylesbury, UK, 180 pp.

Van den Bossche, P. (1988) Preliminary observations of tsetse flies fed on a pig dipped in deltamethrin. *Annales de la Société Belge de Médecine Tropicale*, 68, 159–160.

Van den Bossche, P. (1996) Laboratory bioassays of deltamethrin, topically applied, during the hunger cycle of male *Glossina tachinoides*. *Revue d'Élevage et de Médecine Vétérinaire des Pays Tropicale*, 49, 329–333.

Van den Bossche, P. and Geerts, S. (1988) The effects on longevity and fecundity of *Glossina tachinoides* after feeding on pigs treated with Ivermectin. *Annales de la Société Belge de Médecine Tropicale*, 68, 133–139.

Van den Bossche, P. and Staak, C. (1997) The importance of cattle as a food source for *Glossina morsitans morsitans* in Katete district, Eastern Province, Zambia. *Acta Tropica*, 65, 105–109.

Van der Vloedt, A.M.V., Baldry, D.A.T., Politzar, H., Kulzer, H. and Cuisance, D. (1980) Experimental helicopter applications of decamethrin followed by release of sterile males for the control of riverine vectors of trypanosomiasis in Upper Volta. *Insect Science and its Application*, 1, 105–112.

Van Etten, J. (1982a) Comparative studies on fat reserves, feeding and metabolic strategies of flies from two allopatric populations of *Glossina pallidipes* Austen in Kenya. *Acta Tropica*, 39, 157–169.

Van Etten, J. (1982b) Comparative studies on the diurnal activity pattern in two field and laboratory populations of *Glossina pallidipes*. *Entomologia Experimentalis et Applicata*, 32, 38–45.

Van Etten, J. (1982c) Enzyme polymorphism in populations of the tsetse fly *Glossina pallidipes* in Kenya. *Entomologia Experimentalis et Applicata*, 31, 197–201.

Van Hoeve, K. and Cunningham, M.P. (1964) Prophylactic activity of Berenil against trypanosomes in treated cattle. *Veterinary Record*, 76, 260.

Van Hoeve, K., Onyango, R.J., Harley, J.M.B. and de Raadt, P. (1967) The epidemiology of *Trypanosoma rhodesiense* sleeping sickness in Alego location, Central Nyanza, Kenya II. The cyclical transmission of *Trypanosoma rhodesiense* isolated from cattle to man, a cow and to sheep. *Transactions of the Royal Society of Tropical Medicine and Hygiene*, 61, 684–687.

Van Hoof, L.M.J.J. (1947) Observations on trypanosomiasis in the Belgian Congo. The second Royal Society of Tropical Medicine and Hygiene Chadwick Lecture. *Transactions of the Royal Society of Tropical Medicine and Hygiene*, 40, 728–761.

Van Hoof, L., Henrard, C. and Peel, E. (1937a) Influences modificatrices de la transmissibilité cyclique du *Trypanosoma gambiense* par *Glossina palpalis*. *Annales de la Société Belge de Médecine Tropicale*, 17, 249–272.

Van Hoof, L., Henrard, C. and Peel, E. (1937b) Sur le rôle du porc indigène comme réservoir de *Trypanosoma gambiense*. *Société Belge de Biologie*, 126, Comtes Rendus de la Société de Biologie, pp. 72–75.

Van Hoof, L., Henrard, C. and Peel, E. (1937c) Action de repas médicamenteux sur l'évolution des trypanosomes pathogènes chez la *Glossina palpalis*. *Annales de la Société Belge de Médecine Tropicale*, 17, 385–440.

Van Sickle, J. (1988) Invalid estimates of the rate of population increase from *Glossina* (Diptera: Glossinidae) age distributions. *Bulletin of Entomological Research*, 78, 155–161.

Van Sickle, J. and Phelps, R.J. (1988) Age distributions and reproductive status of declining and stationary populations of *Glossina pallidipes* Austen (Diptera: Glossinidae) in Zimbabwe. *Bulletin of Entomological Research*, 78, 51–61.

Vanderplank, F.L. (1941) Activity of *Glossina pallidipes* and the lunar cycle (Diptera). *Proceedings of the Royal Entomological Society of London (A)*, 16, Parts 4–6, 61–64.

Vanderplank, F.L. (1944) Studies on the behaviour of the tsetse-fly (*Glossina pallidipes*) in the field: Attractiveness of various baits. *Journal of Animal Ecology*, 13, 39–48.

Vanderplank, F.L. (1947a) Some observations on the hunger-cycle of the tsetse-flies *Glossina swynnertoni* and *G. pallidipes* (Diptera) in the field. *Bulletin of Entomological Research*, 38, 431–438.

Vanderplank, F.L. (1947b) Experiments with DDT on various species of tsetse flies in the field and laboratory. *Transactions of the Royal Society of Tropical Medicine and Hygiene*, 40, 603–620.

Vanderplank, F.L. (1947c) Experiments in the hybridisation of tsetse-flies (*Glossina*, Diptera) and the possibility of a new method of control. *Transactions of the Royal Entomological Society*, 98, 1–18.

Vanderplank, F.L. (1948) Experiments in cross-breeding tsetse-flies (*Glossina* species). *Annals of Tropical Medicine and Parasitology*, 42, 131–152.

Verboom, W.C. (1965) The use of aerial photographs for vegetation surveys in relation with tsetse control, and grassland surveys in Zambia. *ITC Publication Number 28, Series B*, 21 pp.

Vey, A. (1971) Recherches sur les champignons pathogènes pour les glossines. Études sur *Glossina fusca congolensis* Newst. et Evans en République Centrafricaine. *Revue d'Élevage et de Médecine Vétérinaire des Pays Tropicale*, 24, 577–579.

Vickerman, K. (1973) The mode of attachment of *Trypanosoma vivax* in the proboscis of the tsetse fly *Glossina fuscipes*: an ultrastructural study of the epimastigote stage of the trypanosome. *Journal of Protozoology*, 20, 394–404.

Vickerman, K. (1974) The ultrastructure of pathogenic flagellates. In: Elliot, K., O'Connor, M. and Walstenholme, G.E.W. (eds) *Trypanosomiasis and Leishmaniasis with Special Reference to Chagas Disease*. Ciba Foundation Symposium No. 20, Amsterdam Association of Scientific Publishers, pp. 171–198.

Vreysen, M.J.B. (1995) Radiation induced sterility to control tsetse flies: the effect of radiation and hybridisation on tsetse biology and the use of the sterile insect technique in integrated tsetse control. PhD thesis, Wageningen University, The Netherlands.

Vreysen, M.J.B. and Van der Vloedt, A.M.V. (1995a) Radiation sterilization of *Glossina tachinoides* Westw. pupae. I. The effect of dose fractionation and nitrogen during irradiation in the mid-pupal phase. *Revue d'Élevage et de Médecine Vétérinaire des Pays Tropicaux*, 48, 45–51.

Vreysen, M.J.B. and Van der Vloedt, A.M.V. (1995b) Radiation sterilization of *Glossina tachinoides* Westw. pupae. II. The combined effects of chilling and gamma irradiation. *Revue d'Élevage et de Médecine Vétérinaire des Pays Tropicaux*, 48, 53–61.

Vreysen, M.J.B., Khamis, I.S. and Van der Vloedt, A.M.V. (1996) Evaluation of sticky panels to monitor populations of *Glossina austeni* (Diptera: Glossinidae) on Unguja island of Zanzibar. *Bulletin of Entomological Research*, 86, 289–296.

Vundla, R.M.W. and Whitehead, D.L. (1985) An enterokinase in the gut of pharate adult of *Glossina morsitans morsitans*. *Acta Tropica*, 42, 79–85.

Waage, J.K. (1981) How the zebra got its stripes – biting flies as selective agents in the evolution of zebra coloration. *Journal of the Entomological Society of South Africa*, 44, 351–358.

Wacher, T.J., Rawlings, P. and Snow, W.F. (1993) Cattle migration and stocking densities in relation to tsetse – trypanosomiasis challenge in The Gambia. *Annals of Tropical Medicine and Parasitology*, 87, 517–524.

Wacher, T.J., Milligan, P.J.M., Rawlings, P. and Snow, W.F. (1994) Tsetse-trypanosomiasis challenge to village N'Dama cattle in The Gambia: field assessments of spatial and temporal patterns of tsetse–cattle contact and the risk of trypanosomiasis infection. *Parasitology*, 109, 149–162.

Wain, E.B., Sutherst, R.W., Burridge, M.J., Pullan, N.B. and Reid, H.W. (1970) Survey for trypanosome infections in domestic cattle and wild animals in areas of east Africa. III. Salivarian trypanosome infections in relation to the *Glossina* distribution in Busoga District, Uganda. *British Veterinary Journal*, 126, 634–641.

Wakelin, D. (1978) Genetic control of susceptibility and resistance to parasitic infection. *Advances in Parasitology*, 16, 219–308.

Wall, R. (1988) Tsetse mating behaviour: effects of hunger in *Glossina morsitans* and *G. pallidipes*. *Physiological Entomology*, 13, 479–486.

Wall, R. (1989a) Ovulation, insemination and mating in the tsetse fly *Glossina pallidipes*. *Physiological Entomology*, 14, 475–484.

Wall, R. (1989b) Sexual responses of males of *Glossina morsitans morsitans* Westwood and *G. pallidipes* Austen (Diptera: Glossinidae) to traps and targets. *Bulletin of Entomological Research*, 79, 335–343.

Wall, R. (1989c) The rôles of vision and olfaction in mate location by males of the tsetse fly *Glossina morsitans morsitans. Medical and Veterinary Entomology,* 3, 147–152.

Wall, R. (1990) Ovarian ageing of tsetse flies (Diptera: Glossinidae) – interspecific differences. *Bulletin of Entomological Research,* 80, 85–89.

Wall, R. and Langley, P.A. (1991) From behaviour to control: The development of trap and target techniques for tsetse fly population management. *Agricultural Zoology Reviews,* 4, 137–159.

Wall, R. and Langley, P.A. (1993) The mating behaviour of tsetse flies (*Glossina*): a review. *Physiological Entomology,* 18, 211–218.

Wall, R., Langley, P.A., Stevens, J. and Clarke, G.M. (1990) Age determination in the old-world screw-worm fly, *Chrysomyia bezziana* by pteridine fluorescence. *Journal of Insect Physiology,* 36, 213–218.

Wallace, F.G. (1962) The trypanosomatid parasites of horseflies with the description of *Crithidia rileyi* n. sp. *Journal of Protozoology,* 9, 53–58.

Wallbanks, K.R., Ingram, G.A. and Molyneux, D.H. (1986) The agglutination of erythrocytes and *Leishmania* parasites by sandfly gut extracts; evidence for lectin activity. *Tropical Medicine and Parasitology,* 37, 409–413.

Ward, R.A. (1968) The susceptibility of *Glossina austeni* to infection with *Trypanosoma brucei. Transactions of the Royal Society of Tropical Medicine and Hygiene,* 62, 672–678.

Ward, R.A. and Bell, L.H. (1972) Susceptibility of *Glossina austeni* and *G. morsitans* to infection with *Trypanosoma congolense. Transactions of the Royal Society of Tropical Medicine and Hygiene,* 66, 308.

Warnes, M.L. (1989) Responses of tsetse flies (*Glossina* spp.) to compounds on the skin surface of an ox: A laboratory study. *Medical and Veterinary Entomology,* 3, 399–406.

Warnes, M.L. (1990a) The effect of host odour and carbon dioxide on the flight of tsetse flies (*Glossina* spp.) in the laboratory. *Journal of Insect Physiology,* 36, 607–611.

Warnes, M.L. (1990b) Field responses of *Glossina morsitans morsitans* Westwood and *G. pallidipes* Austen (Diptera: Glossinidae) to the skin secretions of oxen. *Bulletin of Entomological Research,* 80, 91–97.

Warnes, M.L. (1995) Field studies on the effect of cattle skin secretion on the behaviour of tsetse. *Medical and Veterinary Entomology,* 9, 284–288.

Warnes, M.L. and Finlayson, L.H. (1987) Effect of host behaviour on host preference in *Stomoxys calcitrans. Medical and Veterinary Entomology,* 1, 53–57.

Warnes, M.L. and Maudlin, I. (1992) An analysis of supernumerary or B-chromosomes of wild and laboratory strains of *Glossina morsitans morsitans. Medical and Veterinary Entomology,* 6, 175–176.

Warnes, M.L., Torr, S.J. and Hargrove, J.W. (1995) Current research into the use and deployment of odour-baited targets in Zimbabwe. *ISCTRC 22nd meeting, Kampala,* Publication No. 117, pp. 269–271.

Waterston, J. (1915) Notes on African Chalcidoidea II. *Bulletin of Entomological Research,* 5, 343–372.

Waterston, J. (1916) Chalcidoidea bred from *Glossina morsitans* in Nyasaland. *Bulletin of Entomological Research,* 6, 381–393.

Waterston, J. (1917) Chalcidoidea bred from *Glossina* in the northern territories, Gold Coast. *Bulletin of Entomological Research,* 8, 178–179.

Watson, H.J.C. (1962) The domestic pig as a reservoir of *Trypanosoma gambiense*. *ISCTR 9th meeting, Conakry*, 327.

Weidhaas, D.E. and Haile, D.G. (1978) A theoretical model to determine the degree of trapping required for insect population control. *Bulletin of the Entomological Society of America (ESA Bulletin)*, 24, 18–20.

Weir, J. and Davison, E. (1965) Daily occurrence of African game animals at water holes during dry weather. *Zoologica Africana*, 1, 353–368.

Weiss, J.B. (1995) DNA probes and PCR for diagnosis of parasitic infections. *Clinical Microbiology Reviews*, 8, 113–130.

Weitz, B. (1952) The antigenicity of sera of man and animals in relation to the preparation of specific precipitating antisera. *Journal of Hygiene*, 50, 275–294.

Weitz, J.B. (1956) Identification of bloodmeals of blood sucking arthropods. *Bulletin of the World Health Organisation*, 15, 473–490.

Weitz, B. (1960) Feeding habits of bloodsucking arthropods. *Experimental Parasitology*, 9, 63–82.

Weitz, B. (1963) The feeding habits of *Glossina*. *Bulletin of the World Health Organisation*, 28, 711–729.

Weitz, B.G.F. (1970) Methods for identifying the blood meals of *Glossina*. In: Mulligan, H.W. (ed.) *The African Trypanosomiases*. George Allen and Unwin, London, pp. 416–423.

Weitz, B. and Buxton, P.A. (1953) The rate of digestion of blood meals of various haematophagous arthropods as determined by the precipitin test. *Bulletin of Entomological Research*, 44, 445–450.

Weitz, B. and Glasgow, J.P. (1956) The natural hosts of some species of *Glossina* in East Africa. *Transactions of the Royal Society of Tropical Medicine and Hygiene*, 50, 593–612.

Weitz, B. and Jackson, C.H.N. (1955) The host animals of *Glossina morsitans* at Daga-Iloi. *Bulletin of Entomological Research*, 46, 531–538.

Weitz, B., Langridge, W.P., Bax, P.N. and Lee-Jones, F. (1960) The natural hosts of *Glossina longipennis* Corti and of some other tsetse flies in Kenya. *ISCTRC/CCTA 7th meeting Bruxelles, 1958*, Publication No. 41, 303–312.

Welburn, S.C. and Gibson, W.C. (1989) Cloning of a repetitive DNA from the Rickettsia-like organisms of tsetse flies (*Glossina* spp.). *Parasitology*, 98, 81–84.

Welburn, S.C. and Maudlin, I. (1989) Lectin signalling of maturation of *Trypanosoma congolense* infections in tsetse. *Medical and Veterinary Entomology*, 3, 141–145.

Welburn, S.C. and Maudlin, I. (1990) Haemolymph lectin and the maturation of trypanosome infections in tsetse. *Medical and Veterinary Entomology*, 4, 43–48.

Welburn, S.C. and Maudlin, I. (1991) Rickettsia-like organisms, puparial temperature and susceptibility to trypanosome infection in *Glossina morsitans*. *Parasitology*, 102, 201–206.

Welburn, S.C. and Maudlin, I. (1992) The nature of the teneral state in *Glossina* and its role in the acquisition of trypanosome infection in tsetse. *Annals of Tropical Medicine and Parasitology*, 86, 529–536.

Welburn, S.C., Ellis, D.S. and Maudlin, I. (1987) *In vitro* cultivation of Rickettsia-like-organisms from *Glossina* spp. *Annals of Tropical Medicine and Parasitology*, 81, 331–335.

Welburn, S.C., Maudlin, I. and Ellis, D.S. (1989) Rate of trypanosome killing by lectins in midguts of different species and strains of *Glossina*. *Medical and Veterinary Entomology*, 3, 77–82.

Welburn, S.C., Arnold, K., Maudlin, I. and Gooday, G.W. (1993) Rickettsia-like organisms and chitinase production in relation to transmission of trypanosomes by tsetse flies. *Parasitology*, 107, 141–145.

Welburn, S.C., Maudlin, I. and Molyneux, D.H. (1994) Midgut lectin activity and sugar specificity in teneral and fed tsetse. *Medical and Veterinary Entomology*, 8, 81–87.

Welburn, S.C., Maudlin, I. and Milligan, P.J.M. (1995) *Trypanozoon*: infectivity to humans is linked to reduced transmissibility in tsetse. I. Comparison of human-serum-resistant and human-serum-sensitive field isolates. *Experimental Parasitology*, 81, 404–408.

Wellde, B.T., Hockmeyer, W.T., Kovatch, R.M., Bhogal, M.S. and Diggs, C.L. (1981) *Trypanosoma congolense*: natural and acquired resistance in the bovine. *Experimental Parasitology*, 52, 219–232.

Wellde, B.T., Chumo, D.A., Waema, D., Reardon, M.J. and Smith, D.H. (1989a) A history of sleeping sickness in Kenya. *Annals of Tropical Medicine and Parasitology*, 83, Supplement 1, 1–11.

Wellde, B.T., Waema, D., Chumo, D.A., Reardon, M.J., Adhiambo, A., Olando, J. and Mabus, D. (1989b) The Lambwe valley and its people. *Annals of Tropical Medicine and Parasitology*, 83, Supplement 1, 13–20.

Wellde, B.T., Chumo, D.A., Reardon, M.J., Waema, D., Smith, D.H., Gibson, W.C., Wanyama, L. and Siongok, T.A. (1989c) Epidemiology of Rhodesian sleeping sickness in the Lambwe valley Kenya. *Annals of Tropical Medicine and Parasitology*, 83, Supplement 1, 43–62.

Wellde, B.T., Waema, D., Chumo, D.A., Reardon, M.J., Oloo, F., Njogu, A.R., Opiyo, E.A. and Mugutu, S. (1989d) Review of tsetse control measures taken in the Lambwe valley in 1980–1984. *Annals of Tropical Medicine and Parasitology*, 83, Supplement 1, 119–125.

Wells, E.A. (1972) The importance of mechanical transmission in the epidemiology of Ngana: a review. *Tropical Animal Health and Production*, 4, 74–88.

Westwood, J.O. (1851) Observations on the destructive species of dipterous insects known in Africa under the names of the tsetse, Zimb, and Tsaltsalya, and their supposed connection with the fourth plague of Egypt. *Proceedings of the Zoological Society of London*, 18, 258–270.

Whitehead, D.L. (1981) The effect of phytosterols on tsetse reproduction. *Insect Science and its Application*, 1, 281–288.

Whitelaw, D.D. and Jordt, T. (1985) Colostral transfer of antibodies to *Trypanosoma brucei* in goats. *Annales de la Société Belge de Médecine Tropicale*, 65, 199–205.

Whiteside, E.F. (1949) An experiment in control of tsetse with DDT-treated oxen. *Bulletin of Entomological Research*, 40, 123–134.

Whiteside, E.F. (1958) The maintenance of cattle in tsetse infested country. A summary of four years' experience in Kenya. *Proceedings of the 7th meeting ISCTR, Brussels*, Publication No. 41, 83–90.

Whiteside, E.F. (1960) Recent work in Kenya on the control of drug-resistant cattle trypanosomiasis. *Proceedings of the 8th Meeting of the ISCTR, Jos*, Publication No. 62, 141–154.

Whiteside, E.F. (1962a) The control of cattle trypanosomiasis with drugs in Kenya: methods and costs. *East African Agricultural and Forestry Journal*, 28, 67–73.

Whiteside, E.F. (1962b) Interactions between drugs trypanosomes and cattle in the field. In: Goodwin, L.G. and Nimmo-Smith, R.H. (eds) *Drugs, Parasites and Hosts*. Churchill, London, pp. 116–141.

Whiteside, E.F. (1963) A strain of *Trypanosoma congolense* directly resistant to Berenil. *Journal of Comparative Pathology*, 73, 167–175.

Whitnall, A.B.M. (1932) The trypanosome infections of *Glossina pallidipes* in the Umfolosi Game Reserve, Zululand. *18th Report of the Division of Veterinary Services and Animal Industry, Union of South Africa*, 21.

WHO (1979) *Parasitic Zoonoses*. WHO Technical Report Series No. 637. World Health Organisation, Geneva.

WHO (1995) *Planning Overview of Tropical Diseases Control*. Division of Tropical Diseases, World Health Organisation, Geneva.

Wiedemann, C.R.W. (1830) Aussereuropaische zweiflugelige. *Insekten*, pt II, 253–254.

Wigglesworth, V.B. (1929) Digestion in the tsetse fly: a study of structure and function. *Parasitology*, 21, 288–321.

Wigglesworth, V.B. (1972) *The Principles of Insect Physiology*, 7th Edn. Chapman & Hall, London, 827 pp.

Wijers, D.J.B. (1958a) Factors that may influence the infection rates of *Glossina palpalis* with *Trypanosoma gambiense*. I. The age of the fly at the time of the infected feed. *Annals of Tropical Medicine and Parasitology*, 52, 385–390.

Wijers, D.J.B. (1958b) The importance of the age of *Glossina palpalis* at the time of infective feed with *Trypanosoma gambiense*. *ISCTR/CCTA 7th meeting, Bruxelles*, Publication no. 41, 319–320.

Wijers, D.J.B. (1969) The history of sleeping sickness in Yimbo location (Central Nyanza, Kenya). *Tropical and Geographical Medicine*, 21, 323–337.

Wijers, D.J.B. (1974a) The complex epidemiology of Rhodesian sleeping sickness in Kenya and Uganda. I. The absence of the disease on Mfangano island (Kenya). *Tropical Geography and Medicine*, 26, 58–64.

Wijers, D.J.B. (1974b) The complex epidemiology of Rhodesian sleeping sickness in Kenya and Uganda. II. Observations in Samia (Kenya). *Tropical Geography and Medicine*, 26, 182–197.

Wijers, D.J.B. (1974c) The complex epidemiology of Rhodesian sleeping sickness in Kenya and Uganda. III. The epidemiology in the endemic areas along the lake. *Tropical Geography and Medicine*, 26, 307–318.

Wijers, D.B.J. (1974d) The complex epidemiology of Rhodesian sleeping sickness in Kenya and Uganda. IV. Alego and the other fuscipes areas north of the endemic. *Tropical Geography and Medicine*, 26, 341–351.

Wilkes, J.M., Peregrine, A.S. and Zilberstein, D. (1995) The accumulation and compartmentalisation of isometamidium chloride in *Trypanosoma congolense*, monitored by its intrinsic fluorescence. *Biochemical Journal*, 312, 319–329.

Wilkes, J.M., Mulugeta, W., Wells, C. and Peregrine, A.S. (1997) Modulation of mitochondrial electrical potential: a candidate mechanism for drug resistance in Africa trypanosomes. *Biochemical Journal*, 325, 755–761.

Willemse, L. (1991) A trial of odour baited targets to control the tsetse fly *Glossina morsitans centralis* Machado (Diptera: Glossinidae) in west Zambia. *Bulletin of Entomological Research*, 81, 351–357.

Willemse, L.P.M. and Takken, W. (1994) Odor-induced host location in tsetse flies (Diptera: Glossinidae). *Journal of Medical Entomology*, 31, 775–794.

Willett, K.C. (1955) A special method for the dissection of *Glossina*. *Annals of Tropical Medicine and Parasitology*, 49, 376–383.

Willett, K.C. (1956) An experiment on dosage in human trypanosomiasis. *Annals of the Society of Medical Parasitology*, 50, 75.

Willett, K.C. (1965) Some observations on the recent epidemiology of Sleeping sickness in Nyanza region, Kenya, and its relation to the general epidemiology of Gambian and Rhodesian sleeping sickness in Africa. *Transactions of the Royal Society of Tropical Medicine and Hygiene*, 59, 374–386.

Willett, K.C. (1966) Development of the peritrophic membrane in *Glossina* (tsetse flies) and its relation to infection with trypanosomes. *Experimental Parasitology*, 18, 290–295.

Willett, K.C., McMahon, J.P., Ashcroft, M.T. and Baker, J.R. (1964) Trypanosomes isolated from *Glossina palpalis* and *G. pallidipes* in Sakwa, Kenya. *Transactions of the Royal Society of Tropical Medicine and Hygiene*, 58, 391–396.

Williams, B.G. (1994) Models of trap seeking by tsetse flies: anemotaxis, klinokinesis and edge detection. *Journal of Theoretical Biology*, 168, 105–115.

Williams, B.G., Brightwell, R. and Dransfield, R.D. (1990a) Monitoring tsetse fly populations. II. The effect of climate on trap catches of *Glossina pallidipes*. *Medical and Veterinary Entomology*, 4, 181–193.

Williams, B.G., Dransfield, R.D. and Brightwell, R. (1990b) Monitoring tsetse fly populations. I. The intrinsic variability of trap catches of *Glossina pallidipes* at Nguruman, Kenya. *Medical and Veterinary Entomology*, 4, 167–179.

Williams, B.G., Dransfield, R.D. and Brightwell, R. (1990c) Tsetse fly (Diptera: Glossinidae) population dynamics and the estimation of mortality rates from life-table data. *Bulletin of Entomological Research*, 80, 479–485.

Williams, B.G., Dransfield, R.D. and Brightwell, R. (1992) The control of tsetse flies in relation to fly movement and trapping efficiency. *Journal of Applied Ecology*, 29, 163–179.

Williams, B.G., Campbell, C. and Williams, R. (1995) Broken Houses: science and development in the African savannahs. *Agriculture and Human Values*, Spring 1995, 29–38.

Williams, C.B. (1937) The use of logarithms in the interpretation of certain entomological problems. *Annals of Applied Biology*, 24, 404–414.

Williamson, D.L., Baumgartner, H.H., Mtuya, A.G., Warner, P.V., Tarimo, S.A. and Dame, D.A. (1983a) Integration of insect sterility and insecticides for control of *Glossina morsitans morsitans* Westwood (Diptera: Glossinidae) in Tanzania. I. Production of tsetse flies for release. *Bulletin of Entomological Research*, 73, 259–265.

Williamson, D.L., Dame, D.A., Lee, C.W., Gates, D.B. and Cobb, P.E. (1983b) Integration of insect sterility and insecticides for control of *Glossina morsitans morsitans* Westwood (Diptera: Glossinidae) in Tanzania. IV. Application of endosulfan as an aerosol prior to release of sterile males. *Bulletin of Entomological Research*, 73, 383–389.

Williamson, D.L., Dame, D.A., Gates, D.B., Cobb, P.E., Bakuli, B. and Warner, P.V. (1983c) Integration of insect sterility and insecticides for control of *Glossina morsitans morsitans* Westwood (Diptera: Glossinidae) in Tanzania. V. The impact of sequential releases of sterilised tsetse flies. *Bulletin of Entomological Research*, 73, 391–404.

Wilson, A. (1988) Notes on the control of tsetse fly populations by the application of insecticides to animals. Manuscript for KETRI Workshop, March 1988.

Wilson, A.J., Dar, F.K. and Paris, J. (1973) Serological studies on trypanosomiasis in East Africa. III. Comparison of antigenic types of *Trypanosoma congolense* organisms. *Annals of Tropical Medicine and Parasitology*, 67, 313–317.

Wilson, A.J., Paris, J. and Dar, F.K. (1975) Maintenance of a herd of breeding cattle in an area of high trypanosome challenge. *Tropical Animal Health and Production*, 7, 63–71.

Wilson, A.J., Gatuta, G.M. and Njogu, A.R. (1986) A simple epidemiological method for animal trypanosomiasis to provide relevant data for effective financial decision making. *Veterinary Parasitology*, 20, 261–274.

Wilson, S.G. (1948) The feeding of 'Gammexane' and DDT to bovines. *Bulletin of Entomological Research*, 39, 423–435.

Wilson, S.G. (1953) The control of *Glossina palpalis fuscipes* Newstead in Kenya colony. *Bulletin of Entomological Research*, 44, 711–728.

Wilson, S.G. (1958) Recent advances of *Glossina morsitans submorsitans* in northern Nigeria. *ISCTRC/CCTA 7th meeting Bruxelles*, Publication No. 41, 367–389.

Wilson, V.J. (1972) Observations on the effect of dieldrin on wildlife during tsetse fly *Glossina morsitans* control operations in eastern Zambia. *Arnoldia*, 5, 1–12.

Wilson, V.J. and Roth, H.H. (1967) The effects of tsetse control operations on Common Duiker in eastern Zambia. *East African Wildlife Journal*, 5, 53–64.

Wink, M. (1979) The endosymbionts of *Glossina morsitans* and *G. palpalis*: cultivation experiments and some physiological properties. *Acta Tropica*, 36, 215–222.

Winrock International (1992) *Assessment of Animal Agriculture in Sub-Saharan Africa*. Winrock International Institute for Agricultural Development, Arkansas, USA, 125 pp.

Winterbottom, T.M. (1803) An account of native Africans in the neighbourhood of Sierra Leone. *Sleepy Sickness*, Vol. 2. Whittingham and Co., London, 283 pp.

Wirtz, R.A., Roberts, L.W., Hallam, J.A., Macken, L.M., Roberts, D.R., Buescher, M.D. and Rutledge, L.C. (1985) Laboratory testing of repellents against the tsetse fly *Glossina morsitans. Journal of Medical Entomology*, 22, 271–275.

Wood, D.M. (1987) Oestridae. In: McAlpine, J.F. (ed.) *Manual of Nearctic Diptera*, Vol. 2. Research Branch, Agriculture Canada (Monograph No. 28), pp. 1147–1158.

Wooff, W.R. (1964) The eradication of *Glossina morsitans morsitans* Westw. in Ankole, western Uganda, by dieldrin application. In: *ISCTRC 10th Meeting, Kampala, 1964*. Publication no. 97, ISCTRC, pp. 157–166.

Wooff, W.R. (1967) Consolidation in tsetse reclamation. *ISCTRC 11th meeting, Nairobi*, 141–148.

Wooff, W.R. (1968) The eradication of the tsetse *Glossina morsitans* Westw. and *Glossina pallidipes* Aust., by hunting. *ISCTR 12th meeting, Bangui*, 267–286.

Woo, P.T.K. (1969) The haematocrit centrifuge technique for the detection of trypanosomes in blood. *Canadian Journal of Zoology*, 47, 921–923.

Woo, P.T.K. (1970) Origin of mammalian trypanosomes which develop in the Anterior station of blood-sucking arthropods. *Nature*, 228, 1059–1062.

Woo, P.T.K. (1971) Evaluation of the haematocrit centrifuge and other techniques for the field diagnosis of human trypanosomiasis and filariasis. *Acta Tropica*, 28, 298–303.

Woolhouse, M.E.J., Hargrove, J.W. and McNamara, J.J. (1993) Epidemiology of trypanosome infections of the tsetse fly *Glossina pallidipes* in the Zambezi valley. *Parasitology*, 106, 479–485.

Woolhouse, M.E.J., Bealby, K.A., McNamara, J.J. and Silutongwe, J. (1994) Trypanosome infections of the tsetse fly *Glossina pallidipes* in the Luangwa valley, Zambia. *International Journal for Parasitology*, 24, 987–993.

Woolhouse, M.E.J., McNamara, J.J., Hargrove, J.W. and Bealby, K.A. (1996) Distribution and abundance of trypanosome (subgenus *Nannomonas*) infections of the tsetse fly *Glossina pallidipes* in southern Africa. *Molecular Ecology*, 5, 11–18.

World Bank (1993) *World Development Report (1993) – Investing in Health.* Oxford University Press, Washington, DC.

Wuethrich, B. (1994) Domesticated cattle show their breeding. *New Scientist*, 1926, 16–17.

Wyatt, G.B., Boatin, B.A. and Wurapa, F.K. (1985) Risk factors associated with the acquisition of Sleeping sickness in north Zambia: a case control study. *Annals of Tropical Medicine and Parasitology*, 79, 385–392.

Yagi, A.I. and Razig, M.T. Abdel (1972) Distribution of tsetse flies in southern Darfur District, Sudan. *Bulletin of Epizootic Diseases in Africa*, 20, 287–290.

Yesufu, H.M. (1971) Experimental transmission of *Trypanosoma gambiense* to domestic animals. *Annals of Tropical Medicine and Parasitology*, 65, 341–347.

Yorke, W. and MacFie, J.W.S. (1924) The action of the salivary secretions of mosquitoes and *Glossina tachinoides* on human blood. *Annals of Tropical Medicine and Parasitology*, 18, 103–108.

Yorke, W., Murgatroyd, F. and Hawking, F. (1933) The relation of polymorphic trypanosomes developing in the gut of *Glossina* to the peritrophic membrane. *Annals of Tropical Medicine and Parasitology*, 27, 347–355.

Youdeowei, A. (1975) A simple technique for observing and collecting the saliva of tsetse flies (Diptera, Glossinidae). *Bulletin of Entomological Research*, 65, 65–67.

Youdeowei, A. (1976) Salivary secretion in wild *Glossina pallidipes* Austen. (Diptera, Glossinidae). *Acta Tropica*, 33, 369–375.

Yu, P., Habtemariam, T., Oryang, D., Obasa, M., Nganwa, D. and Robnett, V. (1995) Integration of temporal and spatial models for examining the epidemiology of African trypanosomiasis. *Preventive Veterinary Medicine*, 24, 83–95.

Yvoré, P. (1962) Quelques observations sur l'écologie de deux Glossines du groupe *Fusca* en République Centrafricaine. *ISCTRC/CCTA, 9th meeting, Conakry*, Publication No. 88, 197–204.

Zdárek, J. and Denlinger, D.L. (1991) Wandering behaviour and pupariation in tsetse larvae. *Physiological Entomology*, 16, 523–529.

Zdárek, J. and Denlinger, D.L. (1992a) Eclosion behaviour in tsetse (Diptera: Glossinidae): extrication from the puparium and expansion of the adult. *Journal of Insect Behaviour*, 5, 657–668.

Zdárek, J. and Denlinger, D.L. (1992b) Neural regulation of pupariation in tsetse larvae. *Journal of Experimental Biology*, 173, 11–24.

Zdárek, J. and Denlinger, D.L. (1993) Metamorphosis behaviour and regulation in tsetse flies (*Glossina* spp.) (Diptera: Glossinidae): a review. *Bulletin of Entomological Research*, 83, 447–461.

Zdárek, J., Denlinger, D.L. and Otieno, L.H. (1992) Does the tsetse parturition rhythm have a circadian basis? *Physiological Entomology*, 17, 305–307.

Zdárek, J., Weyda, F., Chimtawi, M.B. and Denlinger, D. (1996) Functional morphology and anatomy of the polypneustic lobes of the last larval instar of tsetse flies, *Glossina* spp. (Diptera: Glossinidae). *International Journal of Insect Morphology and Embryology*, 25, 235–248.

Zerba, E. (1988) Insecticidal activity of pyrethroids on insects of medical importance. *Parasitology Today*, 4, S3-S7.

Zilberstein, D., Wilkes, J., Hirumi, H. and Peregrine, A.S. (1993) Fluorescence analysis of the interaction of isometamidium with *Trypanosoma congolense*. *Biochemistry Journal*, 292, 31–35.

Zillmann, U. and Mehlitz, D. (1979) The natural occurrence of *Trypanozoon* in domestic chicken in the Ivory Coast. *Tropenmedizin und Parasitology*, 30, 244–248.

Zillmann, U., Mehlitz, D. and Sachs, R. (1984) Identity of *Trypanozoon* stocks isolated from man and a domestic dog in Liberia. *Tropenmedizin und Parasitologie*, 35, 105–108.

Zippin, C. (1956) An evaluation of the removal method of estimating animal populations. *Biometrics*, 12, 163–189.

Glossary

Accessory glands	Secretory glands associated with the reproductive system
Achiasmate	No cross-shaped union of chromosomes during nuclear division
Adenotrophic viviparity	Reproduction in which the female gives birth to a larva which developed and has been nourished within the mother's uterus
Alanine	Amidic acid derived from lactic acid
Aposymbiotic	Organisms living together but with no mutually beneficial relationship
Bacteroids	Microorganisms allied to bacteria
Basiconic sensillae	Simple, cone-shaped sensory receptors
Chitin	A strong, fibrous polysaccharide forming the cuticle of insects
Cholinesterase	Enzyme which inactivates acetylcholine, a neuro-transmitter secreted at nerve endings
Chorion	Membrane enclosing embryonic structures
Choriothete	Organ attached to the uterine floor to which the egg is attached
Chymotrypsin	Proteolytic enzyme
Cibarium	Structure formed in the head containing the cibarial pump, which is a modification of the oesophagus for sucking blood, pumping saliva and blood
Corpus allatum	Secretory organ of the insect 'brain'
Corpus cardiaca	Organ of the insect circulatory system equivalent to the heart
Diglycerides	Type of lipid with two fatty acid radicals
Diuresis	Excretion of 'urine'
Ecdysis	Moulting, shedding of the cuticle
Ecdysone	Steroid hormone initiating ecdysis
Eclosion	Emergence of adult insect from the puparium
Egg-tooth	Tooth on the first instar larva for breaking the egg membrane when hatching
Endocuticle	Flexible, inner part of the insect cuticle

548

Endosymbionts	Symbiotic organisms (bacteroids/rickettsia-like organisms) living in the digestive tract
Enterokinase	Enzyme which converts trypsinogen to trypsin
Fibrin	An insoluble protein formed from fibrinogen in blood during clotting
Fibrinolysin	Enzyme causing breakdown of fibrin from the bloodmeal to soluble products
Germarium	Ovary proper – region containing primordial germ cells at the apex of each ovariole
Giant-cell zone	*see* Mycetome
Glucosamine	Any amino acid sugar derived from glucose
Gonotrophic	Regulatory activity of the ovaries
Haematin	Constituent of haemoglobin
Haemocoel	Body cavity, containing blood (haemolymph) in insects and other arthropoda
Haemolymph	Blood fluid of insects
Haemolysin	Substance lysing red blood corpuscles
Hemidesmosomes	Areas of attachment between trypanosomes and tsetse proboscis
Heritability (h^2)	Heritability of a trait is defined as the ratio of additive genetic variance (V_A) to phenotypic variance (V_P), i.e. $h^2 = V_A/V_P$
Imago	Sexually mature adult insect
Incidence	The number of new cases that occur in a known population over a specified period of time
Instar	Immature stage of development of an insect, between two ecdyses, or moults
Juvenile hormone	Hormone that controls the development of larval characteristics in insects
Kinesis	Motion – undirected locomotion
Larviposition	Deposition of a mature, third stage larva
Lectins	A protein which binds to specific carbohydrate component
Lysozyme	Enzyme which destroys or weakens a cell wall leading to its rupture and death
Malpighian tubules	Blind-ending tubular glands with an excretory function opening into anterior hindgut
Meconium	Contents of pre-adult intestine
Micropyle	Pore in egg membrane of the ovum through which sperm enter for fertilization
Milk glands	Modified accessory glands secreting nutrient substances for the developing larvae
Mycetome	A region of specialized cells of the midgut containing symbiotic organisms
Neuroendocrine system	System involving both nervous and endocrine participation
Nulliparous	Females that have not yet given birth
Nurse cells	Cells whose function is to assist the ovum in some way
Oocytes	Cells which undergo meiosis to form the ova

Ookinetes	Same as oocytes
Oogenesis	The process of formation of the ova
Oogonia	Cells of the ovary which undergo mitosis to form oocytes
Optomotor anemotaxis	Response (orientation and flight direction) to visual perception of wind during flight
Orthokinesis	A kinesis in which linear movement is shown
Pharate	Developing adult within the pupal cuticle
Pheromones	Chemical substances released by an insect which affect the behaviour or development of other insects of the same species
Polypneustic lobes	Lobes bearing many respiratory spiracles
Polytene chromosomes	Chromosomes consisting of parallel, identical, exactly paired chromatids
Prevalence	The number of cases of infection in a known population at a designated time
Proline	Amino acid providing energy for flight in tsetse
Proteolytic enzymes	Any enzyme involved in the breakdown of proteins
Proventriculus	Foregut
Ptilinum	Soft vesicular bladder-like membrane between the antennae on the head of cyclorrhaphan Diptera
Pupariation	Process of puparium formation, metamorphosis from larva to puparium
Puparium	Dark brown or blackish cylinder formed by the last larval skin
Pyriproxyfen	A juvenile-hormone mimic
Rickettsia	Group of minute, rod-shaped bacterial obligate parasites of arthropods
Sarcosomes	Interfibrillar granules of insect thoracic muscles
Sebum	Oily lipid substances secreted by sebaceous glands of the skin
Secretogogues	Substances which tend to promote secretion
Spermatheca(ae)	Spherical organ in the female tsetse which receives and stores sperm from the male
Spiracles	External openings of the trachea
Stigmata	Spiracles
Stomatogastric nervous system	Nervous system pertaining to the digestive system (applied to a system of visceral nerves in invertebrates
Supernumerary chromosomes (B-chromosomes)	Additional small chromosomes above the standard chromosome number for a species – may vary in number seasonally
Triglycerides (triglycerol)	Type of lipid with three fatty acid radicals
Uric acid	Complex nitrogen-containing organic compound excreted by insects
Vasodilation	Expansion of blood vessels by relaxation of muscles
Vitellarium	Accessory gland in female reproductive organs which produce the 'egg yolk'

Index

Abortion
  of livestock with trypanosome
      infections 294
  of tsetse 44, 47, 86, 165, 375, 403
Abundance of tsetse 91, 341
Acacia woodland
  and tsetse 88, 102, 350
  A. hockii
    clearance for tsetse control
        387–388
  A. pennata
    as a resting site for tsetse 138
  Acacia/Commiphora bush
    as a habitat of *G. longipennis* 100
Acaricides 374, 378, 418
Accessory glands
  of tsetse 18, 19, 40, 44
  milk glands of female tsetse 41
Acetone
  as an attractant for tsetse 169–177, 365,
      366, 367, 368
  for application of diflubenzuron to
      tsetse 403
  for sterilizing bloodmeal samples 114,
      116
Activity
  of tsetse 31, 37, 53, 117, 137, 139, 291,
      327, 330–331, 366, 403
    and population sampling 57, 66, 67,
        68, 70, 75, 76, 148–150
    circadian 99, 107, 141–142, 145, 326,
        328
    effect of climate 144–145
    effect of light 139, 142–143
    effect of olfactory stimuli 144
    effect of season 143–144
    effect of visual stimuli 144, 362
    night-time 143, 144
    of pregnant flies 130, 145–146
    spontaneous 130, 145

metabolic and enzyme activity 22,
    26, 27, 29, 35, 36, 48, 185, 201, 205,
    206, 210, 213
of humans
  on tsetse 87, 99, 407
  on vegetation types 94
  and feeding 109, 127, 132–134, 137,
      220, 322, 328
  and trypanosome infections in tsetse
      190
  of wild animals in relation to tsetse
      117
  antigenic of bloodmeals 113
Adenosine diphosphate (ADP) 36, 110, 218
Adenosine monophosphate 44, 110
Adenosine triphosphate (ATP) 35, 36, 37,
    110, 218, 296
Adenotrophic viviparity 17, 38
Advances in distribution
  of tsetse 84, 88, 89, 90, 98
  of *T. b. rhodesiense* 238
Aedes aegypti
  trypsin enzyme in 29
Aerial photographs for tsetse surveys 341
Aerial spraying to control tsetse 153, 269,
    345, 346, 351–354, 357, 400, 409,
    410, 414, 415, 416, 419
Aerosols of insecticides 351, 352, 354
Afzelia spp. 95
Age of tsetse 196
  Determination of 153–156
    ovarian 71, 153, 154–155, 329
    pteridine 49, 155–156
    wing-fray 154
  in relation to trypanosome infection
      193–196
  mean age at death 196
Agglutinins in tsetse 195, 206–207
AIDS (HIV) effect on tsetse populations 89
Alanine 32, 35, 36

551